AF894880

Angiotensin Converting
Enzyme Inhibitors

Proceedings of the A. N. Richards Symposium sponsored by the Physiological Society of Philadelphia—May 8-9, 1980.

Angiotensin Converting Enzyme Inhibitors

Mechanisms of Action and Clinical Implications

Edited by
Zola P. Horovitz, Ph.D.
The Squibb Institute for Medical Research
Princeton, New Jersey

WILLIAM BEAUMONT HOSPITAL
MEDICAL LIBRARY
ROYAL OAK, MICHIGAN

Urban & Schwarzenberg
Baltimore-Munich 1981

Urban & Schwarzenberg, Inc.
7 E. Redwood Street
Baltimore, Maryland 21202
USA

Urban & Schwarzenberg
Pettenkoferstrasse 18
D-8000 München 2
Germany

© Urban & Schwarzenberg 1981

All rights including that of translation, reserved. No part of this publication may be reproduced, stored in a retrieval system, or transmitted in any other form or by any means, electronic, mechanical, recording, or otherwise without the prior written permission of the publisher.

Printed in the United States of America

Library of Congress Cataloging in Publication Data

Richards (A. N.) Symposium, Philadelphia, 1980.
Angiotensin converting enzyme inhibitors.

"Proceedings of the A.N. Richards Symposium sponsored by the Physiological Society of Philadelphia, May 8–9, 1980."
Includes index.
1. Angiotensin converting enzyme—Congresses. 2. Enzyme inhibitors—Therapeutic use— Congresses. 3. Hypertension—Chemotherapy—Congresses. 4. Captopril—Congresses. I. Horovitz, Zola P. II. Physiological Society of Philadelphia. III. Title.
QP609.A53R53 1980 615'.71 81–388
ISBN 0–8067–0821–2 AACR1

ISBN 0-8067-0821-2 (Baltimore)

ISBN 3-541-70821-2 (Munich)

Acknowledgments

This symposium was supported by the generous contributions of the following companies to The Physiological Society of Philadelphia:

Abbott Laboratories
American Cyanamid Company
Arnar-Stone Laboratories, Inc.
Berlex Laboratories, Inc., Subsidiary of Schering AG
Boehringer Ingelheim Ltd.
CIBA-GEIGY Pharmaceuticals
Cutter Laboratories, Inc.
E. I. du Pont de Nemours & Company
Hoechst-Roussel Pharmaceuticals Incorporated
Hoffman-La Roche Inc.
E. Lilly
ICI Americas Inc.
3M, Central Research Laboratories
Mead Johnson & Company
Merck Sharpe & Dohme Research Labs.
McNeil Laboratories, Inc.
Ortho Pharmaceutical Corporation
Pfizer Inc.
Schering-Plough Corporation
G. D. Searle and Company
Smith Kline & French Laboratories
Squibb Institute for Medical Research
The Upjohn Company
USV Pharmaceutical Corporation
Wallace Laboratories
Warner Lambert Foundation
Wyeth Laboratories

Contents

Contributors and Participants

Ader, R.
University of California,
San Francisco
Division of Cardiology &
Cardiovascular Research
Institute
San Francisco, CA U.S.A.

Alexander, J. C.
The Squibb Institute for Medical
Research
Princeton, NJ U.S.A.

Amery, A.
Department of
Pathophysiology
University of Leuven
Leuven, Belgium

Antonaccio, M. J.
The Squibb Institute for Medical
Research
Princeton, NJ U.S.A.

Asaad, M.
The Squibb Institute for Medical
Research
Princeton, NJ U.S.A.

Bing, R. F.
University of Leicester
Leicester, UK

Bravo, E.
Cleveland Clinic Foundation
Research Division
Cleveland, OH U.S.A.

Brunner, H. R.
Department of Medicine
Centre Hospitalier Universitaire
Vaudois
1011 Lausanne, Switzerland

Buckley, J. P.
University of Houston
Houston, TX U.S.A.

Carretero, O. A.
Henry Ford Hospital
Detroit, MI U.S.A.

Chatterjee, K.
University of California,
San Francisco
Division of Cardiology &
Cardiovascular Research
Institute
San Francisco, CA U.S.A.

Cheung, H. S.
The Squibb Institute for Medical
Research
Princeton, NJ U.S.A.

Clappison, B. H.
Monash University
Prince Henry's Hospital
Melbourne, Australia

Clough, D.
ICI Pharmaceuticals Division
Mereside, Alderley Park
Macclesfield, Cheshire, UK

Conway, F. J.
John Radcliffe Hospital
Headington, Oxford, UK

Cushman, D. W.
The Squibb Institute for Medical Research
Princeton, NJ U.S.A.

Fagard, R.
Inwendige Geneeskunde
Academisch Ziekenhuis Pellenberg
Pellenberg, Belgium

Ferguson, R.
Jefferson Medical College
Thomas Jefferson University
Philadelphia, PA U.S.A.

Fouad, F. M.
Cleveland Clinic Foundation
Research Division
Cleveland, OH U.S.A.

Ganten, D.
University of Heidelberg, and
German Institute for High
Blood Pressure Research
Heidelberg, Germany

Gavras, H. P.
Boston City Hospital
Boston, MA U.S.A.

Gross, F.
University of Heidelberg
Heidelberg, Germany

Hatton, R.
ICI Pharmaceuticals Division
Mereside, Alderley Park
Macclesfield, Cheshire, UK

Horovitz, Z. P.
The Squibb Institute for Medical Research
Princeton, NJ U.S.A.

Hübner, D.
University of Heidelberg, and
German Institute for High
Blood Pressure Research
Heidelberg, Germany

Jones, J. C.
The Medical Research Council
Blood Pressure Unit
Glasgow, Scotland

Johnston, C. I.
Prince Henry's Hospital
Melbourne, Australia

Kaufman-Bühler, I.
University of Heidelberg, and
German Institute for High
Blood Pressure Research
Heidelberg, Germany

Koffer, H.
Jefferson Medical College
Thomas Jefferson University
Philadelphia, PA U.S.A.

Laragh, J. H.
The New York Hospital
Cornell Medical Center
New York, NY U.S.A.

Lefer, A. M.
Jefferson Medical College
Philadelphia, PA U.S.A.

Lijnen, P.
Hypertension and Cardiovascular
Rehabilitation Unit
University of Leuven
Leuven, Belgium

MacGregor, G. A.
Charing Cross Hospital Medical School
London, UK

Maitra, S. R.
Henry Ford Hospital
Detroit, MI U.S.A.

Markandu, N. D.
Charing Cross Hospital Medical School
London, UK

McCaa, R. E.
University of Mississippi
School of Medicine
Jackson, MS U.S.A.

McKinstry, D. N.
The Squibb Institute for Medical Research
Princeton, NJ U.S.A.

Meyer, J. H.
The Squibb Institute for Medical Research
Princeton, NJ U.S.A.

Morton, J. J.
The Medical Research Council
Glasgow, Scotland

Ondetti, M. A.
The Squibb Institute for Medical Research
Princeton, NJ U.S.A.

Parmley, W. W.
University of California, San Francisco
Division of Cardiology & Cardiovascular
Research Institute
San Francisco, CA U.S.A.

Rockhold, R. W.
University of Heidelberg, and
German Institute for High
Blood Pressure Research
Heidelberg, Germany

Roulston, J. E.
Charing Cross Hospital Medical School
London, UK

Rubin, B.
The Squibb Institute for Medical Research
Princeton, NJ U.S.A.

Russell, G. I.
University of Leicester
Leicester, UK

Sabo, E.
The Squibb Institute for Medical Research
Princeton, NJ U.S.A.

Scicli, A. G.
Henry Ford Hospital
Detroit, MI U.S.A.

Schüll, B.
University of Heidelberg and
German Institute for High
Blood Pressure Research
Heidelberg, Germany

Speck, G
University of Heidelberg and
German Institute for High
Blood Pressure Research
Heidelberg, Germany

Stritar, J.
The Squibb Institute for Medical Research
Princeton, NJ U.S.A.

Swales, J. D.
University of Leicester
Leicester, UK

Tarazi, R. C.
Cleveland Clinic Foundation
Research Division
Cleveland, OH U.S.A.

Thurston, H.
Department of Medicine
University of Leicester
Leicester, UK

Trachte, G. J.
University of Virginia
School of Medicine
Charlottesville, VA U.S.A.

Turini, G. A.
Department of Medicine
Centre Hospitalier Universitaire
Vaudois
1011 Lausanne, Switzerland

Unger, T.
University of Heidelberg and
German Institute for High Blood
Pressure Research
Heidelberg, Germany

Vlasses, P. H.
Jefferson Medical College
Thomas Jefferson University
Philadelphia, PA U.S.A.

Waeber, B.
Department of Medicine
Centre Hospitalier Universitaire
Vaudois
1011 Lausanne, Switzerland

Wong L, P.C.
University of Minnesota
Minneapolis, MN U.S.A.

Yasujima, M.
Monash University
Prince Henry's Hospital
Melbourne, Australia

Zimmerman, B. G.
University of Minnesota
Minneapolis, MN U.S.A.

Introduction

This volume is a compilation of the papers and discussions presented at a meeting in Philadelphia, PA, USA, on May 8–9, 1980. It represents an interdisciplinary interaction of biochemists, physiologists, pharmacologists and clinicians, all of whom have been active in studying the effects of angiotensin converting enzyme (ACE) inhibitors.

Because of the complexities of drug discovery we live with today, breakthroughs representing new types of pharmacological agents are very rare. However, we have a unique opportunity in this volume to explore the implications of a completely new class of agents. The questions raised are obviously numerous. Can these agents help explain the role of the hormone angiotensin in both normal function and pathological conditions? Do these agents work only by lowering effective levels of angiotensin or are there other mechanisms involved? Do ACE inhibitors represent an improved method of treating diseases such as hypertension or congestive heart failure? Is this therapy only symptomatic or can it actually reverse pathological causative factors? These questions are very complex and will probably require decades of research to answer. The text of this volume, though, begins to address these and other questions that the discovery of selective ACE inhibitors has raised.

The first section of this book presents an interesting history of the decade of research that went into finding and proving the merit of ACE inhibitors. It also describes the unique molecular design studies at The Squibb Institute for Medical Research that resulted in the synthesis of captopril, the first orally effective ACE inhibitor. It also includes an excellent, in-depth review of the effects of these agents on various hypertensive animal models. The remaining papers in Section I and throughout Section II address themselves to the possible site and mechanism of action of these new compounds.

Section III describes research studies in animals and man on the effects of angiotensin and ACE inhibitors on hemodynamics and fluid balance. We also find preliminary data on the effect of decreasing the levels of angiotensin in diseases other than hypertension. The papers in Section IV address themselves to the treatment of various types of hypertension with ACE inhibitors.

Each section is followed by discussion between the authors and members of the symposium audience, which numbered over 300 people. This volume closes with a paper by Dr. John Laragh, The A. N. Richards Memorial Lecturer, entitled "The Renin System in Hypertension: A Research Journey," and an excellent symposium summary and overview by Dr. Franz Gross.

My colleagues and I at The Squibb Institute for Medical Research are very proud of being able to provide the scientific and medical communities with a new way to study the renin angiotensin system and possibly a new method to treat cardiovascular diseases. The importance of the information these compounds have generated and the ideas coming from this volume are made more relevant by a letter to the editor in *The Journal of the American Medical Association* (*JAMA*, April 25, 1980, Vol. 243, p. 1631, No. 16) appearing just two weeks before this symposium and entitled "Angiotensin and Hypertension". The first paragraph reads: "I think the time has come when the cliché, 'the mechanisms and cause of essential hypertension are unknown,' can justifiably be altered to, 'it is highly probable that the core of the mechanism of essential, renovascular, and malignant hypertension is angiotensin.' The evidence from blocking the converting enzyme and from peptide blocking agents all but conclusively shows that when angiotensin is unable to function, blood pressure returns to normal or near normal, especially if salt has been depleted." The author was Dr. Irving Page, and although he might be accused of being somewhat biased since he was one of the discoverers of angiotensin, I do believe the results described in this book partially support him. With the availability, in some countries, of the first marketed ACE inhibitor, captopril,* occurring while this book is being printed, the research community may soon be able to make the evidence conclusive and vindicate Dr. Page's belief in the crucial role of angiotensin in the pathogenesis of hypertension.

Finally, I would like to thank the many people who helped make the symposium and this volume possible:

Dr. *Warren Chernick* and the Board of the Physiological Society of Philadelphia.

Drs. *M.J. Antonaccio, J.P. Buckley, R.E. McCaa* and *D.N. McKinstry*, who served as Session Chairpersons and helped edit this book.

Dr. *Morton E. Goldberg*, who originated the idea of this symposium, who served as its Co-Chairman and did such a wonderful job of arranging the facilities and serving as liaison with the Society.

*Capoten®, E. R. Squibb & Sons, Inc.

Mrs. *Linda Tywford* for the long hours she spent in assisting Dr. Goldberg and me in the arrangements for the symposium.

Messrs. *M. Perez, J. Alcantara, R. Roytos* and *A. Warade* for their assistance during the symposium.

And especially Mrs. *Marilyn Stopko* for her help and patience throughout the editing and preparation of this volume.

August, 1980

ZOLA P. HOROVITZ
Princeton, New Jersey

SECTION I

Development: Central and Autonomic Effects

Joseph P. Buckley, *Chairperson*

Angiotensin Converting Enzyme Inhibitors: Evolution of a New Class of Antihypertensive Drugs

David W. Cushman, Hong Son Cheung, Emily F. Sabo and Miguel A. Ondetti
The Squibb Institute for Medical Research
Princeton, New Jersey 08540

INTRODUCTION

The history of the development of antihypertensive drugs that act by inhibiting angiotensin converting enzyme is inextricably linked to the prevailing level of understanding of the structural and catalytic properties of this unusual peptidase. The action of most drugs is initiated by specific attachment to a macromolecular "receptor," the chemical constitution of which is usually unknown. For converting enzyme inhibitors, the drug receptor is the active site of the enzyme, a limited but readily accessible portion of the protein structure that is responsible for specific binding of peptide substrates and for hydrolytic cleavage of the proper peptide bond. Specific binding to the active site of angiotensin converting enzyme has been the guiding principle behind the development of teprotide and captopril, potent competitive inhibitors of this enzyme that have found clinical use as antihypertensive drugs. Development of captopril, in particular, was highly dependent on a general understanding of the chemical nature of the active site of angiotensin converting enzyme, one result of almost three decades of study of its properties.

PROPERTIES OF ANGIOTENSIN CONVERTING ENZYME

In 1954, Skeggs and coworkers found that angiotensin, the pressor product of renin action known then as hypertensin or angiotonin, existed in two forms that were separable by countercurrent distribution. The product obtained from incubations of renin with its plasma substrate in the presence of chloride was named angiotensin II, that obtained in the absence of such a monoanion was called angiotensin I. That angiotensin I lacked direct vasoconstrictor action was demonstrated by studies in artificially perfused kidney (Skeggs et al., 1956) and on the isolated aortic strip (Helmer, 1957). The enzyme responsible for conversion of the inactive angiotensin I to the vasoconstrictor angiotensin II in the presence of chloride ion was partially purified from horse plasma by Skeggs et al. (1956), and given the name angiotensin converting enzyme (originally hypertensin converting enzyme). This enzyme appeared to contain a tightly bound functional metal ion, since it was inhibited by metal-chelating agents such as EDTA, but lost no activity upon dialysis. When the sequences of angiotensins I and II were determined, it was apparent that the converting enzyme hydrolyzed the carboxyl-terminal dipeptide His-Leu from the decapeptide angiotensin I to liberate the

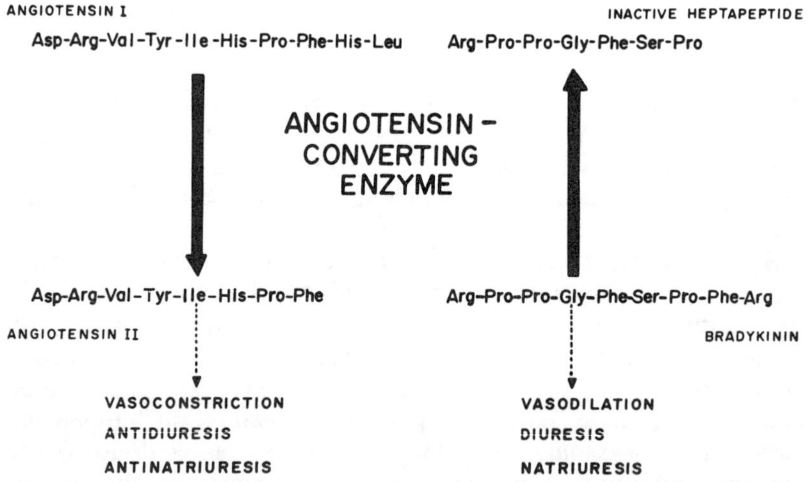

Fig. 1. Two reactions catalyzed by angiotensin converting enzyme (kininase II) that have potential importance in blood pressure regulation. Reprinted with permission from *Enzyme Inhibitors as Drugs*, Merton Sandler, editor, 1980, p. 232. Copyright by The Macmillan Press, Ltd., London.

biologically active octapeptide angiotension II (Fig. 1; Lentz et al., 1956).

Interest in angiotensin converting enzyme lay dormant for the next decade, as Skeggs' group turned their attention to the renin and angiotensinogen (renin substrate) components of the renin-angiotensin system (*see* Skeggs et al., 1966). In the late 1960s and early 1970s, however, several factors contributed to a renaissance of interest in this enzyme and in the role of the renin-angiotensin system in hypertensive disease. Ng and Vane (1967, 1968) showed that functionally important conversion of angiotensin I to angiotensin II took place not in plasma, but in vascular beds of tissues, particularly the lung. Simple tripeptide substrates (Fig. 2) were developed that allowed quantitative and specific spectrophotometric and fluorometric assay of angiotensin converting enzyme activity in tissue extracts (Piquilloud et al., 1970; Cushman and Cheung, 1971a). Angiotensin converting enzyme was shown to be identical with an enzyme (kininase II) that plays a major role in inactivation of the hypotensive peptide bradykinin (Fig. 1; Yang et al., 1970). Finally, as discussed later, the first specific inhibitors were isolated from snake venom.

Substrate Specificity

By 1971, it was obvious to most investigators (see Cushman and Cheung, 1971a; Yang et al., 1970) that angiotensin converting enzyme was not an endopeptidase, but an unusual type of exopeptidase that cleaved the second peptide bond from the carboxyl-terminal end of a variety of peptide substrates including angiotensin I, bradykinin, Hip-His-Leu, and Z-Phe-His-Leu (Fig. 2, Table 1). Results of Ng and Vane (1970) appeared to confirm this mechanism of peptide bond hydrolysis in vivo; they showed that an undecapeptide analog of angiotensin I with an extra histidine (Asp-Arg-Val-Tyr-Ile-His-Pro-Phe-His-His-Leu) was inactivated rather than converted to angiotensin II in the pulmonary vascular beds, presumably because of sequential hydrolysis of the dipeptides His-Leu and Phe-His from the carboxyl-terminal end.

The early 1970s saw increased study of the substrate specificity of angiotensin converting enzyme with simple peptide substrates. As shown in Table 1, a tripeptide with an acylated terminal amino group is the simplest peptide cleaved by the enzyme. However, the tripeptide Z-Phe-His-Leu, analogous to the terminal tripeptide sequence of angiotensin I, binds to the active site of angiotensin converting enzyme as well as the intact decapeptide, as judged by the relative K_m values shown in Table 1. From results shown in Table 1, it can be seen that an aromatic amino acid such as phenylalanine in the antepenultimate

Table 1. Substrate specificity of angiotensin converting enzyme.

	Peptide Structure[a]	K_m (μM)	References
Substrates			
(AI)	Asp-Arg-Val-Tyr-Ile-His-Pro-Phe-His-Leu	50–70	Skeggs et al., 1956
			Cushman and Cheung, 1972
			Das and Soffer, 1975
	Z-Phe-His-Leu	50	Piquilloud et al., 1970
	Bz-Gly-His-Leu	2200	Cushman and Cheung, 1972
			Das and Soffer, 1975
	Bz-Gly-Gly-Gly	3000	Yang et al., 1971
			Dorer et al., 1976
(BK)	Arg-Pro-Pro-Gly-Phe-Ser-Pro-Phe-Arg	0.9	Yang et al., 1970
			Dorer et al., 1974
	<Glu-Lys-Trp-Ala-Pro	—	Cheung and Cushman, 1973
	Bz-Gly-Ala-Pro	30	Cheung et al., 1980
(Met⁵-Enk)	Tyr-Gly-Gly-Phe-Met	—	Erdös et al., 1978
Not Hydrolyzed			
(AII)	Asp-Arg-Val-Tyr-Ile-His-Pro-Phe		Skeggs et al., 1956
	<Glu-Trp-Pro-Arg-Pro-Gln-Ile-Pro-Pro		Cheung and Cushman, 1973
	Z-Gly-Pro-Gly		Elisseeva et al., 1971
	Bz-Gly-Phe-Glu		Elisseeva et al., 1971
	Bz-Gly-Gly-Phe-NH_2		Elisseeva et al., 1971
	Bz-Gly-His		Cushman and Cheung, 1971a
	Phe-His-Leu		Angus et al., 1973

[a]Abbreviations: AI, angiotensin I; BK, bradykinin; Met^5-Enk, Met^5-enkephalin; AII, angiotensin II; Z, benzyloxycarbonyl; Bz, benzoyl; <Glu, pyroglutamyl or pyrrolidone-5-carbonyl

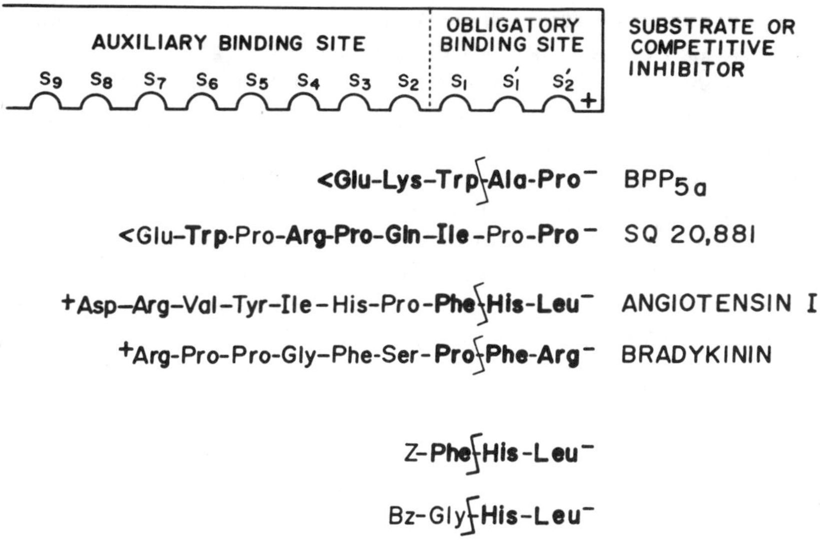

Fig. 2. Comparison of binding of substrates and competitive snake venom inhibitors to the active site of the angiotensin converting enzyme. Circular clefts, labeled S_1 to S_n or S'_1 to S'_2 in both directions from the catalytic center of the enzyme, represent subsites that can potentially interact with the corresponding amino acid residues of peptide substrates or inhibitors. The region of the active site designated the obligatory binding site is essential for binding and proper alignment of both substrates and competitive inhibitors; only the competitive snake venom peptide inhibitors appear to interact significantly with the region of the active site designated the auxiliary binding site. Amino acid residues of substrates or inhibitors thought to interact effectively with either obligatory or auxiliary binding sites are shown in boldface. Abbreviations used are: <Glu, pyrrolidone-5-carbonyl (pyroglutamyl); Z, benzyloxycarbonyl; Bz, benzoyl; BPP_{5a}, bradykinin potentiating peptide 5a (a synthetic venom peptide inhibitor); SQ 20,881, the designation employed by the Squibb Institute for Medical Research for the synthetic nonapeptide from *Bothrops jararaca* venom.

position of a peptide substrate contributes greatly to the overall affinity for the enzyme (Z-Phe-His-Leu vs. Bz-Gly-His-Leu). Peptides such as angiotensin II with a penultimate proline residue, those with a terminal dicarboxylic amino acid such as glutamate, and those lacking an ionizable terminal carboxyl group are not cleaved by the converting enzyme (Table 1). These structural requirements for substrates of angiotensin converting enzyme are quite similar to those observed with carboxypeptidase A of bovine pancreas (Hoffmann and Bergmann, 1940; Hartsuck and Lipscomb, 1971), and provide additional evidence that the converting enzyme is an exopeptidase with a hydrolytic mechanism similar to that of the classical carboxypeptidases.

These early substrate specificity studies and more recent examples included in Table 1 indicate that peptide substrates bind to the active site of angiotensin converting enzyme predominantly via their carboxyl-terminal tripeptide residues. The region of the active site that binds the essential carboxyl-terminal tripeptide sequence of substrates is designated 'obligatory binding site' in Figure 2. The pockets or other discrete areas at this obligatory binding site that specifically interact with the amino acid side chains of substrates are represented by circular clefts and are referred to as 'subsites,' labeled S_1, S_1', and S_2' in Figure 2 in accordance with the nomenclature of Schechter and Berger (1967). Thus, the early substrate specificity studies indicated that subsite S_1 was probably a hydrophobic pocket with a good affinity for aromatic side chains such as that of phenylalanine; that subsite S_1' did not interact effectively with proline; and that subsite S_2' did not tolerate acidic side chains of dicarboxylic amino acids. It was also clear from these early studies that a positively charged group on the active site of angiotensin converting enzyme formed an ionic bond with the negatively charged terminal carboxyl group of the substrate and thus initiated the proper alignment of the terminal amino acid residues of the substrate with subsites at the active site.

Enzymatic and Physicochemical Properties

The availability of simple spectrophotometric and fluorometric assay procedures for assessing activity of angiotensin converting enzyme in crude tissue extracts was essential for accurate determination of tissue distribution of this enzyme (Roth et al., 1969; Cushman and Cheung, 1971b and 1972), for its purification (Yang et al., 1971; Dorer et al., 1972; Cushman and Cheung, 1972; Cheung and Cushman, 1973; Oshima et al., 1974; Nakajima et al., 1973) and for its enzymatic and physicochemical characterization. Biological or radiochemical assays with angiotensin I as substrate used in many pioneering studies (Huggins and Thampi, 1968; Huggins et al., 1970; Bakhle and Reynard, 1971; Alabaster and Bakhle, 1972) suffer from interference by a variety of other peptidases present in extracts and, in the case of the bioassay procedures, from inherent variability.

The pH optimum of angiotensin converting enzyme was found to vary with the substate employed and to be influenced by the presence or absence of chloride ion. With longer peptide substrates such as angiotensin I or bradykinin in the presence of chloride ion, the optimal pH for hydrolytic action of the converting enzyme was about 7.5; with tripeptide substrates such as Z-Phe-His-Leu, Hip-His-Leu, or Hip-Gly-Gly, it was about pH 8.5 (Cushman and Cheung, 1971a; Dorer et al.,

1970, 1972 and 1974; Piquilloud et al., 1970). The effect of chloride ion on rate of hydrolysis of substrates also varied with the particular substrate; the rates of cleavage of angiotensin I, Z-Phe-His-Leu, Hip-His-Leu and Hip-Gly-Gly were all greatly enhanced by chloride ion, while cleavage of bradykinin was activated only slightly (Dorer et al., 1972 and 1974; Piquilloud et al., 1970, Cushman and Cheung, 1971a and 1972; Alabaster and Bakhle, 1972). As discussed in the following section, cleavage of some peptide substrates can be inhibited rather than activated by chloride ion (Cheung and Cushman, 1973; Cheung et al., 1980). Michaelis constants for several substrates of angiotensin converting enzyme are listed in Table 1.

Early studies on inhibition of angiotensin converting enzyme by chelating agents such as EDTA and reactivation by various divalent metal ions suggested that the enzyme contained a tightly bound zinc ion that was essential for the hydrolytic reaction that it catalyzes (Cushman and Cheung, 1971a; Dorer et al., 1970; Yang et al., 1971; Stevens et al., 1972); this was later shown directly by Das and Soffer (1975). Angiotensin converting enzyme in various tissues has been shown to be a glycoprotein consisting of a single polypeptide chain with a molecular weight of 130,000 to 150,000 (see Das and Soffer, 1975; Soffer, 1976; Lanzillo and Fanburg, 1976).

PEPTIDE INHIBITORS FROM SNAKE VENOM

Bradykinin Potentiating Factor

Before 1968, the only known inhibitors of angiotensin converting enzyme were nonspecific metal chelators such as EDTA. However, the key discovery that led eventually to the first potent and specific inhibitors of this enzyme was published three years earlier. Ferreira (1965) showed that a nontoxic ethanol extract of the venom of the Brazilian arrowhead viper, *Bothrops jararaca*, potentiated smooth muscle contraction, hypotension and increased capillary permeability induced by bradykinin, a nonapeptide originally isolated after incubation of plasma with the same venom. Various properties of this 'bradykinin potentiating factor' (BPF), including its inactivation by trypsin, indicated that it was a peptide or mixture of peptides. At least part of the bradykinin potentiating activity of BPF appeared to be due to its inhibition of the enzymatic degradation of bradykinin in plasma (Ferreira, 1966); but several years were to pass before it became clear that angiotensin converting enzyme was the bradykininase inhibited by BPF.

Table 2. Venom peptides of *Bothrops jararaca*.

Alternative Nomenclatures[a]		Structures[b]	References
SQ 20,475	BPP_{5a}	<Glu-Lys-Trp-Ala-Pro	Ferreira et al., 1970b
—	BPP_{6a}	(Glx,Ser,Trp,Pro,Gly,Pro)	Ferreira et al., 1970b
SQ 20,881	BPP_{9a}	<Glu-Trp-Pro-Arg-Pro-Gln-Ile-Pro-Pro	Ondetti et al., 1971; Ferreira et al., 1970b
SQ 20,859	BPP_{10a}	<Glu-Ser-Trp-Pro-Gly-Pro-Asn-Ile-Pro-Pro	Ferreira et al., 1970b
SQ 20,861	BPP_{10b}	<Glu-Asn-Trp-Pro-Arg-Pro-Gln-Ile-Pro-Pro	Ondetti et al., 1971
SQ 20,858	BPP_{10c}	<Glu-Asn-Trp-Pro-His-Pro-Gln-Ile-Pro-Pro	Ondetti et al., 1971; Ferreira et al., 1970b
SQ 20,661	BPP_{11d}	<Glu-Trp-Pro-Arg-Pro-Thr-Pro-Gln-Ile-Pro-Pro	Ferreira et al., 1970b
—	BPP_{12a}	(Glx,Glx,Trp,Pro,Arg,Pro,Ile,Pro,Glx,Trp,Ala,Pro)	Ferreira et al., 1970b
SQ 20,718	BPP_{13a}	<Glu-Gly-Gly-Trp-Pro-Arg-Pro-Gly-Pro-Glu-Ile-Pro-Pro	Ondetti et al., 1971; Ferreira et al., 1970b
—	BPP_{13b}	(Glx,Gly,Gly,Leu,Pro,Arg,Pro,Gly,Pro,Glx,Ile,Pro,Pro)	Ferreira et al., 1970b

[a] The SQ numbers and the general nomenclature of the venom peptides are discussed by Cheung and Cushman, 1973; the BPP (bradykinin potentiating peptide) nomenclature was suggested by Stewart et al., 1971.

[b] Peptides of undetermined structure are shown in parentheses with the component amino acids arranged in a likely sequential order; Glx designates either glutamic acid or glutamine.

The initial testing of BPF for inhibition of angiotensin converting enzyme was based on a mistaken assumption about the relationship of this enzyme to the plasma bradykininase inhibited by BPF. In 1968, the best characterized bradykinin degrading enzyme of plasma was a carboxypeptidase referred to as kininase I (Erdös and Yang, 1966), an enzyme that cleaved the carboxyl-terminal arginine residue from bradykinin, as indicated below. Ng and Vane (1968) noted that this enzyme and angiotensin converting enzyme both cleaved bonds following the sequence Pro-Phe in their peptide substrates, and speculated that they might be identical.

kininase I
↓
Arg-Pro-Pro-Gly-Phe-Ser-Pro-Phe-Arg *bradykin*
Asp-Arg-Val-Tyr-Ile-His-Pro-Phe-His-Leu *angiotensin I*
↑
converting enzyme

Although the reasoning was not quite correct, the decision to test BPF as an inhibitor of angiotensin converting enzyme was a nexus in the history of development of inhibitors of angiotensin converting enzyme. Bakhle (1968) reported that BPF was a potent inhibitor of angiotensin converting enzyme of dog lung homogenate, and the long-delayed purification of the active components of BPF was initiated by two groups of investigators.

Synthetic Venom Peptides

Ferreira et al. (1970b) reported the isolation and amino acid content of nine bradykinin potentiating peptides from *Bothrops jararaca* venom; they determined the amino acid sequence of one of these, the pentapeptide BPP_{5a} (Table 2), which they also synthesized. Ondetti et al. (1971) isolated six peptidic inhibitors of angiotensin converting enzyme from the same venom, and sequenced and synthesized each of these nonapeptide to tridecapeptide inhibitors. All of the peptides isolated by both groups, including three not yet sequenced or synthesized (Table 2), have both converting enzyme inhibitory and bradykinin potentiating activities in vitro and in vivo (Cushman and Cheung, 1972; Cheung and Cushman, 1973; Ferreira et al., 1970a and 1970b; Stewart et al., 1971; Engel et al., 1972; Bakhle, 1972; Greene et al., 1972). Since no universally accepted nomenclature exists for these peptides, we will employ designations for individual peptides used by the investigators who first determined their structures. Five similar peptides isolated from the venom of the Japanese pit viper *Agkistrodon halys blomhoffii* (Kato and Suzuki, 1971) have not been studied extensively.

Table 3. Specificity of action of inhibitors of angiotensin converting enzyme.

System	I_{50} or A_{50} (μM)[a]				
	BPP_{5a}	SQ 20,881	2	3	4 (captopril)
Peptidases					
Angiotensin converting enzyme	0.06	0.55	17	0.2	0.02
Carboxypeptidase A	>300	1800	1500	1500	1500
Carboxypeptidase B	—	520	>1000	>1000	800
Leucine aminopeptidase	1000	>100	>1000	40	5.4
Trypsin	>300	1000	>1000	>1000	>1000
Chymotrypsin	1000	380	>1000	>1000	>1000
Agents Contracting Guinea Pig ileum					
Angiotensin I	<3	0.06	18	0.2	0.02
Bradykinin	0.2	0.002	1	0.02	0.003
Angiotensin II	>50	>30	>500	>500	>500
Acetylcholine	>50	>30	>500	>500	>500

[a]All values are I_{50} (concentrations producing 50% inhibition) except for those describing effects on bradykinin induced contraction of guinea pig ileum, which are A_{50} (concentrations producing 50% augmentation). Values for captopril and SQ 20,881 are according to Cushman et al., 1979, Ondetti et al., 1979a, and Rubin et al., 1978a and 1978b; values for inhibition of carboxypeptidase A were reported by Cushman et al., 1978. Other values are unpublished data obtained by Cushman, Rubin and coworkers.

Activities in vitro and in vivo

The pentapeptide BPP_{5a} (Table 3) was the most potent venom peptide inhibitor of angiotensin converting enzyme of rabbit lung (Cheung and Cushman, 1973; Cushman and Cheung, 1972); but, unlike other venom peptides, it can also be cleaved by the enzyme (most rapidly in the absence of chloride ion) to liberate its terminal dipeptide residue Ala-Pro. The nonapeptide SQ 20,881 was the next most potent inhibitor of the rabbit lung converting enzyme, and is now known to be an even more potent inhibitor of angiotensin converting enzyme isolated from tissues such as guinea pig ileum and rat testis (Cushman et al., unpublished observations), from rat brain (Yang and Neff, 1972) or from human urine (Fisher and Ryan, 1979). In studies with isolated strips of guinea pig ileum, the biologically stable nonapeptide SQ 20,881 was more effective than BPP_{5a} as an inhibitor of contractions induced by angiotensin I or as an augmentor of the contractions induced by bradykinin (Table 3). Similar results were obtained in vivo when the venom peptides were tested for inhibition of the vasopressor action of angiotensin I or for augmentation of the vasodepressor action of bradykinin (Engel et al., 1972; Greene et al., 1972); the pentapeptide BPP_{5a} has a very short duration of activity in such in vivo tests, presumably because of its rapid metabolism by peptidases. SQ 20,881 was also the most effective of the venom peptides for lowering blood pressure in animal models of hypertension, and was eventually shown to be markedly effective for treatment of human hypertension when administered parenterally (Engel et al., 1973; Samuels et al., 1976; Muirhead et al., 1974; Gavras et al., 1974; Johnson et al., 1975).

Mechanism and Specificity of Inhibition

Structure-activity correlations among analogs of BPP_{5a} and SQ 20,881 suggested that these snake venom peptide inhibitors compete with substrates for binding to the active site of angiotensin converting enzyme by binding in the same manner as substrates to the previously discussed obligatory binding site of the enzyme (subsites S_1, S_1', and S_2', in Fig. 2); the increased affinity of the snake venom peptide inhibitors over that normally observed with peptide substrates appeared to be due to additional interactions with an adjacent region at the active site of the enzyme (subsites S_2 to S_9 in Fig. 2) that we referred to as the 'auxiliary binding site' (Cushman et al., 1973; Cheung and Cushman, 1973). Only carboxyl-terminal fragments of BPP_{5a} or SQ 20,881 retained significant enzyme inhibitory activity, and the tripeptide analog Phe-Ala-Pro, analogous to the terminal tripeptide sequence of BPP_{5a}, was a moderately potent inhibitor. Substitution of certain amino acid residues into the carboxyl-terminal tripeptide sequence of the competitive snake venom peptide inhibitors had the same effect on enzyme binding affinity as had been observed earlier with peptide substrates; an antepenultimate aromatic amino acid enhanced binding, and a penultimate proline, terminal dicarboxylic amino acid or substituted carboxyl group decreased binding. The terminal sequence Trp-Ala-Pro or Phe-Ala-Pro appeared to be nearly optimal for binding to the obligatory binding site, whether it was present in an analog of BPP_{5a} or incorporated into an analog of SQ 20,881. Substitution of an alanine residue for the penultimate proline of SQ 20,881, however, makes the resulting analog, like BPP_{5a}, susceptible to cleavage by angiotensin converting enzyme.

As shown in Table 3, neither BPP_{5a} nor SQ 20,881 has any significant inhibitory activity toward other peptidases, including zinc metallopeptidases that are similar in mechanism of action to angiotensin converting enzyme. However, excised guinea pig ileum contracting in response to a variety of agonists added to the bath fluid has proved to be a more convenient test system for demonstrating such specificity of action of converting enzyme inhibitors (Rubin et al., 1978a and 1978b). Thousands of randomly selected compounds, including many potent but nonspecific converting enzyme inhibitors, have been found to have no effect on agonists contracting guinea pig ileum, or to uniformly inhibit all such contractions, usually at high doses. Only specific inhibitors of angiotensin converting enzyme have the profile of activities in the guinea pig ileum test system shown in Table 3 for BPP_{5a} and SQ 20,881: inhibition of contractions elicited by angiotensin I, augmentation of contractions induced by bradykinin and no effect, even at high concentrations, on contractions due to angiotensin II, acetylcholine, prostaglandins, catecholamines or other agonists. The guinea pig ileum test has also been an excellent predictor of in vivo effectiveness of converting enzyme inhibitors.

DESIGN OF NONPEPTIDE INHIBITORS

Although SQ 20,881 (teprotide) is an effective antihypertensive drug, its lack of oral activity would probably restrict its use to diagnostic testing and hypertensive emergencies. Thus, in early 1974, the efficacy of converting-enzyme inhibitors as antihypertensive drugs had been demonstrated, but the development of an orally effective agent for use in chronic therapy seemed, at best, many years away. This gloomy situation was dramatically altered by a new approach to development of converting-enzymes inhibitors, the design of specific active site binding agents, that led in only a year and a half to the first synthesis of captopril, a markedly potent and specific orally active inhibitor.

Hypothetical Active Site Model

As discussed earlier, the substrate specificity and other properties of angiotensin converting enzyme suggested that it was a zinc metallopeptidase similar in mechanism to carboxypeptidase A, an enzyme whose active site had been well characterized by x-ray crystallography and other methods (Quiocho and Lipscomb, 1961). Thus, in our earliest concepts of the active site of angiotensin converting enzyme, it was implicit that the enzyme had a positively charged carboxyl-binding residue, as shown in Figures 2 to 4. It was also clear, although not usually discussed in reference to active site binding of the snake venom peptide inhibitors, that the zinc ion of angiotensin converting enzyme was appropriately located (between subsites S_1 and S_1' in Fig. 2) to participate in hydrolytic cleavage of the peptide bond between antepenultimate and penultimate amino acid residues of the substrate, resulting in release

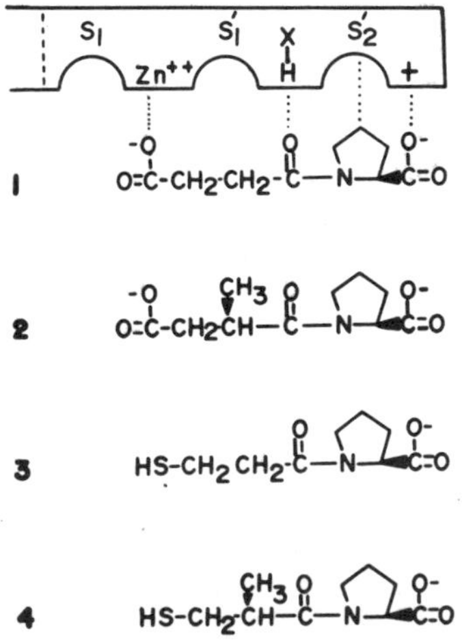

	RABBIT LUNG CONVERTING ENZYME		GUINEA PIG ILEUM	
			AI	BK
	I_{50}(μM)	K_i(μM)	I_{50}(μM)	A_{50}(μM)
1	330	-	440	37
2	22	2.5	18	0.87
3	0.20	0.012	0.30	0.025
4	0.023	0.0017	0.023	0.0032

SQ 14,225 (CAPTOPRIL)

Fig. 3. Structures and activities of important intermediates in the design of captopril, a potent and specific orally active inhibitor of angiotensin converting enzyme. The diagrammatic model of the active site of angiotensin converting enzyme is explained in the legend to Figure 2. Abbreviations used are: X–H, a postulated hydrogen bond donating group at the active site of the enzyme; AI, angiotensin I; BK, bradykinin; I_{50}, concentration producing 50% inhibition; A_{50}, concentration producing 50% augmentation; K_i, inhibitor constant (enzyme inhibitor dissociation constant).

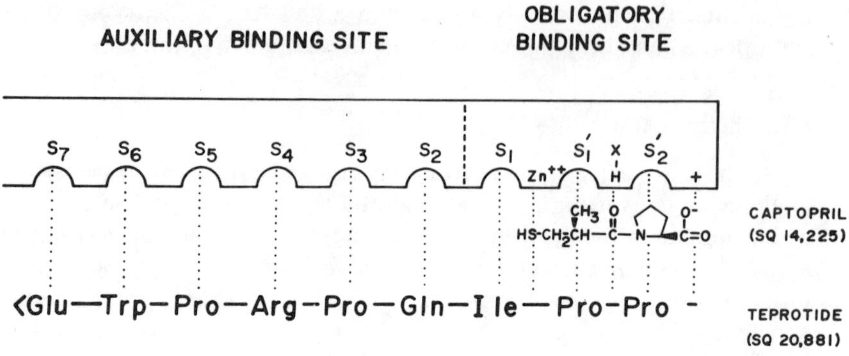

Fig. 4. Comparison of active site binding of two clinically useful inhibitors of angiotensin converting enzyme. Abbreviations used and active site designations are explained in the legends to Figures 2 and 3.

of the dipeptide product. Later, it seemed reasonable to include in our more complete model of the obligatory binding site of the converting enzyme (Figs. 3 and 4) a hydrogen bond donating group (X-H), located between subsites S_1' and S_2', that could bind the terminal, nonscissile peptide bond of a peptide substrate, although evidence for such an interaction did not yet exist (Ondetti et al., 1977a, 1977b and 1979b; Cushman et al., 1977, 1978 and 1979). As described earlier, studies with venom peptides indicated that subsites S_1, S_1', and S_2' had strong affinities for side chains of tryptophan, alanine and proline, respectively.

Biproduct Analog Inhibitors

Byers and Wolfenden (1972 and 1973) had described a potent competitive inhibitor of carboxypeptidase A, D-2-benzylsuccinic acid, which they felt was bound at the active site of this enzyme with ionic and hydrophobic interactions of the 1-carboxyl and benzyl groups corresponding to similar interactions of a typical hydrophobic amino acid product such as L-phenylalanine, but with an additional interaction of the 4-carboxyl group that they thought was analogous to the active site binding interaction of the terminal carboxyl group of the non-amino acid product of carboxypeptidase action.

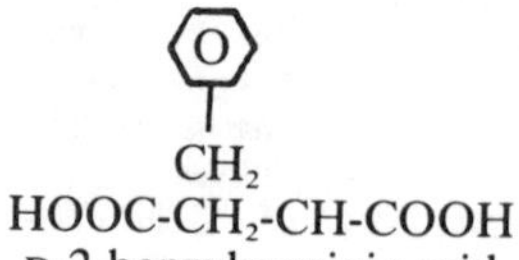

D-2-benzylsuccinic acid

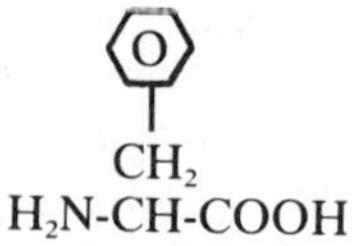

L-phenylalanine

Such an inhibitor thus could be viewed as a 'biproduct analog' or as a 'bisubstrate analog' for the reverse reaction of carboxypeptidase A.

The exact site of binding of the 4-carboxyl group D-2-benzylsuccinic acid to the enzyme turned out to be important for our own studies on inhibition of angiotensin converting enzyme. Byers and Wolfenden hesitated to suggest the zinc ion of the enzyme as this site, because they felt that they had evidence to indicate that the 4-carboxyl group was protonated when bound to the active site of carboxypeptidase A. Bunting and Myers (1974), studying various dicarboxylic acid inhibitors of carboxypetidase A, felt that such compounds were bound as dianions, with one carboxyl group bound to the positively charged carboxyl recognition site on the enzyme and the other to the zinc ion of the enzyme, replacing the water molecule that normally serves as the fourth ligand for the zinc. When we attempted to develop analogous biproduct analog inhibitors of angiotensin converting enzyme (Ondetti et al., 1979a; Cushman et al., 1977), we also assumed that the second carboxyl group of such compounds could interact with the zinc ion of this enzyme (Fig. 3).

Figure 3 shows our initial biproduct analog inhibitor of angiotensin convertig enzyme, succinyl-L-proline (1). This compound is an analog of the dipeptide Gly-Pro, which should have a favorable interaction with subsite S_2' of the enzyme, as deduced from our studies with venom peptide inhibitors, but would not interact with subsite S_1'; we assumed that the succinyl carboxyl group would be properly situated to interact with the zinc ion of angiotensin converting enzyme as diagrammed in Figure 3. As shown in Figure 3, succinyl-L-proline is a fairly weak inhibitor of angiotensin converting enzyme, but it did have the profile of a specific inhibitor in the guinea pig ileum test system: inhibition of angiotensin I–induced contractions, augmentation of bradykinin-induced contractions, but no other agonistic or antagonistic actions.

Optimization of Active Site Binding: The Road to Captopril

Encouraged by the apparent specificity of succinyl-L-proline (1) as a converting enzyme inhibitor, we embarked upon structure-activity studies that eventually led to the synthesis of hundreds of analogs designed to test our hypothetical model and to improve the modest active site binding observed with the prototype biproduct analog inhibitor (Ondetti et al., 1977a, 1977b and 1979b; Cushman et al., 1977, 1978 and 1979). Five postulated enzyme binding interactions of succinyl-L-proline and its analogs formed the basis for these highly directed structure-activity studies: side chain interactions with subsites S_1' and S_2', ionic interaction of the amino acid carboxylate with the positively charged residue of the enzyme, hydrogen bonding of the amide carbonyl with the putative donor X-H and interaction of the succinyl carboxylate or alternate structure with the zinc ion of the enzyme (Figs. 3 and 4).

In agreement with the postulated mode of binding of succinyl-L-proline to the active site of the converting enzyme (Fig. 3), introduction of a substituent in the 2-position of the succcinyl group enhanced inhibitory activity, but only if its steric configuration mimicked that of the penultimate L-amino acid residue of a substrate that normally interacts with subsite S_1' of the enzyme. One of the most effective of such substituents has been the methyl group; incorporation of this side chain (compound 2, Fig. 3) yields a biproduct analog of the dipeptide Ala-Pro, the terminal sequence of the tightly bound snake venom peptide BPP_{5a}. Compound 2 (D-2-methyl-succinyl-L-proline) was 15 to 20 times more inhibitory than succinyl-L-proline, and was the first compound of this series to be shown to be active in vivo after oral administration (Ondetti et al, 1977a). With regard to enhancing inhibitory activity, the most important of the five postulated interactions between compounds such as 2 and the active site of angiotensin converting enzyme turned out to be the binding of the succinyl carboxylate to the zinc ion of the enzyme. When this group was replaced by sulfhydryl, which appeared from existing data on zinc complexes to be a better single ligand for zinc than carboxylate, the resulting compounds (3 and 4) were 1000 times more inhibitory than their carboxylic acid precursors 1 and 2 (Fig. 3). Captopril, compound 4, is an extremely potent competitive inhibitor of angiotensin converting enzyme; its specific binding to the active site of this enzyme (Fig. 4) has not been substantially bettered in testing hundreds of carefully selected analogs.

Confirmation and Refinement of the Active Site Model

The large number of structure-activity correlations obtained with analogs leading to or related to captopril have validated each of the five postulated interactions of these compounds with the enzyme, and have indicated the optimal structural requirements for each interaction. The negatively charged amino acid carboxyl group had been shown to be essential for inhibitory activity, all activity being lost upon its replacement by hydrogen. The side chain of proline (incorporated into the ring structure) is one of the most effective, if not the most effective, for interacting with subsite S_2'; and the methyl group is one of the most effective for interacting with subsite S_1'. Studies with dipeptide product inhibitors (Table 4) suggest that valine and tryptophan side chains, respectively, may interact somewhat more efffectively with subsites S_1'

and S_2' than the analogous side chains of captopril, which correspond to alanine and proline (Cheung et al., 1980). In the case of the tryptophan side chain, however, the difference appears to be more important with the rabbit lung enzyme than with the enzyme from guinea pig ileum. The amide carbonyl, proposed to hydrogen bond with the enzyme, is important for inhibitory activity; compounds in which this group is removed or shifted in position are about 1000-fold less inhibitory. The mercaptopropanoyl chain length has been shown to be optimal;

Table 4. Inhibition of angiotensin converting enzyme by dipeptides.

Dipeptide	I_{50}[a] (μM)	Dipeptide	I_{50}[a] (μM)
Val-Trp	1.6	Gly-Ala	2,000
Ile-Trp	2.0	Tyr-Gly	2,000
Ile-Tyr	3.7	Gly-Leu	2,500
Ala-Trp	10	Ala-Gly	2,500
Arg-Trp	16	Gly-His	3,100
Val-Tyr	22	His-Leu[b]	3,200
Gly-Trp	30	Gly-Arg	3,200
Val-Phe	53	Lys-Gly	3,200
Ala-Tyr	88	Phe-Gly	3,700
Ile-Pro	130	Asp-Ala[b]	3,800
Arg-Pro	180	Gly-Ser	3,800
Ala-Phe	190	Gly-Val	4,600
Gly-Tyr	210	Met-Gly	4,800
Ala-Pro	230	Gly-Lys	5,400
Arg-Phe	230	Gly-Glu	5,400
Leu-Ala[b]	310	Gly-Thr	5,700
Lys-Arg[b]	380	Trp-Gly	5,900
Val-Pro	420	His-Gly	6,300
Gly-Pro	450	Gly-Gln	7,000
Arg-Ala[b]	460	Gly-Gly	7,200
Tyr-Ala[b]	460	Gln-Gly	7,400
Ala-Ala[b]	620	Ser-Gly	8,500
Gly-Phe	630	Leu-Gly	8,800
Phe-Arg[b]	920	Gly-Asp	9,200
Ile-Phe	930	Thr-Gly	9,900
Val-Gly	1,100	Glu-Gly	10,000
Ile-Gly	1,200	Glu-Ala[b]	10,000
Gly-Ile	1,200	Asp-Gly	14,000
Gly-Met	1,400	Pro-Gly	17,000

[a] Most I_{50} values are from Cheung et al., 1980; amino-terminal amino acids and carboxyl-terminal amino acids of dipeptides are thought to interact with subsites S_1' and S_2', respectively.
[b] I_{50} values derived from K_i values of Cheung et al., 1980, and Das and Soffer, 1975, using the formula of Cheng and Prusoff, 1973.

and the sulfhydryl group or another zinc binding group such as phosphonic acid or hydroxamate (Ondetti et al., 1977b and 1979b) is essential for inhibition.

As shown in Table 3, captopril (4) and its precursors (2) and (3) are all specific inhibitors of angiotensin converting enzyme, as judged either from studies on inhibition of other peptidases or from their effects on various agonists contracting guinea pig ileum. The only significant activity unrelated to converting enzyme inhibition was the inhibition of the zinc-containing leucine aminopeptidase by compounds 3 and 4 at concentrations 200 times greater than those required to inhibit angiotensin converting enzyme.

CONCLUSIONS

The postulated active site binding interactions of teprotide (SQ 20,881) and captopril (SQ 14,225) are compared in Figure 4. Although two-dimensional and highly idealized, this model of the active site of angiotensin converting enzyme provides a good summary of existing knowledge of the mechanism of binding of substrates and competitive inhibitors by the enzyme. The best characterized portion of the extended active site is that in the immediate vicinity of the 'catalytic center,' defined by the zinc ion shown in the model and other catalytically important residues that were not shown because they do not appear to play any role in binding of substrates or competitive inhibitors, but only in facilitating peptide bond hydrolysis. Both substrates and inhibitors must bind to this obligatory binding site that encompasses subsites S_1, S_1', and S_2'. Studies with simple substrates and with various classes of competitive inhibitors described in this review have indicated that subsites S_1, S_1' and S_2' are all more or less hydrophobic in nature, interacting most effectively with aromatic or branched chain aliphatic groups. Subsites S_2 to S_n, which constitute the auxiliary binding site that enhances binding of the snake venom peptides, are not well characterized. It is not clear whether or not important interactions occur at each of these subsites, and in no case is the nature of the interaction well characterized.

The effectiveness of captopril as an antihypertensive drug is amply documented in other chapters of this volume. The mechanism of captopril's blood pressure lowering action may be quite complex, since it could potentially involve both decreased circulating or tissue levels of angiotensin II and increased circulating or tissue levels of bradykinin, and could include both direct vasoactivities of these peptides and various secondary effects on aldosterone secretion, sodium ion excretion and release of precursors of prostaglandins and prostacyclin (*see* Cushman and Ondetti, 1980). Such complexity has led some investigators to unwarranted speculation about antihypertensive mechanisms for captopril that are unrelated to inhibition of angiotensin converting enzyme. In addition to the lack of direct evidence for any such alternative mechanism, the logical route of development of captopril militates against such a suggestion. As outlined in Figure 3, captopril was developed via a logical sequence of structural alterations from the prototype compound succinyl-L-proline (1), with an attendant 20,000-fold increase in converting enzyme inhibitory activity. At no point in this development process was the optimization of chemical structure guided by antihypertensive activity, except inasmuch as such activity was expected to parallel converting enzyme inhibition. The logical route of development of captopril not only provides great assurance of its primary mechanism of action, but also introduces a specificity of action that should translate clinically into an excellent therapeutic ratio.

REFERENCES

Alabaster, V.A., and Bakhle, Y.S. 1972. Converting enzyme and bradykininase in the lung. Circ Res 30:II–72–II–84

Angus, C.W., Lee, H.J., and Wilson, I.B. 1972. Angiotensin-converting enzyme and a second dipeptidyl carboxypeptidase from hog plasma. Biochim Biophys Acta 276:228–233

Bakhle, Y.S. 1968. Conversion of angiotensin I to angiotensin II by cell-free extracts of dog lung. Nature 220:919–921

Bakhle, Y.S. 1972. Inhibition of converting enzyme by venom peptides. In *Hypertension 1972*, eds. J. Genest and E. Koiw, pp. 541–547. New York: Springer

Bakhle, Y.S., and Reynard, A.M. 1971. Characteristics of the angiotensin I converting enzyme from dog lung. Nature New Biol 229:187–189

Bunting, J.W., and Myers, C.D. 1974. Reversible inhibition of carboxypeptidase A. II. Inhibition of esterase activity by dicarboxylic acid anions. Can J Chem 52:2053–2063

Byers, L.D., and Wolfenden, R. 1972. A potent reversible inhibitor of carboxypeptidase A. J Biol Chem 247:606–608

Byers, L.D., and Wolfenden, R. 1973. Binding of the biproduct analog benzylsuccinic acid by carboxypeptidase A. Biochemistry 12:2070–2078

Cheng, Y.C., and Prusoff, W.H. 1973. Relationship between the inhibition constant (K_i) and the concentration of inhibitor which causes 50 per cent inhibition of an enzymmatic reaction. Biochem Pharmacol 22:3099–3108

Cheung, H.S., and Cushman, D.W. 1973. Inhibition of homogeneous angiotensin-converting enzyme of rabbit lung by synthetic venom peptides of *Bothrops jararaca*. Biochim Biophys Acta 293:451–463

Cheung, H.S., Wang, F.L., Ondetti, M.A., Sabo, E.F., and Cushman, D.W. 1980. Binding of peptide substrates and inhibitors of angiotensin-converting enzyme. Importance of the COOH-terminal dipeptide sequence. J Biol Chem 255:401–407

Cushman, D.W., and Cheung, H.S. 1971a. Spectrophotometric assay and properties of the angiotensin-converting enzyme of rabbit lung. Biochem Pharmacol 20:1637–1648

Cushman, D.W., and Cheung, H.S. 1971b. Concentrations of angiotensin-converting enzyme in tissues of the rat. Biochim Biophys Acta 250:261–265

Cushman, D.W., and Cheung, H.S. 1972. Studies *in vitro* of angiotensin-converting enzyme of lung and other tissues. In *Hypertension 1972*, eds. J. Genest and E. Koiw, pp. 532–541. New York: Springer

Cushman, D.W., and Ondetti, M.A. 1980. Inhibitors of angiotensin-converting enzyme for treatment of hypertension. Biochem Pharmacol 29:1871–1877

Cushman, D.W., Plušĉec, J., Williams, N.J., Weaver, E.R., Sabo, E.F., Kocy, O., Cheung, H.S., and Ondetti, M.A. 1973. Inhibition of angiotensin-converting enzyme by analogs of peptides from *Bothrops jararaca* venom. Experientia 29:1032–1035

Cushman, D.W., Cheung, H.S., Sabo, E.F., and Ondetti, M.A. 1977. Design of potent competitive inhibitors of angiotensin-converting enzyme. Carboxyalkanoyl and mercaptoalkanoyl amino acids. Biochemistry 16:5484–5491

Cushman, D.W., Cheung, H.S., Sabo, E.F., and Ondetti, M.A. 1978. Design of new antihypertensive drugs: Potent and specific inhibitors of angiotensin-converting enzyme. Prog Cardiovasc Dis 21:176–182

Cushman, D.W., Cheung, H.S., Sabo, E.F., and Ondetti, M.A. 1979. Development of specific inhibitors of angiotensin I converting enzyme (kininase II). Fed Proc 38:2778–2782

Das, M., and Soffer, R.L. 1975. Pulmonary angiotensin-converting enzyme. Structural and catalytic properties. J Biol Chem 250:6762–6768

Dorer, F.E., Skeggs, L.T., Kahn, J.R., Lentz, K.E., and Levine, M. 1970. Angiotensin-converting enzyme: Method of assay and partial purification. Anal Biochem 33:102–113

Dorer, F.E., Kahn, J.R., Lentz, K.E., Levine, M., and Skeggs, L.T. 1972. Purification and properties of angiotensin-converting enzyme from hog lung. Circ Res 31:356–366

Dorer, F.E., Kahn, J.R., Lentz, K.E., Levine, M., and Skeggs, L.T. 1974. Hydrolysis of bradykinin by angiotensin-converting enzyme. Circ Res 34:824–827

Dorer, F.E., Kahn, J.R., Lentz, K.E., Levine, M., and Skeggs, L.T. 1976. Kinetic properties of pulmonary angiotensin-converting enzyme. Hydrolysis of hippurylglycylglycine. Biochim Biophys Acta 129:220–228

Elisseeva, Y.E., Orekhovich, V.N., Pavlikhina, L.V., and Alexeenko, L.P. 1971. Carboxycathepsin: A key regulatory component of two physiological systems involved in regulation of blood pressure. Clin Chim Acta 31:413–419

Engel, S.L., Schaeffer, T.R., Gold, B.I., and Rubin, B. 1972. Inhibition of pressor effects of angiotensin I and augmentation of depressor effects of bradykinin by synthetic peptides. Proc Soc Exp Biol Med 140:240–244

Engel, S.L., Schaeffer, T.R., Waugh, M.H., and Rubin, B. 1973. Effects of the nonapeptide SQ 20,881 on blood pressure of rats with experimental renovascular hypertension. Proc Soc Exp Biol Med 143:483–487

Erdös, E.G., and Yang, H.Y.T. 1966. Inactivation and potentiation of the effects of bradykinin. In *Hypotensive Peptides*, eds. N. Back and F. Sicuteri, pp. 235–250. New York: Springer

Erdös, E.G., Johnson, A.R., and Boyden, N.T. 1978. Hydrolysis of enkephalin by cultured human endothelial cells and by purified peptidyldipeptidase. Biochem Pharmacol 27:843–848

Ferreira, S.H. 1965. A bradykinin-potentiating factor (BPF) present in the venom of *Bothrops jararaca*. Brit J Pharmacol 24:163–169

Ferreira, S.H. 1966. Bradykinin-potentiating factor. In *Hypotensive Peptides*, eds. N. Back and F. Sicuteri, pp. 356–367. New York: Springer

Ferreira, S.H., Greene, L.J., Alabaster, V.A., Bakhle, Y.S., and Vane, J.R. 1970a. Activity of various fractions of bradykinin potentiating factor against angiotensin I converting enzyme. Nature 225:379–380

Ferreira, S.H., Bartelt, D.C., and Greene, L.J. 1970b. Isolation of bradykinin-potentiating peptides from *Bothrops jararaca* venom. Biochemistry 9:2583–2593

Fisher, G.H., and Ryan, J.W. 1979. Superactive inhibitors of angiotensin-converting enzyme. Analogs of BPP_{9a} containing dehydroproline. FEBS Lett 107:273–276

Gavras, H., Brunner, H.R., Laragh, J.H., Sealey, J.E., Gavras, I., and Vukovich, R.A. 1974. An angiotensin converting enzyme inhibitor to identify and treat vasoconstrictor and volume factors in hypertensive patients. New Eng J Med 291:817–821

Greene, L.J., Camargo, A.C.M., Krieger, E.M., Stewart, J.M., and Ferreira, S.H. 1972. Inhibition of the conversion of angiotensin I to II and potentiation of bradykinin by small peptides present in *Bothrops jararaca* venom. Circ Res 30:II–62–II–71

Hartsuck, J.A., and Lipscomb, W.N. 1971. Carboxypeptidase A. In *The Enzymes*, Vol. 3, ed. P.D. Boyer, pp. 1–56. New York: Academic Press

Helmer, O.M. 1957. Differentiation between two forms of angiotonin by means of spirally cut strips of rabbit aorta. Am J Physiol 188:571–577

Hofmann, K., and Bergmann, M. 1946. The specificity of carboxypeptidase. J Biol Chem 134:225–235

Huggins, C.G., and Thampi, N.S. 1968. A simple method for the determination of angiotensin I converting enzyme. Life Sci 7:633–639

Huggins, C.G., Corcoran, R.J., Gordon, J.S., Henry, H.W., and John, J.P. 1970. Kinetics of the plasma and lung angiotensin I converting enzymes. Circ Res 26:I–93–I–108

Johnson, J.G., Black, W.D., Vukovich, R.A., Hatch, F.E., Jr., Friedman, B.I., Blackwell, C.F., Shenouda, A.N., Share, L., Shade, R.E., Acchiardo, S.R., and Muirhead, E.E. 1975. Treatment of patients with severe hypertension by inhibition of angiotensin-converting enzyme. Clin Sci Mol Med 48:53s–56s

Kato, H., and Suzuki, T. 1971. Bradykinin-potentiating peptides from the venom of *Agkistrodon halys blomhoffii*. Isolation of five bradykinin potentiators and the amino acid sequences of two of them, potentiators B and C. Biochemistry 10:972–980

Lanzillo, J.J., and Fanburg, B.L. 1976. The estimation and comparison of molecular weight of angiotensin I converting enzyme by sodium dodecyl sulfate–polyacrylamide gel electrophoresis. Biochim Biophys Acta 439:125–132

Lentz, K.E., Skeggs, L.T., Jr., Woods, K.R., Kahn, J.R., and Shumway, N.P. 1956. The amino acid composition of hypertensin II and its biochemical relationship to hypertensin I. J Exp Med 104:183–191

Muirhead, E.E., Brooks, B., and Arora, K.K. 1974. Prevention of malignant hypertension by the synthetic peptide SQ 20,881. Lab Invest 30:129–135

Nakajima, T., Oshima, G., Yeh, H.S.J., Igic, R., and Erdös, E.G. 1973. Purification of the angiotensin I converting enzyme of the lung. Biochim Biophys Acta 315:430–438

Ng, K.K.F., and Vane, J.R. 1967. Conversion of angiotensin I to angiotensin II. Nature 216:762–766

Ng, K.K.F., and Vane, J.R. 1968. Fate of angiotensin I in the circulation. Nature 218:144–150

Ng, K.K.F., and Vane, J.R. 1970. Some properties of angiotensin converting enzyme in the lung *in vivo*. Nature 225:1142–1144

Ondetti, M.A., Williams, N.J., Sabo, E.F., Pluščec, J., Weaver, E.R., and Kocy, O. 1971. Angiotensin-converting enzyme inhibitors from the venom of *Bothrops jararaca*. Isolation, elucidation of structure, and synthesis. Biochemistry 10:4033–4039

Ondetti, M.A., Rubin, B., and Cushman, D.W. 1977a. Design of specific inhibitors of angiotensin-converting enzyme: New class of orally active antihypertensive agents. Science 196:441–444

Ondetti, M.A., Sabo, E.F., Losee, K.A., Cheung, H.S., Cushman, D.W., and Rubin, B. 1977b. The use of an active-site model in the design of specific inhibitors of angiotensin-converting enzyme. In *Peptides - Proceedings of the Fifth American Peptide Symposium*, eds. M. Goodman and J. Meienhofer, pp. 576–578. New York: John Wiley and Sons

Ondetti, M.A., Condon, M.E., Reid, J., Sabo, E.F., Cheung, H.S., and Cushman, D.W. 1979a. Design of potent and specific inhibitors of carboxypeptidases A and B. Biochemistry 18:1427–1430

Ondetti, M.A., Cushman, D.W., Sabo, E.F., and Cheung, H.S. 1979b. The design of active-site-directed reversible inhibitors of exopeptidases. In *Drug Action and Design: Mechanism-Based Enzyme Inhibitors*, ed. T.I. Kalman, pp. 271–283. New York: Elsevier

Oshima, G., Gecse, A., and Erdös, E.G. 1974. Angiotensin I converting enzyme of the kidney cortex. Biochim Biophys Acta 350:26–37

Piquilloud, Y., Reinharz, A., and Roth, M. 1970. Studies on the agiotensin converting enzyme with different substrates. Biochim Biophys Acta 206:136–142

Quiocho, F., and Lipscomb, W.N. 1971. Carboxypeptidase A: A protein and an enzyme. Adv Protein Chem 25:1–78

Roth, M., Weitzman, A.F., and Piquilloud, Y. 1969. Converting enzyme content of different tissues of the rat. Experientia 25:1247

Rubin, B., Antonaccio, M.J., and Horovitz, Z.P. 1978a. Captopril (SQ 14,225) (*D*-3-mercapto-2-methylpropanoyl-*L*-proline): A novel orally active inhibitor of angiotensin-converting enzyme and antihypertensive agent. Prog Cardiovasc Dis 21:183–194

Rubin, B., Laffan, R.J., Kotler, D.G., O'Keefe, E.H., DeMaio, D.A., and Goldberg, M.E. 1978b. SQ 14,225 (*D*-3-mercapto-2-methylpropanoyl-*L*-proline), a novel orally active inhibitor of angiotensin I converting enzyme. J Pharmacol Exp Therap 204:271–280

Samuels, A.I., Miller, E.D., Jr., Fray, J.C.S., Haber, E., and Barger, A.C. 1976. Renin-angiontensin antagonists and the regulation of blood pressure. Fed Proc 35:2512–2520

Schechter I., and Berger, A. 1967. On the size of the active-site in proteases. I. Papain. Biochem Biophys Res Commun 27:157–162

Skeggs, L.T., Jr., Marsh, W.H., Kahn, J.R., and Shumway, N.P. 1954. The existence of two forms of hypertensin. J Exp Med 99:275–282

Skeggs, L.T., Kahn, J.R., and Shumway, N.P. 1956. The preparation and function of the hypertensin-converting enzyme. J Exp Mcd 103:295–299

Skeggs, L.T., Lentz, K.E., Gould, A.B., Hochstrasser, H., and Kahn, J.R. 1966. Biochemistry and kinetics of the renin-angiotensin system. Fed Proc 26:42–47

Soffer, R.L. 1976. Angiotensin-converting enzyme and the regulation of vasoactive peptides. Ann Rev Biochem 45:73–94

Stevens, R.L., Micalizzi, E.R., Fessler, D.C., and Pals, D.T. 1972. Angiotensin I converting enzyme of calf lung. Method of assay and partial purification. Biochemistry 11:2999–3007

Stewart, J.M., Ferreira, S.H., and Greene, L.J. 1971. Bradykinin-potentiating peptide PCA-Lys-Trp-Ala-Pro. An inhibitor of the pulmonary inactivation of bradykinin and conversion of angiotensin I to II. Biochem Pharmacol 20:1557–1567

Yang, H.Y.T., and Neff, N.H. 1972. Distribution and properties of angiotensin converting enzyme of rat brain. J Neurochem 19:2443–2450

Yang, H.Y.T., Erdös, E.G., and Levin, Y. 1970. A dipeptidyl carboxypeptidase that converts angiotensin I and inactivates bradykinin. Biochim Biophys Acta 214:374–376

Yang, H.Y.T., Erdös, E.G., and Levin, Y. 1971. Characterization of a dipeptide hydrolase (kininase II: angiotensin I converting enzyme). J Pharmacol Exp Therap 177:291–300

The Antihypertensive Effects of Captopril in Hypertensive Animal Models

Bernard Rubin, Michael J. Antonaccio and
Zola P. Horovitz
The Squibb Institute for Medical Research
Princeton, New Jersey 08540

INTRODUCTION

Captopril (D-3-mercapto-2-methylpropanoyl-L-proline; SQ 14,225; Capoten®) is a highly potent, orally active and relatively specific competitive inhibitor of angiotensin converting enzyme (kininase II; E.C.3.4.15.1; ACE), as well as an effective antihypertensive agent in both hypertensive animals and man (Ondetti et al., 1977; Cushman et al., 1978; Rubin et al., 1978a, 1978b, 1978c; Jenkins and McKinstry, 1979; Antonaccio et al., 1979d, 1979e; Antonaccio et al., 1980b; Rubin and Antonaccio, 1980).

Previous experience with the nonapeptide inhibitor of ACE, known as teprotide or SQ 20,881, indicated that it could successfully treat a majority of patients with essential hypertension (Gavras et al., 1974; Case et al., 1977). Unfortunately, teprotide can be employed only parenterally, and the need for an orally effective inhibitor of ACE prompted the search for a compound such as captopril.

Captopril, a nonpeptidic compound, was developed from a program of rational chemical design that made use of a hypothetical model of the active site of ACE as the basis for synthesis of compounds with increasingly greater affinities for the enzyme. The hypothetical active site was based on the assumption that substrate binding and peptide bond cleavage by ACE occurred in a manner similar to that of carboxypeptidase A (Cushman et al., 1977, 1978; Ondetti et al., 1977).

In the last few years, captopril has been undergoing intensive preclinical and clinical investigation in the treatment of high-, normal- and

Table 1. Effects of captopril on blood pressure (BP) of 2 kidney renal hypertensive rats.

Animal Model	Captopril: Dose mg/kg	Captopril: Route	Captopril: Period	Effect	Comments	Reference
Renal Hypertensive (2K–1C) Rats	0.3–1.2	i.v.t.	60′	↓↓		Schölkens et al., 1979
	1	i.v.	≲1D.	↓↓	(2K–2C)	Ferrone et al., 1978
	1	i.v.	≲1D.	↓, ↔		Lee et al. 1979
	1.25	i.v.	≲1D.	↓↓	Anesth., 30′	Marks et al., 1980
	1–30	p.o.	1–2D.	↓↓		Laffan et al., 1978; Rubin et al., 1978b
	12.5	p.o.	≲1D.	↓↓		Thurston et al., 1980
	6;20	i.m.	≲1D.	↓↓		Sen et al., 1979
	30	p.o.	≲D.	↓↓		Antonaccio et al., 1979a,b,e; 1980a, 1979a,b,e; 1980a
	60	p.o.	3D.	↓↓		Ribeiro et al., 1979
	≳45	p.o.	12D.	↓↓		Bengis et al., 1978, 1979; McCaa et al., 1979
	0.08–0.16*	i.p.	12D.	↓↓		Freeman et al., 1979
	6	p.o.	21D.	↓↓		Weed et al., 1979
	100	p.o.	28D.	↓↓		Scicli et al., 1980
	30	p.o.	12M.	↓↓	Survival ↑ Heart Wt ↓ ±HCTZ	Rubin et al., 1978a,b; Antonaccio et al., 1979c; Horovitz et al., 1979

* = per hour
D. = days
M. = months

↑ = <20 mm Hg, <10% increase in BP
↑↑ = ≳20 mm Hg, ≫10% increase in BP
↔ = no change in BP
↓ = <20 mm Hg, <10% decrease in BP
↓↓ = ≲20 mm Hg, ≫10% decrease in BP

low (plasma) renin types of hypertension and secondarily in the treatment of refractory congestive heart failure (*see* Rubin and Antonaccio, 1980).

The actions of inhibitors such as captopril or teprotide on ACE should reduce the conversion of the relatively inactive decapeptide angiotensin I to angiotensin II, a highly potent pressor vasoconstrictor octapeptide that is also aldosterone-stimulating and thus causes sodium and fluid retention. ACE inhibitors also decrease the degradation of kinins, such as bradykinin, which manifest vasodilator, vasodepressor, natriuretic, diuretic and prostaglandin releasing actions. Furthermore, inhibitors of ACE tend to increase plasma renin levels, mainly as the result of inhibition of the negative feedback mechanism by angiotensin II on the renal juxtaglomerular apparatus. The potential for dual actions by captopril or teprotide, however, can complicate the analysis of the mechanism(s) of their antihypertensive actions. Furthermore, the roles of the renin-angiotensin-aldosterone system, as well as that of the kallikrein-kinin-prostaglandin system, in essential hypertension are still steeped in controversy.

This report is primarily restricted to briefly reviewing the effects of captopril on blood pressure in various hypertensive animal models.

HYPERTENSIVE ANIMAL MODELS

Renal Hypertensive Animals

Two-kidney Models

One Clip or 2 Clip Rats The 2 kidney, 1 clip (Goldblatt et al., 1934) hypertensive rat (2K-1C RHR) is a high plasma renin model, typical of renin-angiotensin hypertension in both the initial and established phases for at least several weeks (Gavras et al., 1976; Thurston et al., 1980).

Conscious 2K-1C RHR, at least 6 wk after surgery, were dosed orally for 1 or 2 days with 1 to 30 mg/kg of captopril. Mean blood pressure (MBP) decreased by 10 to 65 mm Hg within the first four hr after dosage and remained depressed at least 6 to 10 hr. These effects represented about 10 to 32% decreases in MBP (Laffan et al., 1978; Rubin et al., 1978b).

As shown in Table 1, in similar short- to intermediate-term studies in 2K-1C RHR, qualitatively similar antihypertensive effects were produced by captopril, given intraventricularly, intraperitoneally, intravenously, intramuscularly or orally in a dose range of 0.08 to 100 mg/kg for periods of 1 hr to as long as one month. The blood pressure decreases were accompanied by decreases in total peripheral resistance and either no change or slight increases in cardiac output (Bengis et al., 1978; Ferrone et al., 1978). Increasing the salt intake of 2K-1C RHR diminished the antihypertensive effect of captopril (Ribeiro et al., 1979).

Freeman et al. (1979) infused captopril 1 day before and for 12 days after renal artery constriction in 2K-1C RHR; captopril prevented a significant rise in blood pressure of conscious 2K-1C RHR for the 12 days after clipping (Table 1).

Long-term oral dosage of 2K-1C RHR with 30 mg/kg/day of captopril for up to 12 mo (Table 1) initially lowered systolic blood pressure (SBP) to about 150 mm Hg and then subsequently to between 170 and 180 mm Hg, whereas control animals maintained SBP of 200 to 210 mm Hg (Fig. 1). The slight degree of tolerance to captopril's antihypertensive effect in this model may have reflected a sympathetic component, since dosage of these rats with guanethidine could further reduce SBP to normotensive levels (circa 130 mm Hg). Hydrochlorothiazide (HCTZ) alone, 6 mg/kg/day p.o., had little or no effect on SBP. Daily dosing

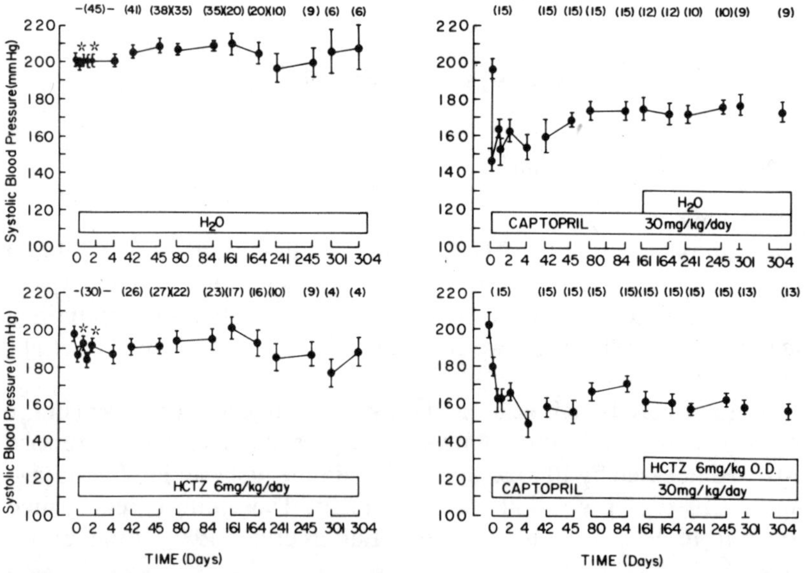

Fig. 1. Effects of daily oral doses of water, captopril and hydrochlorothiazide (HCTZ) on systolic blood pressures of conscious 2 kidney, 1 clip renal hypertensive rats. Numbers in parentheses are numbers of rats tested at each point. Vertical bars at each point are means ± S.E. Reproduced by permission of Z.P. Horovitz, Development of angiotensin converting enzyme inhibitors, 1979, Med J Aust 2 (Spec. Suppl): 3–5; Australasian Medical Publishing Co., Ltd.

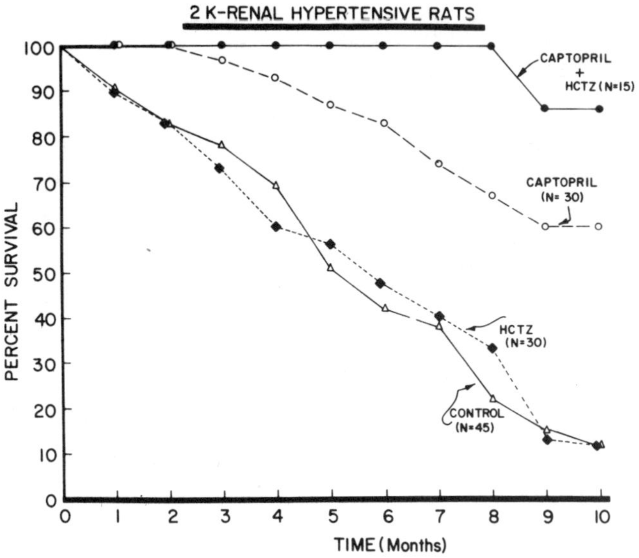

Fig. 2. Survival rates of 2 kidney, 1 clip renal hypertensive rats dosed with water, captopril and/or hydrochlorothiazide (HCTZ) for ten months.

of such rats with both captopril and HCTZ, however, increased the antihypertensive effect of captopril, suggesting that a salt water retention factor was also operating in these animals. The combination dosage of captopril with HCTZ, however, did not normalize SBP, since SBP was maintained at 150 to 160 mm Hg (Rubin et al., 1978a, 1978b; Antonaccio et al., 1979e; Horovitz et al., 1979). The survival rates of these 2K-1C RHR were greater (87%) in those rats treated with both captopril and HCTZ, whereas those treated with captopril alone showed a somewhat lesser (60%) survival rate (Fig. 2). In contrast, control rats as well as rats treated with HCTZ alone showed only 13% incidence of survival. Captopril treated rats also showed less cardiac hypertrophy than did the vehicle treated controls.

Aortic-Coarctated Rats Accelerated severe hypertension, resembling a malignant type of renin-mediated renal hypertension (Carretero, O. A. et al., 1971) was induced in rats within several days after ligation of the aorta midway between the origin of the two renal arteries.

Four to 7 days after surgery, anesthetized or conscious aortic-coarctated rats were infused intravenously with captopril or dosed orally with captopril in short-term studies (Table 2). Blood pressure was decreased by up to 48 mm Hg.

Table 2. Effects of captopril on blood pressure (BP) of coarctated and of 1 kidney renal hypertensive rats.

Animal Model	Captopril Dose mg/kg	Captopril Route	Captopril Period	Effect	Comments	Reference
Renal Hypertensive Aortic-Coarct. Rats	3–10	i.v.(10′)	≲1 D.	↓↓	Anesth.,120′	Laffan et al., 1978 Rubin et al., 1978b
	5;20	p.o.	≲1 D.	↓↓		Sweet and Columbo, 1979
Renal Hypertensive Rats (1K–1 fig. 8)	(2X/WK.)	s.c.	2M.	↓↓		Loyke, 1978
(1K–Low Na+)	0.08*	i.p.	12D.	↓↓		Seymour et al., 1980
Renal Hypertensive Rats (1K–1C)	30	p.o.	4D.	↓↓	Grad. Decr.	Rubin et al., 1978b
	33–73	p.o.	7D.	↓↓	Grad. Decr.	Bengis et al., 1978, 1979
	20	i.m.	<1D.	↓↓		Sen et al., 1979
	0.08–0.16*	i.p.	12D.	↓	(8D.)⟷(8–12D.)	Freeman et al., 1979

* = per hour
D. = days
M. = months

↑ = <20 mm Hg, <10% increase in BP
↑↑ = ≥20 mm Hg, ≥10% increase in BP
↔ = no change in BP
↓ = <20 mm Hg, <10% decrease in BP
↓↓ = ≥20 mm Hg, ≥10% decrease in BP

Two-Wrapped (Perinephritic) Dogs 2 kidney, 2 wrapped (perinephritic) hypertensive dogs (Page, 1939) were prepared and tested by R.R. Vollmer (*see* Rubin et al., 1978b). Four weeks after surgery, the animals were dosed orally with captopril, 31 mg/kg/day, for the next 9 wk (Table 3). Blood pressures in the captopril treated dogs were 25 to 50 mm Hg lower than those dosed orally with lactose placebo. Blood pressures returned gradually to hypertensive levels during the 4 wk after captopril dosage was stopped (Fig. 3). The placebo dosed control hypertensive dogs had normal plasma renin levels, whereas the captopril treated dogs showed 2- to 15-fold increases in plasma renin activity after initiation of dosage with captopril (R.R. Vollmer, personal communication).

Vrobel et al. (1979) studied the effects of captopril on blood pressure and regional blood flow in 1 kidney (1K) perinephritic hypertensive dogs. MBP was reduced slightly by captopril when the dogs were at rest; however, during exercise, captopril produced little or no decrease in blood pressure in this normal plasma renin animal model (Table 3). Captopril did not alter distribution of perfusion to the major organs under either set of conditions.

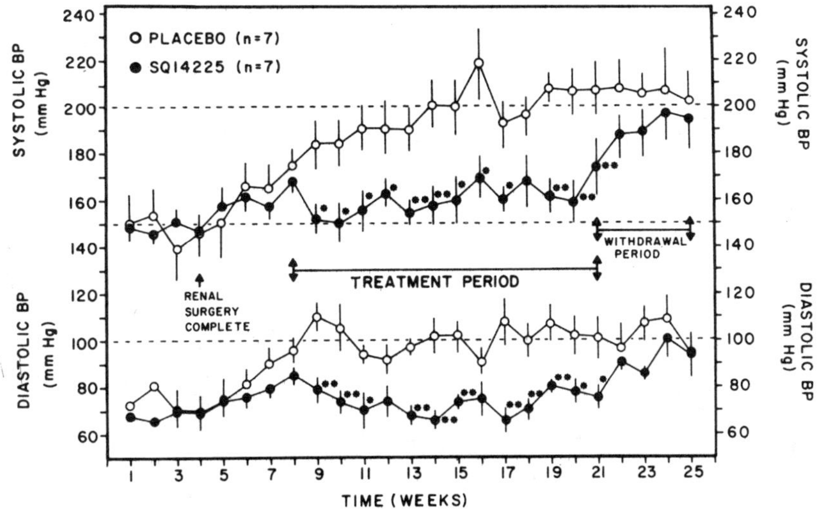

Fig. 3. Systolic and diastolic blood pressures of 2 kidney, 2 wrapped (perinephritic) dogs dosed orally with SQ 14,225 (captopril) 31 mg/kg/day or with lactose placebo from the eighth week through the 20th week. Vertical bars at each point are means ± S.E.; asterisks in captopril treated group indicate significantly lower pressures at $p < 0.001$. Reproduced by permission of B. Rubin, Captopril (SQ 14,225)(D-3-mercapto-2-methylpropanoyl-L-proline): A novel orally active inhibitor of angiotensin converting enzyme and antihypertensive agent, 1978, Progr Cardiovasc Dis 31: 183–194, Grune and Stratton, Inc.

Table 3. Effects of captopril on blood pressure (BP) of 2 kidney and of 1 kidney renal hypertensive dogs.

Animal Model	Captopril Dose mg/kg	Route	Period	Effect	Comments	Reference
Renal Hypertensive Dogs (2K–1C)	0.1–0.2	i.v.	≂1D.	↓		Zimmerman et al., 1980
(2K–2Wrap)	31	p.o.	63D.	↓↓		Vollmer; Rubin et al., 1978b
(1K–1Wrap)	1	i.v.	≂1D.	↓		Vrobel et al., 1979
Renal Hypertensive Dogs (1K–Renal A.	1	i.v.	≂1D.	↓↓ (10–50′)	(120′)	Antonaccio, 1980
Stenosis)	30–35	p.o.	12D.	↓↓ (6D.)	↔(6–12D.)	Watkins et al., 1978
Chronic	0.1	i.v.t.	<1D.	↑		Schölkens et al., 1980 Schölkens, 1979

D. = days
M. = months

↑ = <20 mm Hg, <10% increase in BP
↑↑ = ≥20 mm Hg, ≥10% increase in BP
↔ = no change in BP
↓ = <20 mm Hg, <10% decrease in BP
↓↓ = ≥20 mm Hg, ≥10% decrease in BP

One Kidney Models

One Clip Rats In 1 kidney, 1 clip renal hypertensive rats (1K-1C RHR) maintained on normal sodium diet, hypertension is apparently associated with normal levels of plasma renin activity within a few days postsurgery.

Rubin et al. (1978b) administered captopril, 30 mg/kg/day p.o. for 4 days to conscious 1K-1C RHR, at 6 and 14 wk postsurgery; blood pressures decreased gradually by at least 50 mm Hg (Fig. 4). Similar results after oral or intramuscular dosage of 1K-1C RHR with captopril have been reported (Table 2).

Intraperitoneal infusion of captopril into conscious 1K-1C RHR prevented a significant rise in SBP until the eighth day after clipping, but systolic pressure then increased to hypertensive levels (Table 2) despite continuing infusion of captopril for another 4 days.

One Figure-8 Rats In conscious chronic 1 figure-8 rats (Grollman, 1944) subcutaneous dosage of captopril for 7 wk (Table 2) decreased SBP by 27 mm Hg.

Low-Sodium Rats Intraperitoneal infusion of captopril in conscious sodium deplete 1 kidney rats (without renal artery clipping) for 12 days

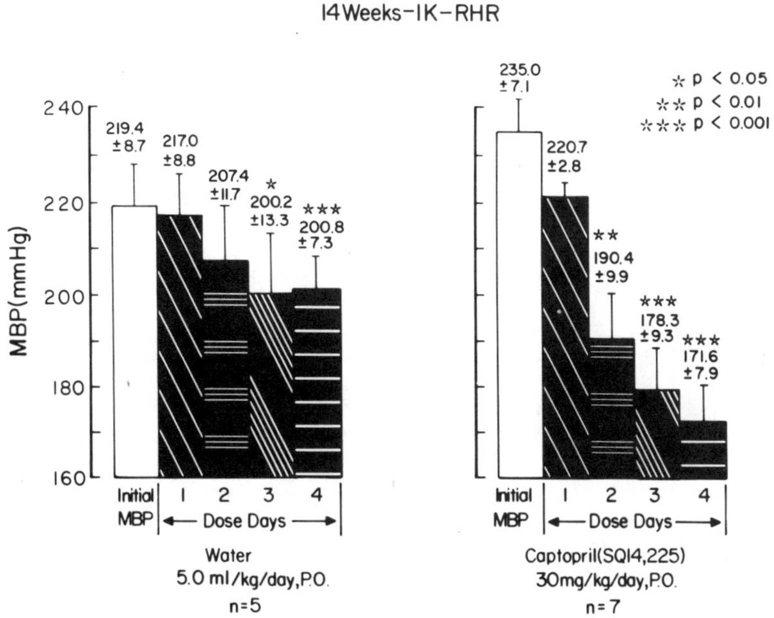

Fig. 4. Direct mean blood pressures (MBP) of conscious 1 kidney, 1 clip hypertensive rats dosed for 4 days with water or captopril 14 weeks after surgery.

Table 4. Effects of captopril on blood pressure (BP) of hypertensive SHR.

Animal Model	Captopril Dose mg/kg	Route	Period	Effect	Comments	Reference
Okamoto-Aoki	0.3–1.2	i.v.t.	20′	↑ ↑ ; ↓		Schölkens et al., 1980; Schölkens, 1979
SHR	2	i.v.t.	60′	↓ ↓		Stamler et al., 1980
(Adult Rats)	2	i.v.	≲1D.	↓ ↓	Anesth.,180′	Hutchinson et al., 1980; Mendelsohn et al., 1979
	0.3–100	p.o.	1–11D.	↓ ↓		Laffan et al., 1978; Rubin et al., 1978b
	3	p.o.	≲1D.	↓		Watanabe and Sokabe, 1979
	100	p.o.	1D.	↓ ↓		Ferrone et al., 1978, 1980
	30;100	p.o.	1–3D.	↓ ↓		Antonaccio et al., 1979a,b; 1980a
	6–100	p.o.	5D.	↓ ↓		Muirhead et al., 1978
	100	p.o.	8D.	↓ ↓		Asaad and Antonaccio, 1979
	100	p.o.	14D.	↓ ↓		Ferrone et al., 1978, 1979, 1980
	100	p.o.	21D.	↓ ↓		Maitra et al., 1979
	100	p.o.	28D.	↓ ↓		Scicli et al., 1980
	10;30	p.o.	42D.	↓ ; ↓ ↓		Koike et al., 1980
	30;100	p.o.	42D.	↓ ↓		Ito et al., 1980
	100	p.o.	70D.	↓ ↓		Samar and Coleman, 1980
	100	p.o.	12M.	↓ ↓	Heart Wt. ↓	Antonaccio et al., 1979d; Horovitz et al., 1979; Ferrone et al., 1979

D. = days
M. = months

↑ = <20 mm Hg, <10% increase in BP
↑ ↑ = ⋝20 mm Hg, ⋝10% increase in BP
↔ = no change in BP
↓ = <20 mm Hg, <10% decrease in BP
↓ ↓ = ⋝20 mm Hg, ⋝10% decrease in BP

(Table 2) abolished the pressure rise previously observed in this model in the absence of treatment with the converting enzyme inhibitor. When dosage with captopril was discontinued, SBP progressively increased during the next week.

Renal Artery Stenosed Dogs Captopril was administered intravenously to conscious Goldblatt 2K-1C hypertensive dogs (Table 3). Renal blood flow increased, renal vascular resistance of the contralateral kidney decreased, and blood pressure decreased slightly.

In 1 kidney conscious renal hypertensive dogs, daily oral dosage of captopril, starting 1 day before renal arterial constriction and then continuing for the next 12 days (Table 3), delayed the onset of hypertension for only the first 6 days after constriction.

In several dogs made hypertensive by unilateral (right) nephrectomy and by a 60% reduction in blood flow to the other kidney, intravenous injection of captopril (Table 3) significantly reduced blood pressure of these conscious animals by about 40 mm Hg (M.J. Antonaccio, personal communication).

Genetic Hypertensive Rats

Okamoto-Aoki rats (SHR)

The Okamoto-Aoki strain (Okamoto and Aoki, 1963) of genetic hypertensive rats (SHR) is a normal plasma renin model of hypertension (Koletsky et al., 1972). Short term (1 or 2 days) dosage orally of 12-week-old SHR with captopril, 0.3 to 100 mg/kg/day, decreased MBP by 23 to 40 mm Hg, usually within 3 hr after dosage. These decreases in MBP represented about 13 to 20% reductions in blood pressure and were slightly slower in onset than those observed in 2K-1C RHR (Laffan et al., 1978; Rubin et al., 1978b). The antihypertensive effect of captopril in the SHR was slightly intensified on the 2nd day as well as on the 11th day.

The antihypertensive effects of short- or intermediate-term dosage with intraventricular, intravenous or oral dosage of captopril, 0.3 to 100 mg/kg, for periods of 20 min to 70 days in SHR were further studied by many investigators (Table 4). Moderate to marked decreases in blood pressure occurred in virtually all of these studies with captopril.

Muirhead et al. (1978) also reported that cardiac output was unchanged by captopril and total peripheral resistance was decreased. When dosage with captopril was stopped, about 2 to 4 days elapsed before MBP returned to predrug hypertensive levels.

Ferrone and Heran (1978) and Ferrone et al. (1980) also examined the acute and subacute hemodynamic effects of captopril in SHR. MBP was acutely lowered by 22 to 32 mm Hg, cardiac output increased and peripheral resistance decreased. Captopril increased organ blood flow, especially in lung and splanchnic organs. In the subacute two week study, organ blood flows were not consistently increased; slight increases in flow were noted in heart, lung and splanchnic organs. All organ vascular resistances tended to decrease, especially so in skeletal muscle, skin and splanchnic organs.

Koike et al. (1980), in their subacute studies, also noted that heart rates showed little or no change, cardiac output tended to increase, and heart/body weight ratios as well as total peripheral resistances decreased. No significant increases occurred in plasma volumes. Cerebral and renal blood flow increased, while blood flow to other organs tended to increase or remained unchanged.

Long-term oral dosage of conscious SHR with captopril, 100 mg/kg/day for 12 to 13 mo, progressively lowered MBP to normotensive levels by the sixth month (Table 4). MBP was decreased by about 70 mm Hg and no tolerance was observed (Fig. 5). Reductions in heart weight–body weight ratios paralleled the decreases in blood pressure (Antonaccio et al., 1979c; Horovitz et al., 1979). Cessation of therapy in these rats after 2 wk, 3, 6 and 12 mo of daily dosing resulted in a prolonged return of blood pressure to predrug levels; the rapidity of this return appeared to be inversely related to the duration of treatment with captopril. SHR that were previously dosed with captopril for 13 mo required 3 mo to return to hypertensive levels (Ferrone et al., 1979).

Several preliminary reports have appeared recently on the antihypertensive and related effects of captopril in combination with hydrochlorothiazide (HCTZ) in SHR. Ferrone and Heran (1980) dosed conscious SHR with captopril; captopril lowered arterial pressure, decreased total peripheral resistance and slightly increased cardiac output. HCTZ alone had no consistent hemodynamic effects. Addition of HCTZ to captopril, however, enhanced captopril's effects on blood pressure, total peripheral resistance, and cardiac output; organ vascular resistances (heart, brain, lung, muscle, skin, kidney and splanchnic) were also decreased more than was observed with captopril alone (Table 5). Chan et al. (1980) and Lai et al. (1980) also found that there was a synergistic antihypertensive effect in combination oral dosage of SHR with captopril and HCTZ; a similar qualitative pattern of synergistic antihypertensive effects also occurred in the SHR after bilateral ureteral ligation (Table 5).

Daily oral doses of captopril were administered to conscious weanling or young SHR for 4 to 22 wk by Sen et al. (1980), Ferrone and Antonaccio (1979), Forslund et al. (1980), and Crofton et al., (1979). Sen et al. (1980) reported that captopril prevented cardiac hypertrophy and

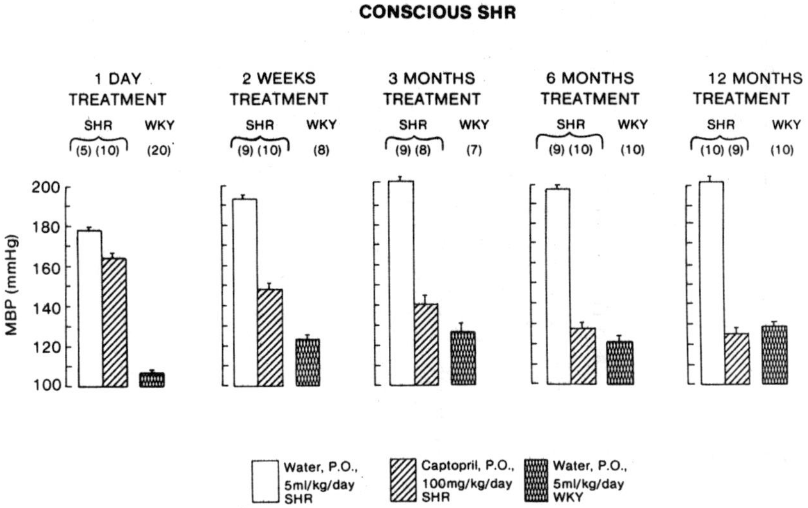

Fig. 5. Direct mean blood pressures (MBP) of conscious SHR after 1 day to 12 months oral dosage with water or captopril. WKY was dosed only with water.

hypertension. Ferrone and Antonaccio (1979) also found that captopril treated rats did not develop hypertension during the 4- to 16-week period of dosage; however, within one week after discontinuation of captopril, the usual development of hypertension occurred. Forslund et al. (1980) obtained qualitatively similar results in preventing the development of hypertension.

Crofton et al. (1979) gave captopril to young SHR; captopril treated rats, at the end of 4 wk of treatment, had SBP which were 38 mm Hg lower than those of control SHR. Sen et al. (1980) also treated young SHR for 6 wk with either captopril or with both captopril and hydrochlorothiazide. Blood pressures were decreased about 55 and 70 mm Hg respectively and heart weights were reduced, in comparison with non–drug treated controls (Table 5).

Stroke-Prone (spSHR)

Watanabe and Sokabe (1979) administered captopril orally to conscious SHR and to the more plasma renin dependent stroke-prone SHR (spSHR) (Okamoto et al., 1974; Matsunaga et al., 1975). One hour after dosage, MBP was decreased only 18 mm Hg in SHR, whereas in spSHR blood pressure was decreased 45 mm Hg (Table 5).

Unger et al. (1979, 1980) administered captopril to 6-month-old spSHR for 1 mo; blood pressure was decreased by 62 mm Hg. The pressor responses to intraventricular injections (i.v.t.) of angiotensin I

Table 5. Effects of captopril on blood pressure (BP) of hypertensive SHR and spSHR.

Animal Model	Captopril Dose mg/kg	Captopril Route	Captopril Period	Effect	Comments	Reference
Okamoto-Aoki SHR	100	p.o.	2D.	↓↓	±HCTZ	Chan et al.; Lai et al., 1980
(Adult Rats)	100	p.o.	14D.	↓↓	±HCTZ	Ferrone and Heran, 1980
Okamoto-Aoki SHR	15–30	p.o.	28D.	↓↓		Crofton et al., 1979
(Weanling & Young Rats)	30–100	p.o.	42D.	↓↓	Heart Wt. ↓	Sen et al., 1980
		p.o.	4M.	↓↓	±HCTZ	Ferrone and Antonaccio, 1979
		p.o.	3–5–1/2M.	↓↓		Forslund et al., 1980
spSHR	3	p.o.	⩽1D.	↓↓		Watanabe and Sokabe, 1979
	50	p.o.	1M.	↓↓	I.V.T.:AI,BK	Unger et al., 1979, 1980
	30	p.o.	3M.	↓↓	(Age⩾1–1/2M.)	Watanabe et al., 1980
	100	p.o.	4M.	↓↓	(Age⩾4M.)	Watanabe et al., 1980

D. = days
M. = months

↑ = <20 mm Hg, <10% decrease in BP
↑↑ = ⩾20 mm Hg, ⩾10% increase in BP
↔ = no change in BP
← = <20 mm Hg, <10% decrease in BP
↓ = ⩾20 mm Hg, ⩾10% decrease in BP
↓↓

and of bradykinin were each unaltered after such chronic treatment with captopril.

Watanabe et al. (1980) treated conscious young and adult spSHR with captopril for 3 and 4 mo. SBP was decreased about 35 to 50 mm Hg by captopril. Heart weight–body weight ratios were decreased in the captopril-treated groups (relative to water dosed controls). Treatment with captopril decreased the incidence of vascular disease in the young and its severity in the adult rats, respectively, and tended to decrease the incidence of cerebral stroke in the adult spSHR (Table 5).

Several studies have been reported recently on the blood pressure effects of i.v.t. administered captopril in hypertensive as well as in normotensive animals. Schölkens (1979) and Schölkens et al. (1979) reported that captopril i.v.t. produced 20-min pressor responses (+20 and +13%) in normotensive and in spontaneous hypertensive rats, followed by a decrease in blood pressure. In chronic 2K-1C RHR, captopril i.v.t. did not cause a significant initial increase in blood pressure; however, a lasting (>60 min) decrease in blood pressure of 17% developed. In chronic renal hypertensive dogs, i.v.t. dosage of captopril increased blood pressure and heart rate. On the other hand, Stamler et al. (1980) found that i.v.t. captopril in conscious SHR was more hypotensive (−25 mm Hg) than was the same i.v. dose of captopril (−16 mm Hg); the effect lasted about 1 hr by either route (Tables 1, 3, 4).

In normotensive rats (Schölkens, 1979; Schölkens et al., 1979) dosage with captopril i.v.t. reduced the pressor response to i.v.t. injection of angiotensin I but not of angiotensin II and augmented the pressor response to i.v.t. bradykinin. Intraventricular injection of captopril in chloralose-anesthetized cats had been previously reported by Vollmer and Boccagno (1977) to also reduce the pressor response to i.v.t. angiotensin I but not to angiotensin II. Unger et al. (1979, 1980) also found that either short-term i.v. dosage or one month's oral dosage of spSHR with captopril did not alter the pressor responses to i.v.t. angiotensin I or to i.v.t. bradykinin. Similar negative results were found by Vollmer and Boccagno (1977) on the pressor responses to i.v.t. angiotensin I and II after acute i.v. dosage of cats with captopril.

New Zealand GH Rats

This strain of genetic hypertensive rat, originally described by Smirk and Hall (1958) typically manifests a lowered, if not normal, plasma renin level (Folkow and Hallbäck, 1977).

In conscious GH rats, a single oral dose of captopril decreased MBP by about 35 mm Hg within the first few hours (Rubin et al., 1978). In conscious GH rats injected i.v. with captopril, blood pressure decreased

Table 6. Effects of captopril on blood pressure (BP) of hypertensive GH and DOCA + salt rats.

Animal Model	Captopril Dose mg/kg	Route	Period	Effect	Comments	Reference
New Zealand GH Rats	2	i.v.	60′	↓↓ (10′)		Andrews, 1979
	30	p.o.	1D.	↓↓		Rubin et al., 1978
	0.01–1	i.v.	<60′	↓↓	Anesth.	Phelan and Clark, 1979
	0.01–1	i.a.	<60′	↓↓	Anesth.	Phelan and Clark, 1979
Metacorticoid Hypertensive Rats (DOCA + NaCl;1K)	1.25	i.v.	<30′	↓↓	Anesth.	Marks et al., 1980
	1	i.p.	30′	↓↓		Dollery and Miyamori, 1980
	5	p.o.	2–3M.	↔		Tikkanen et al., 1980
	10–100	p.o.	4D.	↔		Rubin et al., 1978b
	10	p.o.	14D.	↔		Dean and Ingham, 1978
	15–25	p.o.	21D.	↔		Douglas et al., 1979

D. = days
M. = months

↑ = <20 mm Hg, <10% increase in BP
↑↑ = ⩾20 mm Hg, ⩾10% increase in BP
↔ = no change in BP
↓ = <20 mm Hg, <10% increase in BP
↓↓ = ⩾20 mm Hg, ⩾10% decrease in BP

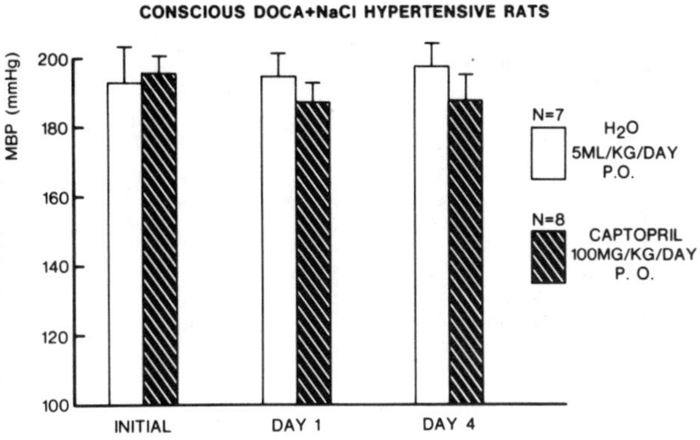

Fig. 6. Direct mean blood pressures (MBP) of conscious DOCA-salt hypertensive rats after 1 to 4 days oral dosage with water or captopril. Blood pressures shown are the averages of the first 4 hr recordings on each day.

by 12% within 10 min after the injection; recovery was complete within 1 hr (Andrews, 1979). Phelan and Clark (1979) injected captopril i.v. (jugular vein) or i.a. (femoral artery, perfused hind limb at constant rate) in chloralose-anesthetized GH rats. Captopril i.v. caused an immediate dose-related decrease in blood pressure of 10 to 50% because of a delay coil in the circuit; when captopril entered the hind limb, perfusion pressure fell 20 to 45%. Blood pressure and perfusion pressure returned to predrug levels 10 to 60 min later. When captopril was injected i.a., an immediate fall was followed within 5 min by a second, more pronounced fall in perfusion pressure (Table 6).

Metacorticoid Hypertensive Rat

Dosage of rats with DOCA and maintenance on increased sodium intake for 4 to 6 wk results in a self-sustaining hypertension (Friedman et al., 1951).

In conscious low renin, high kallikrein activity (Margolius et al., 1972), DOCA-salt hypertensive rats, oral dosage of captopril alone for 4 days (Fig. 6) (Rubin et al., 1978b) or for 14 to 21 days (Dean and Ingham, 1978; Douglas et al., 1979) did not result in any consistent decreases in blood pressure. Furthermore, similar oral dosage of conscious rats with captopril did not prevent the development of DOCA-salt induced hypertension (Dean and Ingham, 1978; Douglas et al., 1979; Tikkanen et al., 1980).

On the other hand, in pentobarbital anesthetized DOCA-salt hypertensive rats, i.v. captopril transiently decreased blood pressure by about 20 mm Hg (Marks et al., 1980). Dollery (1980) injected captopril i.p. into conscious DOCA-salt hypertensive rats and also transiently decreased SBP by 23 mm Hg (Table 6).

Table 7. Effects of captopril on blood pressure (BP) of miscellaneous hypertensive rat models.

Animal Model	Captopril Dose mg/kg	Captopril Route	Captopril Period	Effect	Comments	Reference
Miscellaneous Angiotensin II +NaCl (Rats)	30–60	p.o.	32D.	↓↓		
Renoprival Hypertensive +NaCl (Rabbits)	—	—	—	↔	Sodium-Vol. Expanded	Muirhead et al., 1979
Na+ Deficiency +Methylprednisone (Rats)	1	i.v.	<1D.	↓↓		Elijovich and Krakoff, 1979

D. = days
M. = months

↑ = <20 mm Hg, <10% increase in BP
↑↑ = ⩾20 mm Hg, ⩾10% increase in BP
↔ = no change in BP
↓ = <20 mm Hg, <10% decrease in BP
↓↓ = ⩾20 mm Hg, ⩾10% decrease in BP

Miscellaneous Hypertensive Models

Orally administered captopril for 1 mo decreased blood pressure by about 20 mm Hg in angiotensin II–salt induced hypertensive intact rats; this model represents a sodium-volume expanded, renin-angiotensin-aldosterone suppressed hypertensive state (Muirhead et al., 1979). In another sodium-volume expanded hypertensive setting (in which the kidneys were excised), such as renoprival hypertension of the saline-loaded rabbit, captopril failed to prevent the hypertensive state (Muirhead et al., 1979).

In sodium deficient rats made hypertensive by dosage with methylprednisolone i.m., intravenous dosage with captopril reduced SBP by at least 20 mm Hg (Elijovich and Krakoff, 1979) (Table 7).

DISCUSSION

It is apparent that captopril generally decreases blood pressure in hypertensive animals. In some studies this effect can be independent of pretreatment levels of plasma renin activity. Similar reports have been reported clinically (Gavras et al., 1978; Bravo and Tarazi, 1979; Johnston et al., 1979). Other clinical reports, however, have indicated a correlation between degree of hypotensive effect and precaptopril plasma renin levels (Asplund et al., 1979; Case et al., 1978).

In hypertensive animal models dosed with captopril, total peripheral resistance decreased and cardiac output was either unchanged or increased slightly. Renal blood flow increased, diuresis, natriuresis and decreased aldosterone levels occurred in most animal studies after dosage with captopril. Furthermore, except in salt depleted animals, captopril produced little or no tachycardia (see Rubin and Antonaccio, 1980; Antonaccio et al., 1980b; Atkinson and Robertson, 1979).

Similar to the situation that occurred with teprotide a number of years ago, the question again arises with captopril: are all of its antihypertensive effects directly or indirectly attributable to inhibition of ACE or kininase II? At the other extreme, are there so-called nonspecific hypotensive mechanisms independent of inhibition of ACE, as suggested by Jandhyala et al. (1978) and by Vollmer et al. (1978), also operational in the antihypertensive action of captopril? Furthermore, several factors which may contribute to captopril's antihypertensive activity include: (1) direct vasodilator effects, including reductions in sensitivity to autacoid agonists, or possible alterations in vascular Na^+/K^+ ion flux; (2) inhibition of sympathetic ganglionic, neuronal and adrenergic activity or inhibitory effects on related receptors; (3) inhibitory effects on the brain renin-angiotensin system after peripheral dosage with captopril; (4) systemic or local accumulation of bradykinin; (5) direct or indirect release of vasodepressor prostaglandins; (6) decreases in aldosterone levels, accompanied by natriuresis and diuresis; and (7) equal or greater effects on vascular renin levels than on plasma renin levels.

Most of these factors have been reviewed by Rubin et al. (1978b), Horovitz (1980), Rubin and Antonaccio (1980), and Antonaccio et al. (1980b, 1981). The reports cited above have emphasized that captopril *is* a potent inhibitor of ACE or kininase II. Captopril does not appear to affect sympathetic or parasympathetic ganglionic, neuronal or adrenergic activity. Little or no evidence exists that captopril is either a direct vasodilator or that it reduces vascular sensitivity to various agonists such as angiotensin II, norepinephrine, K+, etc., or that it acutely reduces either the pressor effects of norepinephrine, serotonin, tyramine, vasopressin or the depressor effects of isoproteronol or prostaglandin E_2 (Rubin et al., 1978b; Rubin and Antonaccio, 1980). One report by Okuno et al. (1979) indicated that captopril at 3 to 81 μg/ml reduced the contractile effects of norepinephrine on the mesenteric vascular bed of the rat; the relevance of this effect at these concentrations is uncertain, since rats dosed orally with captopril at 30 mg/kg showed peak blood concentrations of 5 to 7 μg/ml of captopril and human subjects dosed orally with 100 mg of captopril showed maximum blood concentrations of less than 1 μg/ml of captopril (B.H. Migdalof, personal communication). Another recent report by Ito et al. (1980) indicated that the Na+ and K+ content of aortas excised from SHR were decreased after 6 wk of dosage of SHR with captopril (30 to 100 mg/kg/day p.o.); the pertinence of this observation is unclear and awaits further investigation.

No evidence in favor of central effects by peripherally administered captopril has been obtained in animals. The brain renin-angiotensin system has not been shown to be affected by oral or intravenous doses of captopril. Intraventricular (i.v.t.) injections of captopril, however, will inhibit the pressor responses to i.v.t. angiotensin I but not to angiotensin II; the pressor responses to i.v.t. bradykinin can be augmented.

The evidence for possible localized accumulation of bradykinin is mixed. Some studies suggest an increase or even decrease in bradykinin levels, other suggest little or no change. Endogenous kinins, if increased, could promote the systemic or local release of prostaglandins, of which only some are vasodepressor, diuretic and natriuretic. Furthermore, there is an ever-growing body of data that suggest that increased arteriolar wall renin activity may be more pertinent or relevant than the levels of plasma renin activity in contributing to hypertension, especially in low and normal renin models of hypertension. Hence, previous data and conclusions based primarily on degrees of correlation or lack of correlation between *plasma* renin activity and hypotensive effects of captopril may need to be reevaluated (Rubin and Antonaccio, 1980; Horovitz, 1980; Antonaccio et al., 1980b, 1981).

It still appears, therefore, that captopril owes its antihypertensive activity primarily, if not totally, to inhibition of ACE or kininase II with reference to the direct or indirect consequences of that inhibition. The obvious complexities concerning captopril's mechanism(s) of antihypertensive action emphasize the continuing challenge to further clarification of these mechanism(s).

SUMMARY AND CONCLUSIONS

Captopril is antihypertensive in most rodent and canine hypertensive animal models, including a) renovascular (Goldblatt, Grollman, Page-type) animals, and b) genetic spontaneous hypertensive animal models, including SHR, spSHR and GH animals, as well as c) other miscellaneous models such as sodium-volume expanded models and sodium deficient, glucocorticoid treated animals.

Little or no consistent and long-lasting antihypertensive effects are produced by captopril alone in metacorticoid hypertensive animals, such as DOCA-salt treated rats.

Most of the antihypertensive activity of captopril still appears to be directly or indirectly related to the consequences of its inhibitory effect on ACE (kininase II). Special consideration should be given to captopril's effects on the vascular renin-angiotensin system, among other possible renal and cardiovascular factors.

REFERENCES

Andrews, D.I. 1979. Effects of saralasin and captopril on the blood pressure of conscious genetically hypertensive and normotensive rats. Med J Aust 2 (Spec Suppl):7–9

Antonaccio, M.J., Harris, D., Goldenberg, H., High, J.P., and Rubin, B. 1979a. The effects of captopril, propanolol, and indomethacin on blood pressure and plasma renin activity in spontaneously hypertensive and normotensive rats. Proc Soc Exp Biol Med 162:429–433

Antonaccio, M.J., High, J.P., Rubin, B., and Schaeffer, T. 1979b. Contribution of the kidneys but not adrenal glands to the acute antihypertensive effects of captopril in spontaneously hypertensive rats. Clin Sci 57(Suppl 5):127s–130s

Antonaccio, M.J., Rubin, B., Horovitz, Z.P., Laffan, R.J., Goldberg, M.E., High, J.P., Harris, D.N., and Zaidi, I. 1979c. Effects of chronic treatment with captopril (SQ 14,-225), an orally active inhibitor of angiotensin I-converting enzyme, in spontaneously hypertensive rats. Jpn J Pharmacol 29:285–294

Antonaccio, M.J., Rubin, B., Horovitz, Z.P., Mackaness, G., and Panesevich, R. 1979d. Long term efficacy of captopril (SQ 14,225) in 2-kidney renal hypertensive rats. Clin Exp Hypertens 1:505–519

Antonaccio, M.J., Waugh, H., Rubin, B., High, J., and Schaeffer, T. 1979e. Roles of the sympathetic nervous and renin-angiotensin systems in the development of two-kidney, one-clip rat hypertension. Circulation 59/60(Suppl II):175

Antonaccio, M.J., Asaad, M.M., High, J., Schaeffer, T., and Rubin, B. 1980a. The role of vascular renin and kidneys in the antihypertensive action of captopril in spontaneously hypertensive rats (SHR). Fed Proc 34(Abs):813

Antonaccio, M.J., Asaad, M., Rubin, B., and Horovitz, Z.P. 1981. Captopril: Factors involved in its mechanism of action. In *Angiotensin Converting Enzyme Inhibitors: Mechanisms of Action and Clinical Implications,* ed. Z.P. Horovitz, in press. Baltimore: Urban and Schwarzenberg

Antonaccio, M.J., Rubin, B., and Horovitz, Z.P. 1980b. Effects of captopril in animal models of hypertension. Clin Exp Hypertens, 2:613–637

Asaad, M.M., and Antonaccio, M.J. 1979. The effects of captopril (SQ 14,225) on vascular tissue renin in spontaneously hypertensive (SHR) and WKY-normotensive (NTR) rats; relationship to systolic blood pressure (SBP) and plasma renin activity (PRA). Pharmacologist 21(Abs):212

Asplund, J., Carlberg, B., and Ohman, P. 1979. A converting enzyme inhibitor in the treatment of essential hypertension. Proc 6th Sci Mtg Int Soc Hypertens, Goteborg (Abs):121

Atkinson, A.B., and Robertson, J.I.S. 1979. Captopril in the treatment of clinical hypertension and cardiac failure. Lancet 2:836–839

Bengis, R.G., and Coleman, T.G. 1979. Antihypertensive effect of prolonged blockade of angiotensin formation in benign and malignant, one- and two-kidney Goldblatt hypertensive rats. Clin Sci 57:53–62

Bengis, R.G., Coleman, T.G., Young, D.B., and McCaa, R.E. 1978. Long-term blockade of angiotensin formation in various normotensive and hypertensive rat models using converting enzyme inhibitor (SQ 14,-225). Circ Res 43(Suppl I):45–53

Bravo, E.L., and Tarazi, R.C. 1979. Converting enzyme inhibition with an orally active compound in hypertensive man. Hypertension 1:39–46

Carretero, O.A., Kuk, P., and Piwonska, S. 1971. Role of the renin-angiotensin system in the pathogenesis of severe hypertension in rats. Circ Res 29:654–663

Case, D.B., Wallace, J.M., Keim, H.J., Weber, M.A., Sealey, J.E., and Laragh, J.H. 1977. Possible role of renin in hypertension as suggested by renin-sodium profiling and inhibition of converting enzyme. N Engl J Med 296:641–646

Case, D.B., Atlas, S.A., Laragh, J.H., Sealey, J.E., Sullivan, P.A., and McKinstry, D.N. 1978. Clinical experience with blockade of the renin-angiotensin-aldosterone system with an oral converting enzyme inhibitor (SQ 14,225; captopril) in hypertensive patients. Progr Cardiovasc Dis 21:195–206

Chan, P.S., Ronsberg, M.A. and Cervoni, P. 1980. Synergistic antihypertensive activity of captopril and hydrochlorothiazide in spontaneously hypertensive rats (SHR). Fed Proc 39(Abs):813

Crofton, J.T., Share, L., and Horovitz, Z.P. 1979. The effect of SQ 14,225 on systolic blood pressure and urinary excretion of vasopressin in the developing spontaneously hypertensive rats. Hypertension 1:462–467

Cushman, D.W., Cheung, H.S., Sabo, E.F., and Ondetti, M.A. 1977. Design of potent competitive inhibitors of angiotensin-converting enzyme. Biochemistry 16:5484–5491

Cushman, D.W., Cheung, H.S., Sabo, E.F., and Ondetti, M.A. 1978. Design of new antihypertensive drugs: potent and specific inhibitors of angiotensin converting enzyme. Progr Cardiovasc Dis 21:176–182

Dean, H.G., and Ingham, S. 1978. The effect of SQ 14,225 on fluid intake in DOCA/salt hypertensive rats. Br J Pharmacol 64(Abs): 390P–391P

Dollery, C.T., and Miyamori, I. 1980. Indomethacin and the hypotensive action of captopril in DOCA salt hypertensive rats. Br J Pharmacol 68(Abs):117P–118P

Douglas, B.H., Langford, H.G., and McCaa, R.E. 1979. Response of mineralocorticoid hypertensive animals to an angiotensin I converting enzyme inhibitor. Proc Soc Exp Biol Med 161:86–87

Elijovich, F., and Krakoff, L.R. 1979. Absolute sensitivity to angiotension II and converting enzyme inhibition in glucocorticoid hypertension. Clin Res 27(Abs):592A

Ferrone, R.A., and Antonaccio, M.J. 1979. Prevention of the development of spontaneous hypertension in rats by captopril (SQ 14,225). Eur J Pharmacol 60:131–137

Ferrone, R., and Heran, C.L. 1978. Hemodynamic effects of SQ 14,225 and guanethidine in spontaneously hypertensive rats. Circulation 57/58(Suppl II):144

Ferrone, R., and Heran, C. 1980. Effects of hydrochlorothiazide (HCTZ) on the hemodynamic responses of SHR to captopril(C). Fed Proc 39(Abs):1190

Ferrone, R.A., Kardon, M.B., and Walsh, G.M. 1978. Systemic hemodynamic effect of converting enzyme inhibitor (CEI, SQ 14,225) during acute renovascular hypertension. Fed Proc 37(Abs):718

Ferrone, R.A., Antonaccio, M.J., and High, J.P. 1979. Effects of captopril on blood pressure of spontaneously hypertensive rats (SHR) during the pre and established stage of hypertension. Fed Proc 38(Abs):738

Ferrone, R.A., Heran, C.L., and Antonaccio, M.J. 1980. Comparison of the acute and chronic hemodynamic effects of captopril and guanethidine in spontaneous hypertensive rats. Clin Exp Hypertens 2:247–272

Folkow, B.U.G., and Hallbäck, M.I.L. 1977. Physiopathology of spontaneous hypertension in rats. In *Hypertension, Physiopathology and Treatment,* eds. J. Genest, E. Koiw and O. Kuchel, pp. 507–528. New York: McGraw-Hill

Forslund, T., Tikkanen, I., and Fyhrquist, F. 1980. Prevention of hypertension in spontaneously hypertensive rats with captopril. Proc 7th Mtg Int Soc Hypertens, New Orleans (Abs):35

Freeman, R.H., Davis, J.O., Watkins, B.E., Stephens, G.A., and DeForrest, J.M. 1979. Effects of continuous converting enzyme blockade on renovascular hypertension in the rat. Am J Physiol 236:F21–F24

Friedman, S.M., Friedman, C.L., and Nakashima, M. 1951. Sustained hypertension following the administration of desoxycorticosterone acetate. J Exp Med 93:361–371

Gavras, H., Brunner, H.R., Laragh, J.H., and Vukovich, R.A. 1974. An angiotensin converting enzyme inhibitor to identify and treat vasoconstrictor and volume factors in

hypertensive patients. N Engl J Med 291:817–821
Gavras, H., Oliver, J.A., and Cannon, P.J. 1976. Interrelations of renin, angiotensin II, and sodium in hypertension and renal failure. In *Annual Reviews of Medicine,* eds. W.P. Kreger, C.H. Coggins and E.W. Hancock, pp. 485–521. Palo Alto: Annual Reviews, Inc.
Gavras, H., Brunner, H.R., Turini, G.A., Kershaw, G.R., Tifft, C.P., Cuttelod, S., Gavras, I., Vukovich, R.A., and McKinstry, D.N. 1978. Antihypertensive effect of the oral angiotensin converting enzyme inhibitor SQ 14,225 in man. N Engl J Med 298:991–995
Goldblatt, H., Lynch, J., Hanzal, R.F., and Summerville, W.W. 1934. Studies in experimental hypertension I. The production of persistent elevation of systolic blood pressure by means of renal ischemia. J Exp Med 59:347–379
Grollman, A. 1944. Simplified procedure for inducing chronic renal hypertension in the mammal. Proc Soc Exp Biol Med 57:102–104
Horovitz, Z.P. 1980. Pharmacology and mechanism of action of inhibitors of the renin-angiotensin system. In Drug Development and Evaluation, Vol. 4, *Pharmacology and Clinical Use of Angiotensin I Converting Enzyme Inhibitors,* eds. F. Gross and R.K. Liedtke, pp. 9–16. Stuttgart: Gustav Fischer Verlag
Horovitz, Z.P., Antonaccio, M.J., Rubin, B., and Panasevich, R.E. 1979. Influence of various antihypertensive agents on lifespan of renal hypertensive rats. Br J Clin Pharmacol 7(Suppl 2):2435–2485
Hutchinson, J.S., Mendelsohn, F.A.O., and Doyle, A.E. 1980. Hypotensive action of captopril and saralasin in intact and anephric spontaneously hypertensive rats. Hypertension 2:119–124
Ito, K., Koike, H., Miyamoto, M., and Urakawa, N. 1980. Long-term blockade of angiotensin-converting enzyme alters passive ion transport of vascular smooth muscle. Life Sci 26:1023–1027
Jandhyala, B.S., Washington, G.F., and Lokhandwala, M.F. 1978. Influence of inhibition of angiotensin I converting enzyme with SQ 14,225 on the arterial blood pressure of mongrel dogs. Res Commun Chem Pathol Pharmacol 22:257–265
Jenkins, A.C., and McKinstry, D.N. 1979. Review of clinical studies of hypertensive patients treated with captopril. Med J Aust 2(Spec Suppl):32–37
Johnston, C.I., Millar, J.A., McGrath, B.P., and Matthews, P.G. 1979. Long-term effects of captopril (SQ 14,225) on blood pressure and hormone levels in essential hypertension. Lancet 2:493–496
Koike, H., Ito, K., Miyamoto, M., and Nishino, H. 1980. Effects of long-term blockade of angiotensin converting enzyme with captopril (SQ 14,225) on hemodynamics and circulating blood volume in spontaneously hypertensive rats. Hypertension 2:299–303
Koletsky, S., Shook, P., and Rivera-Velez, J. 1972. Absence of hyperactive renal humoral pressor system in spontaneously hypertensive rats. In *Spontaneous Hypertension,* ed. K. Okamoto, pp. 199–202. Tokyo: Igaku Shion, Ltd.
Laffan, R.J., Goldberg, M.E., High, J.P., Schaeffer, T.R., Waugh, M.H., and Rubin, B. 1978. Antihypertensive activity in rats of SQ 14,225, an orally active inhibitor of angiotensin converting enzyme. J Pharmacol Exp Ther 204:281–288
Lai, F.M., Tanikella, T., Herzlinger, H., Chan, P.S., and Cervoni, P. 1980. Studies on the mechanism of the enhancement of the antihypertensive activity of captopril by a diuretic in spontaneously hypertensive rats. Fed Proc 39(Abs):496

Lee, T.C., Kushiro, T., Gassic J.-P., Lupu, A.N., and Maxwell, M.H. 1979. The kallikrein-kinin system (KKS) in blood pressure homeostasis. Proc 6th Sci Mtg Int Soc Hypertens, Goteberg(Abs):42

Loyke, H.F. 1978. The effect of converting enzyme and angiotensin II blockade in chronic experimental hypertension. Proc 7th World Congress of Cardiology, Tokyo, (Abs):320

McCaa, R.E., McCaa, C.S., Bengis, R.G., and Guyton, A.C. 1979. Role of aldosterone in experimental hypertension. J Endocrinol 81:69P–78P

Maitra S., Miyazaki, S., Scicli, A.G., and Carretero, O.A. 1979. Role of mineralocorticoid in the chronic antihypertensive effect of SQ 14,225. Circulation 59/60(Suppl II):228

Margolius, H.S., Geller, R.G., DeJong, W., Pisano, J.J., and Sjoerdsma, A. 1972. Urinary kallikrein excretion in hypertension. Circ Res 30/31(Suppl II):125–131

Marks, E.S., Bing, R.F., Thurston, H., and Swales, J.D. 1980. Vasodepressor property of the converting enzyme inhibitor captopril (SQ 14,225): the role of factors other than renin-angiotensin blockade in the rat. Clin Sci 58:1–6

Matsunaga, M., Yamamoto, J., Hara, A., Ogino, K., Yamori, Y., and Okamoto, K. 1975. Plasma renin and hypertensive vascular complications, an observation in the stroke-prone spontaneously hypertensive rats. Jpn Circ J 39:1305–1311

Mendlesohn, F.A.O., Hutchinson, J.S., and Doyle, A.F. 1979. Effect of captopril in spontaneously hypertensive rats: persistence of its hypotensive action after nephrectomy as evidence for hyperactivity of the tissue renin-angiotension system. Med J Aust 2(Spec Suppl):5–7

Muirhead, E.E., Prewitt, R.L., Jr., Brooks, B., and Brosius, W.L., Jr. 1978. Antihypertensive action of the orally active converting enzyme inhibitor (SQ 14,225) in spontaneously hypertensive rats. Circ Res 43(Suppl I):53–59

Muirhead, E.E., Brooks, B., and Brosius, W.L. 1979. Captopril in sodium-volume expanded hypertension. Fed Proc 38(Abs):1439

Okamoto, K., and Aoki, K. 1963. Development of a strain of spontaneously hypertensive rats. Jpn Circ J 27:282–293

Okamoto, K., Yamori, Y., and Nagaoka, A. 1974. Establishment of the stroke-prone spontaneously hypertensive rat (SHR). Circ Res 34/35(Suppl 1):143–153

Okuno, T., Kondo, K., Konishi, K., Saruta, T., and Kato, E. 1979. SQ 14,225 attenuates the vascular response to norepinephrine in the rat mesenteric arteries. Life Sci 25:1343–1350

Ondetti, M.A., Rubin, B., and Cushman, D.W. 1977. Design of specific inhibitors of angiotensin-converting enzyme: new class of orally active antihypertensive agents. Science 196:441–444

Page, I.H. 1939. The production of persistent arterial hypertension by cellophane perinephritis. J A M A 113:2046–2048

Phelan, E.L., and Clark, D.W.J. 1979. Effect of captopril on blood pressure and vascular resistance in normotensive and genetically hypertensive rats. Med J Aust 2(Spec Suppl):9–10

Ribeiro, A.B., Melo, G.M., Franco, R.J.S., Marson, O., Kohlmann, O., Jr., and Ramos, O.L. 1979. Renal hypertension, diet salt and the antihypertensive action of SQ 14,225 (SQ). Proc 6th Sci Mtg Int Soc Hypertens, Goteborg(Abs):165

Rubin, B., and Antonaccio, M.J. 1980. Captopril. In *Pharmacology of Antihypertensive Drugs,* ed. A. Scriabine, pp. 21–42. New York: Raven Press

Rubin, B., Antonaccio, M.J. Goldberg, M.E., Harris, D.N., Itkin,

A.G., Horovitz, Z.P., Panasevich, R.E., and Laffan, R.J. 1978a. Chronic antihypertensive effects of captopril (SQ 14,225), an orally active angiotensin I-converting enzyme inhibitor, in conscious 2-kidney renal hypertensive rats. Eur J Pharmacol 51:377–388

Rubin, B., Antonaccio, M.J., and Horovitz, Z.P. 1978b. Captopril (SQ 14,225) (D-3-mercapto-2-methylpropanoyl-L-proline): a novel orally active inhibitor of angiotensin-converting enzyme and antihypertensive agent. Prog Cardiovasc Dis 21:183–194

Rubin, B., Laffan, R.J., Kotler, D.G., O'Keefe, E.H., DeMaio, D.A., and Goldberg, M.E. 1978c. SQ 14,225 (D-3-mercapto-2-methylpropanoyl-L-proline), a novel orally active inhibitor of angiotensin I-converting enzyme. J Pharmacol Exp Ther 204:271–280

Samar, R.E., and Coleman, T.G. 1980. Effects of sodium restriction and antihypertensive therapy in spontaneously hypertensive rats (SHR).Fed Proc 39(Abs):1193

Schölkens, B.A. 1979. Intraventricular injection of the angiotensin converting enzyme inhibitor SQ 14,225 in rats with experimental hypertension. IRCS Med Sci 7:720

Schölkens, B.A., Steinbach, R., and Jung, W. 1979. Intracerebroventricular injection of an angiotensin I converting enzyme inhibitor (SQ 14,225) in animals with experimental hypertension. Arch Pharmacol 307(Suppl):R46

Scicli, A.G., Maitra, S.R., Miyazaki, S., and Carretero, O.A. 1980. Role of mineralocorticoids in the antihypertensive effect of converting enzyme inhibitor. Fed Proc 39(Abs):1189

Sen, S., Smeby, R.R., Bumpus, F.M., and Turcotte, J.G. 1979. Role of renin-angiotensin system in chronic renal hypertensive rats. Hypertension 1:427–434

Sen, S., Tarazi, R.C., and Bumpus, F.M. 1980. Effect of converting enzyme inhibitor (SQ 14,225) on myocardial hypertrophy in spontaneously hypertensive rats. Hypertension 2:169–176

Seymour, A.A., Davis, J.O., Freeman, R.H., DeForrest, J.M., Rowe, B.P., Stephens, G.A., and Williams, G.M. 1980. Hypertension produced by sodium depletion and unilateral nephrectomy: a new experimental model. Hypertension 2:125–129

Smirk, F.H., and Hall, W.H. 1958. Inherited hypertension in rats. Nature 182:727–728

Stamler, J.F., Brody, M.J., and Phillips, M.I. 1980. The central and peripheral effects of captopril (SQ 14,225) on the arterial pressure of the spontaneously hypertensive rat. Brain Res 186:499–503

Sweet, C.S., and Columbo, J.M. 1979. Cardiovascular properties of antihypertensive drugs in a model of severe renal hypertension. J Pharmacol Meth 2:223–239

Thurston, H., Bing, R.F., Marks, E.S., and Swales, J.D. 1980. Response of chronic renovascular hypertension to surgical correction or prolonged blockade of the renin-angiotensin system by two inhibitors in the rat. Clin Sci 58:15–20

Tikkanen, I., Fyhrquist, F., Tikkanen, T., Miettinen, A., Törnroth, T., and Forslund, T. 1980. Captopril prevents nephritis-DOCA-NaCl hypertension but is ineffective in 1-kidney-DOCA-NaCl hypertension. Proc 7th Sci Mtg Int Soc Hypertens, New Orleans(Abs):135

Unger, T., Rockhold, R.W., Bönner, G., Vecsei, P., and Ganten, D. 1979. Long-term oral administration of converting enzyme inhibitor (SQ 14,225) in spontaneously hypertensive rats. Arch Pharmacol 307(Suppl):R46

Unger, T., Rockhold, W., Bönner, G., Rascher, W., and Ganten, D.

1980. Effects of different orally active inhibitors of the converting enzyme in spontaneously hypertensive rats. In Drug Development and Evaluation, Vol. 4, *Pharmacology and Clinical Use of Angiotensin I Converting Enzyme Inhibitors,* eds. F. Gross and R.K. Liedtke, pp. 23–29. Stuttgart: Gustav Fischer Verlag

Vollmer, R.R., and Boccagno, J.A. 1977. Central cardiovascular effects of SQ 14,225, an angiotensin converting enzyme inhibitor in chloralose anesthetized cats. Eur J Pharmacol 45:117–125

Vollmer, R.R., Boccagno, J.A., Harris, D.N., and Murthy, V.S. 1978. Hypotension induced by inhibition of angiotensin-converting enzyme in pentobarbital-anesthetized dogs. Eur J Pharmacol 51:39–45

Watanabe, T.X., and Sokabe, H. 1979. Acute vasodepressor effect of D-3-mercapto-2-methylpropanoyl-L-proline (SQ 14,225) in the stroke-prone substrain of spontaneously hypertensive rats (SHRSP). Jpn J Pharmacol 29:133–135

Watanabe, T.X., Kawashima, K., Saito, K., and Sokabe, H. 1980. Effects of long-term treatments with captopril on blood pressure and renin activity in stroke-prone spontaneously hypertensive rats. Jpn J Pharmacol, in press

Watkins, B.E., Davis, J.O., Freeman, R.H., Stephens, G.A., and DeForresst, J.M. 1978. Effects of the oral converting enzyme inhibitor (SQ 14,225) on one-kidney hypertension in the dog. Proc Soc Exp Biol Med 157:245–249

Weed, W.C., Vaughan, E.D. Jr., and Peach, M.J. 1979. Prolongation of the saralasin responsive state of two-kidney, one clip Goldblatt hypertension in the rat by the orally administered converting enzyme inhibitor captopril (SQ 14,225). Hypertension 1:8–12

Zimmerman, B.G., Mommsen, C., and Kraft, E. 1980. Renal vasodilatation caused by captopril in conscious normotensive and Goldblatt hypertensive dogs. Proc Soc Exp Biol Med, 164:459–465

Effects of Angiotension Converting Enzyme Inhibitors on the Brain

Thomas Unger, Robin W. Rockhold, Irene Kaufmann-Bühler, Dorsay Hübner, Bernd Schüll, Georg Speck and Detlev Ganten

Department of Pharmacology, University of Heidelberg and German Institute for High Blood Pressure Research 6900 Heidelberg, West Germany

INTRODUCTION

Converting enzyme inhibitors (CEI) have proved to be powerful antihypertensive agents of great theoretical, practical and clinical interest. They were originally conceived as inhibitors of a carboxydipeptidase named converting enzyme (EC 3.4.15.1), which converts the biologically inactive decapeptide angiotensin I (ANG I) to the effector octapeptide angiotensin II (ANG II) of the renin-angiotensin system (RAS). Inhibition of the generation of the pressor peptide ANG II in the plasma was thought to be the exclusive cause for the antihypertensive action of CEI.

Recent results mainly obtained with the CEI SQ 14,225 (captopril) raised the following questions:

1. Is the inhibition of the conversion of ANG I to ANG II in plasma in all cases sufficient to explain the blood pressure lowering effect of CEI?
2. Does local ANG II generation in extrarenal tissue such as arterial wall or brain participate in blood pressure regulation and are the orally active CEI effective at these tissue sites?
3. Do CEI also act as inhibitors of other peptidases apart from converting enzyme and do they effectively interfere with the biosynthesis and degradation of other peptides such as kinins, substance P and opioid peptides?

These questions are addressed in this chapter by a discussion of a) the specificity of CEI and their possible inhibitory action on peptidases other than converting enzyme; b) the specificity of converting enzyme, i.e., the effects of some biologically active peptides (as possible substrates) on enzyme activity; c) the role of these peptides in the brain for blood pressure regulation; d) the penetration of CEI into the brain, and e) the central effects of CEI on blood pressure in spontaneously hypertensive rats.

CONVERTING ENZYME (EC 3.4.15.1) IN THE BRAIN

Converting enzyme is present in the brain at concentrations lower than in lung tissue but higher than in most other organs (Ganten et al., 1975a; Roth et al., 1969). High concentrations were found in the striatum, the pituitary gland, the cerebellum (Yang and Neff, 1972), in the caudate nucleus (Poth et al., 1975), the plexus chorioideus (Arregui and Iversen, 1978), and in microvessels of the brain (Gimbrone et al., 1979). Converting enzyme is also present in the cerebrospinal fluid (CSF) of rats and dogs (Schelling et al., 1980). Subcellular fractionation studies showed that the enzyme is mainly found in the microsomal fraction of synaptosomes (Yang and Neff, 1972; Benuck and Marks, 1978). The fact that CEI such as captopril can block the biological effects elicited by intracranial injections of renin or ANG I, whereas CEI do not block the central ANG II blood pressure responses, clearly demonstrates that the converting enzyme is active in vivo in brain tissue (Fitzsimons et al., 1977; Reid and Day, 1977; Mann et al., 1979; Schölkens et al., 1980).

OTHER COMPONENTS OF THE RAS IN THE BRAIN

The other components of the RAS (angiotensinogen, renin, angiotensin and angiotensinases) (Fig. 1) have been demonstrated to be present in the brain (Ganten and Speck, 1978). Angiotensin II–like immunoreactive material has been localized in various parts of the brain, with particularly high density in the median eminence and in other hypothalamic nuclei, in the medulla oblongata, in the substantia gelatinosa of the spinal cord and in the sympathetic lateral column (Fuxe et al., 1976; Ganten et al., 1978). Renin-like immunoreactive material has recently

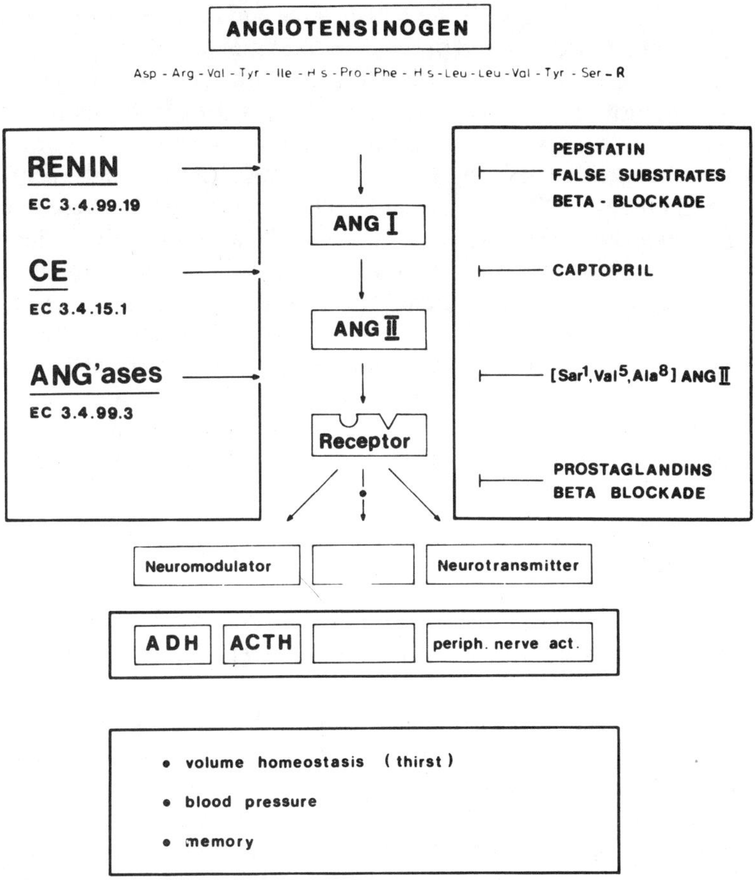

Fig. 1. General outline of the renin-angiotensin system. On the left are the enzymes involved in the biosynthesis of the effector peptide angiotensin II from its high molecular weight precursor angiotensinogen via the inactive intermediate peptide precursor angiotensin I; inhibitors of the enzymes and receptor blockers are shown on the right. The possible modes of action, effects on hypothalamic hormones and the overall regulatory effects of angiotensin are indicated on the lower part.

been localized in the peri- and paraventricular nuclei, nucleus supraopticus and cerebellum of rats and mice (Fuxe et al., 1980). In the CSF, angiotensinogen is present in high concentrations (Printz et al., 1978; Schelling et al., 1980), ANG I and ANG II-like material was also found in the CSF (Ganten and Speck, 1978). Thus, the RAS is the only peptide system of which all components, including the peptide synthetizing and degrading enzymes, have been characterized in the brain.

The central effects of angiotensin have been extensively studied. They include increase of blood pressure, stimulation of drinking behavior with selective salt uptake (Avrith and Fitzsimons, 1980), release of antidiuretic hormone (ADH) and adrenocorticotrophic hormone (ACTH) from the pituitary gland and stimulation of catecholamine synthesis and turnover (for review see Fuxe et al., 1980; Ganten and Speck, 1978; Phillips et al., 1979; Severs and Daniels-Severs, 1973). These effects are mediated by specific ANG II receptors localized mainly in the vicinity of the brain ventricles and which can be blocked by competitive ANG II receptor antagonists (Bennett and Snyder, 1976; Sirett et al., 1977; Simpson et al., 1978; Phillips et al., 1979). Thus, stimulation of endogenous brain angiotensin biosynthesis by intracerebroventricular (i.c.v.) administration of renin or the application of angiotensin into the brain is followed by a concerted output of different neurohumoral effects that can all by themselves contribute to the elevation of blood pressure (Ganten et al., 1979).

ROLE OF BRAIN ANGIOTENSIN AND OTHER PEPTIDES IN BLOOD PRESSURE CONTROL

In recent years evidence has accumulated suggesting that peptide systems in the brain, particularly the RAS, may play a functional role in pathophysiological mechanisms of blood pressure control. Most of these data have been obtained in the spontaneously hypertensive rat (SHR), which is generally believed to be the best available experimental model for human essential hypertension. Thus, it was found that renin activity in young SHR is significantly higher in a number of noradrenergic brain nuclei (A1, A5, A6, nucleus tractus solitarii) known to be involved in blood pressure control and brain catecholamine turnover (Ganten et al., 1980). In addition, it has been shown that these rats were supersensitive to central angiotensin (Hoffman et al., 1976) as well as to enkephalins (Schaz et al., 1980) and substance P (Unger et al., 1980), and may possess an increased number of angiotensin receptors in the brain (Stamler et al., 1980). In some strains of SHR, increased CSF angiotensin levels have been reported (Ganten et al., 1975b). In addition, it has been reported by several investigators independently that central application of the angiotensin receptor antagonist saralasin by injection or long-term infusion over several days lowered blood pressure in SHR (Phillips et al., 1975; Wickre et al., 1978; Clough et al., 1979). This effect persisted after nephrectomy (Phillips et al., 1977). Intravenous administration of the inhibitor did not decrease but rather increased blood pressure in these rats (Mann et al., 1978). This is characteristically seen in hypertensive states with low plasma renin, as is the case for SHR of the stroke-prone strain (SHR-sp).

Taking these data together, it appears that stimulation of the brain RAS is the hallmark of central peptidergic stimulation (Ganten et al., 1979) contributing to the development and maintenance of high arterial blood pressure in SHR (Fig. 2). A correlate in man is found in a recently published report by Kaneko et al. (1979), who observed increased angiotensin-like activity in the CSF of hypertensive patients.

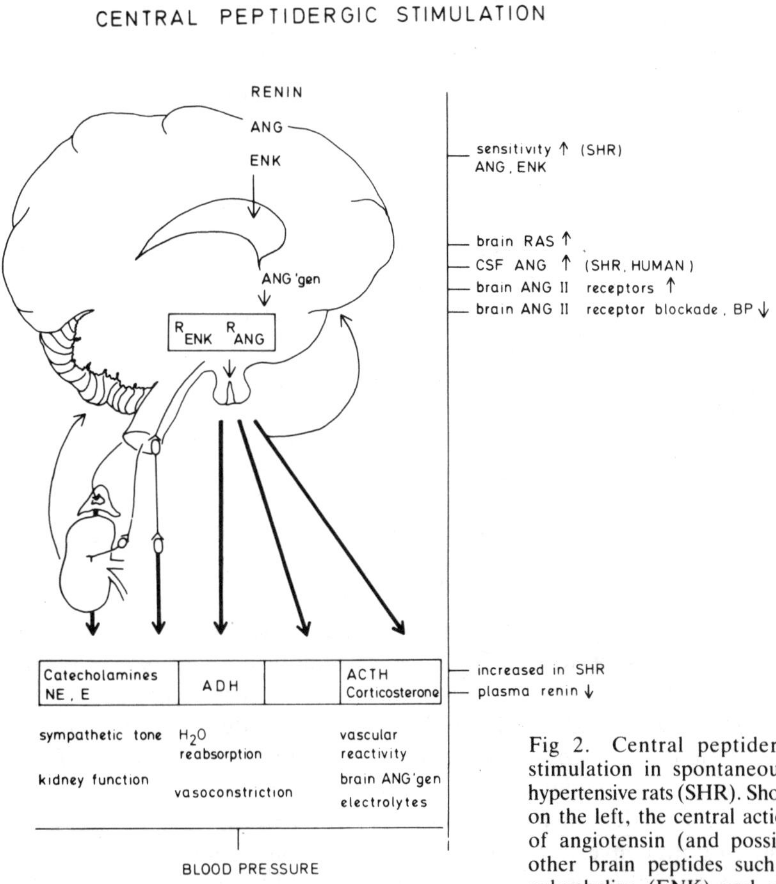

Fig 2. Central peptidergic stimulation in spontaneously hypertensive rats (SHR). Shown on the left, the central actions of angiotensin (and possibly other brain peptides such as enkephalins (ENK) and substance P) on blood pressure via stimulation of the sympathetic nervous system and release of hypothalamic and pituitary hormones. Summarized on the right, the findings suggesting a contribution of brain angiotensin and other peptides to pathophysiological mechanisms in hypertension.

INTERACTIONS OF CEI WITH OTHER ENZYME-PEPTIDE SYSTEMS IN THE BRAIN

Angiotensin converting enzyme is identical with kininase II, a bradykinin degrading enzyme (Yang et al., 1970). Bradykinin-like immunoreactive material has been demonstrated in the brain (Hori, 1968; Correa et al., 1979). Cells containing bradykinin were found in the hypothalamus with dense clusters in the vicinity of the periventricular and dorsomedial nuclei (Correa et al., 1979). In contrast to its peripheral depressor effects, bradykinin increases blood pressure when injected into the CSF of rats (Lambert and Lang, 1970; Correa and Graeff, 1974). In the periphery, potentiation of the depressor action of kinins by inhibiting kininase II is a mechanism which could explain in part the antihypertensive properties of CEI. CEI can also potentiate the pressor effects of centrally administered bradykinin. Although this effect is obviously not contributory to the blood pressure lowering action of CEI, it has to be taken into account as part of the complex actions that CEI may have on the brain.

More important, in terms of possible central antihypertensive actions, seems to be the fact that converting enzyme has properties in common with an enkephalin degrading dipeptidyl-carboxypeptidase (Benuck and Marks, 1979), although a more specific enkephalinase is apparently not identical with converting enzyme (Swerts et al., 1979b). Enkephalins are widely distributed in the brain (Hökfelt et al., 1977) and, besides other biological functions, the opioid peptides act on the cardiovascular system in a rather complex way, depending on the length of the peptide chain and on the site of application or action, respectively. For instance, leucine-enkephalin is pressor when injected intravenously or into the lateral or fourth brain ventricle (Bolme et al., 1978; Schaz et al., 1980). On the other hand, the long-lived analogue D-Ala-Met-enkephalin also exhibits depressor effects similar to β-endorphin (Bolme et al., 1978; Bellet et al., in press). In SHR peripheral tissues such as the coeliac ganglion, the superior cervical ganglion and the adrenal gland have been found to contain less Met-enkephalin-like activity than normotensive controls (Di Giulio et al., 1979) and the opiate receptor antagonist naloxone increases blood pressure when administered in states of shock hypotension (Holaday and Faden, 1978).

It has been demonstrated in vitro that the CEI SQ 20,881 and captopril inhibit both converting enzyme activity and enkephalinase activity from mouse and rat brain preparations (Swerts et al., 1979b), although both inhibitors had about 1000-fold higher inhibitory potencies against converting enzyme (Table 1). Captopril has also been demonstrated to inhibit the Met-enkephalin inactivation by a membrane bound dipep-

Table 1. Some similarities and dissimilarities between converting enzyme-like brain peptidases.

	Enkephalinases	Converting Enzyme	Substance P–Degrading Enzyme
Distribution in the brain			
striatum	+ + +[b]	+ + +[b]	
cortex	+ +	+	
cerebellum	+	+ +	
Subcellular localization	Membranes[h]	Microsomes[f g] Membranes[h]	Cytoplasma[d]
Purification	Partial[e h]	Partial[e g h]	No[d]
Cl-Activation	No[a] Little[h]	Yes[g]	No[d]
Inhibited by			
EDTA	Yes[g]	Yes[d]	No[d]
PCMBA	—	No[d]	Yes[d]
Inhibitors (IC_{50})			Particulate-Soluble
Captopril	1×10^{-5}M[b] K_i: 10^{-7}M[h]	2×10^{-8}[b]	25% Inhib. No inhib.[d]
SQ 20,881	1×10^{-4}M[b]	2×10^{-7}M[b]	30% Inhib. 5×10^{-6}M[d]
Leu-ENK	1×10^{-5}M[b]	1×10^{-3}M[b]	
Met-ENK	1×10^{-6}M[b]	2×10^{-3}M[b]	
Substance P	—	5×10^{-5}M[c]	

References:
[a] Swerts et al., 1979a
[b] Swerts et al., 1979b
[c] McGeer and Singh, 1979
[d] Lee et al., 1979
[e] Gorenstein and Snyder, 1979
[f] Yang and Neff, 1972
[g] Benuck and Marks, 1978
[h] Benuck and Marks, 1979
Leu-ENK: Leucine-enkephalin; Met-ENK: methionine-enkephalin; PCMBA: p-chloromercuribenzoate;

tidyl-carboxypeptidase from rat brain claimed to be identical with converting enzyme (Benuck and Marks, 1979) (Table 1).

Evidence for interference of CEI with enkephalin metabolism in vivo was reported by Stine et al. (1979) who demonstrated that captopril (900 μg i.c.v.) inhibited the degradation of ^{3}H-Met-enkephalin in the rat striatum and caused prolonged analgesia. The latter effect could be prevented by intraperitoneal injection (10 mg/kg) of naltrexone.

In view of these findings, it can be assumed that CEI interfere in vivo with the degradation of opioid peptides. Thus, potentiation of their cardiovascular effects due to inhibition of opioid peptide degradation in the periphery or in the brain could be a mechanism involved in the antihypertensive actions of CEI.

Another peptide, the undecapeptide substance P, has recently been found to inhibit converting enzyme activity from rat striatal homogenates (McGeer and Singh, 1979) (Table 1). In the assay used, substance P was approximately 100 times more active than leucine-enkephalin as an inhibitor of the hydrolysis of a synthetic substrate, Hippuryl-Histidyl-Leucine (Hip-His-Leu), by striatal homogenates. In another study (Lee et al., 1979), the CEI SQ 20,881 was found to be a potent inhibitor of the soluble substance P-degrading activity from rat brain homogenates, while having only a weak action on the particulate enzyme fraction (Table 1). Captopril did not inhibit the soluble enzyme but had a weak inhibitory action on the particulate enzyme. The substance P-degrading enzyme differed from converting enzyme in terms of chloride ion requirement, subcellular distribution and sensitivity to various inhibitors (Table 1).

Substance P is a widely distributed peptide in the brain (Hökfelt et al., 1978). It is a potent vasodepressor agent when injected intravenously (von Euler and Gaddum, 1931), but has marked pressor effects upon administration into the lateral brain ventricles (Unger et al., 1980). Application directly into the nucleus tractus solitarii was followed by hypotension and bradycardia, compatible with baroreceptor reflex activation, and it was suggested that substance P is the transmitter at the first synapse of the baroreceptor reflex (Haeusler and Osterwalder, 1980).

Whether or not CEI do inhibit the breakdown of this peptide in vivo, and whether this has consequences for cardiovascular control and the antihypertensive action of captopril, e.g., stimulation of the baroreceptor reflex, remains to be investigated.

EFFECTS OF CEI ON BRAIN CONVERTING ENZYME

Experiments to investigate whether CEI can penetrate the blood brain barrier and whether they can inhibit brain converting enzyme following peripheral administration were first performed by Vollmer and Boccagno (1977). In the chloralose anaesthetized cat, the CEI captopril was given intravenously (i.v.), and the pressor responses to i.c.v. infused

ANG I (310 ng/kg) were tested. Intravenous doses of 310 μg/kg and 3.1 mg/kg of captopril did not inhibit the central (i.c.v.) ANG I pressor responses, while markedly diminishing the blood pressure increases to i.v. administered ANG I.

Conversely, when captopril was applied i.c.v., the pressor effects of i.c.v. infused ANG I (310 ng/kg) were significantly inhibited at doses of 3.1 μg/kg and above. High doses of 310 μg/kg and 3.1 mg/kg i.c.v. were found to block the pressor responses not only to i.c.v. but also to i.v. injected ANG I. The pressor responses to i.v. and i.c.v. ANG II (200 ng/kg) were not affected by i.v. or i.c.v. application of the CEI.

The authors concluded that captopril was a potent inhibitor of ANG I conversion in the brain as well as in the peripheral circulation. Since i.v. administration of the CEI had no effect on converting enzyme activity in the brain, penetration of the compound from the periphery to the brain through the blood-CSF barrier was questioned. However, entry of the CEI from the CSF to the peripheral blood was observed at higher doses.

A similar study was conducted by Mann et al. (1979) in SHR since, in these animals, increased blood pressure may alter the permeability of the blood-brain barrier. Application of captopril by gavage and with the drinking fluid overnight (total intake approximately 25 mg/kg) for 24 hr blocked the pressor responses to i.v. ANG I (50 ng) for some hours, but had no effect on the response to ANG I infused into the lateral brain ventricle.

Infusion of cumulative doses of the CEI (1, 10, 100 μg/min for 30 min) into the brain ventricles of α-chloralose-anaesthetized SHR-sp inhibited the pressor responses to i.c.v. ANG I (100 ng/min, 10 min infusion). In harmony with the results of Vollmer and Boccagno (1977), it was concluded that peripherally administered captopril did not reach concentrations in CSF high enough to inhibit the central ANG I responses even in hypertensive animals.

Although blood pressure can be lowered in SHR through inhibition of the brain RAS by competitive ANG II receptor antagonists, Mann et al. (1979) did not observe a blood pressure fall following i.c.v. application of the CEI in the anaesthetized rats. This raised the question whether inhibition of kinin degradation could lead to an accumulation of kinins in the brain and whether this could counteract the blood pressure lowering effect of inhibited central ANG II formation.

The possibility that central administration of CEI would affect the brain mediated cardiovascular action of kinins was therefore investigated in our laboratory. Following infusion of the CEI captopril (1 μg) into the lateral brain ventricle (i.c.v.) of SHR, there was a threefold potentiation of the pressor response to i.c.v. bradykinin (2 μg) (Fig. 3). The

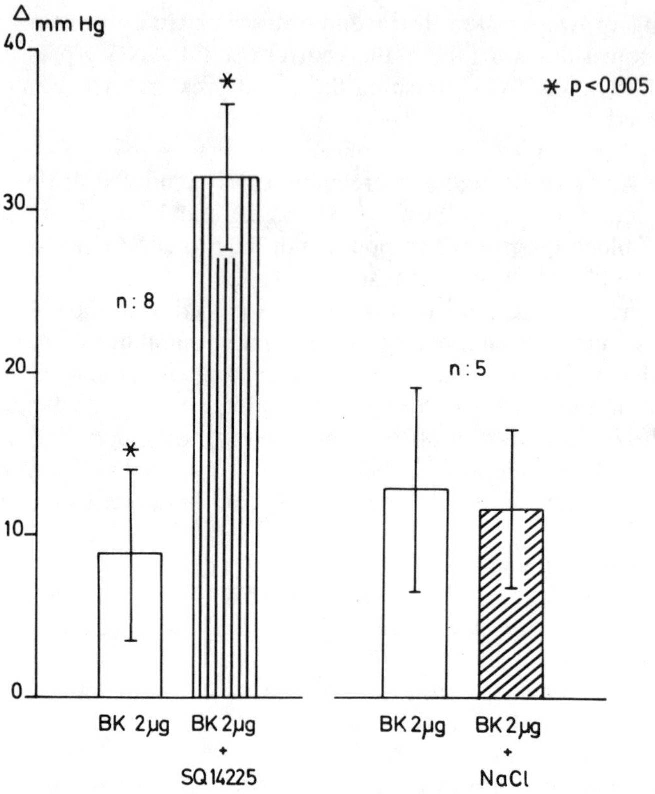

Fig. 3. Blood pressure increases due to bradykinin (BK) injected into the lateral brain ventricle (i.c.v.) before (open columns) and after (shaded columns) i.c.v. administration of 1 μg captopril (SQ 14,225) or vehicle alone (NaCl) in spontaneously hypertensive rats. Shown are changes in mean arterial blood pressure ($\bar{x} \pm$ SEM).

lack of concomitant potentiation of the vasodepressor responses to i.v. bradykinin indicated that the CEI had not penetrated into the periphery at the dose used. However, in a control study in normotensive Sprague-Dawley rats, we observed a reversal of the blood pressure responses to i.c.v. bradykinin (10 μg), namely, a significant blood pressure decrease when the animals were pretreated i.c.v. with a large dose (500 μg) of the CEI (Fig. 4). At this dose, captopril penetrated from the CSF into the peripheral blood as shown by the inhibition of the pressor responses to i.v. ANG I and the potentiation of the depressor responses to i.v. bradykinin (Fig. 5a,b).

A possible explanation for the observed blood pressure response to i.c.v. bradykinin following the high i.c.v. dose of captopril could therefore be that kininase inhibition in the brain facilitated the penetration of i.c.v. administered bradykinin into the peripheral blood, where its depressor effect was potentiated by circulating CEI equally derived from the CSF.

Potentiation of the pressor responses to i.c.v. bradykinin (2 μg) by captopril (2 μg i.c.v.) in normotensive Wistar rats has also been reported by Kondo et al. (1979) in a recent study.

These findings demonstrate the presence of central kininase activity in vivo and indicate that the enzyme can be inhibited by CEI in the brain as well as in the periphery. These data furthermore stress the necessity to carefully exclude effects due to penetration of compounds from the brain to the peripheral blood when characterizing the central action of peptides or CEI.

BLOOD PRESSURE EFFECTS OF CEI IN SPONTANEOUSLY HYPERTENSIVE RATS

Administration of the CEI captopril into the brain ventricles has recently been reported to lower blood pressure in SHR (Schölkens et al., 1979; Hutchinson et al., 1980; Stamler et al., 1980). In one study (Stamler et

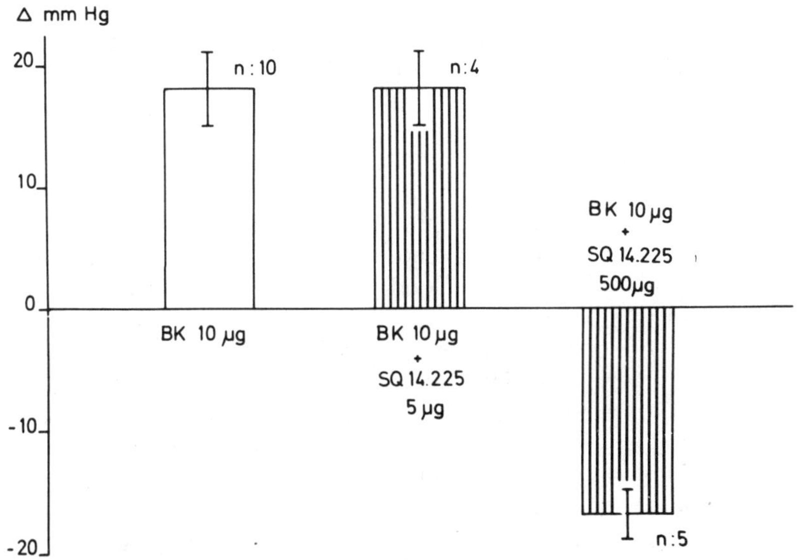

Fig. 4. Blood pressure responses to i.c.v. bradykinin (BK) before (open columns) and following (shaded columns) 5μg or 500 μg captopril (SQ 14,225) i.c.v. in Sprague-Dawley rats. Shown are changes in mean arterial blood pressures ($\bar{x}$ ± SEM).

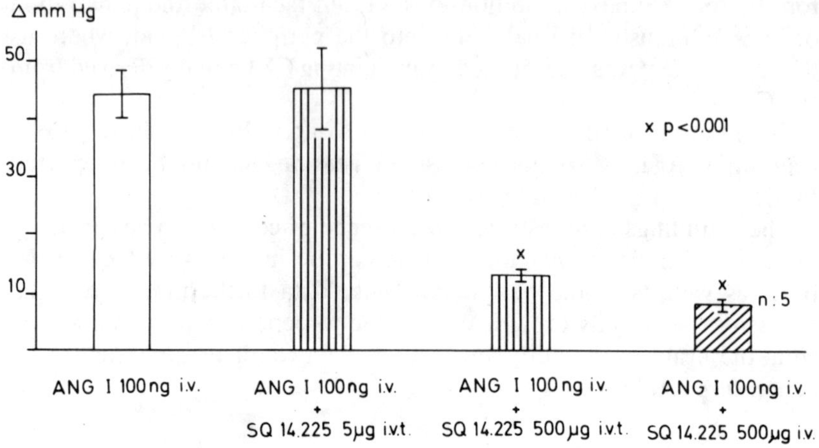

Fig. 5. Shown are changes in mean arterial blood pressure ($\bar{x}$ ± SEM). *Upper panel:* Blood pressure increases due to intravenous (i.v.) angiotensin I before (open columns) and after (shaded columns) central (i.v.t. = i.c.v.) application of captopril at two different doses (5 µg and 500 µg) in Sprague-Dawley rats. Low dose captopril i.c.v. (5 µg) does not lead to converting enzyme inhibition in the peripheral blood (first shaded column). At 500 µg i.c.v., captopril penetrates from the CSF into the peripheral blood and inhibits the pressor responses to angiotensin I (second shaded column) as seen after i.v. injection of the inhibitor.

Lower panel (page 67): Blood pressure decreases (Δmm Hg, ordinate) following intravenous (i.v.) injections of increasing doses of bradykinin (BK, abscissa). ●——● control i.v. BK injections; *– – –* i.v. BK following 500 µg SQ 14,225 centrally (i.v.t.); □– – –□ i.v. BK following 500 µg SQ 14,225 i.v. Note shifts to the left of the BK dose-response curves following both routes of high dose SQ 14,225 application.

al., 1980), a dose of 75 µg captopril per rat was injected either into the lateral brain ventricle or intravenously. Application via either route caused a fall of mean arterial blood pressure. These effects were significantly smaller following i.v. (−16.3 ± 4.8 mm Hg) than i.c.v. (−25.3 ± 4.7 mm Hg) injection of the CEI.

In another study (Hutchinson et al., 1980), a dose of 2 mg/kg captopril was given i.c.v. to SHR and normotensive Wistar Kyoto (WKY) controls. A rapid hypotensive response to i.c.v. captopril was observed in conscious SHR that was maximal after 2 hr (−29 ± 4 mm Hg) and lasted for more than 7 hr. Intravenous injection of the CEI (2 mg/kg) to SHR was followed by a transient fall of blood pressure that was less marked than the i.c.v. effect. In the normotensive control rats, no decrease of blood pressure was observed following i.c.v. injection of the same dose (2 mg/kg) of the CEI.

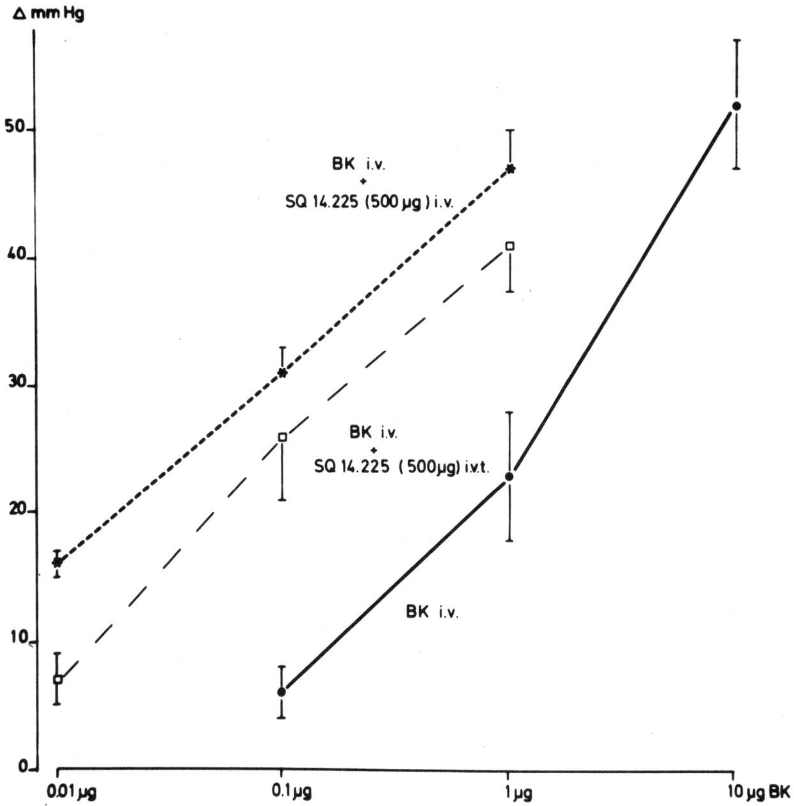

These results demonstrate that, in SHR, blood pressure can indeed be lowered by application of CEI to the brain via the i.c.v. route. However, since high doses of captopril were applied i.c.v. in both studies, penetration of the compound from the CSF into the peripheral blood cannot be excluded, and peripheral mechanisms may have contributed substantially to the hypotensive effect observed.

We therefore conducted experiments designed to test the central and peripheral inhibition of converting enzyme simultaneously following different i.c.v. doses of captopril and to correlate the effects on converting enzyme inhibition with changes in blood pressure (Unger et al., in press).

In the first experiment, SHR-sp and normotensive Wistar Kyoto rats were injected i.c.v. with 500 μg of the CEI captopril (Fig. 6). In SHR-sp, i.c.v. injection of 500 μg captopril elicited a biphasic blood pressure response. An initial increase (maximum +54 ± 5 mm Hg mean arterial

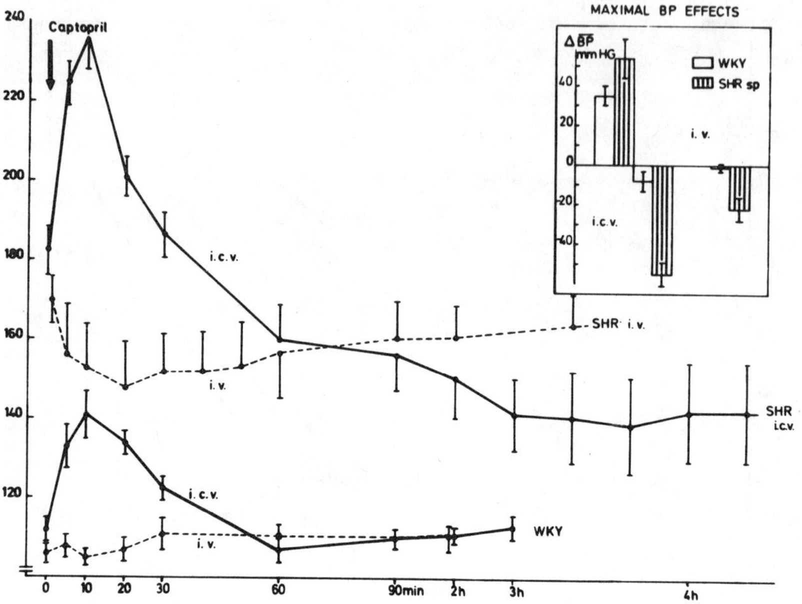

Fig. 6. Changes in mean arterial blood pressure ($\bar{x} \pm$ SEM) in response to a high dose of 500 μg of the converting enzyme inhibitor captopril (●——● i.c.v.; ●- - -● i.v.) in spontaneously hypertensive rats (SHR) and in Wistar-Kyoto rats (WKY). The insert depicts maximal blood pressure effects following both routes of captopril administration in SHR-sp and in WKY.

pressure) of rapid onset and about 30 min duration was followed by a blood pressure decrease (maximum −55 ± 6 mm Hg below control level) that lasted for 4 hr or longer. The pressor responses to i.c.v. ANG I (100 ng) were completely blocked immediately after the injection of the inhibitor. This effect lasted at least 24 hr in all rats; in some animals, central blockade was still present 48 hr after the injection. The pressor responses to i.v. ANG I (100 ng) were reduced by 80% as early as 3 min after the central injection of the CEI. They returned gradually to control values, which were almost reached after 5 hr.

Intravenous injection of the same captopril dose in SHR-sp caused a small transient blood pressure decrease (maximum −20.8 ± 6.1 mm Hg) of less than 1 hr duration. Blockade of the pressor responses to i.v. ANG I was almost complete 5 min after i.v. captopril injection and lasted for 2 hr.

In WKY, i.c.v. injection of 500 μg captopril had only blood pressure increasing effects (similar duration but smaller than in SHR-sp); no significant decreases occurred. Following i.v. injection with the CEI, blood pressure remained unaltered.

A similar pattern of blood pressure changes following high dose captopril i.c.v. in spontaneously hypertensive and normotensive rats was observed by Schölkens et al. (1979).

In a second experiment (Fig. 7), a dose of 5 μg captopril was injected i.c.v. in SHR-sp. This time, only decreases of blood pressure were observed (maximum −12.8 ± 1.3 mm Hg mean arterial pressure), which started 2 to 3 min after the CEI injection and lasted for about 20 min. The central (i.c.v.) pressor responses to ANG I (100 ng) were partially blocked, with a maximum of about 70% inhibition after 7 min,

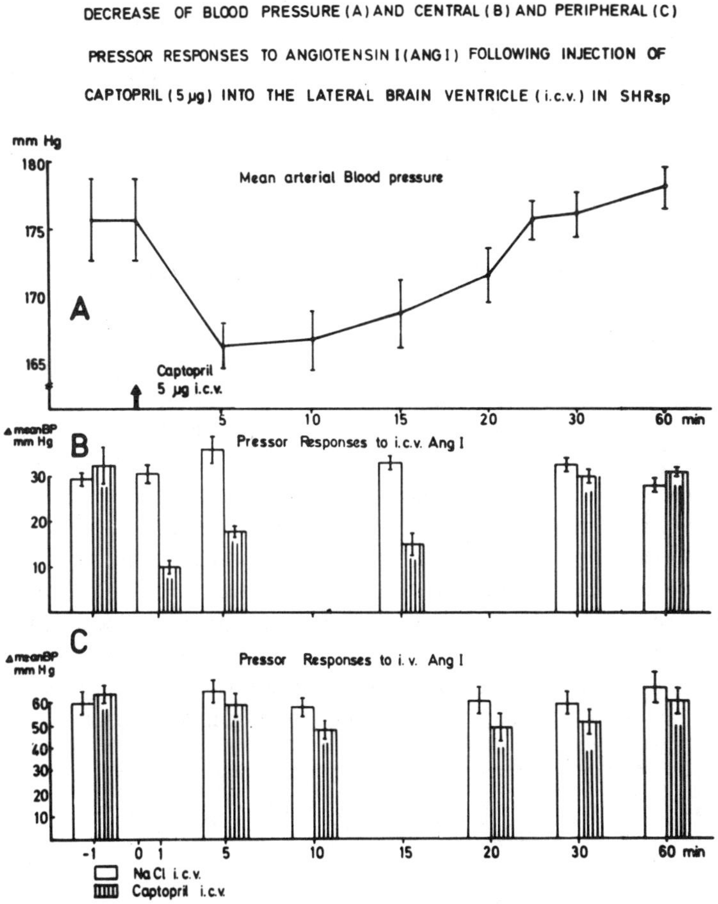

Fig. 7. Effects of i.c.v. captopril (5 μg) on systemic blood pressure (A) and on the angiotensin induced pressor responses following central (i.c.v.) (B) and peripheral (i.v.) (C) angiotensin I (ANG I) administration. Experiments were carried out in spontaneously hypertensive rats. Shown are changes in mean arterial blood pressure ($\bar{x}$ ± SEM).

and returned gradually to control values within 60 min. The pressor responses to i.c.v. and i.v. ANG I were studied in two groups of SHR-sp: one group had been injected i.c.v. with 5 μg captopril and the other one with the same volume of the vehicle (0.9% NaCl) only. Thus, tachyphylactic effects due to repeated ANG I injections at short time intervals could be excluded. The i.v. ANG I pressor effects in the group injected with 5 μg captopril i.c.v. were at no time significantly different from those injected i.c.v. with NaCl.

The following conclusions can be drawn from the experiments described above:

1. Blood pressure decreases in response to i.c.v. administered captopril in SHR but not in normotensive WKY.
2. Low doses of i.c.v. captopril (5 μg) cause converting enzyme inhibition in the CSF and in the brain but not in the peripheral blood. The fall of blood pressure appears, therefore, to be centrally mediated. Since inhibition of ANG II formation in the CSF and blood pressure decrease have similar time courses both effects can be correlated.
3. High doses of i.c.v. captopril (e.g., 500 μg) entail marked and sustained blood pressure decreases. At these doses, the CEI leaks from the CSF into the peripheral blood and causes prolonged inhibition of converting enzyme in the periphery. However, it is unlikely that the hypotensive effect is due to peripheral converting enzyme inhibition alone, since i.v. administration of the same doses of captopril had little effect on blood pressure, although converting enzyme in the plasma was blocked almost completely.
4. In SHR and WKY, high doses of captopril injected i.c.v. elicit an initial rise of blood pressure, the cause of which remains unclear. An unspecific toxic pressor effect, possibly due to acidity of the drug solution, was ruled out in control experiments. It could thus be speculated that the pressure rise is due to transiently elevated endogenous kinins in the brain following central converting enzyme (kininase II) blockade, but this needs further investigation.

CENTRAL EFFECTS OF CHRONIC ORAL TREATMENT WITH CEI

The studies described above demonstrate that CEI can interfere with a number of pressor and depressor peptide systems in the brain and that CEI can exert an antihypertensive action when applied directly to the brain. The data available from acute experiments, however, do not favor the idea that brain mediated antihypertensive mechanisms would play a significant role following acute oral or i.v. administration of the drug, since oral treatment overnight or i.v. injections of captopril did not alter the pressor responses to i.c.v. ANG I (Vollmer and Boccagno, 1977; Mann et al., 1979) indicating that this CEI did not penetrate the blood-brain barrier to an extent leading to inhibition of converting enzyme in the CSF. Furthermore, in a radioautographic study in rats, no radioactivity could be detected in the brain following i.v. injection of captopril-^{35}S (Heald and Ita, 1977). However, these experiments did not exclude the possibility that penetration of the blood-brain barrier and accumulation at central sites not accessible for CSF ANG I could occur following prolonged oral administration of CEI.

This question was addressed in several studies in our laboratory. Spontaneously hypertensive rats were used as the experimental animal model of hypertension for three reasons: a) the resemblance of this animal type of hypertension to human essential hypertension; b) the possible dependence of high blood pressure on central peptidergic mechanisms; and c) the fact that, in these rats, blood pressure can be lowered by central application of CEI.

In the first experiment (Unger et al., 1979a), captopril (50 mg/kg/day) was given orally in the drinking fluid for 4 wk to 6-month-old SHR-sp. At the end of the treatment period, the rats were challenged with i.v. and i.c.v. injections of ANG I and bradykinin. Blood pressure was effectively lowered in the treated animals, the pressor responses to i.v. ANG I were inhibited and the depressor effects of i.v. bradykinin were potentiated. There was, however, no change in the pressor responses to i.c.v. ANG I and i.c.v. bradykinin.

In a second experiment (Unger et al., 1979b), 7-week-old SHR-sp and WKY were given orally the CEI YS 980 for 3 months. This CEI bears close structural similarity to captopril but is characterized by a higher lipid solubility. It was therefore assumed that penetration of the blood-brain barrier might be easier for this compound than for captopril.

Blood pressure was kept low (slightly above WKY control levels) in the treated SHR-sp, while the treated WKY had a slight decrease in systolic but not in diastolic blood pressure. Intravenous injections with ANG I and bradykinin demonstrated similar inhibition of the plasma converting enzyme in treated SHR-sp and WKY. In SHR-sp, central (i.c.v.) injection of the two peptides was followed by unchanged responses to bradykinin and slightly (but statistically not significant) diminished responses to ANG I. In WKY, the pressor effects of i.c.v. ANG I were significantly attenuated (Fig. 8), demonstrating for the first time inhibition of central converting enzyme after prolonged oral treatment with a CEI.

Since the inhibitory effect on ANG I conversion was small, and no effect was observed on the central action of kinins despite the high solubility of the CEI YS 980, it was argued that the testing of i.c.v. peptide effects may not be a sensitive enough bioassay for the penetration of CEI into the brain tissue.

Hence, in a third experiment, the effects of chronic oral treatment with CEI on brain renin and converting enzyme activity in brain tissue were measured by direct assays of the two enzymes.

SHR-sp weanlings were treated orally with captopril (50 mg/kg/day). Their parent SHR-sp had already been given the drug 2 wk before mating; treatment of their mothers had been continued during pregnancy and lactation period. In the treated SHR-sp offspring, hypertension did not occur throughout their life span. At the age of 6 months, the rats were sacrificed and renin and converting enzyme activity were measured in different areas of the brain (medulla oblongata, hypothalamus, striatum, cerebellum, cortex, plexus chorioideus), in peripheral tissues (aorta, kidney, adrenal gland, lung) and in the plasma. They were compared with those of strictly age- and sex-matched SHR-sp controls.

Measurement of Renin Following CEI Treatment

The effects of chronic oral treatment with captopril on renin activity in brain areas known to be involved in blood pressure control are shown in Figure 9.

Renin concentrations in the medulla oblongata, hypothalamus and neurohypophysis were significantly increased irrespective of whether or not the animals had been nephrectomized (NX) 24 hr before sacrifice. The nephrectomy experiments were performed to exclude contamination of brain tissue with kidney and plasma renin.

Renin concentration was also increased in the plasma and in some peripheral tissues (kidney, adrenals) but not in the lung and in the aorta.

These results are interpreted to mean that tissue renin concentrations increased as a consequence of converting enzyme blockade and inhibition of ANG II generation, since decreased ANG II levels have been shown to lead to a rise of renin by interruption of the negative feedback loop. This has been demonstrated by several authors including ourselves for plasma and kidney renin (Davis and Freeman, 1976; Peach, 1977), and has also been reported to occur in aortic tissue (Asaad and Antonaccio, 1979) following 8 days of oral treatment with CEI.

The increase of renin concentration in medulla oblongata, hypothalamus and neurohypophysis suggests inhibition of converting enzyme activity and, thereby, decreased ANG II feedback inhibition in these

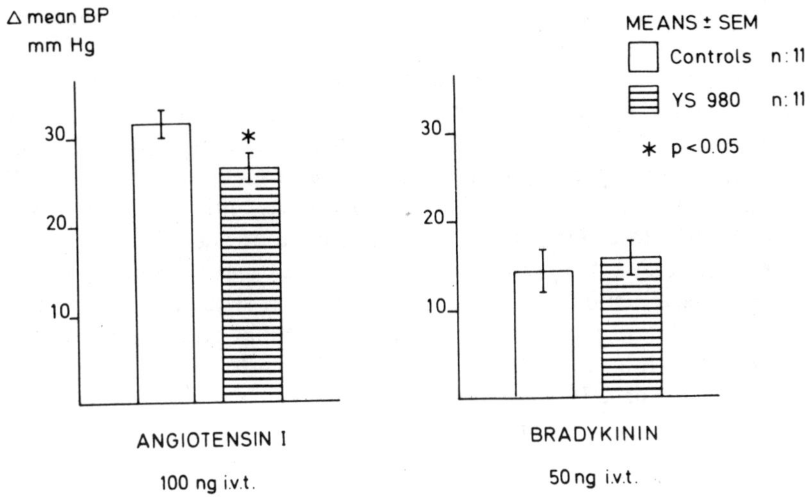

Fig. 8. Blood pressure effects of intraventricular (i.v.t. = i.c.v.) angiotensin I and bradykinin following 12 weeks of oral treatment with the converting enzyme inhibitor YS 980 (50 mg/kg/day) in Wistar-Kyoto rats.

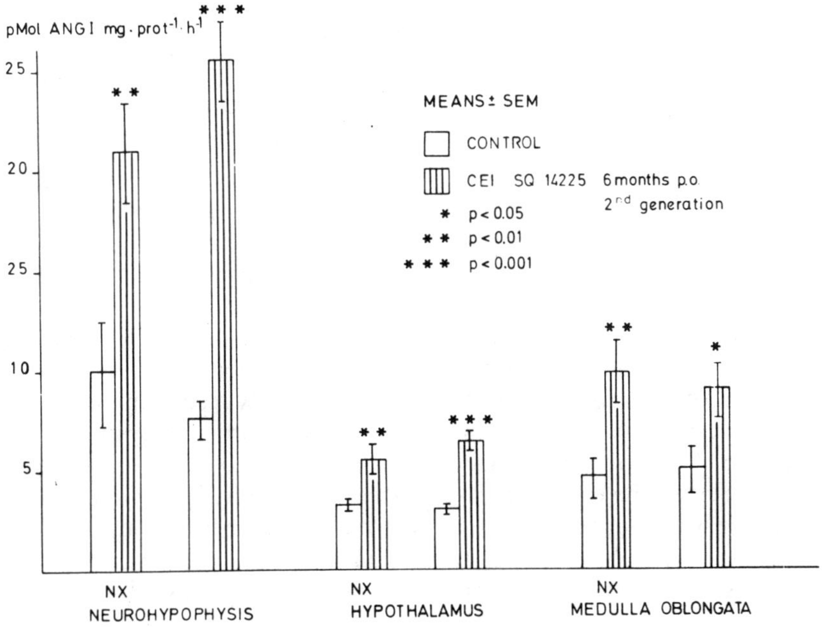

Fig. 9. Effects of chronic oral treatment with the converting enzyme inhibitor captopril (SQ 14,225) on tissue renin concentration in different brain areas in spontaneously hypertensive rats (n:10 per group). NX: nephrectomy.

blood pressure controlling brain areas by chronic oral treatment with CEI.

The finding of unchanged renin concentration in the aorta in our experiments is at variance with the results reported by Asaad and Antonaccio (1979) and does not seem to favor their hypothesis that inhibition of ANG II in vascular tissue is a major factor contributing to the antihypertensive mechanism of CEI. However, this discrepancy could also be due to differences in the time course of oral CEI treatment in the two studies. It may well be that different mechanisms of the antihypertensive action of CEI come into play in sequential order. Thus, the acute blood pressure decreasing effect could predominantly be associated with an interference with the plasma RAS, while converting enzyme (kininase II) inhibition in peripheral tissues such as the vascular walls would become important at a later stage of the treatment period, and penetration of the CEI into the brain with inhibition of central ANG II formation could be an even later event.

Measurement of Converting Enzyme Following CEI Treatment

Direct measurement of converting enzyme activity was performed by a modified method of Depierre and Roth (1975). In this procedure, the dipeptide l-histidyl-L-leucine (His-Leu) is cleaved by the enzyme from a synthetic substrate (carbobenzoxy-Phe-His-Leu) and is detected fluorimetrically after reaction with o-phthaldialdehyde.

In the CEI-treated SHR-sp, changes in converting enzyme activity were found in the brain (medulla oblongata and hypothalamus), in the lung and in the plasma. Surprisingly, increases and not decreases of the enzyme were measured in these tissues and in plasma. The biochemical results of enzyme measurement were thus in clear contradiction to the effective blockade of ANG I to ANG II conversion in the circulation in vivo during oral CEI treatment. The most probable explanation for this unexpected rise of converting enzyme activity resides in the finding that the inhibitor enzyme complex dissociates during processing of the sample. A thorough investigation into the method of converting enzyme measurement revealed that, e.g., increased converting enzyme activity was measured in the plasma after it had been frozen for even a very short time, but that decreased converting enzyme activity was measured in the same plasma of captopril treated rats when the sample was assayed freshly without freezing. It thus appears that mere freezing dissociates the enzyme inhibitor complex, leading to the discrepancy between biologically and biochemically measured converting enzyme activity. It follows that the biochemically measured increases in converting enzyme "activity" were, in fact, measurements of increased converting enzyme protein concentration in plasma and brain. This would indicate that converting enzyme is regulated in a product-dependent way, i.e., ANG II suppresses converting enzyme synthesis and, in captopril treated rats with blockade of ANG II generation, converting enzyme synthesis is stimulated. The effects of oral CEI treatment on brain renin and converting enzyme can thus be explained by an interruption of the negative feedback loop. These findings are a strong argument for central effects of captopril.

In summary, it has been shown that peptides such as angiotensin, kinins, substance P, and opioid peptides can participate in central mechanisms of blood pressure regulation. These peptides are substrates for converting enzyme and for closely related dipeptide carboxypeptidases which are also inhibited by CEI. Central administration of CEI leads to a brain mediated decrease of blood pressure in SHR-sp.

The testing of biological effects of i.c.v. injected peptides was revealed to be an insensitive bioassay for the penetration of CEI into the brain, but measurements of renin and converting enzyme indicate that chronic oral treatment with CEI leads to an increased synthesis of both enzymes in brain tissue.

We conclude that CEI do have central effects, and that these may contribute to the antihypertensive action of CEI by interference with central peptidergic blood pressure regulating systems.

REFERENCES

Arregui, A., and Iversen, L.L. 1978. Angiotensin-converting enzyme: Presence of high activity in chorioid plexus of mammalian brain. Europ J Pharmacol 52:147–150

Asaad, M.M., and Antonaccio, M.J. 1979. The effect of captopril (SQ 14225) on vascular tissue renin in spontaneously hypertensive (SHR) and WKY-normotensive (NTR) rats: Relationship to systolic blood pressure (SBP) and plasma renin activity (PRA). Pharmacologist 21:356, 212

Avrith, D.B., and Fitzsimons, J.T. 1980. Increased sodium appetite in the rat induced by intracranial administration of components of the renin-angiotensin system. J Physiol 301:349–364

Bellet, M., Elghozi, J.L., Meyer, P., Pernollet, M.O., and Schmitt, H. Central cardiovascular effects of narcotic analgesics and enkephalins in rats. Br J Pharmacol, in press

Bennett, J.P., Jr., and Snyder, S.H. 1976. Angiotensin II binding to mammalian brain membranes. J Biol Chem 251:7423–7430

Benuck, M., and Marks, N. 1978. Subcellular localization and partial purification of a chloride dependent angiotensin I converting enzyme from rat brain. J Neurochem 30:729–734

Benuck, M., and Marks, N. 1979. Co-identity of brain angiotensin converting enzyme with a membrane bound dipeptidyl carboxypeptidase inactivating met-enkephalin. Biochem Biophys Res Commun 88:215–221

Bolme, P., Fuxe, K., Agnati, L.F., Bradley, R., and Smythies, J. 1978. Cardiovascular effects of morphine and opioid peptides following intracisternal administration in chloralose-anaesthetized rats. Europ J Pharmacol 48:319–324

Clough, D., Rettig, R., Langmann, E., Rockhold, R., Unger, Th., and Ganten, D. 1979. Effects of chronic central angiotensin II blockade in spontaneously hypertensive rats. Pflüger's Arch 319(Suppl):R8

Correa, F.M.A., and Graeff, F.G. 1974. Central mechanisms of the hypertensive action of intraventricular bradykinin in the unanaesthetized rat. Neuropharmacology 13:65–75

Correa, F.M.A., Innis, R.B., Uhl, G.R., and Snyder, S.H. 1979. Bradykinin-like immunoreactive neuronal systems localized histochemically in rat brain. Proc Natl Acad Sci USA 76: 1489–1493

Davis, J.O., and Freeman, R.H. 1976. Mechanisms regulating renin release. Physiol Rev 56:1–56

Depierre, D., and Roth, M. 1975. Fluorimetric determination of dipeptidyl-carboxypeptidase (angiotensin I-converting enzyme). Enzyme 19:65–70

Euler, U.S. von, and Gaddum, J.H. 1931. An unidentified depressor substance in certain tissue extracts. J Physiol (London) 72:74–87

Fitzsimons, J.T., Epstein, A.N., and Johnson, A.K. 1977. The peptide specificity of receptors for angiotensin-induced thirst. *Central Actions of Angiotensin and Related Hormones*. New York: Pergamon Press, p. 405

Fuxe, K., Ganten, D., Hökfelt, T., and Bolme, P. 1976. Immunohistochemical evidence for the existence of angiotensin II-containing nerve terminals in the brain and spinal cord of the rat. Neurosci Lett 2:229–234

Fuxe, K., Andersson, K., Ganten, D., Hökfelt, T., and Eneroth, P. 1980. Evidence for the existence of an angiotensin II-like immunoreactive central neuronal system and its interactions with the central catecholamine pathways. *Enzymatic Release of Vasoactive Peptides*. New York: Raven Press, p. 161

Fuxe, K., Ganten, D., Hökfelt, T., Locatelli, V., Poulsen, K., Stock, G., Rix, E., and Taugner, R. 1980. Renin-like immunocytochemical activity in the rat and mouse brain. Neurosci Lett 18:245–250

Ganten, D., and Speck, G. 1978. The brain renin-angiotensin system: A model for the synthesis of peptides in the brain. Biochem Pharmacol 27:2379–2389

Ganten, D., Ganten, U., Schelling, P., Boucher, R., and Genest, J. 1975a. The renin and isorenin-angiotensin systems in rats with experimental pituitary tumors. Proc Soc Exp Biol Med 148:568–572

Ganten, D., Hutchinson, J.S., and Schelling, P. 1975b. The intrinsic brain isorenin-angiotensin system in the rat: Its possible role in central mechanisms of blood pressure regulation. Clin Sci Mol Med 48:265s–268s

Ganten, D., Fuxe, K., Phillips, M.I., Mann, J.F.E., and Ganten, U. 1978. The brain isorenin-angiotensin system: Biochemistry, localization and possible role in drinking and blood pressure regulation. *Frontiers in Neuroendocrinology, Vol. 5.* New York: Raven Press, p. 61

Ganten, D., Unger, Th., Rockhold, R., Schaz, K., and Speck, G. 1979. Central peptidergic stimulation: Its possible contribution to blood pressure regulation. *Radioimmunoassay of Drugs and Hormones in Cardiovascular Medicine.* Amsterdam: Elsevier, p. 33

Ganten, D., Speck, G., Meyer, D., Loos, H.-E., Schelling, P., Rettig, R., and Unger, Th. 1980. Biochemical and functional aspects of brain renin. *Enzymatic Release of Vasoactive Peptides.* New York: Raven Press, p. 31

Gimbrone, M.A., Majeau, G.R., Atkinson, W.J., Sadler, W., and Cruise, S.A. 1979. Angiotensin-converting enzyme activity in isolated brain microvessels. Life Sci 25:1075–1084

Di Giulio, A.M., Yang, H.-Y.T., Fratta, W., and Costa, E. 1979. Decreased content of immunoreactive enkephalin-like peptide in peripheral tissues of spontaneously hypertensive rats. Nature 278:646–647

Gorenstein, C., and Snyder, S.H. 1979. Two distinct enkephalinases: solubilization, partial purification and separation from angiotensin converting enzyme. Life Sci 25:2065–2070

Haeusler, G., and Osterwalder, R. 1980. Evidence for a transmitter function of substance P at the first synapse of the baroreceptor reflex. Naunyn Schmiedeberg's Arch Pharmacol 311(Suppl):R46

Heald, A.F., and Ita, C.E. 1977. Distribution in rats of an inhibitor of angiotensin converting enzyme (SQ 14225), as studied by whole-body autoradiography and liquid scintillation counting. Pharmacologist 19:129

Hoffman, W.E., Phillips, M.I., and Schmid, P.G. 1976. Central angiotensin II-induced responses in spontaneously hypertensive rats. Am J Physiol 232:H426–H433

Hökfelt, T., Elde, R., Johansson, O., Terenius, L., and Stein, L. 1977. The distribution of enkephalin-immunoreactive cell bodies in the rat central nervous system. Neurosci Lett 5:25–31

Hökfelt, T., Elde, R., Johansson, O., Ljungdahl, A., Schultzberg, M., Fuxe, K., Goldstein, M., Nilsson, G., Pernow, P., Terenius, L., Ganten, D., Jeffcoate, S.L., Rehfeld, J., and Said, S. 1978. Distribution of peptide-containing neurons. *Psychopharmacology: A Generation of Progress.* New York: Raven Press, p. 39

Holaday, J.W., and Faden, A.J. 1978. Naloxone reversal of endotoxin hypotension suggests role of endorphins in shock. Nature 275:450–452

Hori, S. 1968. The presence of bradykinin-like polypeptides, kinin-releasing and destroying activity in brain. Jap J Pharmacol 18:772

Hutchinson, J.S., Mendelsohn, F.A.O., and Doyle, A.E. 1980. Blood pressure responses of consious normotensive and spontaneously hypertensive rats to intracerebroventicular and peripheral administration of captopril. Hypertension 2:546–550

Kaneko, Y., Ohnishi, T., Fujishima, S., Tanaka, T., and Umemura, S. 1979. Relation of angiotensin II concentrations in cerebrospinal fluid to systemic blood pressure levels in essential hypertension. Proceedings of the 6th Scientific Meeting of the International Society of Hypertension, Göteborg

Kondo, K., Okuno, T., Saruta, T., and Kato, E. 1979. Central and peripheral effects of bradykinin and prostaglandin E_2 on blood pressure in conscious rats. Naunyn Schmiedeberg's Arch Pharmacol 308:111–115

Lambert, G.A., and Lang, W.J. 1970. The effects of bradykinin and eledoisin injected into the cerebral ventricles of conscious rats. Europ J Pharmacol 9:383–386

Lee, C.M., Arregui, A., and Iversen, L.L. 1979. Substance P degradation by rat brain peptidases: Inhibition by SQ 20881. Biochem Pharmacol 28:553–556

Mann, J.F.E., Phillips, M.I., Dietz, R., Haebara, H., and Ganten, D.

1978. Effects of central and peripheral angiotensin blockade in hypertensive rats. Am J Physiol 234:H629–H637

Mann, J.F.E., Rascher, W., Dietz, R., Schömig, A., and Ganten, D. 1979. Effects of an orally active converting-enzyme inhibitor, SQ 14225, on pressor responses to angiotensin administered into the brain ventricles of spontaneously hypertensive rats. Clin Sci 56:585–589

McGeer, E.G., and Singh, E.A. 1979. Inhibition of angiotensin converting enzyme by substance P. Neurosci Lett 14:105–108

Peach, M.J. 1977. Renin-angiotensin system: Biochemistry and mechanisms of action. Physiol Rev 57: Vol 2

Phillips, M.I., Phipps, J., Hoffman, W.E., and Leavitt, M. 1975. Reduction of blood pressure by intracranial injection of angiotensin blocker in spontaneously hypertensive rats. Physiologist 18:350

Phillips, M.I., Mann, J.F.E., Haebara, H., Hoffman, W.E., Dietz, R., Schelling, P., and Ganten, D. 1977. Lowering of hypertension by central saralasin in the absence of plasma renin. Nature 270:445–447

Phillips, M.I., Weyhenmeyer, J., Felix, D., Ganten, D., and Hoffman, W.E. 1979. Evidence for an endogenous brain renin-angiotensin system. Fed Proc 38:2260–2266

Poth, M.M., Heath, R.G., and Ward, M. 1975. Angiotensin-converting enzyme in human brain. J Neurochem 25:83–85

Printz, M.P., Printz, J., and Gregory, T.J. 1978. Identification of angiotensinogen in animal brain homogenates. Relationship to a possible angiotensin I-generating pathway. Circ Res 43(I):21–27

Reid, I.A., and Day, R.P. 1977. Interactions and properties of some components of the renin angiotensin system in rat brain. *Central Actions of Angiotensin and Related Hormones*. New York: Pergamon Press, p. 267

Roth, M., Weitzman, A.F., and Piquilloud, Y. 1969. Converting enzyme content of different tissues of the rat. Experientia 25:1247

Schaz, K., Stock, G., Simon, W., Schlör, K.H., Unger, Th., Rockhold, R., and Ganten, D. 1980. Enkephalin effects on blood pressure, heart rate and baroreceptor reflex. Hypertension, 2:395–407

Schelling, P., Ganten, U., Sponer, G., Unger, Th., and Ganten, D. 1980. Components of the renin-angiotensin system in the cerebrospinal fluid of rats and dogs with special consideration of the origin and the fate of angiotensin II. Neuroendocrinology 31:297–308

Schölkens, B.A., Steinbach, R., and Jung, W. 1979. Intracerebroventricular injection of an angiotensin I converting enzyme inhibitor (SQ 14225) in animals with experimental hypertension. Naunyn Schmiedeberg's Arch Pharmacol 307 (Suppl):R46

Schölkens, B.A., Steinbach, R., and Jung, W. 1980. Induction of angiotensin formation in brain: Biological effects and pharmacological interferences. *Enzymatic Release of Vasoactive Peptides*. New York: Raven Press p. 365

Severs, W.B., and Daniels-Severs, A.E. 1973. Effects of angiotensin on the central nervous system. Pharmacol Rev 25:415–449

Simpson, J.B., Epstein, A.N., and Camardo, J.S., Jr. 1978. Localization of receptors for the dipsogenic action of angiotensin II in the subfornical organ of rats. J Compar Physiol Psychol 92:581–608

Sirett, N.E., McLean, A.S., Bray, J.J., and Hubbard, J.I. 1977. Distribution of angiotensin II receptors in rat brain. Brain Res 122:299–312

Stamler, J.F., Brody, M.J., and Phillips, M.I. 1980. The central and peripheral effects of captopril (SQ 14225) on the arterial pressure of the spontaneously hypertensive rat. Brain Res 186:499–503

Stine, S., Yang, H.–Y., and Costa, E. 1979. Intraventricular Captopril inhibits in situ metabolism of ^{3}H Met-enkephalin and induces naltrexone sensitive analgesia. Pharmacologist 21:286A

Swerts, J.P., Perdrisot, R., Malfroy, B., and Schwartz, J.C. 1979a. Is "enkephalinase" identical with "angiotensin converting enzymes"? Europ J Pharmacol 53:209–210

Swerts, J.P., Perdrisot, R., Patey, G., de la Baume, S., and Schwartz, J.C. 1979b. "Enkephalinase" is distinct from brain "angiotensin converting enzyme." Europ J Pharmacol 57:279–281

Unger, Th., Rockhold, R.W., Bönner, G., Vecsei, P., and Ganten, D. 1979a. Long-term oral administration of converting enzyme inhibitor SQ 14225 in spontaneously hypertensive rats. Naunyn Schmiedeberg's Arch Pharmacol 307(Suppl):R46

Unger, Th., Rockhold, R.W., Schaz, K., Vecsei, P., Bönner, G., Rascher, W., and Ganten, D. 1979b. A novel orally active converting-enzyme inhibitor YS 980: Effects on blood pressure in spontaneously hypertensive rats. Clin Sci 57:157s—160s

Unger, Th., Rockhold, R.W., Yukimura, T., Rettig, R., and Ganten, D. 1980. Blood pressure and heart rate responses to centrally administered substance P are increased in spontaneously hypertensive rats. Clin Sci 59:000–000

Unger, Th. Kaufman-Bühler, I., Schölkens, B.A., and Ganten, D. Brain-converting enzyme inhibition. A possible mechanism for the antihypertensive action of captopril in spontaneously hypertensive rats. Europ J Pharmacol, in press

Vollmer, R., and Boccagno, J.A. 1977. Central cardiovascular effects of SQ 14225, an angiotensin-converting enzyme inhibitor in chloralose-anaesthetized rats. Europ J Pharmacol 45:117–125

Wickre, C., McDonald, M.J., Bau, D., Moffitt, B., and Aumann, S.K. 1978. Blood pressure reduction in the spontaneously hypertensive rat induced by chronic intracerebroventricular infusion of angiotensin antagonist. Endocrinol Soc 60:207

Yang, H.-Y.T., and Neff, N.H. 1972. Distribution and properties of angiotensin converting enzyme of rat brain. J Neurochem 19:2243–2250

Yang, H.-Y.T., Erdös, E.G., and Levin, Y. 1970. A dipeptidyl carboxypeptidase that converts angiotensin I and inactivates bradykinin. Biochim Biophys Acta 214:374

Angiotensin, Interaction with Homeostatic Reflexes

J. Conway,[1] R. Hatton and D. Clough
Biology Department, ICI Pharmaceuticals Division, Mereside Alderley Park, Macclesfield, Cheshire, United Kingdom

INTRODUCTION

Blood pressure is maintained by the sympathetic nervous system and by angiotensin, and it is known that the two systems interact with each other (Zimmerman and Gomez, 1965; Khairallah et al., 1971; Malick and Nasjletti, 1976; Zimmerman and Kraft, 1979). Although these interactions have only been studied in isolated preparations. There must, however, be a complex interplay of the two systems as they are brought into action—for example, with change of posture and with sodium restriction. The relative roles of the two systems under these circumstances have not been adequately delineated. The availability of agents like saralasin and converting enzyme inhibitors, which antagonize the activity of angiotensin, make it possible to examine the role of the hormone in controlling blood pressure under varying conditions.

In the experiments to be described the importance of angiotensin in maintaining blood pressure has been studied during sodium depletion, which activates the renin-angiotensin system, and during lower-body negative pressure, which activates the sympathetic nervous system.

SODIUM DEPLETION

Progressive sodium depletion has been induced in unanaesthetized dogs by diet and the periodic use of diuretic to raise plasma renin activity

[1] Present address: Cardiac Department, John Radcliffe Hospital, Headington, Oxford, United Kingdom.

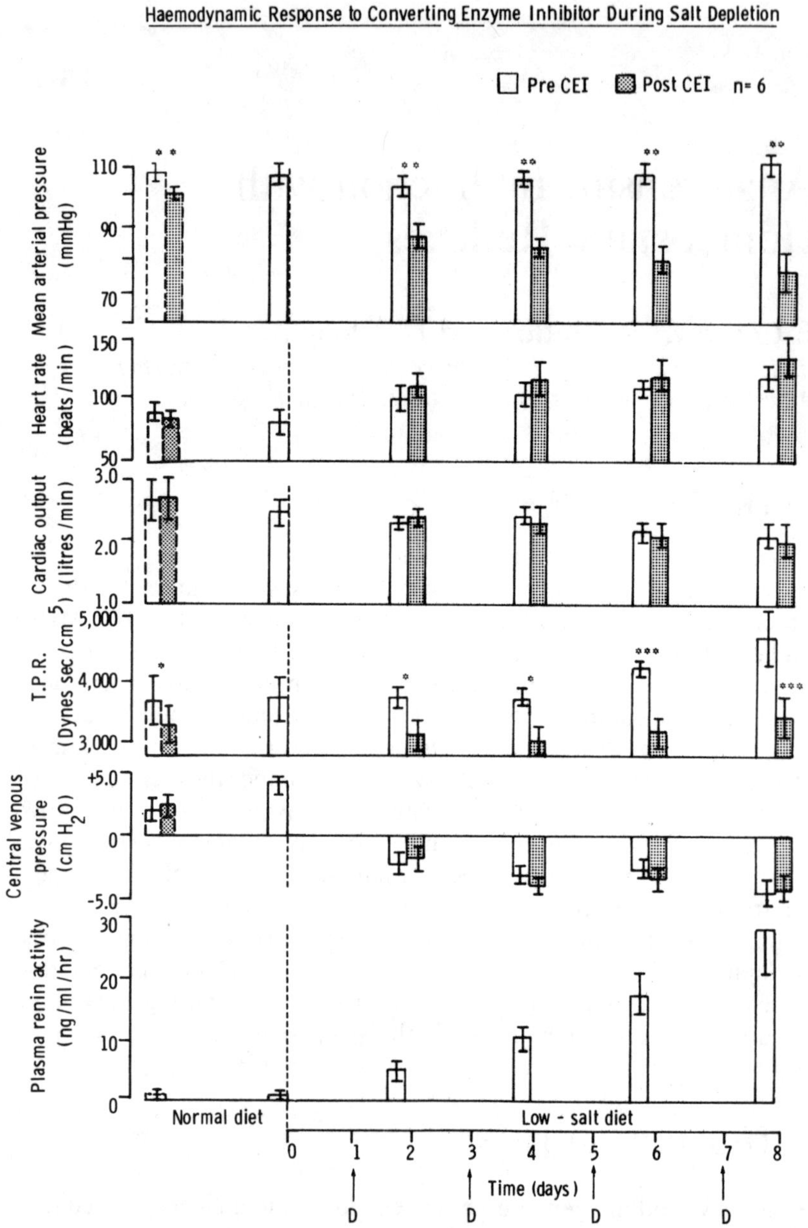

Fig. 1. The hemodynamic effect of converting enzyme inhibitor in dogs subjected to increasing levels of sodium depletion. Note that although blood pressure does not fall with sodium depletion, the increasing effect of CEI shows that blood pressure becomes increasingly dependent on angiotensin. From Conway et al., Hypertension 1:402–409, 1979, with permission.

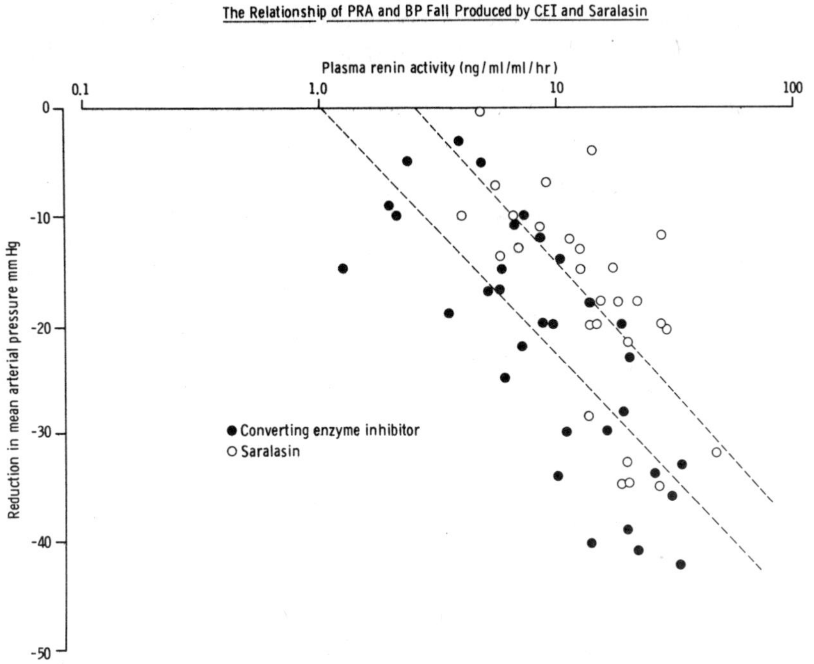

Fig. 2. The fall in blood pressure in response to CEI or saralasin in relation to the prevailing level of plasma renin activity.

(PRA) from 1 to about 27 ngAI/ml/hr. Resting blood pressure did not fall with this regimen, but the administration of an angiotensin converting enzyme inhibitor (nonapeptide, Beckman) (CEI) demonstrated that blood pressure is dependent on circulating angiotensin (Fig. 1).

Administration of CEI initiates a fall in pressure that is proportional to the prevailing PRA level (Fig. 2). Projection of this line to its origin demonstrates that the level of PRA at which the angiotensin generated affects blood vessels is 1.06 ngAI/ml/hr. This is the median of the normal range for the resting dog on a normal sodium intake.

The same experiment performed with saralasin demonstrates a similar relationship with the prevailing level of PRA, but blood pressure is always higher after blocking the action of angiotensin with this agent, with the extrapolated line now cutting the renin axis at 2.6 ngAI/ml/hr. One can speculate on the reasons for the difference between CEI and SAR, but the point to be made here is that with each agent PRA levels just above the level seen at rest generate sufficient angiotensin to directly affect blood pressure.

Since angiotensin is a potent vasoconstrictor peptide, the direct action of this hormone would adequately explain the findings. However, further experimentation has indicated that this may not be the entire explanation.

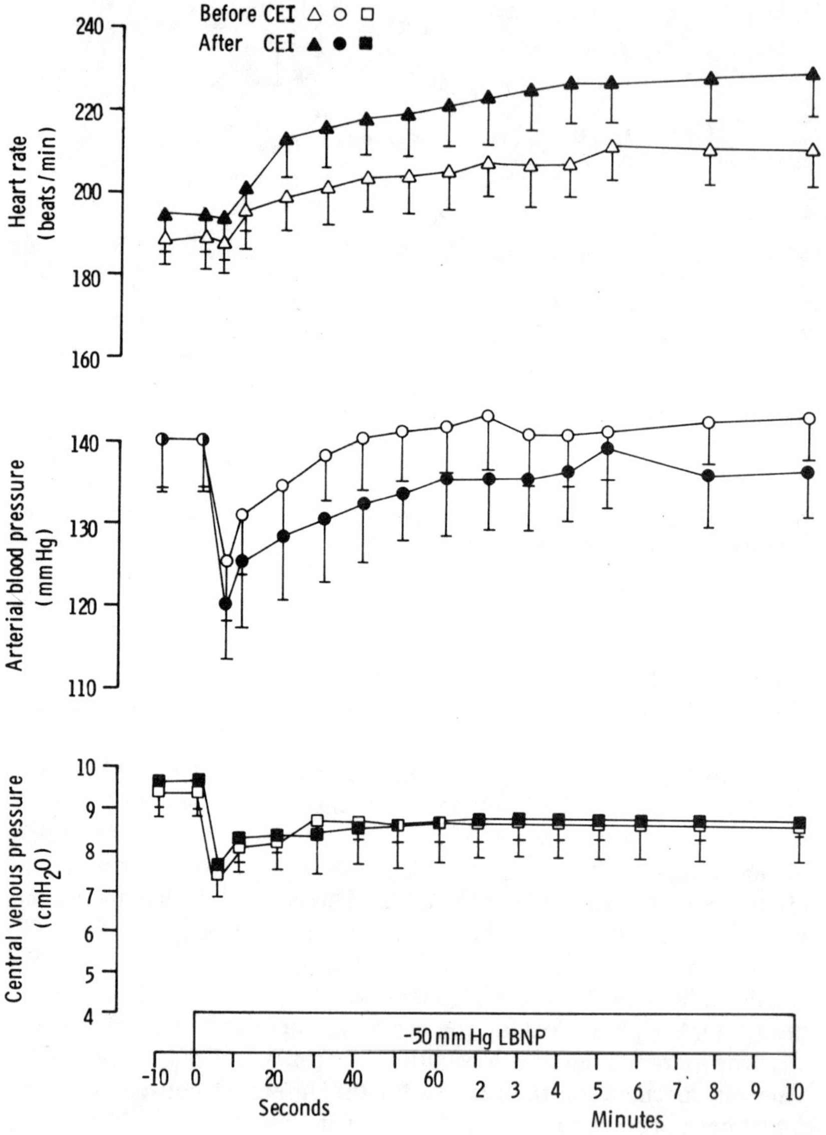

Fig. 3. The effect of inhibition of the action of angiotensin on the blood pressure response to lower-body negative pressure. From Hatton et al., 1980 (in press), with permission.

INTERACTION BETWEEN ANGIOTENSIN AND THE SYMPATHETIC NERVOUS SYSTEM

Changes in posture have long been used to test the efficiency of homeostatic reflexes. This can conveniently be duplicated experimentally by applying negative pressure (−50 mm Hg) to the lower part of the body. This causes sequestration of blood in the lower extremities, a fall in central venous pressure and blood pressure. The blood pressure is restored by the activity of the sympathetic reflexes to the initial level in approximately 30 seconds even though suction is still maintained.

In a group of cats anaesthetized with chloralose, with a mean PRA level of 2.9 ± 0.3 ngAI/ml/hr, this negative-pressure procedure was performed before and after dosing with CEI (Fig. 3). At this level of PRA, angiotensin could not be shown to affect blood pressure, since CEI had no more than a transient effect on the level of pressure. However, the application of lower-body negative pressure after blocking the action of angiotensin led to a greater fall in pressure and an obvious delay in the restoration of blood pressure—in fact, the control level was not restored (Fig. 3). However, the change in central venous pressure and its partial reflex restoration was unaffected by CEI. (These findings could also be duplicated with saralasin.) Thus, blockade of the action of angiotensin led to a reduction in the efficacy of the blood pressure restoration reflex, which is believed to depend primarily upon the activity of the sympathetic nervous system (Hatton et al., 1980). It can be seen from Figure 3 that at the level of renin prevailing in these experiments there was insufficient circulating angiotensin II to affect blood vessels. CEI therefore had no effect on blood pressure; nevertheless, after the action of angiotensin was blocked, the reflex response to lower body suction was clearly impaired.

This interaction with sympathetic reflexes can be illustrated in a more dramatic fashion if CEI is administered after the reflex has had an opportunity to express itself. If the circulatory system is allowed to restabilize after application of lower-body negative pressure, the administration of CEI will cause an immediate and sustained fall in blood pressure (Fig. 4), although there will be no change in central venous pressure. Saralasin produces results that are essentially similar, indicating that the effect of CEI is due to its effect on angiotensin conversion and not on bradykinin degradation.

The ability of angiotensin to affect the release and reuptake of norepinephrine has long been known (Khairallah et al., 1971). It has also been shown to have central actions which augment sympathetic tone (Scroop and Lowe, 1969; Joy and Lowe, 1970; Ferrario et al., 1972).

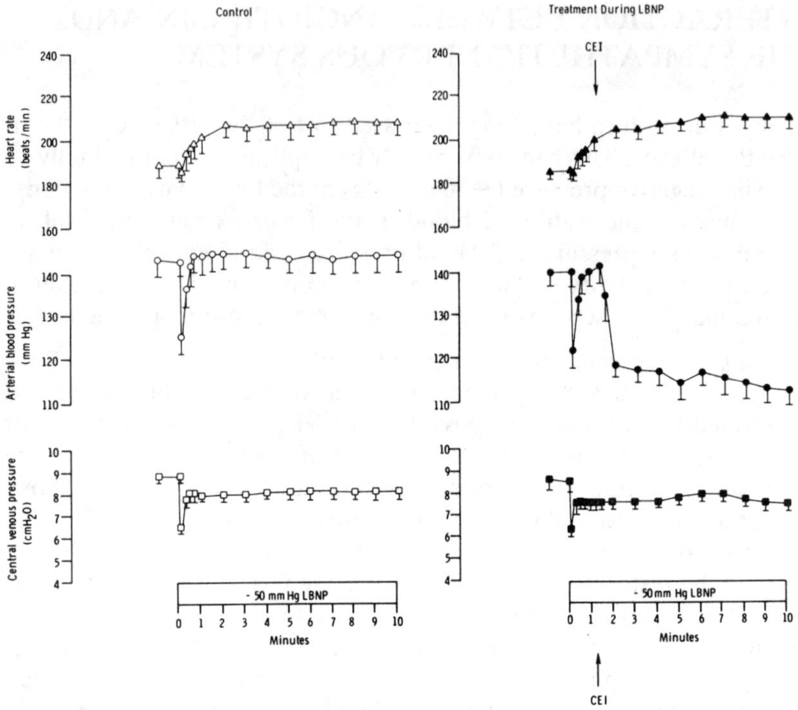

Fig. 4. The effect of the administration of CEI during the application of lower-body negative pressure. From Hatton et al., 1980 (in press), with permission.

Our experiments demonstrate the augmentation of a homeostatic reflex by angiotensin. Furthermore, it has been shown that this becomes apparent when prevailing level of PRA (2.5 ngAI/ml/hr) is below the level required to generate sufficient angiotensin to have an effect on blood pressure (Fig. 3). When the reflex had already been activated, the effect of angiotensin blockade was more striking. This probably explains the well-known observation that saralasin or CEI may not lower blood pressure after moderate sodium depletion in a subject in the horizontal position, but they will readily do so in subjects who are upright (Sancho et al., 1976).

CONCLUSIONS

These studies indicate that interference with the action of angiotensin will show as an interference with sympathetic reflexes. Even when the level of angiotensin is below the concentration at which it exerts a direct constrictor action on blood vessels, it apparently affects reflex function. Converting enzyme administration at this stage therefore must produce vasodilation of neurogenic origin.

REFERENCES

Conway, J., Hatton, R., Keddie, J., and Dawes, P. 1979. Role of angiotensin in the control of blood pressure during sodium depletion. Hypertension 1:402–409

Ferrario, C.M., Gildenberg, P.O., and McCubbin, J.W. 1972. Cardiovascular effects of angiotensin mediated by the central nervous system. Circulation Research 30:257–262

Hatton, R., Clough, D.P., Adigun, S.A., and Conway, J. 1980. Functional interaction between angiotensin and sympathetic reflexes. In press.

Joy, M.D., and Lowe, R.D. 1970. Evidence that the area postrema mediates the central cardiovascular response to angiotensin II. Nature 228:1303–1304

Khairallah, P.A., Davila, D., Papanicolaou, N., Glende, N.M., and Meyer, P. 1971. Effect of angiotensin infusion on catecholamine uptake and reactivity in blood vessels. Circ Res 28 & 29, Supp 2, II–96–II–106

Malik, K.U., and Nasjletti, A. 1976. Facilitation of adrenergic transmission by locally generated angiotensin II in rat mesenteric arteries. Circ Res 38:26–30

Sancho, J., Burton, J., Barger, A.C., and Haber, E. 1976. The role of the renin-angiotensin-aldosterone system in cardiovascular homeostasis in normal human subjects. Circulation 53:400–405

Scroop, G.E., and Lowe, R.D. 1969. Efferent pathways of the cardiovascular response to vertebral artery infusion of angiotensin in the dog. Clin Sci 37:605–619

Zimmerman, B.G., and Gomez, J. 1965. Increased response to sympathetic stimulation in the cutaneous vasculature in presence of angiotensin. Int J Neuropharm 4:185–193

Zimmerman, B.G., and Kraft, E. 1979. Blockade by saralasin of adrenergic potentiation induced by renin-angiotensin system. J Pharmacol Exp Ther 210:101–105

Discussion

D.M. ACKERMAN: Dr. Cushman, one of the enzymes you did not mention being tested for inhibition by these compounds was carbonic anhydrase, which is a zinc containing enzyme. Have any of these compounds been looked at for inhibition of carbonic anhydrase activity?

D.W. CUSHMAN: Yes, Dr. Karen Bush tested captopril as an inhibitor of carbonic anhydrase and alcohol dehydrogenase, both of which are zinc containing enzymes. She found no inhibition of these enzymes by captopril. The only problem with specificity studies of this type is that every new enzyme tested requires a different assay method. I would not have expected much inhibition of a zinc containing enzyme like carbonic anhydrase, since the zinc ion of that enzyme plays a different functional role from that in angiotensin converting enzyme (ACE) or in other zinc containing peptidases such as carboxy peptidase A or leucineaminopeptase.

UNIDENTIFIED: You mentioned an assay of plasma angiotensin converting enzyme activity using a tripeptide substrate. Can you adapt that assay in plasma of patients treated with captopril and is there an effect of plasma storage during prolonged time under plasma angiotensin converting enzyme activity?

D.W. CUSHMAN: With respect to the first question, you can adapt the Hip-His-Leu assay to measure activity in human plasma, using the fluorometric assay originally developed by Roth and co-workers, as we and others have done. Friedland and Silverstein published such a method that was nothing more than Marc Roth's fluorometric assay procedure applied to our substrate and assay conditions. With regard to the second question, I don't think there is any problem with the plasma angiotensin converting enzyme if you store samples, but there is no doubt that captopril itself is unusually unstable in plasma. This is where Jacques Genest's group probably went wrong in their recent paper where they found "no inhibition" of plasma converting enzyme by captopril at a time when blood pressure was still decreased. They measured the plasma converting enzyme activity of samples collected at various times after captopril administration. But, if you don't acidify plasma very rapidly, captopril is oxidized to the disulfide dimer very readily, a reaction that is probably catalyzed by metal ions or some other factor in plasma. In aqueous solution, captopril is very stable at neutral pH, but in plasma it is quickly oxidized. The converting enzyme, from our experience, should be stable for a relatively long period of time. We have kept it for as long as two years in the refrigerator without much loss of activity.

B.G. ZIMMERMAN: While you were talking about converting enzymes in various sites, I was wondering if you found that there is some difference in blocking converting enzyme in different tissues? For instance, in the kidney there seems to be some inability of captopril to inhibit as well as it does in the lung. Is this perhaps your experience as well?

D.W. CUSHMAN: There are apparently certain differences in specificity of converting enzymes from different tissues. The guinea pig ileum enzyme, for instance, is inhibited to a different degree by certain analogs of captopril than is the rabbit lung enzyme; but the converting enzymes that we have tested from different tissues are all inhibited about equally by captopril itself. Baranowski and coworkers claimed that the rat kidney angiotensin converting enzyme was not inhibited by captopril or SQ 20,881. However, they used very crude extracts and a fluorometric assay with angiotensin I as substrate. I'm fairly certain that, under these conditions, they were not measuring activity of the converting enzyme, but rather the release of the two histidine residues of angiotensin I, along with the other amino acids liberated by a variety of peptidases. When they used hippuryl-histidylleucine as a substrate for that same kidney enzyme, its properties were the same as those that we and other investigators have reported. Recently Erdös has purified the human kidney enzyme and found it to be essentially the same as that obtained from other tissues and other species.

M.J. ANTONACCIO: I would like to ask Dr. Ganten a question and suggest perhaps an alternative mechanism for the increase in central converting enzyme after captopril. I think you showed very carefully and very clearly that under the best of circumstances and for prolonged periods of dosage times, you cannot inhibit the pressor response to centrally administered angiotensin I after peripherally administered captopril. I would suggest that this is in fact a very sensitive method for measuring inhibition. I would like to draw an analogy of what was described perhaps 10 years ago by many authors on the sympathetic nervous system. If one, for instance, gives ganglion blockers peripherally—or any other sympatholytic agent, for that matter, that does not get into the brain—one gets a very dramatic turn-on of tyrosine hydroxylase in the brain. It is not because these compounds get into the brain and do something to the enzyme. There is some alternative method of allowing these systems to turn on through reflex activation, etc. Isn't it highly possible that this is analogously what happens to the central converting enzyme activity? That captopril does not affect it directly, but by lowering pressure or by doing something peripherally that you get some kind of compensatory turn-on of central angiotensin converting enzyme?

D. GANTEN: I think I agree with you that we have not shown that the antihypertensive effect of converting enzyme inhibitors is due to central inhibition of converting enzyme. I did not want to say that was the message, as it may very well be secondary. I think, however, that with the renin experiments we found that the increased renin activity in the brain, despite nephrectomy, is some indication that the plasma angiotensin system apparently is not more involved. The enzyme measurement is some indication that we do affect a system in the brain which is related to blood pressure and which has some specific relationship to the converting enzyme inhibition. I agree with you there is a lot of work to do and I know that auto-radiographic studies show that converting enzyme inhibitor does not easily penetrate but these are acute studies and I don't think chronic studies are available and the auto-radiographic studies, of course, are much less sensitive than the enzyme measurements in the brain. I would still maintain, though I appreciate your alternative explanation, that there is a very good possibility of penetration and central effects and I would not agree that the CNS angiotensin I conversion is a sensitive assay. I don't think it is.

T. UNGER: Just one comment to the experiment Dr. Ganten was talking about. Using long-term, oral treatment with converting enzyme inhibitor in spontaneously hypertensive rats, we did not see increases of arterial wall renin despite the increases in the brain. But we saw increases in the plasma and in the kidney.

S. EHRREICH: I would like to ask Dr. Conway if he ever tried his reflex heart rate experiments to determine "set point" changes with the converting enzyme inhibitor in hypertensive animals. And, secondly, his explanation that elevation in renin would be necessary since the converting enzyme inhibitor elevates renin on its own, wouldn't you expect that it would be active almost all the time? Another question for Dr. Rubin is about his experiment with the dose-response relationship and time-action curves of ACE inhibition in the hypertensive rats. Has he tried to determine whether there was any correlation either on the dose-response or the time-action curve with either elevation of plasma renin or converting enzyme inhibition or whatever. In other words, in those particular animal models, does the time course of the response curve seem to indicate that there is indeed an effect on the enzyme which correlates with blood pressure at those particular time points, and doses.

F.J. CONWAY: In answer to your first question, we have not raised plasma renin activity in any other way except sodium depletion so our experiments are confined to that particular model. In regard to the rise in plasma renin activity it should be remembered that these are acute experiments and there was insufficient time for PRA to rise significantly. In the control experiment we were trying to show that converting enzyme inhibitor was not having action of its own and did need the presence of renin in order to shift the curve.

B. RUBIN: I think I can answer in terms of the dose-response curves that you mentioned. As Dr. Ehrreich may recall, we published in the Journal of Pharmacology and Experimental Therapeutics in 1978 that, in studying the effects of captopril in the renal hypertensive model and in the SHR model, we did not get parallel dose-response curves. Whether that will stand the test of time until somebody else attempts to confirm or refute it I don't know, unless somebody else in the audience perhaps has some information bearing on this. The other questions were about the plasma renin activity?

S. EHRREICH: My question was whether there was a relationship between the blood pressure effects you saw in the experimental animals with the drug at various doses or various times, with either an elevation of plasma renin or converting enzyme inhibition. In other words, can you show that the blood pressure responses are related to what you expect the drug to be doing or to enzyme inhibition?

B. RUBIN: I don't think we have the kind of pinpoint data to which you are referring. At this time, I am unable to make a positive or negative statement to that effect. I think that there definitely is a difference in the dose-hypotensive response curves in the two types of models, so I can only answer part of your question. Reports have appeared suggesting that in the SHR or Goldblatt rat, blood pressure is still decreased for some time after withdrawal of captopril and in spite of increasing *plasma* converting enzyme activity.

F.J. CONWAY: You are giving doses of captopril from—I think—0.3 to 100 mg/kg to your spontaneously hypertensive rats. Would you be able to draw a curve of converting enzyme inhibition in the plasma over that dose range?

B. RUBIN: I don't believe that at the time we were doing this we had enough data on the plasma renin activities or converting enzyme inhibition activity versus dose to which you refer. The question is well taken. As indicated previously, plasma renin activities do not necessarily correlate with captopril's antihypertensive effects. However, in conscious normotensive dogs dosed acutely with captopril, increases in plasma renin activities appeared to be correlated in degree and/or duration with the hypotensive effects.

F. FYHRQUIST: This seems to be the first occasion on which induction of ACE is mentioned. I would like to confirm and extend the data presented by Professor Ganten. We have observed that there is induction of angiotensin converting enzyme production by captopril. We can confirm that there is an increase of ACE in serum and plasma from hypertensive patients treated with captopril. This can be demonstrated by first dialyzing the plasma or serum sample. In spontaneously hypertensive rats (Okamoto Aoki strain) treated wtih captopril (30 mg/kg body weight daily) for 12 or 24 wk (Fig. 1) there was

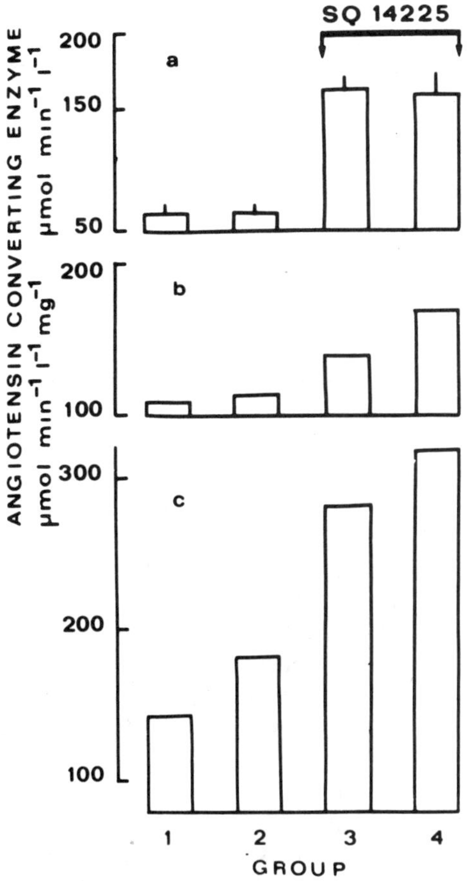

Fig. 1. Mean values of angiotensin converting enzyme in serum (a), in purified pulmonary plasma membranes (b) (per mg of membrane protein), and in enzyme solubilized from membranes with Triton X-100 (c), experiment with Okamoto Aoki rats (n = 40, 10 in each group).

an increase of serum ACE with a factor of 2.5 to 3. These rats (n = 40) were divided into four groups of 10 rats in each group. Group IV received captopril all the time, Group III for the last 12 wk, Group II for weeks 0–12, and Group I was control. When ACE was purified in pulmonary plasma membranes (panel b), or subsequently eluted from membranes with Triton X–100 (panel c), there was an increase in pulmonary membrane ACE concentration in animals treated with captopril. This increase is more marked in animals treated for 24 wk than in those treated for 12 wk only. We think that these data prove that at least in the Okamoto rat there is marked induction of pulmonary angiotensin converting enzyme by captopril. The same we have seen in Wistar rats. This seems to be true in humans too, as judged by increasing serum ACE levels during captopril therapy.

R. LIPICKY: I think this is directed towards Dr. Rubin. I noticed that the effects of captopril in terms of the time course of the duration of effect differs when single acute doses were being administered as opposed to when it was administered chronically wherein the effect lasted hours; after chronic administration it lasted weeks. Is there some explanation for that, and could it be that it is related in some way to the induction of the angiotensin converting enzyme that Dr. Fyhrquist has just been talking about?

B. RUBIN: Theoretically, if the converting enzyme activity is increasing by induction, the effects of captopril should decrease and we don't see anything of this sort except in the renal hypertensive rat, where the hypotensive effect wanes slightly as time goes on. In the SHR, the hypotensive effect, however, becomes more and more predominant; furthermore my colleagues tell me this also happens in man. You get greater intensification of effect over time as mentioned in one of the early reports by Case; a progressive, continuing antihypertensive effect was observed in both a group of renal hypertensive patients as well as in essential hypertensive patients. I can't reconcile these effects solely with the increase in plasma converting enzyme activity that has been mentioned previously. As you continue to dose with captopril and then stop, it takes a longer period of time for the blood pressure to bounce back. There is no real pinpoint answer as to the reason as yet, unless we invoke any of the other mechanisms that were postulated or presented or are going to be presented during the rest of this symposium. I don't believe, from everything we are learning as we go along, that the plasma renin activity is a key factor in monitoring effect. I think what we have to do is get down to the sites of action. After all, the PRAs that we are measuring are circulating in the blood stream. They can be very discordant from those in vascular tissue or possibly even the brain tissue. I think that this is a continuing open question and I believe that some of the other reports that are to be presented are actually going to give some indications of that.

Z.P. HOROVITZ: A number of questions were asked and obviously we don't have all the answers but I think some comments are appropriate. To get to the comment on the duration of effect and also the return. There really isn't any really good correlation. It would be nice to say that when the blood pressure falls you measure exact correlation with renin, converting enzyme activity or even renin substrate, angiotensinogen, which is also important. But, you can't do that. The correlations are not that exact even when the blood pressure returns after you stop captopril. Dr. Rubin showed that if you give captopril acutely for one day, a few days, a week or so, the blood pressure response returns reasonably quickly in two or three days and I think that is the case with most anti-hypertensive agents. The blood pressure does not come back in minutes or hours. It takes a little while. It also appears that after chronic administration, especially in the spontaneous hypertensive rat, it takes a longer time for the blood pressure to return to its hypertensive level. We do not have any good correlations with the biochemistry. It appears that the enzymes, and I think the response to angiotensin I, comes back a lot faster than the blood pressure in the animal. The plasma renin goes way up and then tends to come down towards control levels although in most cases not back to pre-captopril levels, even though you continue to give the same dose of captopril. Now I think that reflects back on the real problem which I don't think anyone has an answer to and that's what happens to these enzymes once you disturb the system. Dr. Ganten has brought up the fact that the converting enzyme activity may indeed increase and Dr. Fyhrquist has supported that. We have some evidence that the substrate angiotensinogen activity tends to decrease while captopril is being given and I think Dr. Fyhrquist may even have some data on that point. So obviously when you interfere with that system, probably anywhere along the line, these enzymes try to adapt. We don't know yet how much of the control of the blood pressure is related over time to the changes in those enzymes.

G.A. MACGREGOR: We were surprised to find no inhibition of converting enzyme activity in patients and normal subjects who had taken captopril. We wondered if this might be due to the length of time the samples had been stored in the deep freeze. We therefore looked at converting enzyme activity in patients and normals who had been taking captopril chronically at increasing times after storage in deep freeze at −20° C. We found that there was undetectable converting enzyme activity as compared to the control value before captopril, when samples were assayed within 2 to 4 hr of taking the blood, but there then was a progressive but varaiable increase in converting-enzyme activity with storage time; so that, after 40 days of storage in a deep freeze at −20° C, the converting enzyme activity had returned to control levels (Fig. 2). I should like to ask Dr. Cushman whether there is any way of stabilizing samples so that captopril does not dissociate from the converting enzyme. If this was possible, it would make the measurement more practical and useful.

D.W. CUSHMAN: I am not sure whether or not I have the explanation for your results. Are these blood samples from patients treated with captopril that should have had an inhibitory level of captopril at the time of collection?

G.A. MACGREGOR: Yes, these are all. These people were on captopril several days or months who then had blood taken and the plasma was then stored for the time on the slide, before measurement.

D.W. CUSHMAN: Well, I don't think anybody has the exact answer to your question, but from what I understand it could very well be that the captopril is "going to pot" during storage of your samples. It is not an unstable compound, but it is capable of being oxidized to the disulfide dimer, which under in vitro conditions would not be inhibitory. The other possiblity, that I don't really believe strongly but will mention, is that converting enzyme activity in extracts may increase with time under certain conditions; we have seen this with brain extracts. This could be similar to the results that John Laragh and others have seen with renin. Perhaps there is a proconverting enzyme that becomes activated by plasma proteases, or alternatively, an endogenous inhibitor of plasma converting enzyme that becomes destroyed over a period of time. Under your conditions, it would be

CAPTOPRIL

Increase in Angiotensin converting enzyme activity with storage time.

Age of sample	Angiotensin converting enzyme activity expressed as a percentage of pre-captopril value (+ ISD)
<4 hours	0.99 ± 1.39 (n=5)
18-24 hours	4.62 ± 1.38 (n=4)
1-20 days	41.0 ± 21.0 (n=7)
21-40 days	62.6 ± 24.0 (n=7)
>40 days	98.5 ± 9.4 (n=23)

Fig. 2. Increase in angiotensin converting enzyme with storage time in plasma samples from subjects taking captopril.

hard to decide absolutely whether the changes were merely due to loss of inhibition by inactivation of captopril, or to some change that takes place in the enzyme itself. This is my way of saying that I really don't have any satisfactory explanation, but only a warning to look for instability of captopril under some conditions.

J.H. LARAGH: I was just going to say that apropos of the question raised a minute ago that I don't think we have the whole answer, but there are a lot of reasons that could be put forth to explain why the blood pressure does not come down abruptly, which do not rule out a renin basis. This is fairly characteristic in patients even after nephrectomy and also occurs after giving captopril as we and others have observed. The antihypertensive effect isn't fully achieved in many patients till about 8 or 10 days. There is an immediate response and then a rebound of the pressure in the next 3 to 6 days going sometimes back up to control levels. The final antihypertensive expression can take up to 10 days to observe even though the inhibition on the renin system may be instantaneous. It seems that you can have a patient who is "reset" so that there is a transient defense of the baseline blood pressure, perhaps by the baroreceptor system. Accordingly, too, it is not surprising that when you stop captopril after an animal or patient has been on it for months that the blood pressure does not rebound instantly. In the latter case you have to recover your sodium balance, the renin system has to go back to work and potassium balance also shifts and this has been observed fairly often. So I think we do have some explanations.

D.W. CUSHMAN: I think that what John Laragh says is true, but there is also another possibility. A lot of people are suggesting that some of the side effects of captopril might be due to possible formation of mixed disulfides within various proteins in blood or other tissues. It is possible that if you give it long enough you saturate all kinds of sites in the body with captopril in its mixed disulfide form, and many such sites might be located near the functional angiotensin converting enzyme on vascular endothelium. It might take quite a while to eliminate such stores of captopril from the body. Of course, this is pure speculation.

D. GANTEN: We have made the observation that converting enzyme in the brain is inhibited for a much, much longer time than plasma. If you give a single dose of converting enzyme inhibitor into the brain, the enzyme is inhibited for 48 hr sometimes, whereas in the plasma, angiotensin I conversion is restarted in a very short time. So I think the central effects might be one explanation, not the full explanation, for the prolonged effects of converting enzyme inhibitor. My second point is that the increase of converting enzyme following captopril treatment of course raises the question of the regulation of converting enzyme activity and we have investigated this by looking into simple negative feedback mechanisms of angiotensin II–converting enzyme, and if you choose angiotensin II there is no decrease of converting enzyme. If you nephrectomize the animals and take angiotensin II away there is no increase of converting enzyme. So negative feedback which we were operating on in the beginning to explain our hypothesis seems not to be true and this was the third point which I think has to be considered. We are, all of us, measuring converting enzyme by synthetic substrate and, as we have seen, various other peptides can interfere with the converting enzyme activity measurement. We are probably not measuring one enzyme. We are measuring a mixture of enzymes, and the lack of feedback in the renin-angiotensin system on converting enzyme measurements indicates to us that other peptide systems have to be investigated.

P. VELLETRIE: I have some data I'd like to mention that is in agreement with Dr. Conway's observation concerning heart rate and mean arterial pressure during ACE inhibition. I've looked at experimental renal hypertension in a two kidney, one clip model in the cat and have noted, first of all, during the slow rise in arterial pressure a positive correlation with heart rate, an increase in heart rate during the disease. During captopril reversal I've noticed bradycarida, a significant bradycardia, along with the drop in blood pressure. When I looked at the slopes relating heart rate to blood pressure, it seemed to be due to a change in sensitivity rather than a change in set point, as the slope of the line during captopril therapy was much greater than during the development of the disease.

D. SOFIA: Dr. Cushman. Is the free sulfhydryl group necessary to impart converting enzyme inhibitory activity to captopril? And, if not, it has been implicated in side effect manifestations at least in the clinic. Is there a reason to have the free sulfhydryl group on captopril?

D.W. CUSHMAN: Yes. Well, at least until someone comes up with a suitable alternative. If, as I pointed out in my talk, the sulfhydryl of captopril is replaced by a hydroxyl all of the inhibitory activity is lost; if it is converted to the methyl sulfide derivative, all activity is lost. Although the disulfide dimer is active in vivo, it is completely inactive in vitro, and is presumably reduced in vivo. Of course we would like to have a nonsulfhydryl compound with properties as good as those of captopril. I don't think, however, that there is any guarantee that a non-sulfhydryl compound would be less toxic. It might be more toxic. I think that the sulfhydryl has been given a bad rap. It has been stated that captopril is like penicillamine, and must therefore be bad. Well, it is even more similar to cysteine, a common component of our diet. I think it is going to take years to prove exactly how much of a problem the sulfhydryl represents. One thing is for certain, however—if it weren't for the sulfhydryl, we wouldn't have captopril. Of course we are looking for inhibitors acting by a variety of mechanisms that might be as good as captopril without requiring the presence of a sulfhydryl group. We've already started to see some compounds of this type in the literature. There was an abstract in the latest Federation Proceedings by Galardy and Fernandez showing that N-phosphyl-alanyl-proline is as potent as captopril as an inhibitor of angiotensin converting enzyme. This compound was administered in vivo by infusion, suggesting that it may not have a very long duration of action; but I think it has been apparent for a long time that compounds like N-phospheryl derivatives can bind very tightly to angiotensin converting enzyme or to similar zinc-containing peptidases such as thermolysin. We've looked at a great number of such nonsulfhydryl compounds, but haven't yet come up with one with properties in vitro and in vivo that we consider good enough to initiate clinical studies.

C.I. JOHNSTON: I think that one must be careful about correlating the return of blood pressure with the pharmacokinetics of any antihypertensive drug. Blood pressure after prolonged drug treatment takes a long time to come back to normal, regardless of the antihypertensive drug used. This has been shown both in man and also in the spontaneously hypertensive rat. The reasons are multiple and include resetting of the baroreceptors together with vascular remodeling in the tissues. Then the return of blood pressure after chronic therapy really bears very little relationship to the blood level or to drug metabolism and whether the drug is still present in the body or not.

F. FYHRQUIST: Mr. Chairman, since Dr. Horovitz already mentioned the angiotensinogen effect, could I show a slide (Fig. 3)? This work was done in eight healthy humans, who received captopril 25 mg three times daily for 1 wk. The plasma renin substrate (angiotensinogen) level decreased rapidly; in fact, within 1 hr we observed a significant decrease. The rapid decrease of plasma renin substrate concentration within 1 hr was surprising. There was a further decrement after 24 hr and even more so after 1 wk of captopril treatment. This effect we have seen also in clinical studies on patients with essential hypertension, treated with captopril for up to 12 months. There is an individually varying, clear reduction of plasma renin substrate concentration in some patients to 80% from starting levels. The possible clinical relevance of this we don't know, but one could speculate that it adds to the effect of captopril.

M.J. ANTONACCIO: Let's go back to one of the very first questions asked. I think Dr. Ehrreich asked whether the antihypertensive effect of captopril was related to plasma renin activity. I think this question is critical. Several studies have shown that if one deals with renal hypertension alone, one can relate the antihypertensive activity of captopril very clearly to plasma renin activity. Once you get away from that model, when you start dealing with spontaneous hypertensive rats for instance, it is very clear that there is no relationship with the plasma renin activity and ability of these drugs to lower pressure. Also, in chronic renal hypertension where plasma renin activity returns to normal, or in models of hypertension that involve the kidneys, such as two kidney, two wrapped, where plasma renin activity never rises, captopril is effective. I think it is very clear that you can sometimes get associations between the antihypertensive effect and plasma renin but this is not necessarily the case.

F.W. WOLFF: It's interesting that the sulfhydryl group was brought up as a particular site of activity for converting enzyme inhibitor activity. It sort of reminds one of two questions. One, are there any obser-

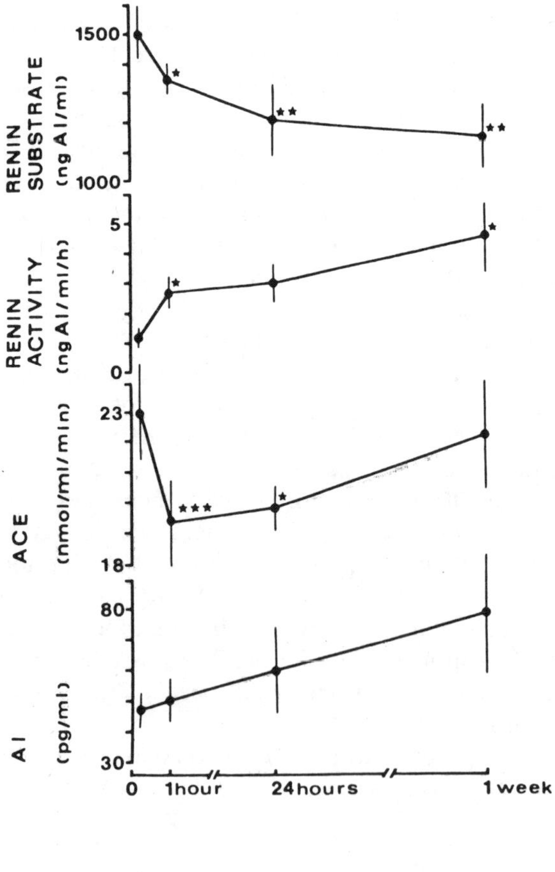

Fig. 3. Effect of captopril, 25 mg × 3 for one week, in healthy humans, on plasma renin substrate, renin activity, serum angiotensin converting enzyme, and angiotensin I (A I) concentration. Mean ± SEM. Difference from zero value: (*) = $p<0.05$, (**) = $p<0.01$, (***) = $p<0.001$.

vations on the blood level and kinetics of converting enzyme between normal individuals and individuals with severe hypertension? Quite apart from the action of the inhibitor? Have any observations been made? Second question—there are a lot of drugs about for hypertension. We don't really quite know how they act and some of these drugs actually have very powerful active sulfhydryl groups or sulfamyl groups. Have any observations been made with some of the so-called vasodilators? Do they have an action on converting enzymes?

D.W. CUSHMAN: Yes. We have looked at a very large number of drugs of every therapeutic classification. Most had no effect on angiotensin converting enzyme in vivo, although certain drugs such as propranolol produce some inhibition at very high concentrations, such as 10^{-3} M.

F.W. WOLFF: I guess you looked at diazoxide. . .

D.W. CUSHMAN: We have tested diazoxide, and it does not inhibit. But I think it's important to point out that it takes a lot more than just a sulfur or sulfhydryl to produce a good inhibitor of angiotensin converting enzyme. For example, 2-mercaptoethanol is about the simplest compound you can think of that is soluble and contains a single sulfhydryl group, and it doesn't touch angiotensin converting enzyme. Some compounds, with two sulfhydryl groups that can chelate the zinc, will inhibit; but captopril has only one sulfhydryl group, and that interaction alone would not be enough to make it a good inhibitor. It is the relatively strong interaction of the sulfhydryl of captopril with the enzyme in combination with several other interactions that produces its great overall affinity.

F.W. WOLFF: Have any observations been made on converting enzymes in normal individuals compared to those with severe hypertension?

D.W. CUSHMAN: I don't know of any, but it could be. We haven't done any work in that direction.

W.L. MATIER: Can you comment on the relative dose of captopril that blocks conversion of angiotensin I compared to the antihypertensive dose. By recollection, you are using about 50 mg/kg to get a good antihypertensive effect. What is the dose to block conversion of angiotensin I to angiotensin II?

D. GANTEN: You mean in the rat and the dog, or man? We have some studies in the rat quite extensively and we find that in the rat you need much more than in man. The dose of 50 mg/kg is a total blocking dose for peripheral conversion of angiotensin I to angiotensin II.

D.W. CUSHMAN: Could I just make one comment? That is misleading because we don't really know to what extent you have to block angiotensin converting enzyme before you begin to get some lowering of blood pressure. Do you only have to block 50% of the activity? I would speculate that you have to achieve a 90 to 99% inhibition to obtain even a weak antihypertensive response. The other question is how do you measure such inhibition? If you test for inhibition of angiotensin I pressor response, you are probably measuring inhibition of conversion in the lung vascular endothelium. At least that would be a good guess. But what is the localization of the converting enzyme activity that you are inhibiting when you are lowering blood pressure? Is it converting enzyme within the vascular wall? Within the kidney? It may take a considerably higher dose of drug to penetrate some of these sites. I certainly don't think one should expect a direct correlation between 50% inhibition of converting enzyme as measured by our usual in vivo tests and 50% lowering of blood pressure. It seems reasonable that it should take a much higher concentration of the drug to lower the blood pressure.

D. GANTEN: Could I make one comment on that? It is my recollection that in clinical studies you have like a hundredfold difference in the dose of captopril that would block conversion of angiotensin I to II by greater than 90%, whereas you need a huge dose of captopril to lower blood pressure.

D.W. CUSHMAN: Well, some people think you don't need as big a dose as is being given. Would you like to comment on that, Dr. Laragh?

J.H. LARAGH: We have recently been experimenting with 1 mg orally and we can lower the blood pressure fairly consistently with that dose of captopril—1/50 of what we started with.

D. GANTEN: Just a comment to what Dr. Cushman said on the availability of endogenous versus exogenous angiotensin to the inhibitors. We have done extensive studies with saralasin and we find that to inhibit endogenous angiotensin you need about 10 times more than to inhibit exogenous angiotensin. So I think the availability of angiotensin which you infuse intravenously compared to the angiotensin which is endogenously produced is completely different and the assay which we have, and that is one of the arguments why we don't believe that our CSF assay is as sensitive an assay for endogenous angiotensin conversion, is not comparable at all exogenously versus endogenously.

SECTION II

Mechanism of Action

Michael J. Antonaccio, *Chairperson*

Role of Kinins in the Pharmacological Effects of Converting Enzyme Inhibitors[1]

Oscar A. Carretero,[2] A. Guillermo Scicli and
Subir R. Maitra
Henry Ford Hospital
Detroit, Michigan 48202

INTRODUCTION

Kinins are potent vasodilator peptides released by enzymes known as kininogenases from precursors, present in both plasma and lymph, known as kininogens or kallikrein substrate. The main kininogenases are plasma and glandular kallikrein, although other enzymes such as trypsin, plasmin and uropepsin are also capable of releasing kinins (Carretero and Scicli, 1980).

The naturally occurring kinins in mammals are bradykinin, lys-bradykinin (kallidin) and met-lys-bradykinin (Fig. 1). Once formed, kinins are broken down rapidly by enzymes called kininases which are found in body fluids and tissues. The half-life of kinins is approximately 15 sec. There are several kininases; some split off the terminal aminoacids (exopeptidases), while others split the kinin molecule at other sites (endopeptidases) (Erdös and Yang, 1970). The best known of the endopeptidases is a peptidyl-dipeptidase that catalyzes the hydrolytic removal of carboxyl terminal dipeptide from kinins and from angiotensin I (Yang et al., 1970). This enzyme is known as kininase II or angiotensin I converting enzyme (EC 3.4.15.1). It participates in blood pressure

[1] Supported in part by NIH Grants HL 15839 and HL 24650
[2] This work was done during the tenure of an established investigatorship of the American Heart Association (Dr. Carretero).

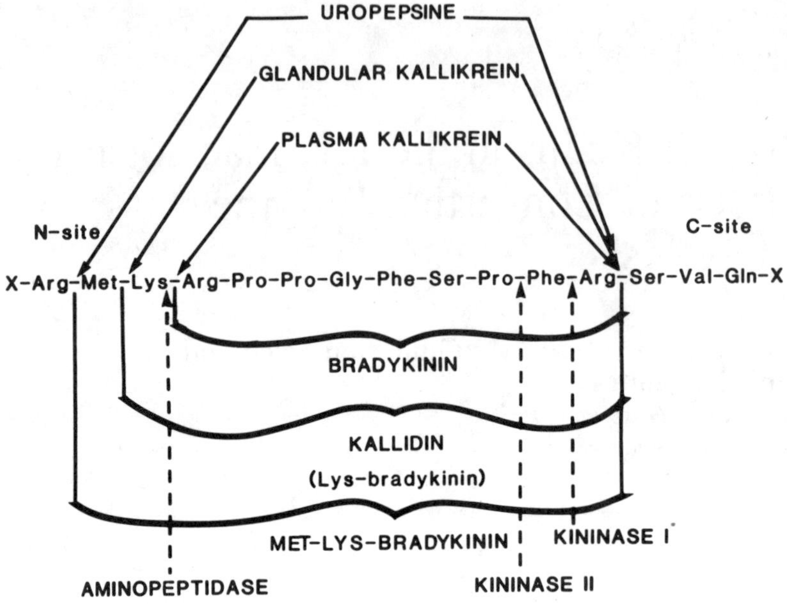

Fig. 1. Structure of bradykinin, kallidin and met-lys-bradykinin. Arrows in the upper part indicate sites of cleavage by plasma and glandular kallikrein and uropepsine from kininogen (kallikrein substrate). Arrows in the lower part indicate sites of cleavage by kininase I, kininase II (angiotensin converting enzyme) and aminopeptidase from kinins.

regulation by converting the inactive angiotensin I to active vasopressor angiotensin II and may also participate in the regulation of blood pressure by inactivating the potent vasodepressor kinins.

Converting enzyme inhibitors (CEI) have been developed that are capable of reducing blood pressure in experimental and human hypertension (e.g., teprotide and captopril). This antihypertensive effect may be due to a decreased concentration of angiotensin II and/or to an increased concentration of kinins in plasma and tissues. There is considerable evidence indicating that in certain experimental and clinical situations the blockade of the renin-angiotensin-aldosterone system is important in mediating the antihypertensive effect of CEI. Although the participation of the kinins in the antihypertensive effect of CEI is not well established, it has been suggested by: a) situations in which the antihypertensive effect of CEI could not be accounted for by a blockade of the renin-angiotensin system; b) an increase in kinin concentrations in body fluids after the administration of CEI; and c) the prevention of the antihypertensive effect of CEI by inhibition of the kallikrein-kinin system. Furthermore, some of the effects of kinins during the administration of CEI could be mediated by prostaglandins, since these peptides have been reported to stimulate the synthesis of these fatty acids (Terragno et al., 1972).

SITUATIONS IN WHICH THE ANTIHYPERTENSIVE EFFECT OF CEI COULD NOT BE ACCOUNTED FOR BY A BLOCKADE OF THE RENIN-ANGIOTENSIN SYSTEM

Converting enzyme inhibitors have been reported to have acute antihypertensive effects in patients with essential hypertension, in nephrectomized patients after dialysis, in spontaneously hypertensive rats (SHR), nephrectomized rats and in DOCA-salt hypertensive rats (Case et al., 1978; de Bruyn et al., 1980; Muirhead et al., 1978; Marks et al., 1979; Miyamori et al., 1980). In these situations, since angiotensin II has not been implicated in the pathogenesis of the hypertension, a blockade of the renin-angiotensin system would not explain the antihypertensive effect of CEI. Thus, an increase in kinins may be responsible for the antihypertensive effect of CEI.

In the DOCA-salt model, the renin-angiotensin system is greatly suppressed while urinary excretion of kallikrein is increased. Therefore, in this model it is possible that the reported antihypertensive effect of one of the CEI, captopril, is mediated by an increase in kinin concentration and not by a blockade of angiotensin II formation. However, it has not been confirmed that captopril has an antihypertensive effect in either unanesthetized or anesthetized DOCA-salt hypertensive rats in which the blood pressure was measured directly from a catheter chronically implanted in the aorta. In contrast, when systolic blood pressure was measured indirectly by the tail-cuff method, a significant decrease in the blood pressure was observed (Table 1).

The discrepancy between the changes in blood pressure observed by these two methods is difficult to explain. Blood flow in the rat's tail varies widely and, if the CEI redistributed blood flow to other organs, this could produce a lower pulse amplitude which could be detected by the indirect method as a decrease in the systolic blood pressure (Borg and Viberg, 1980). Furthermore, it has been reported that the tail-cuff method is not very reliable to determine acute changes in blood pressure (Bunag, 1977). Thus, the changes in blood pressure observed in DOCA-salt hypertension when the blood pressure is measured by the tail-cuff method may be artefacts, due to changes in blood flow distribution. Marks et al. (1979) reported an acute decrease in the direct mean blood

Table 1. Effect of captopril (50 mg/kg) in DOCA-Salt Hypertensive Rats.

	N	Before	After	Significance
Mean Blood Pressure (unanesthetized, p.o. CEI)	8	138±6	135±6	n.s.
Mean Blood Pressure (anesthetized, i.v. CEI)	5	109±6	99±8	n.s.
Systolic Blood Pressure (unanesthetized, p.o. CEI)	9	170±5	152±4	P<0.01
Systolic Blood Pressure (unanesthetized, i.p. CEI)	9	163±4	150±4	P<0.01

N = Number of rats; Mean ± SEM; n.s. = not significant.

pressure of anesthetized DOCA-salt hypertensive rats after the administration of very low doses of captopril (1.25 mg/kg body weight); the reason for the discrepancy between their results and the authors' is not clear. Recently, Miyamori et al. (1980) also reported an acute decrease in the systolic blood pressure of DOCA-salt hypertensive rats after low doses of captopril (1 mg/kg body weight); however, the changes they observed may have been artefacts since they use the tail-cuff method (see above). Nevertheless, the changes observed with the tail-cuff method suggest that CEI may have a pharmacological effect on blood flow distribution which is difficult to explain by a blockade of the formation of angiotensin II.

It is more difficult to exclude the participation of the renin-angiotensin-aldosterone system during the chronic administration of CEI. Although angiotensin II may not participate directly in the pathogenesis of hypertension, it still could have an important function, through the release of aldosterone, in maintaining sodium and water balance, and, as a consequence, in the homeostasis of the blood pressure. The authors have evidence that this is the case in SHR (Maitra et al., 1980). Although kinins may be responsible for the antihypertensive effect of CEI in situations in which the decrease in blood pressure is not explained by a blockade of angiotensin, other possibilities cannot be excluded. For example, the CEI may block the formation of angiotensin II in the arterial wall and in the brain. These extrarenal renin-angiotensin systems may play an important role in the homeostasis of the blood pressure (Ganten et al., 1976; Antonaccio et al., this volume). Furthermore, the possibility that the CEI does more than block the formation of angiotensin II and destroy kinins should not be readily dismissed.

INCREASE IN KININ CONCENTRATIONS IN BODY FLUIDS AFTER THE ADMINISTRATION OF CEI

The blood or plasma concentration of kinins has been reported to increase, to remain unchanged and to decrease after the administration of CEI (Swartz et al., 1979; Swartz et al., 1980; Vinci et al., 1979a; Hulthen and Hokfelt, 1978). These discrepancies can be explained by the unreliability of the methods to determine kinins in blood or plasma (Carretero et al., 1980a). In addition, varied kinin concentrations in untreated normal humans have been found by different investigators (Binia et al., 1963; Mashford and Roberts, 1972; Talamo et al., 1969; Wintroub et al., 1973; Vinci et al., 1978; Hulthen and Borge, 1976), ranging from the picogram to the nanogram (Table 2). Thus, presently, it is not possible to state whether kinins are increased in peripheral blood after the administration of CEI.

Table 2. Concentrations of kinins in Human Blood.

Author	Collection Method	Processing Steps	Kinins (ng/ml)
Binia et al. (1963)*	50 ml blood Directly in ethanol	Ethanol Supernat., evap. Ether and buthanol ext. Water ext.	1–3
Mashford et al. (1972)	5 ml blood Plastic syringe 1,10-phenanthroline Ethanol	Ethanol, evap. Ether ext.	0.07 ± 0.01
Talamo et al. (1969)	Plastic syringe Hexadimethrine bromide EDTA	2 ml plasma TCA precip. Amberlite IRC–50	<3.0 <1.5
Wintroub et al. (1973) Vinci et al. (1978)	Plastic syringe Hexadimethrine bromide EDTA	2 ml plasma TCA precip. QAE-sephadex IRC–50	5.00 ± 2.00 2.95 ± 1.75
Hulthen et al. (1976)	6 ml blood Plastic syringe 1,10-phenanthroline Ethanol	Ethanol supernat., evap. Dichloromethane, evap.	0.22 ± 0.03
Scicli et al. (Unpublished)	5 ml blood Directly in ethanol	Ethanol, boiled, evap. Ether ext. QAE-sephadex IRC–50	0.028 ± 0.004

Mean ± S.E.M.
*Measured by bioassay.

There is better evidence that kinins are increased intrarenally after the administration of CEI. For example, using the stop-flow technique the author found a significant increase in kinins in the distal nephron after the administration of CEI (Scicli et al., 1978a). Similarly, there are reports that kinins increase in urine after the administration of CEI (McCaa, 1979; Nasjletti et al., 1975; Scicli et al., 1978b). An increase of kinins in the kidney could produce an increase in sodium excretion. For example, during CEI administration, the authors found that kinins increase in urine and that their excretion significantly correlated with the urinary excretion of sodium ($r = 0.86$) (unpublished data). Increased intrarenal concentrations of kinins could produce natriuresis either by decreasing tubular sodium reabsorption in the distal nephron (Kauker, 1980) or by influencing intrarenal blood flow distribution (Carretero and Scicli, 1980). Natriuresis combined with a lack of angiotensin II formation could explain part of the antihypertensive effect of CEI. However, this mechanism is probably more important in the chronic rather than in the acute antihypertensive effect of this drug.

The authors have studied whether the effect of CEI in the distribution of renal blood flow is due to mechanisms other than blockade of the angiotensin II formation. Angiotensin II was blocked at the level of the receptor with an infusion of an angiotensin antagonist and, during the infusion, CEI was administered. The CEI produced a significant increase (44 ± 10%) in inner cortical blood flow. However, this increase was not blocked by kinin antibodies which were able to block the vasodepressor effect of both bradykinin and kallikrein. These results would indicate that the changes in blood flow distribution were either produced by a further blockade of the renin-angiotensin system or by kinins formed at a site not reached by the antibodies, such as the lumen of the distal nephron. It could be that kinins in this area of the nephron stimulate the release of PGE_2 by the collecting duct and by the interstitial cells of the medulla. PGE_2 would then reach the inner cortex of the kidney through the ascending Henle's loop where it would produce vasodilatation (Williams et al., 1977).

The authors have also studied the effects of CEI during the blockade of the renin-angiotensin system by infusing an angiotensin antagonist in pregnant nephrectomized rabbits. The antagonist decreased the blood pressure, thus suggesting that utero-placental fetal renin plays an important role in blood pressure homeostasis during pregnancy (Albertini et al., 1980). Although the CEI did not produce a further decrease of the blood pressure, it increased the uterine and placental blood flow significantly. These increases in blood flow were blocked by kinin antibodies, which indicates that kinins mediate the increase in blood flow in the utero-placental complex produced by the CEI.

The effect of kinins may be mediated by prostaglandins, since PGE_2 in the inferior vena cava decreased during the infusion of the angiotensin antagonist but increased during the administration of CEI. This increase in PGE_2 was blocked by kinin antibodies (Seino et al., 1980). It is enticing to postulate that the orally active CEI may be useful in treating toxemia of pregnancy, since in this situation the uterine and placental blood flow are decreased. However, the toxic and teratogenic effects of this drug on the fetus have yet to be determined.

In conclusion, after the administration of CEI, there is no clear evidence that kinins increase in peripheral blood, but there is evidence that kinins increase locally in tissues such as the kidney, uterus and placenta. Thus, some of the effects of CEI in blood flow redistribution may be mediated by kinins.

PREVENTION OF THE ANTIHYPERTENSIVE EFFECT OF CEI BY INHIBITION OF THE KALLIKREIN-KININ SYSTEM

Prevention of the antihypertensive effect of CEI by blockade of the kallikrein-kinin system may provide the most direct evidence that kinins contribute to this effect, provided that the inhibitors used are specific. Blockade may be accomplished either by inhibiting the enzymes which release kinins (kininogenases) or by blocking kinins before they reach the receptors. Presently, there is no specific inhibitor of the two major kinogenases, which are plasma and glandular kallikrein. Although aprotinin (Trasylol®) has frequently been used to inhibit both plasma and glandular kallikrein, it is a polyvalent inhibitor which inhibits many other proteolytic enzymes (Vogel and Werle, 1970). Thus, the effects observed after aprotinin administration could be due either to the inhibition of kallikreins or other enzymes not involved in the generation of kinins in vivo. Nevertheless, it has recently been reported that aprotinin prevented the acute antihypertensive effect of captopril in essential hypertension patients with low and normal renin but not in those with high plasma renin activity (Overlack et al., 1980).

According to Sen et al. (1979), aprotinin did not prevent the antihypertensive effect of captopril in chronic 2 kidney, 1 clip renovascular hypertensive rats. However, in Sen's study, the antihypertensive effect was 50% blocked by aprotinin; before treatment with aprotinin, captopril decreased the blood pressure by 163 mm Hg (249 ± 14 to 86 ± 6 mm Hg) while, after pretreatment with aprotinin, it decreased the blood pressure by only 82 mm Hg (210 ± 10 to 128 ± 5 mm Hg). These data suggest that in severe 2 kidney, 1 clip renovascular hypertension part of the antihypertensive effect of captopril may be due to an increase in kinins. However, since aprotinin is an unspecific serine protease inhibitor, an effect unrelated to the blockade of kininogenases (kinin generating enzymes) cannot be excluded.

At present, though many kinin analogs have been synthesized in the search for specific inhibitors of kinins, no useful antagonist has yet been found. However, the authors have recently developed high titer antibodies against kinins. Although these antibodies were developed against lys-bradykinin, they also cross-react with bradykinin and met-lys-bradykinin, the three naturally occurring kinins in mammals (Carretero and Scicli, 1980). In rats and rabbits, these antibodies block the depressor effect of exogenously administered bradykinin and glandular kallikrein without altering the vasodepressor effect of nitroprusside, thus indicating that the inhibition of the vasodepressor effect of bradykinin and kallikrein is specific (Fig. 2).

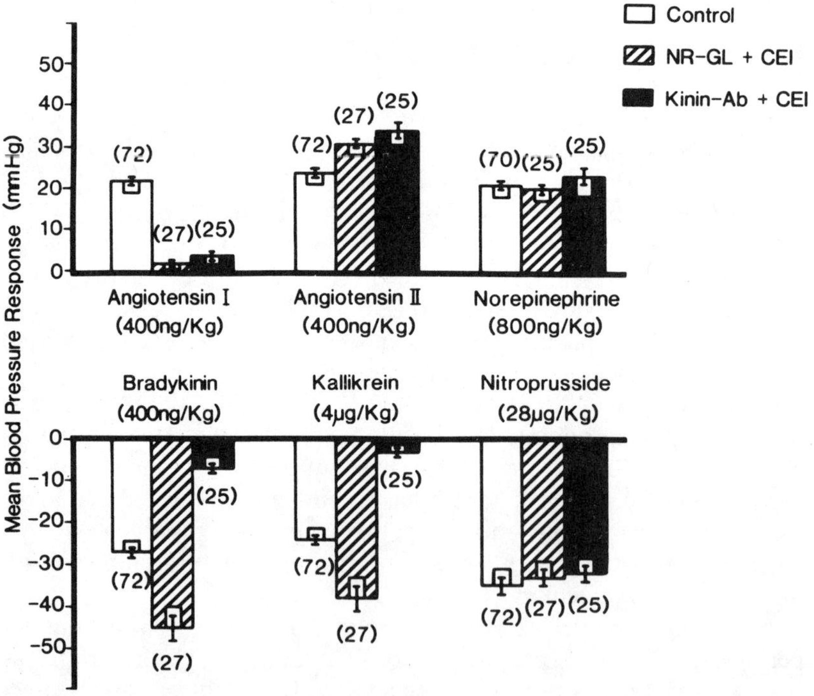

Fig. 2. Pressor response to angiotensin I, angiotensin II and norepinephrine (top) and depressor response to bradykinin, kallikrein and sodium nitroprusside (bottom). White columns indicate untreated group (control); dashed columns indicate group treated with normal rabbit globulins (NR–GL) plus converting enzyme inhibitor (CEI) black columns indicate group treated with antikinin globulins (Kinin-Ab) plus CEI.

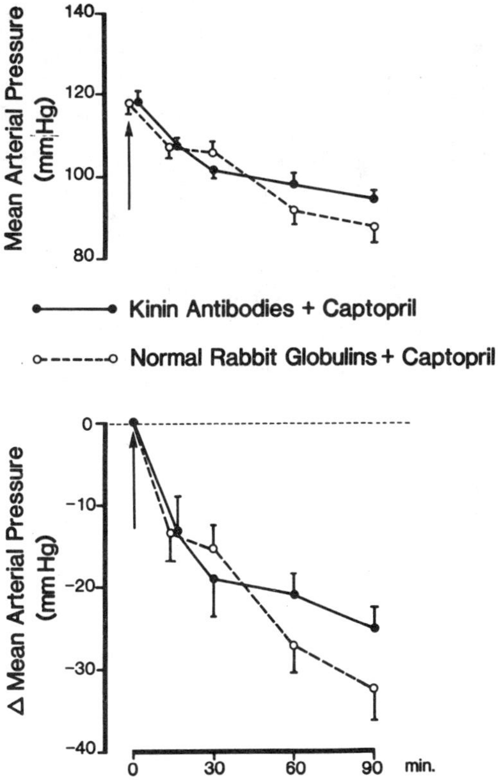

Fig. 3. Effect of the CEI (captopril) on mean blood pressure in the sodium depleted normal rats pretreated with either antikinin globulins or normal rabbit globulins. Δ indicates changes in blood pressure from control period and numbers on the x-axis indicate time in minutes from administration of CEI.

The authors have used these antibodies to determine whether part of the acute antihypertensive effect of CEI is due to kinins (Carretero et al., 1980b). For this, normotensive rats on a low sodium diet, SHR and 2 kidney, 1 clip chronically hypertensive rats on a normal sodium diet were studied. The acute vasodepressor effect of captopril was determined in these three groups of rats after the administration of either semipurified normal rabbit globulins or antikinin globulins. The doses of CEI used (captopril 100 mg/kg of body weight) almost completely blocked the pressor effect of angiotensin I, while the vasodepressor effect of bradykinin and kallikrein was greatly potentiated. The doses of antikinin globulin used completely blocked the depressor effect of both bradykinin and kallikrein. Neither of these two treatments significantly affect the pressor response to angiotensin II and L-norepinephrine and the vasodepressor response to nitroprusside (Fig. 2).

In the normotensive sodium depleted rats treated with either normal rabbit globulins or antikinin globulins, the CEI produced a significant

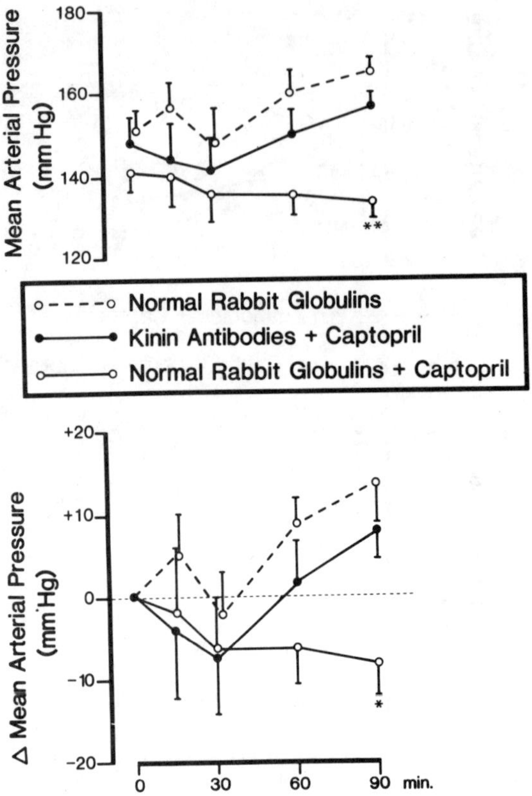

Fig. 4. Effect of the CEI (captopril) on mean blood pressure in SHR pretreated with either normal rabbit globulins or antikinin globulins. Open circle/broken line indicates control group pretreated with normal rabbit globulins but not injected with CEI. Δ indicates changes in blood pressure from control period and numbers on the x-axis indicate time in minutes from administration of CEI. *$P<0.01$; **$P<0.001$.

decrease ($P<0.01$) in blood pressure. The decrease in blood pressure in these two subgroups was similar ($P<0.05$). Figure 3 shows the changes in blood pressure for these two groups.

In the SHR treated with normal rabbit globulins, the CEI produced a small decrease in blood pressure which reached statistical significance 90 min after CEI administration. In the SHR treated with antikinin globulins and CEI, the blood pressure did not decrease. The difference in the changes in blood pressure between these two subgroups was significant ($P<0.01$). In the SHR injected with normal rabbit globulins plus 5% dextrose instead of CEI, the blood pressure tended to increase. Figure 4 shows the changes in blood pressure for these three subgroups.

In the 2 kidney, 1 clip hypertensive rats treated with normal rabbit globulins 90 min after CEI administration, the blood pressure decreased 21 ± 4.5 mm Hg ($P<0.01$). In the subgroup treated with antikinin globulins 90 min after CEI, blood pressure decreased only 4.4 ± 4.1

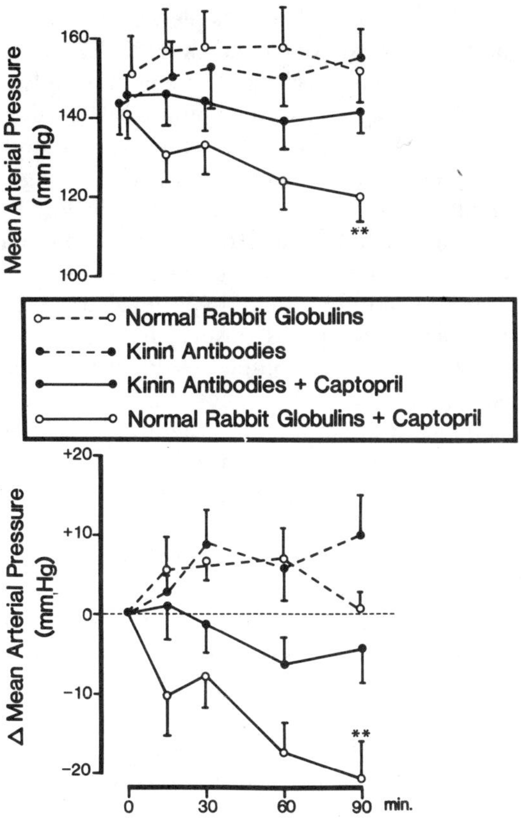

Fig. 5. Effect of CEI (captopril) on mean blood pressure in the 2 kidney, 1 clip hypertensive rats pretreated with either normal rabbit globulins or antikinin globulins. Open circle/broken line and black circle/broken line indicate control groups pretreated with either normal rabbit globulins or antikinin globulins (respectively), but not injected with CEI. Δ indicates changes in blood pressure from the control period. Numbers on the x-axis indicate time in minutes after administration of CEI. $^{**}P<0.02$.

mm Hg ($P>0.05$). The difference in the blood pressure decrease between these two subgroups was significant ($P<0.02$). In the two subgroups which were treated with either normal rabbit globulins or antikinin globulins and which did not receive the CEI, blood pressure had a tendency to increase; however, these changes were not significant. Figure 5 shows the changes in blood pressure for these four subgroups.

The antikinin globulins did not alter the acute hypotensive effect of the CEI in the sodium depleted rats. Thus, during sodium depletion, the antihypertensive effect of the CEI may be due to the blockade of the conversion of angiotensin I to II and not to an increase in the concentration of kinins. This model was selected for study since, in humans, it has been reported that blood kinins are increased during sodium depletion (Vinci et al., 1979b). Accordingly, it was expected that an inhibition of kinin destruction by the CEI would be an important part of its acute vasodepressor mechanism. However, this clearly is not the case in the sodium depleted rat. McCaa (1979), using a different approach, also concluded that the hypotensive actions of CEI are due to the decrease in circulating angiotensin II and not to the accumulation of circulating kinins in sodium depleted dogs.

In the SHR, the acute decrease in blood pressure produced by the intravenous administration of CEI, although statistically significant, was small. This effect was completely blocked by the kinin antibodies, suggesting that the acute effect of this drug may be due to an increase in circulating and/or tissue kinins.

In the 2 kidney, 1 clip hypertensive model, a significant acute antihypertensive effect was observed after the CEI was administered. This effect was almost completely blocked by the antikinin globulins. This result was unexpected, since in this model renin seems to be an important pathogenetic factor. Yet, a significant part of the acute antihypertensive effect of the CEI seemed to be mediated by kinins.

In the SHR and 2 kidney, 1 clip hypertensive rats injected with normal rabbit globulins and in the 2 kidney model injected with antikinin globulins plus 5% dextrose instead of CEI, blood pressure tended to increase; however these changes were not significant. These increases may have been due, in part, to the administration of serum proteins (≈150 mg) which could have produced a small increase in plasma volume and cardiac output. In the 2 kidney model, the blood pressure increase produced by the antikinin globulins was no greater than that produced by the normal rabbit globulins. This suggests that, in this model, blood kinins under basal conditions (no CEI treatment) are not as important in the regulation of the blood pressure. However, kinins seem to play an important role as naturally occurring vasodepressors when their destruction is blocked by the CEI, thus suggesting that the formation and destruction of these vasodilator peptides is an equilibrium situation, and that when this steady state situation is altered, kinins become important in the regulation of blood pressure. The studies with the kinin antibodies neither support nor exclude the possibility that kinins, as local hormones, are important in the antihypertensive effect of the CEI. For example, in the kidney, it has been reported that kinins are formed in the lumen of the distal nephron (Scicli et al., 1978a)where, directly or through the release of prostaglandins, they may participate in the regulation of water and electrolyte excretion and/or in the regulation of the local blood flow to this part of the nephron (Carretero and Scicli, 1980). Yet, the antikinin globulins will not reach this part of the nephron, since due to their high molecular weight they will not be filtered by the glomeruli. However, the CEI have been reported to increase the concentration of kinins in this area of the nephron (Scicli et al., 1978a). Similarly, due to the blood-brain barrier, the antikinin globulins will not reach the central nervous system where the presence of kinins has been reported (Corrêa et al., 1979). Thus, the antikinin globulins probably block only circulating kinins and not tissue kinins.

The blockade of the depressor effect of kinin and kallikrein by antikinin globulins started to dissipate 2 hr after the antibodies had been administered and, 4 to 6 hr later, it had almost completely disappeared. This may be due in part to the fact that antibodies against kinins have a small, but significant cross-reaction with plasma kininogen (kallikrein substrate)(Carretero et al., 1976). The kininogens and the antibodies form a high molecular weight complex which is probably removed rapidly by the reticulo-endothelial system.

Results obtained from using antibodies to block vasoactive substances should be interpreted with caution, since the antigen-antibody complex may cause anaphylactoid reactions, release of histamine and activation of the plasma prekallikrein and complement systems, all of which could decrease the blood pressure independent of the blockade of vasoactive hormones. In the authors' study, such a decrease could only cause a false negative result, since it is the study of the inhibition of the hypotensive effect of the CEI. Thus, the inhibition of the depressor effect of the CEI in the SHR and the 2 kidney, 1 clip models by antikinin globulins cannot be explained by secondary reactions to the antigen-antibody complex formation. The studies with the kinin antibodies suggest that the acute antihypertensive effect of the CEI in the sodium depleted rats is probably due to inhibition of the conversion of angiotensin I to II, while in the SHR and 2 kidney, 1 clip models it is probably due in part to an increase in kinin concentrations secondary to the inhibition of kininase II.

Part of the effect of kinins after CEI may be mediated by prostaglandins, since PGE_2 and its metabolites have been reported increased after the administration of CEI (Swartz et al., 1980; Abe et al., 1980). Furthermore, in essential hypertensive patients, pretreatment with indomethacin, a prostaglandin synthetase inhibitor, partially blocked the antihypertensive effect of captopril (Abe et al., 1980; Cratz et al., 1979).

In conclusion, kinins directly and/or through the release of prostaglandins appear to mediate some of the acute pharmacological effects of CEI. However, there is no direct evidence suggesting that kinins mediate the chronic effects of CEI. The mechanisms by which CEI acutely and chronically decreased the blood pressure are probably various, since evidence indicates that they are quantitatively and qualitatively different in various experimental and clinical situations.

SUMMARY

The antihypertensive effect of converting enzyme inhibitors (CEI) may be due to the blockade of either the formation of the vasoconstrictor peptide, angiotensin II, or the inhibition of the destruction of the potent vasodilator peptides, kinins. Evidence indicates that the blockade of angiotensin II formation is an important mechanism in the antihypertensive effects of CEI. However, the role of kinins in this effect is not well established. Captopril, an orally active CEI, is an effective antihypertensive not only in high renin hypertension, but also in clinical and experimental models of hypertension in which the renin-angiotensin system has not been implicated as a pathogenic factor. Thus, the effects of CEI may be mediated either by an increase in the concentration of kinins or by a yet undetermined mechanism. Reports on the concentration of kinins in blood or plasma after the administration of CEI differ as to whether it is increased, unchanged or decreased. These conflicting results may be a consequence of methodological problems in the determination of kinins in blood or plasma. Kinins in the urine have been reported consistently increased after the administration of CEI which indicates an increase in the concentration of these peptides in the renal tissue. This increase in kinins may participate in the antihypertensive effect of CEI by altering renovascular resistance and by increasing sodium and water excretion. Teprotide (SQ 20,881), a CEI, has been shown to induce increases in both utero-placental blood flow and in plasma PGE_2 in pregnant, nephrectomized rabbits. Since these increases are blocked by kinin antibodies, part of the effect of CEI on the blood flow may be mediated by an increase in kinins directly and/or through the release of prostaglandins.

It has also been reported that aprotinin, a nonspecific inhibitor of kallikrein and other proteases, blocks the acute antihypertensive effect of captopril in patients with low and normal renin essential hypertension. Further, in SHR and in 2 kidney, 1 clip renovascular hypertensive rats, the acute antihypertensive effect of this CEI is almost completely blocked by high titer antibodies against kinins. These antibodies do not alter the vasodepressor effect of the CEI in sodium depleted normotensive rats. There are also reports that PGE_2 and its metabolites are increased in patients with essential hypertension after the administration of captopril. In addition, the antihypertensive effect of captopril is partially blocked by indomethacin, a prostaglandin synthesis inhibitor.

In conclusion, these data suggest that some of the pharmacological effects of the CEI are due to an increase in kinin concentration that directly, or through the release of prostaglandins, could affect local and peripheral vascular resistance and sodium and water excretion.

REFERENCES

Abe, K., Ito, T., Sato, M., Imai, Y., Sato, K., and Yoshinaga, K. 1980. The role of prostaglandin in antihypertensive mechanism of captopril in low renin hypertension. In *Proceedings of the 7th Scientific Meeting of the International Society of Hypertension* (Abs):1

Albertini, R., Seino, M., Scicli, A.G., and Carretero, O.A. 1980. Uterine-placental renin in the regulation of blood pressure in the pregnant rabbit. Am J Physiol, in press

Binia, A., Fasciolo, J.C., and Carretero, O.A. 1963. A method for the estimation of bradykinin in blood. Acta Physiol Lat Amer 13:101–109

Borg, E., and Viberg, A. 1980. Role of heating in non-invasive blood pressure measurements in rats. Acta Physiol Scand 108:73–75

Bunag, R.D. 1977. Blood pressure measurement in rats. In *Proceedings of a Workshop On Blood Pressure Measurement in Hypertensive Animal Models,* pp 21–31. Bethesda: U.S. Department of Health, Education and Welfare

Carretero, O.A., and Scicli, A.G. 1980. The renal kallikrein-kinin system. Am J Physiol 238:F247–F255

Carretero, O.A., Scicli, A.G., and Nasjletti, A. 1980a. The glandular kallikrein-kinin system: Methodology for its measurement. In *Hypertension Research: Methods and Models,* in press

Carretero, O.A., Miyazaki, S., and Scicli, A.G. 1980b. The role of kinins in the acute antihypertensive effect of the converting enzyme inhibitor (captopril). Hypertension, in press

Carretero, O.A., Oza, N.B., Piwonska, A., Ocholik, T., and Scicli, A.G. 1976. Measurement of urinary kallikrein activity by kinin radioimmunoassay. Biochem Pharmacol 25:2265–2270

Case, D.B., Atlas, S.A., Laragh, J.H., Sealey, J.E., Sullivan, P.A., and McKinstry, D.N. 1978. Clinical experience with blockade of the renin-angiotensin-aldosterone system by an oral converting enzyme inhibitor (SQ 14,225, captopril) in hypertensive patients. Prog Card 21:195–206

Corrêa, R.M.A., Innis., R.B., Uhl, G.R., and Snyder S.H. 1979. Bradykinin-like immunoreactive neuronal systems localized histochemically in rat brain. Proc Nat Acad Sci US 76(3):1489–1493

Cratz, F.R., Swartz, S.L., Hollenberg, N.K., Moore, T.J., Dluhy, R.G., Levine, L., and Williams, G.H. 1979. Role of prostaglandins in the hypotensive response to captopril in essential hypertension. Clin Res 27:592A (Abs)

de Bruyn, J., Man in 't Veld, A., Wenting, G., Boomsma, F., and Schalekamp, M. 1980. Converting enzyme inhibitor affects blood pressure equally in renovascular hypertension, essential hypertension and the anephric state. In *Proceedings of the 7th Scientific Meeting of the International Society of Hypertension* (Abs):25

Erdös, E.G., and Yang, H.Y.T. 1970. Kininases. In Bradykinin, Kallidin and Kallikrein, *Handbook of Experimental Pharmacology,* ed. Ervin G. Erdös, 25:289–323 Heidelberg, Springer-Verlag

Ganten, D., Schelling, P., Vecsei, P., and Ganten, U. 1976. Iso-Renin of extrarenal origin. Am J Med 60:760–772

Hulthen, U.L., and Borge, T. 1976. Determination of bradykinin in blood by a sensitive radioimmunoassay. Scand J Clin Lab Invest 36:833–839

Hulthen, U.L., and Hökfelt, B. 1978. The effect of the converting enzyme inhibitor SQ 20,881 on kinins, renin-angiotensin-aldosterone and catecholamines in relation to blood pressure in hypertensive patients. Acta Med Scand 204:497–502

Kauker, M.L. 1980. Bradykinin action on the efflux of luminal ^{22}Na in the rat nephron. J Pharm Exp, in press

Marks, E.S., Bing, R.F., Thurston, H., and Swales, J.D. 1979. Vasodepressor property of the converting enzyme inhibitor captopril (SQ 14,225): the role of factors other than renin-angiotensin blockade in the rat. Clin Sci 58:1–6

Maitra, S.R., Scicli, A.G., and Carretero, O.A. 1980. Role of mineralocorticoid in the chronic antihypertensive effect of converting enzyme inhibitor. Hypertension, in press

Mashford, M.L., and Roberts, M.L. 1972. Determination of blood kinins levels by radioimmunoassay. Biochem Pharmacol 21:2727–2735

McCaa, R.E. 1979. Studies in vivo with angiotensin I converting enzyme (kininase II) inhibitors. Fed Proc 38:2783–2787

Miyamori, I., Brown, M.J., and Dollery, C.T. 1980. The effect of captopril on blood pressure and plasma renin activity of DOCA-salt hypertensive rats. Inhibition of the hypotensive response by indomethacin. Clin Exp Hypert, in press

Muirhead, E.E., Prewitt, R.L. Jr., Brooks, B., and Brosius W.L. Jr. 1978. Antihypertensive action of the orally active converting enzyme inhibitor (SQ 14,225) in spontaneously hypertensive rats. Circ Res 43 (1, Suppl I): I 53–I 59

Nasjletti, A., Colina-Chourio, J., and McGiff, J.C. 1975. Disappearance of bradykinin in the renal circulation of dogs. Circ Res 37:59–65

Overlack, A., Stumpe, K.O., and Sealey, J.E. 1980. Altered blood pressure (BP) and renin (PRA) responses to converting-enzyme (CE) inhibition after aprotinin (AP) induced kallikrein-kinin-system (KKS)-blockade. In *Proceedings of the 7th Scientific Meeting of the International Society of Hypertension* (Abs):100

Seino, M., Albertini, R., Scicli, A.G., and Carretero, O.A., 1980. Isorenin, kinins, and prostaglandins (PGs) in the regulation of blood pressure and utero-placental blood flow during pregnancy. In *Proceedings of the 7th Scientific Meeting of the International Society of Hypertension* (Abs):121

Sen, S., Smeby, R.R., Bumpus, F.M., and Turcotte, J.G. 1979. Role of renin-angiotensin system in chronic renal hypertensive rats. Hypertension 1(4):427–434

Scicli, A.G., Gandolfi, R., and Carretero, O.A. 1978a. Site of formation of kinins in the dog nephron. Am J Physiol 234(1):F35–F40

Scicli, A.G., Rabito, S., Albertini, R., and Carretero, O.A. 1978b. Lack of feedback between urinary kinins and kallikrein in the anesthetized and saline expanded dog. Physiologist 21(4):106

Swartz, S.L., Williams, G.H., Hollenberg, N.K., Levine, L., Dluhy, R.G., and Moore, T.J. 1980. Captopril-induced changes in prostaglandin production. J Clin Invest 65:1257–1264

Swartz, S.L., Williams, G.H., Hollenberg, N.K., Moore, T.J., and Dluhy, R.G. 1979. Converting enzyme inhibition in essential hypertension: The hypotensive response does not reflect only reduced angiotensin II formation. Hypertension 1:106–111

Talamo, R.C., Haber, E., and Austen, K.F. 1969. A radioimmunoassay for bradykinin in plasma and synovial fluid. J Lab Clin Med 74:816–827

Terragno, N.A., Lonigro, A.J., Malik, K.U., and McGiff, J.C. 1972. The relationship of the renal vasodilator action of bradykinin to the release of prostaglandin E-like substances. Experientia 28:437–439

Vinci, J.M., Gill, J.R., Bowden, R.E., Pisano, J.J., Izzo, J.L., Radfar, N.,

Taylor, A.A., Bartter, F.C., and Keiser, H.R. 1978. The kallikrein-kinin system in Bartter syndrome and its response to prostaglandin synthetase inhibition. J Clin Invest 61:1671–1682

Vinci, J.M., Horwitz, D., Zusman, R.M., Pisano, J.J., Catt, K.J., and Keiser, H.R. 1979a. The effect of converting enzyme inhibition with SQ 20,881 on plasma and urinary kinins, prostaglandin E, and angiotensin II in hypertensive man. Hypertension 1:416–426

Vinci, J.M., Zusman, R.M., Izzo, J.L., Bowden, R.E., Horwitz, D., Pisano, J.J., and Keiser, H.R. 1979b. Relationship of urinary and plasma kinins to sodium-retaining steroids and plasma renin activity in normal man. Circ Res 44:228–237

Vogel, R., and Werle, E. 1970. Kallikrein Inhibitors. In Bradykinin, Kallidin and Kallikrein, *Handbook of Experimental Pharmacology,* ed. Ervin G. Erdös, 25:213–249 Heidelberg, Springer-Verlag

Williams, W.M., Frolich, J., Nies, A.S., and Oates, J.A. 1977. Urinary prostaglandins: site of entry into renal tubular fluid. Kidney Intern 11:256–260

Wintroub, B.A., Spraag, S., Stechschulte, D.J., and Austen, K.F. 1973. Characterization and immunoassay for components of the kinin generating system. In *Mechanisms in Allergy Mediated Hypersensitivity,* eds. T.L. Goodfriend, A. Sehan and R. Orange, pp. 495–512 Marcel Dekker, Inc., New York

Yang, H.Y.T., Erdös, E.G., and Levin, L. 1970. A dipeptidyle carboxypeptidase that converts angiotensin I and inactivates bradykinin. Biochem Biophys Acta 214:374

The Kallikrein-Kinin System and Angiotensin Converting Enzyme Inhibition in Hypertension[1]

C.I. Johnston, M. Yasujima and B.H. Clappison
Department of Medicine, Monash University
Prince Henry's Hospital
Melbourne 3004, Australia

INTRODUCTION

Kinins are formed by the action of the enzyme kallikrein on the substrate kininogen. In general, plasma kallikreins generate bradykinin while organ kallikreins produce lysl-bradykinin or kallidin (Levinsky, 1979). However the production, origin and metabolic clearance for circulating blood kinins is not well understood. Some of the possible factors regulating blood bradykinin are depicted in Figure 1. Angiotensin converting enzyme (ACE; kininase II) is a peptidyl dipeptidase and one of the enzymes responsible for the inactivation of bradykinin (Erdös, 1977). This enzyme is found in low concentrations in the plasma but in high concentration in the pulmonary vasculature, endothelial cells and the kidney (Oparil, Koerner and O'Donoghue, 1979).

The recent introduction of specific and potent inhibitors of ACE (Ondetti, Rubin and Cushman, 1977) have provided an important tool to use in the investigation of the kallikrein-kinin system. Captopril (SQ 14,225) has been shown to inhibit bradykinin degradation both in vitro and in vivo (Rubin et al., 1978). Murthy, Waldron and Goldberg (1978) demonstrated that captopril enhanced the hypotensive effect of intravenous bradykinin in conscious rabbits. Although captopril's effect on the renin-angiotensin system in humans has been extensively studied,

[1] These studies were supported by grants in aid from the National Health and Medical Research Council and the National Heart Foundation of Australia.

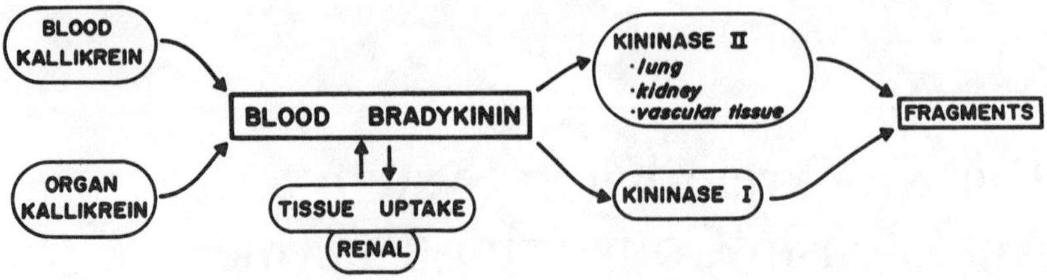

Fig. 1. Factors regulating blood bradykinin.

its effects in potentiating exogenous bradykinin in man have never been reported and its effects on endogenous circulating bradykinin have given conflicting results.

CIRCULATING BLOOD BRADYKININ

Blood contains large amounts of kininogen (kallikrein substrate) together with vast quantities of inactive kallikreins. Even the most stringent sampling techniques may activate kallikrein and thus lead to the formation of bradykinin. Hence it has been difficult to definitely establish whether bradykinin exists in the circulation and its absolute concentration. Original methods for blood kinins used bioassay techniques (Brocklehurst and Zeitlin, 1967; Zacest and Mashford, 1967; Binia, Fasciolo and Carretero, 1963; Nasjletti, Colina-Chourio and McGiff, 1975), but more recently radioimmunoassay methods have been employed (Goodfriend and Ball, 1969; Talamo, Haber and Austen, 1969; Johnston, Matthews and Dax, 1976; Hulthén and Borge, 1976; Williams and Hollenberg, 1977; Johnston et al., 1980a). Most techniques use the method of Mashford and Roberts (1972) to stop the reaction and extract kinins from whole blood. This consists of collecting blood into room temperature absolute ethanol (1:4 vol), reextracting the precipitate with 75% ethanol, washing with diethyl ether and evaporating to dryness. This gives satisfactory recoveries of $97.6 \pm 8.4\%$($n = 20$), immunological identity to standard bradykinin (Sandoz, Basle) with intraassay and interassay variability of 4.4 and 10.2% respectively.

Although blood levels of bradykinin in the past have varied by several orders of magnitude, most recently reported levels in humans have been between 0.2 and 2.0 ng/ml (Johnston et al., 1979; Vinci et al., 1979; Williams and Hollenberg, 1977; Hulthen and Hökfelt, 1978; Mersey et al., 1979).

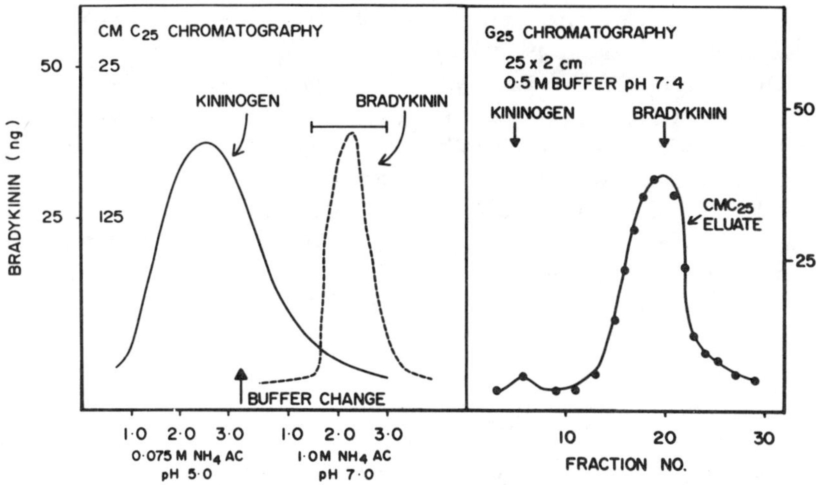

Fig. 2. (a) Elution pattern of ethanolic extract of whole blood on CM C25 Sephadex. Kininogen was determined by incubating column eluates with purified urinary kallikrein or trypsin and measuring the kinins produced by a bradykinin radioimmunoassay. (b) Elution profile of bradykinin immunoreactivity from G25 Sephadex chromatography illustrating that kininogen has been completely separated from bradykinin. The bradykinin immunoreactivity eluates identically to synthetic bradykinin.

Recently the authors have modified the extraction procedure to include an ion exchange chromatographic step with CM Sephadex C25 to eliminate kininogen and other nonspecific immunoreactive substances. Figure 2 (left) illustrates the elution profile of an ethanolic blood extract from such a column, showing clear separation of kininogen and kinins. The right panel of Figure 2 demonstrates that the bradykinin immunoreactive eluate from the ion exchange column has the same chromatographic properties as pure synthetic bradykinin on Sephadex G25 gel chromatography. The ethanolic blood extract after chromatographic purification was shown to be resistant to trypsin digestion but destroyed by chymotrypsin. This enzymatic pattern of degradation was identical to that of synthetic bradykinin.

To further evaluate the measurement of blood bradykinin, an anaesthetized dog was infused either intravenously or into the left ventricle at varying rates with exogenous bradykinin (30–2400 ng/kg/min), and blood samples were obtained from various positions in the arterial and venous side of the circulation. The results are depicted in Figure 3 and show that blood bradykinin levels rose proportionally to the rate of infusion with concomitant hemodynamic changes, and that blood bradykinin was cleared from the circulation across the pulmonary vascular bed and the renal and peripheral vascular bed.

Blood bradykinin has been shown in humans to vary with salt balance (Wong et al., 1975), posture (Wong et al., 1975; Hulthén, Lecerof and Hökfelt, 1978) and to be low (Williams and Hollenberg, 1977) or normal (Mersey et al., 1979) in essential hypertension.

Blood bradykinin concentration is also increased in patients with Bartter's syndrome (Vinci et al., 1978) and patients on cardiopulmonary bypass (Pang et al., 1979).

BLOOD BRADYKININ LEVELS AFTER ANGIOTENSIN CONVERTING ENZYME INHIBITION

Inhibition of angiotensin converting enzyme should lead in vivo to a decrease in angiotensin II concentration, an increase in the angiotensin I level and a corresponding increase in blood bradykinin with a decrease in the metabolic degradation products of bradykinin. Although the changes in the renin-angiotensin system have been extensively and repeatably documented, changes in the kallikrein-kinin system have been little studied and the results are conflicting.

Animal Studies

Matthews and Johnston (1979a) have shown that after angiotensin converting enzyme inhibition by captopril in rats there were dose dependent changes in the renin-angiotensin system and also increases in blood bradykinin levels. However, there was a clear dissociation between the effect of captopril on these two systems. A significant increase in blood bradykinin levels was only demonstrated at the highest dose levels of captopril (30 mg/kg), whereas significant rises in angiotensin I occurred even with mg/kg body weight of captopril (Table 1). Furthermore, the increase in blood bradykinin was only transitory (Matthews and Johnston, 1979b).

In the only other studies in which blood kinins have been measured after captopril administration (20 mg/kg), McCaa, Hall and McCaa (1978) reported a twofold increase in blood kinins in sodium deplete dogs.

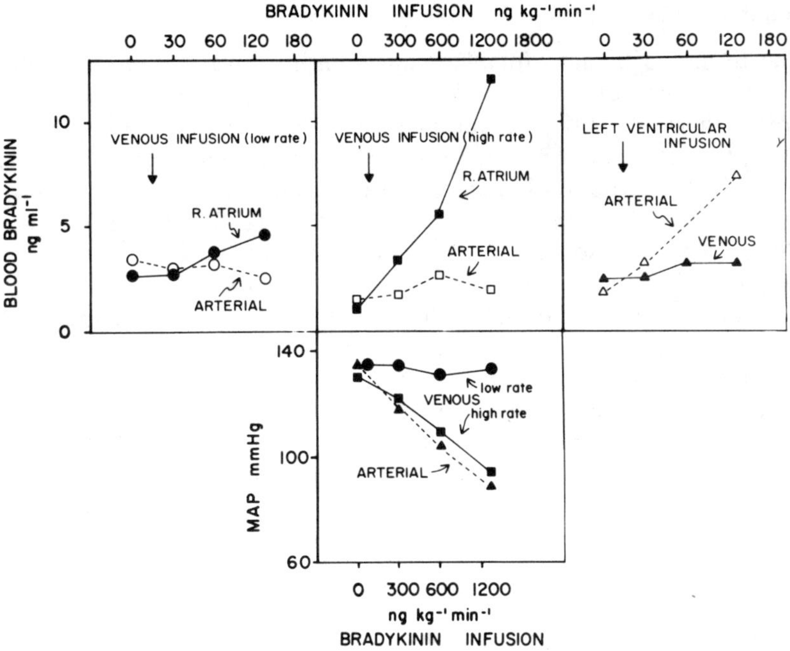

Fig. 3. (a and b) Blood bradykinin levels in an anaesthetised dog measured in samples from the right atrium (●,■) and peripheral artery (○,□) after increasing intravenous infusion rates of bradykinin. (c) Blood bradykinin levels measured in peripheral vein (▲) and artery (Δ) after increasing infusion rates of bradykinin into the left ventricle of a dog. (d) Mean arterial blood pressure in anaesthetised dogs after increasing bradykinin infusion intravenously (●,■) or into the left ventricle (▲).

Table 1. Changes in circulating levels of blood angiotensin I (AI), plasma renin activity (PRA) and blood bradykinin (BK) in rats related to increasing doses of Captopril administered by gastric gavage.

	Captopril mg/Kg				
	0	1	3	15	30
Blood AI pg/ml⁻¹	62 ±12	420* ±62	580* ±78	658* ±58	1242* ±238
PRA ng/ml⁻¹/hr⁻¹	3.8 ±0.9	18.4* ±1.4	21.6* ±2.1	64* ±9	150* ±25
Blood BK ng/ml⁻¹	1.91 ±0.19	1.72 ±0.24	2.86 ±0.64	2.42 ±0.54	4.85* ±0.86

Values are mean ± ISEM
$*P < .05$

Human Studies

The results in humans following angiotensin converting enzyme inhibition are also contradictory. Williams and Hollenberg (1977) have reported that sodium deplete normals showed no increase in plasma bradykinin levels after teprotide (SQ 20,881), but sodium deplete patients with essential hypertension did have significant increases in plasma bradykinin after ACE inhibition. In another publication (Mersey et al., 1979), in which 5 sodium restricted normals were studied, there was a very transient small increase 5 min after SQ 20,881, but plasma bradykinin returned to normal at 10 min. Swartz et al. (1979) found an increase only after 30μg/kg of SQ 20,881 and not 100 or 300 μg/kg in hypertensive patients and in a more recent extensive report the same group (Hollenberg et al., 1979) could find no increase in plasma bradykinin in 16 salt restricted patients with essential hypertension. In contrast, Hulthén and Hökfelt (1978) reported a small fall in blood

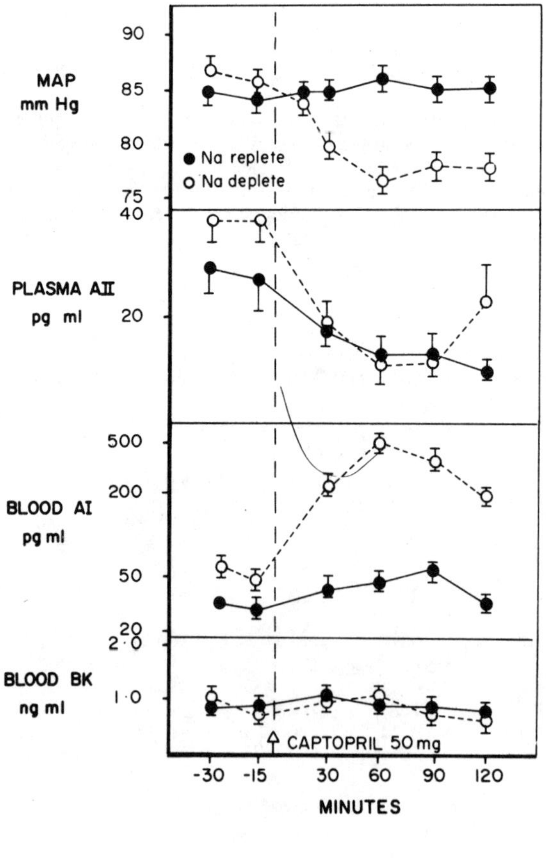

Fig. 4. Changes in mean arterial blood pressure (MAP), plasma angiotensin II (AII), blood angiotensin I (AI) and bradykinin (BK) in 8 sodium replete normal subjects (●) and 6 sodium deplete normal subjects (○) after 50 mg of captopril.

kinins 60 min after the administration of SQ 20,881 to sodium replete hypertensive patients. Millar and Johnston (1979) showed no acute change in blood bradykinin concentration in normal subjects on an unrestricted sodium diet up to 240 min after a single oral dose of 50 or 250 mg of captopril. Similarly Johnston et al. (1979) could demonstrate no increase in blood bradykinin in a group of essential hypertensives given captopril as long term therapy.

Salt Replete and Deplete Normal Subjects

We have recently extended these observations by studying normal subjects who have been sodium depleted by 5 days of a 10 mEq sodium diet together with administration of frusemide on the first 2 days. The sodium deplete subjects showed a small fall in their blood pressure with greater falls in plasma AII and rises in AI and renin. Again, however, there was no alteration in blood bradykinin levels (Fig. 4).

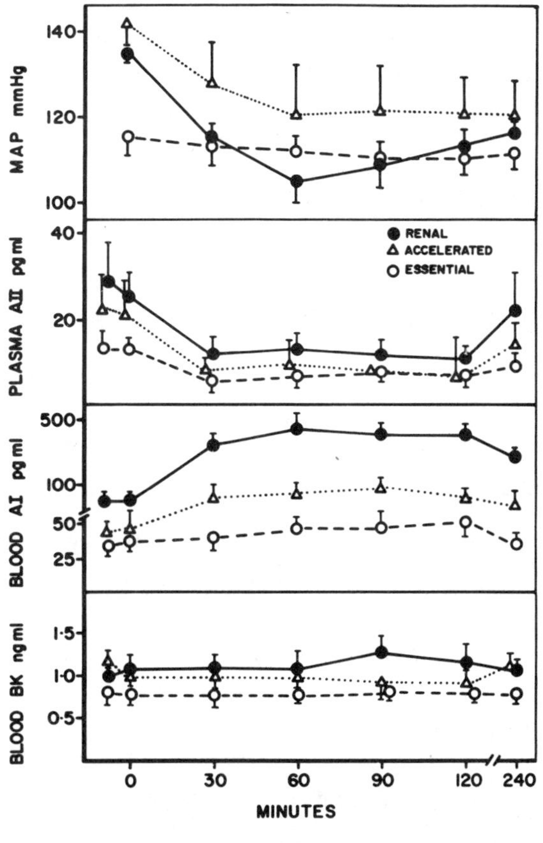

Fig. 5. Effect of captopril on mean arterial blood pressure (MAP), plasma angiotensin II (AII), blood angiotensin I (AI) and blood bradykinin (BK) in 14 patients with renal hypertension (●), 5 patients with accelerated hypertension (Δ) and 16 patients with essential hypertension (●).

Hypertensive Patients

The acute endogenous hormone responses to captopril have also been studied in 3 groups of hypertensive patients (Table 2). Captopril 12.5 to 25 mg orally was followed by a fall in plasma angiotensin II, rise in plasma angiotensin I and renin, but no change in blood bradykinin in the 3 groups. However, there were marked quantitative differences (McGrath et al., 1980) (Fig. 5). The patients with renovascular hypertension also had large falls in systolic and diastolic blood pressure which correlated with thefall in plasma AII and rise in plasma renin (Johnston et al., 1980). Although some patients with essential hypertension had a small fall in blood pressure this could be shown to be due to sodium deficiency from prior and concomitant diuretic therapy (Fig.6).

These studies in sodium deplete subjects and patients with renal hypertension suggest that part of the acute hypotensive response to captopril is due to inhibition of the renin-angiotensin system. In essential hypertension where there was a good long term hypotensive response to captopril, there was no correlation to preexisting renin levels or falls in plasma AII, suggesting that ACE is lowering blood pressure by some other mechanism than inhibition of the renal renin-angiotensin system (Gavras et al., 1978; Johnston et al, 1979).

There may be several explanations for failure to observe an increase in circulating blood bradykinin levels in man. First, the doses of captopril used are much lower; second, the affinity of ACE for bradykinin is 30- to 40-fold less than for angiotensin I. ACE (kininase II) is only one

Table 2. Basal lying systolic (SBP) and diastolic blood pressure (DBP), plasma angiotensin II (AII), plasma renin concentratin (PRC), blood angiotensin I (AI) and bradykinin (BK) in patients with essential, accelerated and Renal hypertension (HT).

		Lying		Plasma AII	PRC	Blood AI	Blood BK
	n	SBP mm Hg	DPB mm Hg	pg/ml	ng/ml/hr	pg/ml	ng/ml
Essential HT	16	155 ±10	99 ±3	13 ±4	1.42 ±0.58	33 ±10	1.03 ±0.22
Accelerated HT	5	204 ±11	114 ±7	22 ±9	1.59 ±0.57	41 ±13	1.08 ±9.13
Renal HT	14	183 ±6	109 ±3	27 ±8	3.03 ±0.55	82 ±15	1.04 ±0.13

Values are mean ± ISEM

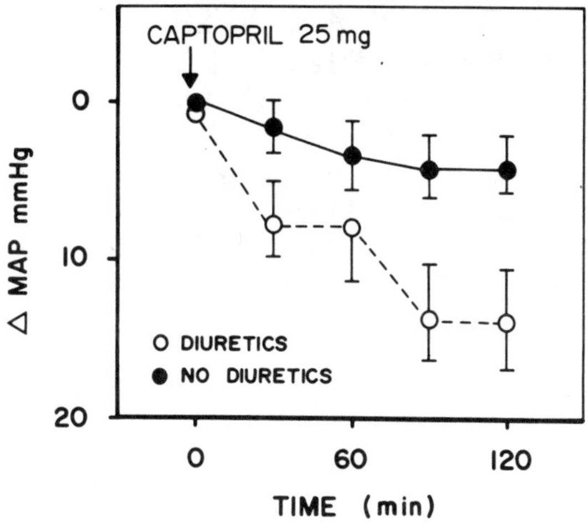

Fig. 6. Changes in mean blood pressure after 25 mg of captopril orally in 10 patients with essential hypertension who had received no prior therapy (●), compared to the fall in blood pressure in 6 patients with essential hypertension who had received diuretic therapy (○).

pathway for the metabolic clearance of bradykinin, and compensatory increases may occur in the alternative bradykinin degradative pathways after captopril. Last, a decrease in bradykinin generation may simultaneously occur after captopril.

EVIDENCE FOR INVOLVEMENT OF THE KALLIKREIN-KININ SYSTEM IN THE HYPOTENSIVE RESPONSE TO ANGIOTENSIN CONVERTING ENZYME INHIBITION

Most of the evidence for involvement of the kallikrein-kinin system in the hypotensive response to inhibition of angiotensin converting enzyme is indirect. Captopril lowers the blood pressure in both anephric humans and animals, (Jandhyala, Washington and Lokhandwala, 1978; Vollmar et al., 1978; man in'T Veld, Wenting and Schalekamp, 1979). After AII receptor blockade with an angiotensin competitive antagonist, captopril is still associated with a hypotensive response (Thurston and Swales, 1978). It is effective in the spontaneously hypertensive rat. Last, after ACE inhibition with SQ 20,881 in patients with essential hypertension, infusion of angiotensin to return blood pressure to its previous level produced plasma angiotensin levels considerably greater than their control AII levels, suggesting some other factor than AII blockade was important in the hypotensive response (Swartz et al., 1979).

EFFECTS OF ANGIOTENSIN CONVERTING ENZYME INHIBITION ON THE RENAL KALLIKREIN-KININ SYSTEM

ACE is found in high concentration in the kidney (Erdös, 1978). Furthermore, the kallikrein and kinins that are excreted in the urine are thought to arise from the kidney (Levinsky, 1979) and to reflect the intrarenal concentration. In a recently concluded study (Clappison et

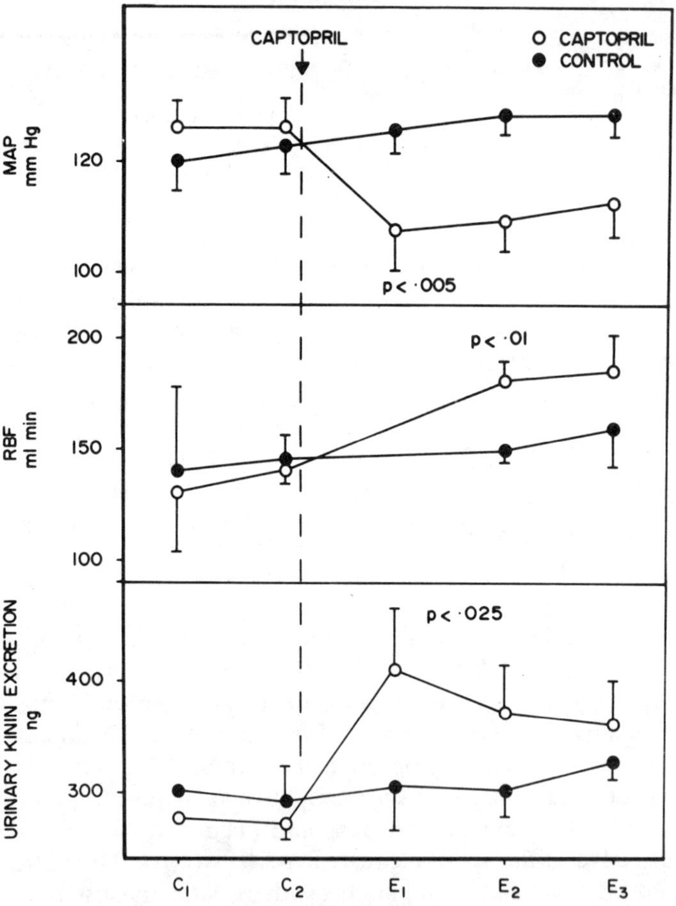

Fig. 7. Effect of captopril on mean arterial pressure (MAP), renal blood flow (RBF) and urinary kinin excretion in a group of 9 dogs (○) given 1.5 mg/kg of captopril compared to 6 control dogs (●) given only vehicle during 2 control 15-min periods (C1 and C2) and 3 experimental 15-min periods (E1, E2 and E3).

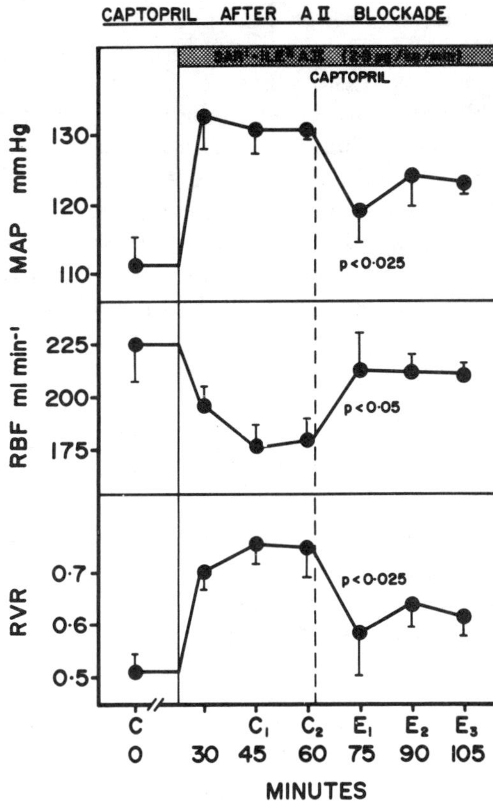

Fig. 8. Changes in mean arterial pressure (MAP), renal blood flow (RBF) and renal vascular resistance (RVR) in a group of 6 anaesthetized dogs infused with the competitive angiotensin II receptor antagonist sar[1]ile[8] angiotensin II and then given 1.5 mg/kg of captopril.

al., 1980; Johnston et al., 1980) administration of captopril to anaesthetized dogs was associated with a fall in mean arterial blood pressure, an increase in renal blood flow, no alteration in glomerular filtration rate but a marked increase in urinary kinin excretion (Fig.7). This increase in renal kinin excretion occurred in the absence of any change in renal arterial or venous blood bradykinin concentration and without change in concurrent urinary kallikrein excretion. To determine whether the renal dilatation associated with ACE inhibition was the result of a decrease in angiotensin II formation or an increase in intrarenal bradykinin concentration, a further series of dogs was studied. In these animals, the physiological effects of angiotensin II were blocked by the continuous administration of the competitive angiotensin receptor antagonist sar[1]ile[8] AII at a dose of 2.5 ng/kg/min intravenously. Blockade of AII receptors was established in each animal by showing a 30-fold shift in the blood pressure dose-response curve to AII and a 100-fold shift in the renal vasoconstrictor dose response to AII. Despite complete AII antagonism,

ACE inhibition by captopril was still associated with a fall in the arterial blood pressure and renal vasodilation (Fig.8). Bradykinin is a potent renal vasodilator; thus, it is tempting to speculate that the renal vasodilation following ACE inhibition by captopril is related to the increase in urinary kinin excretion and that this increase reflects an increase in intrarenal kinins associated with kininase II inhibition. Similar increases in urinary kinins in the dog after captopril have been reported by Olsen and Arrigoni-Martelli (1979). However definitive evidence for this hypothesis must await development of specific antagonists to bradykinin. Unfortunately aprotinin does not inhibit dog renal kallikrein.

ANGIOTENSIN AND BRADYKININ RECEPTOR CHANGES FOLLOWING CAPTOPRIL

Many peptide hormone receptors are modulated by endogenous hormone levels. Increases in plasma angiotensin concentration have been shown to be associated with an increase in adrenal angiotensin II receptors without a change in receptor affinity (Aguilera and Catt, 1978). Millar, Casley and Johnston (1980) have recently demonstrated that with continuous infusions of captopril into rats over 7 days there was a fall in plasma angiotensin II together with a decrease in specific adrenal angiotensin II receptors by direct ligand binding studies. This decrease in adrenal angiotensin II receptors could account for some of the decrease in aldosterone seen after captopril administration.

Much less is known about bradykinin receptors (Odya, Goodfriend and Pena, 1980) but Matthews, Yasujima and Johnston (1980) have recently developed a ligand binding technique employing ^{125}I tyrosine^{8} bradykinin to directly study specific bradykinin receptors in cell membrane fractions from rat uterine smooth muscle. These bradykinin binding sites were shown to be specific, saturable and rapidly reversible with an appropriate K_D and affinity. Sprague Dawley female rats were infused intravenously for up to 7 days with either bradykinin 0.5 μg/kg/min or captopril 80 μg/kg/hr by mini–osmotic pumps (Alza Corp.) implanted into the peritoneal cavity. Daily blood pressures were recorded, and at days 2 and 7, groups of rats were sacrificed. A membrane preparation, prepared by differential ultracentrifugation from a uterine smooth muscle homogenate, was isolated for direct binding studies of bradykinin receptors. Infusion of bradykinin was associated with a transient fall in systolic blood pressure at day 2, with return to normal levels at day 7 (Fig.9). A temporally related decrease and rise occurred in specific bradykinin receptors. Following captopril infusion there was a similar but more persistent decrease in systolic blood pressure, again associated with a prolonged depression in the number of bradykinin receptors (Fig.9). There was no change in bradykinin receptor affinity from a K_D of $0.71 \pm 0.2 \times 10^{-9}$. This fall in the number of bradykinin receptors in uterine smooth muscle may reflect changes in vascular smooth muscle bradykinin receptors. In contrast to angiotensin II receptors, bradykinin receptors appear to be positively regulated by bradykinin levels. The decrease in receptor numbers in uterine smooth muscle after captopril administration may reflect an increase in the local tissue level of bradykinin and may contribute to the hypotensive effect of angiotensin converting enzyme inhibition.

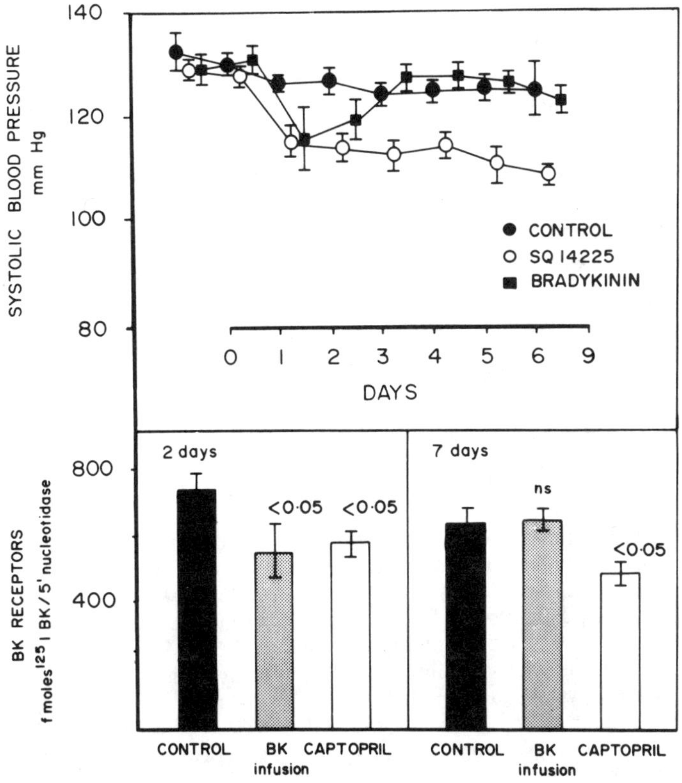

Fig. 9. Top panel: Systolic blood pressure in a group of control rats (●), rats infused intravenously with 0.05 μg/kg/min of bradykinin (■) and rats infused intravenously with captopril 80 μg/kg/hr (○) for 7 days. Bottom Panel: Changes in specific bradykinin receptor numbers measured directly by radioactive ligand binding to a plasma membrane isolated from uterine smooth muscle at days 2 and 7 of the bradykinin and captopril infusions.

CONCLUSIONS

Although the biochemical evidence that angiotensin converting enzyme and kininase II are the same enzyme is conclusive (Soffer, 1978), the pharmacological published evidence that captopril inhibits kininase II and thus accentuates bradykinin's biological action is small.

Angiotensin converting enzyme inhibition by captopril leads to easily identified and appropriate changes in the renin-angiotensin system. There is a fall in endogenous plasma angiotensin II and a rise in plasma angiotensin I and renin. The results of the effect on circulating blood bradykinin levels are conflicting and inconclusive. At very high dose levels of captopril, far in excess of the dose needed for ACE inhibition, blood bradykinin does increase. However, at therapeutically effective dose levels of captopril that lower blood pressure in hypertensive man, no consistent increase in blood bradykinin can be demonstrated.

Captopril produces marked renal vasodilation in animals and man and this is associated with increased urinary kinin excretion, suggesting captopril is an effective inhibitor of tissue converting enzymes. The renal vasodilation is mediated in part through changes in the intrarenal kallikrein-kinin system.

Captopril is also associated with changes in adrenal angiotensin II receptors and in specific bradykinin receptors in smooth muscle. These changes may contribute to the hypotensive action of angiotensin converting enzyme inhibition.

REFERENCES

Aguilera, C., and Catt, K.J. 1978. Regulation of aldosterone secretion by the renin-angiotensin system during sodium restriction in rats. Proc Natl Acad Sci 75,8:4057–4061

Binia, A., Fasciolo, J.C., and Carretero, O.A. 1963. Method for the estimation of bradykinin in blood. Acta Physiol Lat Am 13:101–109

Brocklehurst, W.E., and Zeitlin, I.J. 1967. Determination of plasma kinin and kininogen levels in man. J Physiol 191:417–426

Clappison, B.H., Millar, J.A., Casley, D.J., Anderson, W.P., and Johnston, C.I. 1980. Renal, adrenal, and vascular changes during inhibition of converting enzyme with captopril. Clin & Exp Physiol & Pharmacol, 7:493–498.

Erdös, E.G. 1977. The angiotensin I converting enzyme. Fed Proc 36:1760–1765

Goodfriend, T.L., and Ball, D.L. 1969. Radioimmunoassay of bradykinin: chemical modification to enable use of radioactive iodine. J Lab Clin Med 73:501–511

Gavras, H., Brunner, H.R., Turini, G.A., Kershaw, G.R., Tifft, C.P., Cuttelod, S., Gavras, I., Vukovich, R.A., and McKinstry, D.N. 1978. Antihypertensive effect of the oral angiotensin converting-enzyme inhibitor SQ 14225 in man. N Engl J Med 298:991–995

Hollenberg, N.K., Swartz, S.L., Passan, D.R., and Williams, G.H. 1979. Increased glomerular filtration rate after converting-enzyme inhibition in essential hypertension. N Engl J Med 301:9–12

Hulthén, U.L. and Borge, T. 1976. Determination of bradykinin in blood by a sensitive radioimmunoassay. Scand J. Clin Lab Invest 36:833–839

Hulthén, L., and Hökfelt, B. 1978. The effect of the converting enzyme inhibitor SQ20,881 on kinins, renin-angiotensin-aldosterone and catecholamines in relation to blood pressure in hypertensive patients. Acta Med Scand 204:497–502

Hulthén, U.L., Lecerof, H., and Höfelt, B. 1978. Effect of upright tilting on kinins as compared to renin activity in the renal venous blood from patients with essential hypertension. Acta Med Scand 203:411–414

Jandhyala, B.S., Washington, G.F., and Lokhandwala, M.F. 1978. Influence of inhibition of angiotensin I converting enzyme with SQ14225 on the arterial blood pressure. Res Comm in Chem Path & Pharm22(Abs):257–265

Johnston, C.I., Matthews, P.G. and Dax, E. 1976. Effects of dietary sodium, diuretics and hypertension on renin and kallikrein. In *Systemic Effects of Antihypertensive Agents,* ed. Mohinder P. Sambhi, p 323

Johnston, C.I., McGrath, B.P., Millar, J.A., and Matthews, P.G. 1979. Long-term effects of captopril (SQ14225) on blood-pressure and hormone levels in essential hypertension. The Lancet ii,493–495

Johnston, C.I., Clappison, B., McGrath, B.P., Matthews, P.G., Millar, J.A., and Anderson, W.P. Kallikrein, kinins and blood pressure-effects of angiotensin converting enzyme inhibition. In:*Progress in Biochem Pharmacology,* eds. G.Stokes, S.Karger, Basle, in press

Johnston, C.I., Millar, J.A., Casley, D.J., McGrath, B.P., and Matthews, P.G. 1980a. Hormonal responses to angiotensin blockade: comparison between receptor antagonism and converting enzyme inhibition. Circ Res 46 1:128–134

Levinsky, N.G. 1979. The renal kallikrein-kinin system. Circ Res 44:441–451

McCaa, R.E., Hall, J.E., and McCaa, C.S. 1978. The effects of angiotensin I-converting enzyme inhibitors on arterial blood pressure and urinary sodium excretion. Circ Res 43(Suppl)1:I32–I39

McGrath, B.P., Matthews, G.P., and Johnston, C.I. 1980. Acute changes in blood pressure and vasoactive hormones after captopril in hypertensive patients. Clin & Exp Physiol & Pharmac 7:487–492

Man in 'T Veld., Wenting, G.J., and Schalekamp, M.A.D.H. 1979. Does captopril lower blood pressure in anephric patients? British Medical Journal 1:110

Mashford, M.L., and Roberts, M.L. 1972. Determination of blood kinin levels by radioimmunoassay. Biochem Pharmacol 21:2727–2735

Matthews, P.G., and Johnston, C.I. 1979a. Changes in endogenous angiotensin and bradykinin after inhibition of converting enzyme (kininase II). Med J Australia 2(Suppl):xii–xiv

Matthews, P.G., and Johnston, C.I. 1979b. Responses of the renin-angiotensin system and kallikrein-kinin system to sodium and converting enzyme inhibition (SQ14225). In *Kinnis II:* Systemic Proteases and Cellular Function, eds. Setsuro Fujii, Hiroshi Moriya and Tomoji Suzuki, pp 447–457. Plenum Publishing Corporation

Matthews, P.G., McGrath, B.P., and Johnston, C.I. 1979. Hormonal changes with long-term converting enzyme inhibition by captopril in essential hypertension. Clin Sci & Mol Med, 57 (Suppl) 135s–138s

Matthews, P.G., Yasujima, M., and Johnston, C.I. 1980. Kinin receptors: studies on uterine smooth muscle using a radio-labelled ligand. *Proc. VI International Congress Endocrinology* (Abs):742

Mersey, J.H., Williams, G.H., Emanuel, R., Dluthy, R.G., Wong, P.Y., and Moore, T.J. 1979. Plasma bradykinin levels and urinary kallikrein excretion in normal renin essential hypertension. Journal C1 Endocrinology and Metabolism 48:642–647

Millar, J.A., and Johnston, C.I. 1979. Sequential changes in circulating levels of angiotensin I and II, renin, and bradykinin after captopril. Med J Australia 2 (Suppl):xv–xvii

Millar, J.A., Casley, D.J., and Johnston, C.I. Adrenal angiotensin II receptors and vascular reactivity to angiotensin II in the rat during continuous inhibition of angiotensin converting enzyme. Clin Science, in press

Murthy, V.S., Waldron, T.L., and Goldberg, M.E. 1978. The mechanism of bradykinin potentiation after inhibition of angiotensin-converting enzyme by SQ14,225 in conscious rabbits. Circ Res 43(Suppl):I40–I45

Nasjletti, A., Colina-Chourio, J., and McGiff, J.C. 1975. Disappearance of bradykinin in the renal circulation of dogs: effects of kininase inhibition. Circ Res 37:59–65

Odya, C.E., Goodfriend, T.L., and Pena, C. 1979. Bradykinin receptor-like binding studied with iodinated analogues. Biochem Pharmacol 29:175–188

Olsen, U.B., and Arrigoni-Martelli, E. 1979. The effects of kininase II inhibition by SQ14225 on kidney kallikrein-kinin and prostaglandins systems in conscious dogs. European J Pharmacology 54:229–234

Ondetti, M.A., Rubin, B., and Cushman, D.N. 1977. Design of specific inhibitions of angiotensin converting enzyme. New class of orally active antihypertensive agents. Science 196:441–444

Oparil, S., Koerner, T., and O'Donoghue. J.K. 1979. Mechanism of angiotensin I converting enzyme inhibition by SQ20,881 (<Glu-Trp-Pro-Arg-Pro-Gln-Ile-Pro-Pro) in vivo. Hypertension 1:13–22

Pang, L.M., Stalcup, S.A., Lipset, J.S., Hayes, C.J., Bowman, F.O., and Mellins, R.B. 1979. Increased circulating bradykinin during hypothermia and cardiopulmonary bypass in children. Circulation 60:1503–1507

Rubin, R., Laffan, R.J., Kotler, D.G., O'Keefe, E.H., Demaio, D.A., and Goldberg, M.E. 1978. SQ14,225 (D-3-Mercapto-2-methylpropanoyl-L-proline), a novel orally active inhibitor of angiotensin I-converting enzyme. J Pharmacol & Exp Therapeutics 204:271–280

Soffer, R.L. 1976. Angiotensin-converting enzyme and the regulation of vasoactive peptides. Ann Rev Bioch 43:73–94

Swartz, S.L., Williams, G.H., Hollenberg, N.K., Moore, T.J., and Dluhy, R.G. 1979. Converting enzyme inhibition in essential hypertension: the hypotensive response does not reflect only reduced angiotensin II formation. Hypertension 1:106–111

Talamo, R.C., Haber, E., and Austen, K.F. 1969. A radioimmunoassay for bradykinin in plasma and synovial fluid. J Lab Clin Med V 74:816–827

Thurston, H., and Swales, J.D. 1978. Converting enzyme inhibitior and saralasin infusion in rats. Circ Res 42:588–592

Vinci, J.M., Gill, J.R., Bowden, R.E., Pisano, J.J., Izzo, J.L., Radfar, N., Taylor, A.A., Zusman, R.M., Bartter, F.C., and Keiser, H.R. 1978. The kallikrein-kinin system in Bartter's Syndrome and its response to prostaglandin synthetase inhibition. J Clin Invest 61:1671–1682

Vinci, J.M., Zusman, R.M., Izzo, J.L., Bowden, R.E., Horwitz, D., Pisano, J.J. and Keiser, H.R. 1979. Human urinary and plasma kinins. Relationship to sodium-retaining steroids and plasma renin activity. Circ Res 44:228–237

Vollmer, R.R., Boccagno, J.A., Harris, D.M., and Murthy, V.S. 1978. Hypotension induced by inhibition of angiotensin converting enzyme in pentobarbital-anesthetized dogs. European J Pharmacol 51:39–45

Williams, G.H., and Hollenberg, N.K. 1977. Accentuated vascular and endocrine response to SQ 20881 in hypertension. New Engl J Med 297:184–188

Wong, P.Y., Talamo, R.C., Williams, G.H., and Colman, R.W. 1975. Response of the kallikrein-kinin and renin-angiotensin systems to saline infusion and upright posture. J Clin Invest 55:691–701

Zacest, R., and Mashford, M.C. 1967. Blood bradykinin in the human. Aust J Exp Biol Med Sci 45:89–95

The Vasodepressor Mechanism of Converting Enzyme Inhibitors

H. Thurston, R.F. Bing, G.I. Russell and J.D. Swales
Department of Medicine
University of Leicester
Leicester, United Kingdom

INTRODUCTION

Over the past decade a mass of evidence has been accumulated in support of a major role for the renin-angiotensin system in the control of arterial blood pressure and aldosterone secretion (Laragh and Sealey 1973; Davis et al., 1974). However, many of these studies have depended upon measurements of the circulating levels of renin and angiotensin II that may reflect neither the concentration at the receptor sites nor the degree of vasoconstriction produced by such local hormones in the peripheral arterioles. Unlike many other endocrine systems, where the ablation of the organ of secretion may reveal the major function of its hormone, the renin-angiotensin system cannot be eliminated without removing the kidneys and thereby disrupting secretion of a number of vasoactive substances and fluid balance. Recently, several peptides have been synthesised for the pharmacological blockade of the renin-angiotensin system at one of two main sites: conversion of angiotensin I into angiotensin II (inhibition of converting enzyme) and angiotensin II receptor binding (blockade by competitive angiotensin II antagonists). Two agents which have been utilized in this context are teprotide, the nonapeptide converting enzyme inhibitor (SQ 20881) (Engel et al., 1972) and the octapeptide angiotensin II antagonist, sarcosine^1 alanine^8 (sar^1 ala^8) angiotensin II, saralasin (P113) (Pals et al., 1971). However, both of these peptides must be administered by intravenous infusion and have a relatively short half-life. Captopril (SQ 14,225; D3—mercapto—2 methyl—propanoyl—L—proline) a new orally active converting enzyme inhibitor has been shown to lower blood pressure in man (Gavras et al., 1978) and hypertensive rats (Bengis et al., 1978; Muirhead et al., 1978). In common with teprotide (Engel et al., 1972), (Ondetti et al., 1977) this drug also potentiates the vasodepressor action of bradykinin (Murthy et al., 1977). Moreover, the blood pressure response to all these renin-angiotensin inhibitors is discrepant, with a much greater fall occurring with converting enzyme blockade than with antagonist administration (Case et al., 1976; Thurston and Swales, 1978). This difference may reflect the intrinsic agonist action of saralasin (Pals et al., 1972; Streeten et al., 1975) or the bradykinin potentiating action of converting enzyme inhibitors (Engel et al., 1972; Murthy et al., 1977; Thurston and Swales, 1978). Thus, since the relative importance of the effects upon angiotensin II as opposed to the effects upon bradykinin potentiation are not known, the fall in blood pressure observed with the administration of converting enzyme inhibitor cannot be confidently attributed to the angiotensin II blocking action alone.

Despite these uncertainties, many workers still utilize the degree of blood pressure reduction as a means of assessing the role of the renin-angiotensin system in maintenance of high blood pressures. Such an approach is open to major criticism since no account is taken of: a) the potential vasodepressor properties of the inhibitors and b) the stimulation of compensating mechanisms which may maintain falling blood pressure initially sustained by angiotensin II. Among such homeostatic agents is renin hypersecretion; although saralasin (Brown et al., 1976), teprotide (Case et al., 1977) and captopril (Gavras et al., 1978) all cause a marked rise in plasma renin levels, it would seem unlikely these have significant effects in the presence of a gross excess of antagonist or inhibitor.

The purpose of this review is to examine the mechanism whereby converting enzyme inhibitors produce their hypotensive effects, with particular reference to the role of the renin-angiotensin system, but also to evaluate evidence in support of other actions including potentiation of bradykinin.

PROPERTIES OF CONVERTING ENZYME INHIBITORS

These were initially described in the venom of the viper *Bothrops jararaca* (Greene et al., 1972), and several peptides have been synthesised (Engel et al., 1972). The inhibitors block the action of a peptidyl carboxy-peptidase which is responsible for angiotensin I conversion (Erdös, 1976), which is also the enzyme responsible for the breakdown of most bradykinin (Ferreira et al., 1977). Angiotensin converting enzyme inhibitors therefore prevent angiotensin II generation and could cause accumulation of the blood pressure lowering hormone bradykinin (Greene et al., 1972). Two of these inhibitors will be considered in detail, the nonapeptide teprotide (SQ 20,881) and captopril (SQ 14,-225).

CONVERTING ENZYME INHIBITORS AND THE RENIN-ANGIOTENSIN SYSTEM

Angiotensin II, the active octapeptide of the renin-angiotensin system, is produced by the enzymatic removal of a dipeptide from the physiologically largely inactive decapeptide angiotensin I (Peart, 1965) by angiotensin I converting enzyme (Ng and Vane, 1967). Both teprotide and captopril effectively block the generation of angiotensin II in vivo (Engel et al., 1972; Ondetti et al., 1977) and administration of captopril leads to a rapid fall in plasma angiotensin II levels (Atkinson et al., 1979) associated with a substantial rise in plasma renin and angiotensin I.

The pressor response to angiotensin I—but not angiotensin II—is markedly decreased by oral administration of captopril in man (Ferguson et al., 1977); teprotide inhibits the pressor response to angiotensin I in the rat (Engel et al., 1972; Thurston and Laragh, 1975) and dog (Miller et al., 1975). Although the pulmonary circulation was thought to be the main site of angiotensin I conversion, studies using teprotide on isolated vascular beds have demonstrated significant local generation of angiotensin II in the coronary (Britten et al., 1973), mesenteric (De Salvo et al., 1971), renal and hind limb vasculatures (Aiken and Vane 1970). Moreover, Oparil et al., (1979) have shown that extrapulmonary enzyme is more resistant to blockade with teprotide than the pulmonary enzyme. When teprotide or captopril is administered to animals with intact kidneys a reduction of angiotensin I pressor response is associated with an increase in angiotensin II pressor reactivity (Thurston and Laragh, 1975; Marks et al., 1980) (Fig.1), whereas no change in angiotensin II pressor response occurred in nephrectomized rats (Fig.2). These observations are in keeping with the concept that prior receptor occupancy by endogenous angiotensin II is an important factor in determining the pressor response to exogenous infused angiotensin II (Thurston and Laragh, 1975; Marks et al., 1979). Thus, studies of converting enzyme blockade of angiotensin I pressor responses must be carried out in nephrectomized animals because of the effect of endogenous angiotensin II upon responsiveness.

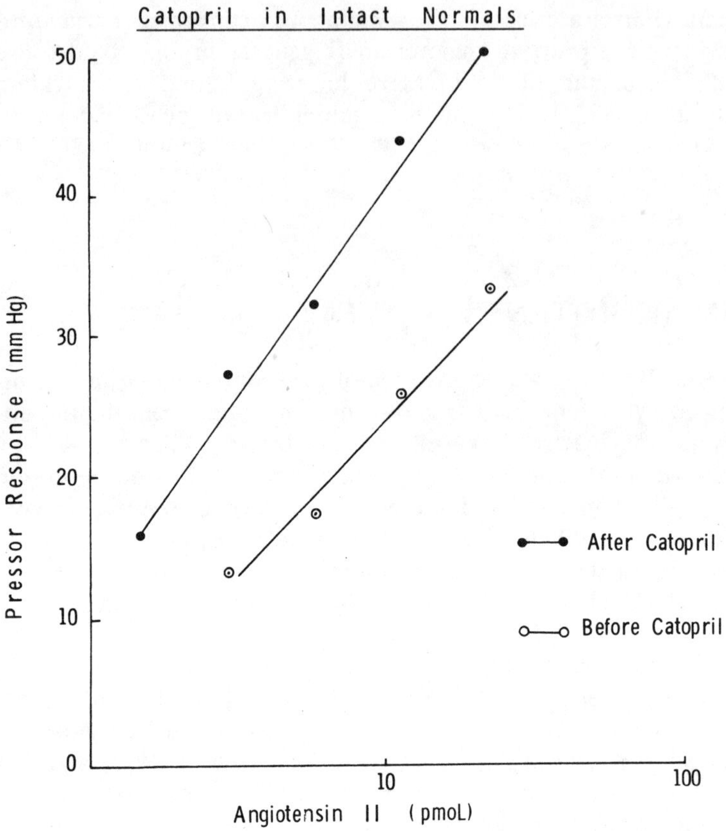

Fig. 1. Effect of captopril (250 μg iv) on the angiotensin II dose-response curve in intact rats.

Acute hypertension can be induced in nephrectomized rats by renin infusion, which causes a maximal increase in blood pressure within 2 to 3 min, after which blood pressure slowly declines (Bing and Neilson, 1973a). The rise in blood pressure can be readily prevented by saralasin infusion (Bing and Neilson, 1973a; Thurston and Swales, 1973). Similarly, teprotide administration prevents the rise of blood pressure during renin infusion and rapidly restores blood pressure to basal levels within minutes (Thurston, 1976) (Fig.3).

This evidence indicates that both these inhibitors block the action of converting enzyme activity in a variety of species in vivo, prevent the pressor response to angiotensin I, enhance responsiveness to angiotensin II and reverse renin infusion hypertension. Furthermore, since teprotide and captopril have little effect on the blood pressure of nephrectomized

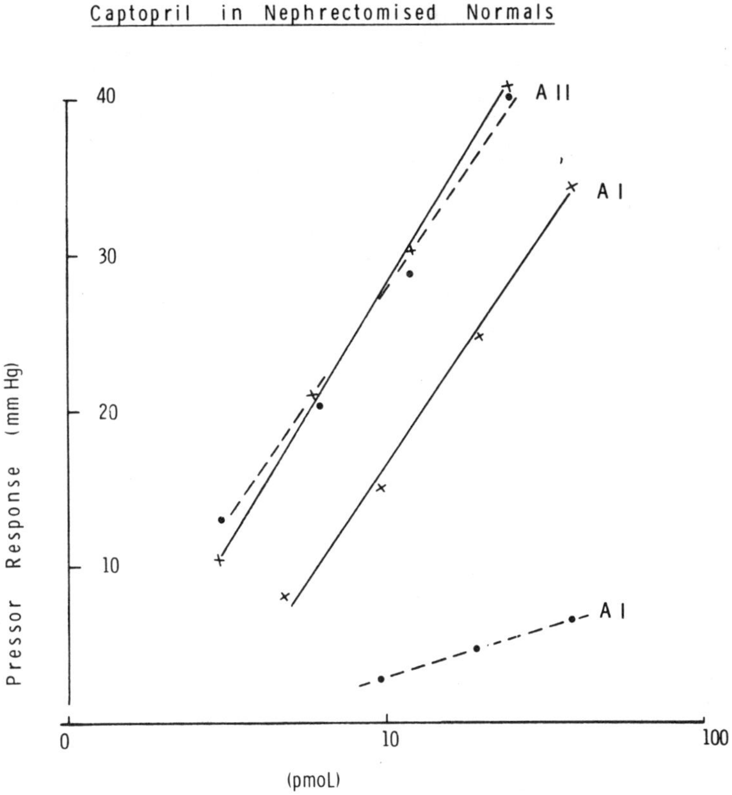

Fig. 2. Angiotensin I and II dose-response curves before (continuous line) and after (broken line) captopril (250 μg iv) in nephrectomised rats.

Table 1. Plasma renin concentration, initial blood pressure and fall in blood pressure with captopril (250 μg iv) in anaesthetized normal and hypertensive rats (all values mean ± SEM).

Experimental Groups	Plasma Renin Concentration (ng/ml/hr)	Initial Blood Pressure (mm Hg)	Fall in Blood Pressure (mm Hg)	*P*
Normals	134.5 ± 17.3	103.7 ± 13.3	16.2 ± 3.7	<0.01
Normal Nephrectomy	1.3 ± 0.3	73.7 ± 6.0	7.8 ± 2.5	<0.05
DOCA-Salt Hypertension	3.2 ± 1.6	151.4 ± 6.3	20.3 ± 2.2	<0.01
Acute 2 Kidney Goldblatt Hypertension	368.3 ± 53.0	156.7 ± 8.0	82.8 ± 13.9	<0.01
Chronic 2 Kidney Goldblatt Hypertension	199.2 ± 19.0	160.3 ± 8.8	40.5 ± 8.4	<0.01

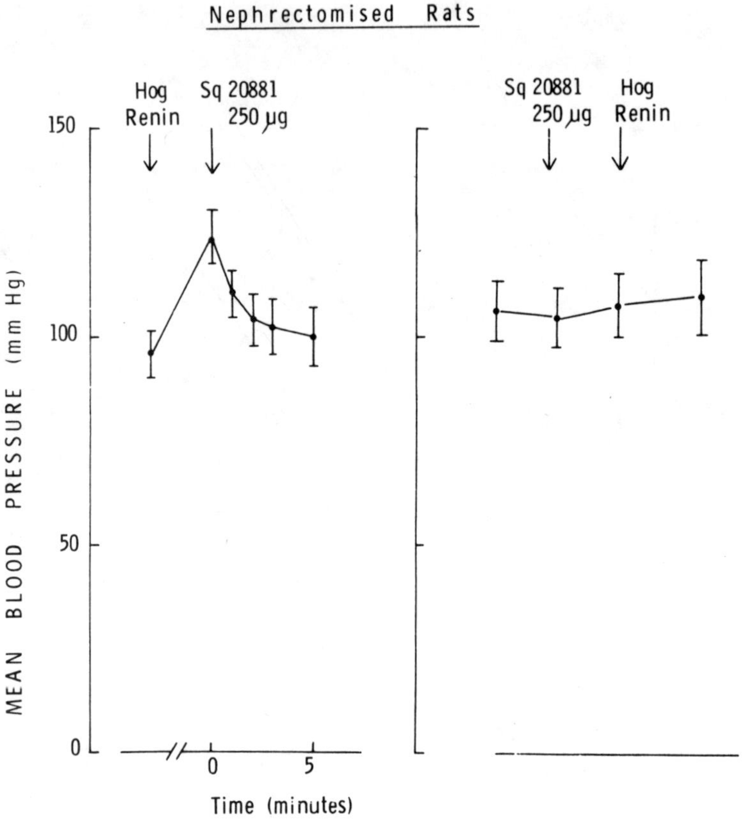

Fig. 3. Effect of teprotide on nephrectomised rats infused with hog renin (0.2 µ).

rats per se (Thurston and Laragh, 1975; Marks et al., 1980) (Table 1)the main mechanism in these conditions may be inhibition of angiotensin II generation. However, in another study captopril reduced the blood pressure of salt depleted, but not salt repleted, anephric subjects (Man in t'Veld, 1980), suggesting an important nonrenal mechanism.

CONVERTING ENZYME INHIBITORS AND BRADYKININ

The identity of converting enzyme of the renin-angiotensin system with kininase II, which degrades bradykinin, suggests that teprotide and captopril might also lead to an increase in circulating bradykinin levels.

However, there are several pathways for bradykinin metabolism (Erdös and Yang, 1970), and furthermore, bradykinin has a higher affinity for converting enzyme than angiotensin I (Dorer et al., 1974). Therefore converting enzyme inhibitor may have a different action upon the two hormone cascades. In an in vitro assay, the ratio of angiotensin I inhibition to bradykinin potentiation was 40 for teprotide and 7 for captopril (Haber, 1978—data calculated from Ondetti et al., 1977).

Teprotide (Engel et al., 1972) and captopril (Murthy et al., 1977; Rubin et al., 1977) potentiate the vasodepressor response to bradykinin in the rat. The authors have administered both teprotide and captopril to bradykinin infused nephrectomized rats and observed substantial blood pressure fall (Thurston and Swales, 1978; Marks et al., 1980). Thus, a steady 4 μg/min infusion of bradykinin was barely vasodepressor, producing blood pressure fall from 72.2 ± 2.83 (mean ± SEM) mm Hg to 65.9 ± 4.62 mm Hg ($P > 0.05$); but when teprotide was given, further fall to 43.2 ± 6.93 mm Hg occurred ($P < 0.01$). Similarly, captopril produced a marked vasodepressor response in bradykinin infused rats (Fig.4) (Marks et al., 1980). A comparison of the effects of

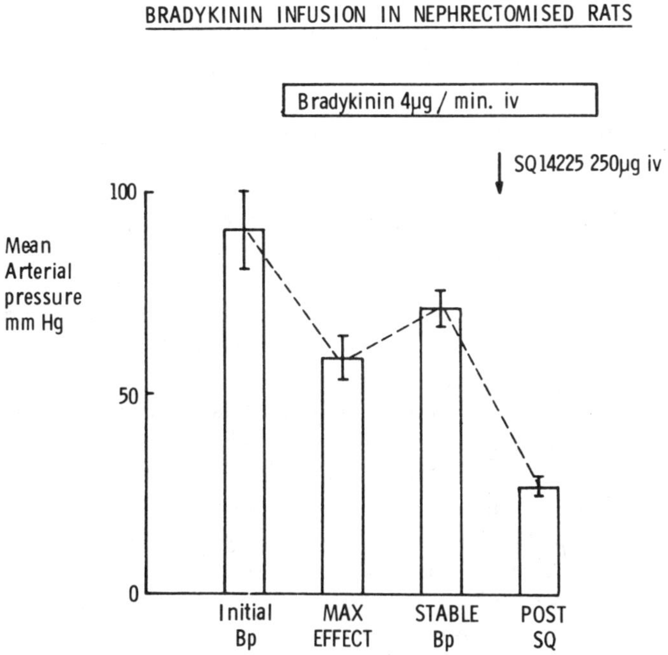

Fig. 4. Effect of captopril on mean blood pressure of nephrectomised rats infused with bradykinin (4 μg/min intravenously) (reproduced with permission of Clinical Science from Marks et al., 1980)

teprotide and captopril in this way would suggest greater bradykinin potentiation by the latter, since final blood pressure with doses of captopril equal to the doses of teprotide was 26.4 ± 1.83 mm Hg. Other studies have shown that bradykinin potentiation by captopril was greater in intact than in nephrectomized rabbits (Murthy et al., 1978). Moreover, pretreatment with indomethacin reduced the vasodepressor response to bradykinin in nephrectomized to a greater extent than in intact rabbits. Thus, renal prostaglandins appear to play a role in bradykinin induced hypotension. In view of this, the lack of blood pressure lowering effect of teprotide in nephrectomized subjects cannot be used as evidence against a significant bradykinin vasodepressor mechanism.

CONVERTING ENZYME INHIBITORS AND BLOOD PRESSURE

In common with the antagonist saralasin, converting enzyme inhibitors teprotide and captopril have been administered in a variety of physiological and hypertensive states in an attempt to determine the role of angiotensin II in blood pressure control. However, where both types of renin-angiotensin blockade have been employed, the results are somewhat discrepant. Thus, in salt depleted rats (Thurston and Swales, 1978), dogs (Miller et al., 1975) and man (Case et al., 1976), although each peptide produced a fall in blood pressure, converting enzyme inhibition had the greater effect. Conversely, in the salt replete state, saralasin more often exhibited a pressor action (Bing and Nielson, 1973b; Case et al., 1976) but teprotide produced a small depressor response (Thurston and Laragh, 1975). Moreover, there was a high correlation between initial renin and the hypotensive action of saralasin (Case et al., 1976) but in one case no such relationship was found in patients treated with captopril (Gavras et al., 1978).

In early (less than 6 weeks) Goldblatt 2 kidney, 1 clip hypertensive rats, administration of saralasin (Brunner et al., 1971; Thurston and Swales, 1974; MacDonald et al., 1975), teprotide (Thurston and Swales, 1977; Thurston et al., 1979) and captopril (Bengis et al., 1978) caused significant falls in blood pressure. However, rats with chronic renovascular hypertension showed little response to saralasin infusion (Thurston and Swales, 1974; Gavras et al., 1975), although captopril produced a significant fall in blood pressure (Thurston et al., 1980). In other studies, when captopril was given to a variety of normal and hypertensive rats (Table 1), a fall in pressure was noted in DOCA-salt hypertension (Marks et al., 1980), whereas saralasin produced a rise in blood pressure

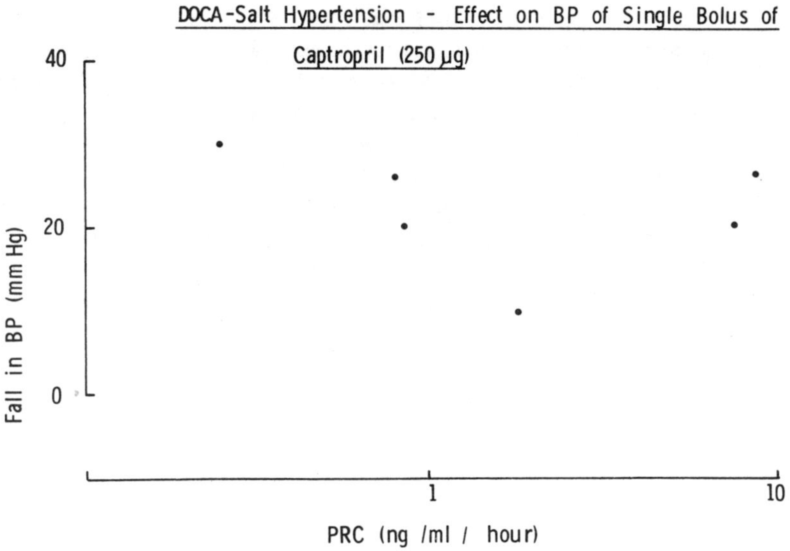

Fig. 5. Captopril induced fall in blood pressure and plasma renin concentration in DOCA hypertensive rats.

in this model (Gavras et al., 1975). In addition, the blood pressure responses to captopril showed little correlation with the plasma renin concentration (Fig.5). Captopril decreased the blood pressure in spontaneously hypertensive rats. (Muirhead et al., 1978; Antonaccio et al., 1979). Furthermore, bilateral nephrectomy attenuated the vasodepressor response to captopril in both DOCA-salt (Marks et al., 1980) and spontaneously hypertensive rats (Antonaccio et al., 1979). Thus captopril lowers the blood pressure in hypertensive rat models in which the renin-angiotensin system is thought not to be implicated in maintaining the raised blood pressure.

Antonaccio et al. (1979) suggested that the blood pressure of spontaneously hypertensive rats may be maintained in part by angiotensin generated by renin within the vascular wall. However, Barrett et al. (1978) showed normal levels of vascular renin in these animals, and furthermore the concentration declined as the blood pressure rose. The authors confirmed this finding and showed low levels in the DOCA-salt rat (unpublished findings), whereas in acute Goldblatt 2 kidney hypertension the levels were more than 10 times greater than in normal rats (Thurston et al., 1978). Therefore, it is unlikely that the depressor response to converting enzyme inhibition in experimental models with low circulating renin levels can be attributed to an action upon vascular renin.

Several hypotheses have been put forward to explain some of these anomalies. First, the intrinsic agonist action of saralasin may limit the extent of blood pressure fall with renin-angiotensin blockade (Jaeger et al., 1978); second, bradykinin potentiation caused by converting enzyme inhibitors could enhance their vasodepressor effects (Thurston and Swales, 1978); and, third, angiotensin II may exert its action by a slow pressor mechanism which may be revealed only by prolonged blockade of the renin-angiotensin system (Reiger et al., 1977) and which may not be related directly to circulating angiotensin levels.

The main support for the importance of the intrinsic agonist activity of saralasin lies in the failure of teprotide to reduce the blood pressure of anephric man (Case et al., 1976) and rats (Thurston and Laragh, 1975). However, as we have seen, captopril does produce a small depressor response in the rat (Marks et al., 1980) and in anephric subjects (Man in t'Veld et al., 1980). On the other hand, bradykinin potentiation

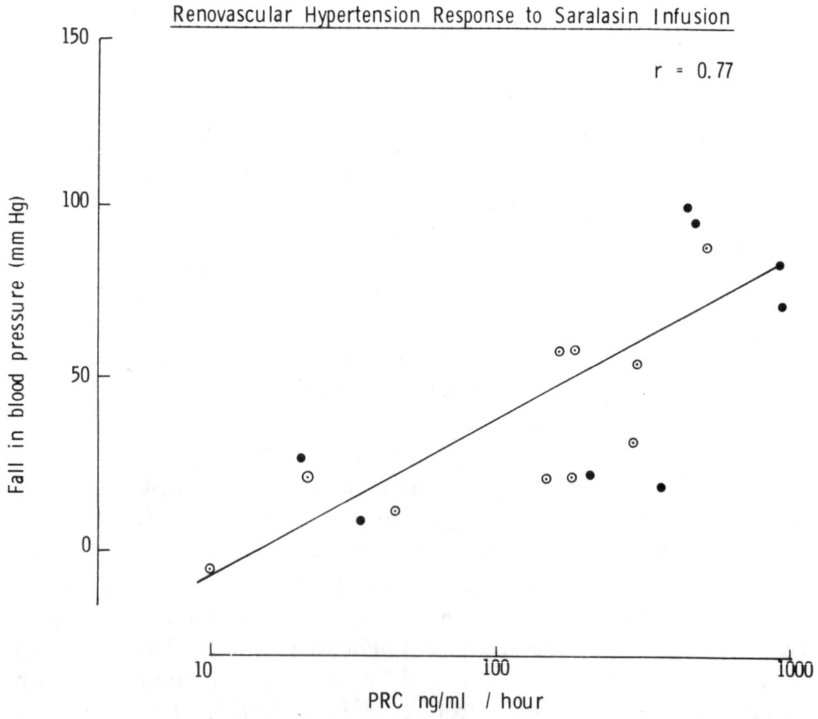

Fig. 6. Relationship between maximum fall in blood pressure with saralasin infusion (10 μg/min) and initial plasma renin concentration in early (closed circles) and chronic (open circles) Goldblatt 2 kidney hypertensive rats.

by converting enzyme inhibitors was attenuated in nephrectomized rabbits, and it appears that the renal prostaglandin system is required for the full vasodepressor action of bradykinin (Murthy et al., 1980). Thus, this would readily explain the marked depressor response to captopril in intact but not nephrectomized spontaneous hypertensive rats. We have recently examined the reported 'slow response' to renin-angiotensin inhibition by prolonged (12 hr) infusion of saralasin and captopril in both early (6 wk) and chronic (4 mo) Goldblatt 2 kidney hypertension. After the immediate fall in blood pressure, which was complete within the first hour, no further progressive reduction in blood pressure was observed with either inhibitor (Bing et al., 1981) (Table 2). Moreover, the fall in blood pressure with saralasin was highly correlated with plasma renin (Fig.6) but no such correlation existed between captopril and blood pressure fall (Fig.7). Thus, we have been unable to demonstrate a slow response to captopril or saralasin of the type reported.

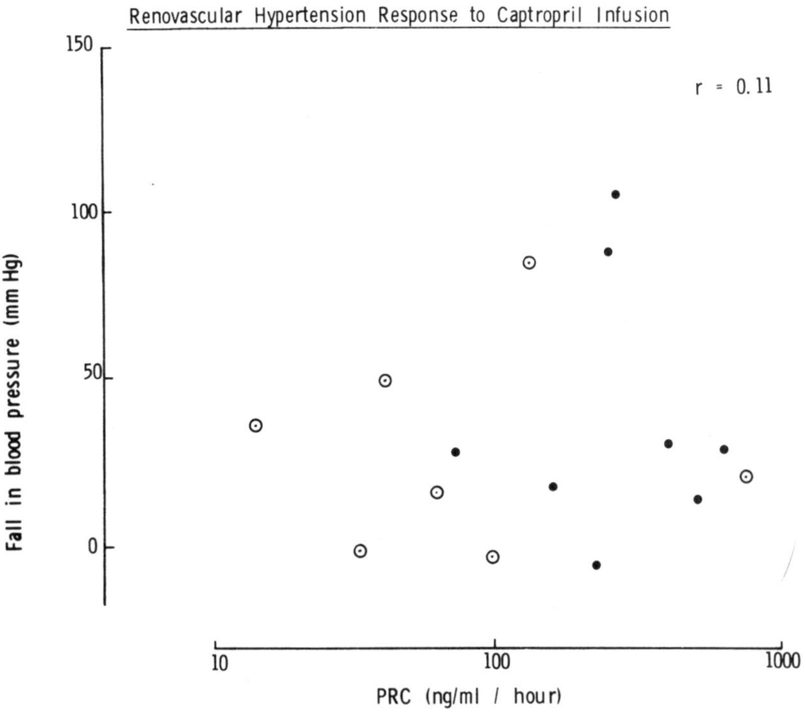

Fig. 7. Fall in blood pressure with captopril infusion (8.3 μg/min) and initial plasma renin concentration in early (closed circles) and chronic (open circles) Goldblatt 2 kidney hypertensive rats.

Table 3. Mean arterial blood pressure before and after saralasin infusion alone or combined with teprotide.

Experimental Groups	N	Blood Pressure (mmHg) Initial	After Saralasin	*P*	After Saralasin and Teprotide	*P*
Salt Loaded Normal	7	120.1 ± 8.29	107.9 ± 7.29	<0.01	96.9 ± 9.88	<0.05
Salt Restricted Normals	7	129.9 ± 1.86	104.2 ± 3.66	<0.01	59.0 ± 6.09	<0.01
Salt Restriction 1 hr Postnephrectomy	7	87.7 ± 7.44	81.5 ± 4.12	NS	81.7 ± 7.02	NS
Renin Infused Nephrectomised	7	141.0 ± 7.54*	85.7 ± 9.80	<0.001	81.7 ± 7.02	NS
Goldblatt 2 Kidney Hypertension	8	175.3 ± 9.3	123.1 ± 17.8	<0.01	112.0 ± 19.10	<0.05
Goldblatt 2 Kidney Hypertension one hr Post-nephrectomy	10	137.8 ± 11.7	124.2 ± 13.1	<0.05	122.3 ± 15.4	NS

*Basal blood pressure before renin infusion 82.4 ± 9.81)

Table 2. Effect of prolonged infusion of saralasin (10 μg/min) or captopril (8.3 μg/min) on blood pressure in conscious early and chronic Goldblatt 2 kidney hypertensive rats.

	N	Blood Pressure (mm Hg Mean ± SEM)		
		Initial	30 min	12 hr
Early Hypertension (duration 36.6 ± 3.8 days)				
Saralasin	8	215 ± 5.5	164 ± 12.7 *P* <0.02	170 ± 12.2
Captopril	8	184 ± 4.1	157 ± 107 *P* <0.05	159 ± 8.5
Chronic Hypertension (duration 151 ± 8.4 days)				
Saralasin	8	195 ± 9.7	181 ± 12.5 *P* <0.1	183 ± 8.8
Captopril	8	184 ± 5.3	158 ± 8.5 *P* <0.05	154 ± 6.8

COMBINED ANGIOTENSIN II AND CONVERTING ENZYME BLOCKADE

When teprotide was given to rats in which the renin-angiotensin system had been blocked with saralasin, an additional fall in blood pressure was observed in all animals with intact kidneys (Thurston and Swales, 1978) (Table 3). Conversely, where rats were pretreated with teprotide, saralasin administration did not cause a further depressor response. Similar results (Fig.8) were obtained when captopril was administered to saralasin-pretreated salt depleted rats (Marks et al., 1980). Since saralasin effectively blocks the renin-angiotensin system, the additional vasodepressor response to teprotide and captopril cannot be explained by angiotensin II inhibition. Furthermore, the additional blood pressure lowering action was absent one hour after bilateral nephrectomy in salt depleted rats. Teprotide caused no change in blood pressure in rats infused with renin and blocked with saralasin (Thurston and Swales, 1978).

In Goldblatt 2 kidney hypertensive rats infused with saralasin, teprotide produced a further fall in blood pressure which was similar to that observed in salt loaded rats (Thurston and Swales, 1978) (Table 3), but bilateral nephrectomy abolished this additional vasodepressor effect

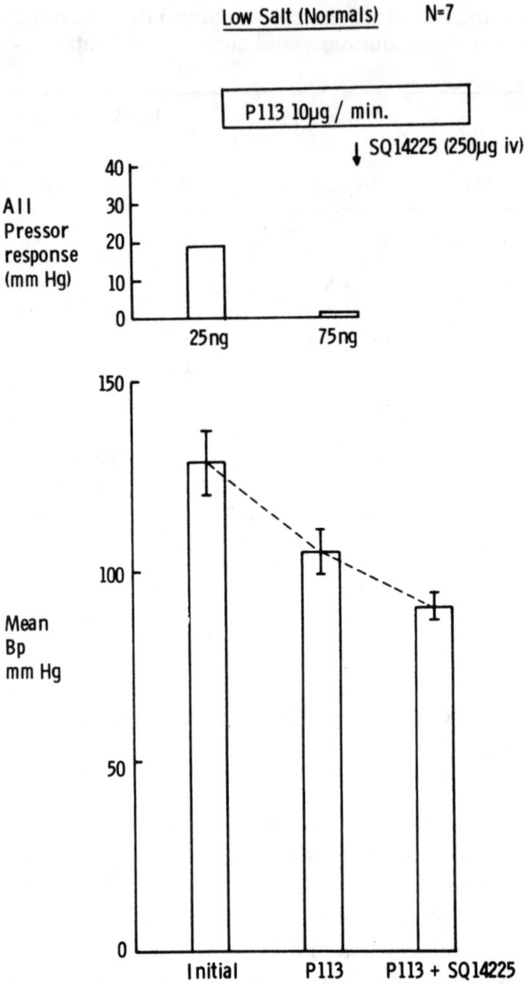

Fig. 8. Pressor response to angiotensin II (AII) and mean arterial blood pressure in seven salt restricted normal rats infused with saralasin alone or in combination with captopril (reproduced with permission of Clinical Science from Marks et al., 1980).

(Thurston and Swales, 1977). Bradykinin infusion into nephrectomised rats restored the vasodepressor action of both teprotide (Thurston and Swales, 1978) and captopril (Marks et al., 1980). Moreover, when Goldblatt 2 kidney hypertensive rats were pretreated with the kallikrein inhibitor trasylol, the fall in blood pressure with captopril was reduced (Sen et al., 1979), and when trasylol was given to normal or Goldblatt 2 kidney hypertensive rats pretreated with saralasin and captopril a rise in blood pressure occurred (Kushiro et al., 1979). Thus the additional vasodepressor response to converting enzyme blockade may be mediated via the kallikrein-kinin system.

CONCLUSIONS

The present information strongly supports the view that converting enzyme inhibitors lower blood pressure more effectively than angiotensin II antagonists. Part of the fall in blood pressure induced by these inhibitors depends upon blockade of the renin-angiotensin system, but an additional action is involved. Thus, these inhibitors produce a vasodepressor response in situations where the renin-angiotensin system plays no part in maintaining blood pressure. Since converting enzyme inhibitors act on the metabolic pathway of bradykinin, the additional action may involve bradykinin potentiation.

ACKNOWLEDGEMENT

We wish to thank Dr. Z.P. Horovitz of the Squibb Institute for Medical Research, USA, for the generous gifts of teprotide and captopril, and Norwich Pharmaceuticals, USA, for saralasin. This work was supported by the National Kidney Research Fund.

SUMMARY

1. The converting enzyme inhibitors teprotide and captopril effectively block the renin-angiotensin system and in addition markedly potentiate the vasodepressor response to bradykinin infusion.
2. Converting enzyme inhibitors were more effective than the competitive antagonist saralasin in lowering the blood pressure of normal, acute and chronic Goldblatt 2 kidney hypertensive rats and hypertensive patients.
3. Captopril lowers blood pressure of nephrectomised rats and anephric subjects, DOCA salt and spontaneous hypertensive rats, although the plasma renin-angiotensin system does not maintain the blood pressure in these situations.
4. When the converting enzyme inhibitors were given to saralasin pretreated rats, an additional fall in blood pressure was observed in all animals with intact kidneys.
5. Converting enzyme inhibitors lower blood pressure by an action additional to angiotensin II blockade. This mechanism may involve the potentiation of bradykinin.

REFERENCES

Antonaccio, M.J., High, J.P., Rubin, B. and Schaeffer, T. 1979. Contribution of the kidneys but not adrenal glands to the acute anti-hypertensive effects of captopril in spontaneously hypertensive rats. Clin Sci 57:127s–130s

Aiken, J.W., and Jane, J.R. 1970. Renin-angiotensin system: inhibition of converting enzyme in isolated tissues. Nature (Lond) 228:30–34

Atkinson, A.B., Brown, J.J., Fraser, R., Leslie, B., Lever A.F., Morton, J.J., and Robertson, J.I.S. 1979. Captopril in hypertension with renal artery stenosis and in intractable hypertension: acute and chronic changes in circulating concentrations of renin, angiotensin I and II and aldosterone, and in body composition. Clin Sci 57:139s–143s

Barrett, J.D., Eggena, P., and Sambhi, M.P. 1978. Partial characterization of aortic renin in the spontaneously hypertensive rat and its interrelationship with plasma renin, blood pressure and sodium balance. Clin Sci Mol Med 55:261–270

Bengis, R.G., Coleman, T.G., Young, D.B., and McCaa, R.E. 1978. Long term blockade of angiotensin formation in various normotensive and hypertensive rat models using converting enzyme inhibitor (SQ 14225). Circ Res 43: Suppl 1 45–53

Bing, J., and Nielson, K. 1973a. Cause of the prolonged pressor action of renin in nephrectomized rats. Acta Path Microbiol A 81:247–253

Bing, J., and Nielson, K. 1973b. Role of the renin system in normo- and hypertension. Acta Path Microbiol Scand A 81:254–259

Bing, R.F., Russell, G.I., Swales, J.D., and Thurston, H. 1981. A comparison of unclipping and prolonged renin-angiotensin blockade on blood pressure in 2 kidney 1 clip hypertension in the rat. J. Lab. Clin. Med. (in press)

Britton, S., and DiSalvo, J. 1973. Effect of angiotensin I to angiotensin II on hind limb and coronary vascular resistance. Am J Physiol 225:1226–1231

Brown, J.J., Brown, W.C.B., Fraser, R., Lever, A.F., Morton, J.J., Robertson, J.I.S., Agabiti-Rosei, E., Tree, M., and Trust, P.M. 1976. Effects of infusing angiotensin II antagonist saralasin on arterial blood pressure and on plasma concentrations of aldosterone, renin and angiotensin II in normal and selected hypertensive patients. In: *Systemic Effects of Antihypertensive Agents,* ed. MP Sambhi, pp 499–512. New York: Stratten

Brunner, H., Kirshman, J.D., Sealey, J.E., and Laragh, J.H. 1971. Hypertension of renal origin: evidence for two different mechanisms. Science 174:1344–1346

Case, D.B., Wallace, J.M., Keim, H.J., Weber, M.A., Drayer, J.I.M., White, R.P., Sealey, J.E., and Laragh, J.H. 1976. Estimating renin participation in hypertension: superiority of converting enzyme inhibitor over saralasin. Amer J Med 61:790–796

Case, D.B., Wallace, J.M., Keim, H.J., Weber, M.A., Sealey, J.E., and Laragh, J.H. 1977. Possible role of renin in hypertension as suggested by renin-sodium profiling and inhibition of converting enzyme. N Engl J Med 296:641–646

Davis, J.O., Freeman, R.H., Johnson, J.A., and Spielman, W.S. 1974. Agents which block the action of the renin-angiotensin system. Circ Res 34:279–285

De Salvo, J., and Montefusco, C.B.

1971. Conversion of angiotensin I to angiotensin II in the canine mesenteric circulation. Am J Physiol 221:1576–1579

Davis, J.O., Freeman, R.H., Johnson, J.A., and Spielman, W.S. 1974. Agents which block the action of the renin-angiotensin system. Circ Res 34:279–285

Dorer, F.E., Kahn, J.R., Lentz, K.E., Levine, M., and Skeggs, L.T. 1974. Hydrolysis of bradykinin by angiotensin converting enzyme. Circ Res 34:824–827

Engel, S.L., Schaeffer, T.R., Gold, B.I., and Rubin, B. 1972. Inhibition of pressor effects of angiotensin I and augmentation of depressor of bradykinin by synthetic peptides. Proc Soc Exp Biol Med 140:240–244

Erdös, E.G., and Yang, H.Y.T. 1970. Kininases. In *Bradykinin, Kallidin and Kallikrein, Handbook of Experimental Pharmacology,* ed. Erdös, E.G. 24: pp 289–323. Berlin: Springer Verlag

Erdös, E.G. 1976. Conversion of angiotensin I to angiotensin II. Am J Med 60:749–759

Ferguson, R.K., Turini, G.A., Brunner, H.R., Gavras, H., and McKinstry, D.N. 1977. A specific orally active inhibitor of angiotensin converting enzyme in man. Lancet 1:775–778

Ferreira, S.H., and Bakhle, Y.S. 1977. Inactivation of bradykinin and related peptides in the lung. In *Metabolic Functions of the Lung,* ed. Y.S. Bakhle and J.R. Vane, p 33. New York: Marcel Dekker

Gavras, H., Brunner, H.R., Laragh, J.H., Darracott-Vaughan, E., Koss, M., Cote, L.J., and Gavras 1. 1975. Malignant hypertension resulting from deoxycorticosterone acetate and salt excess. Circ Res 36:300–309

Gavras, H., Brunner, H.R., Thurston, H., and Laragh, J.H. 1975. Reciprocation of renin dependency with sodium volume dependency in renal hypertension. Science 188:1316–1317

Gavras, H., Brunner, H.R., Turni, G.A., Kershaw, G.R., Tifft, G.P., Cuttelod, S., Gavras, I., Vukovich, R.A., and McKinstry, D.N. 1978. Antihypertensive effect of the oral angiotensin converting enzyme inhibitor SQ 14225 in man. N Eng J Med 298:991–995

Greene, L.J., Camargo, A.C.M., Krieger, E.M., Stewart, J.M., and Ferreira, S.H. 1972. Inhibition of the conversion of angiotensin I to II and potentiation of bradykinin by small peptides present in *Bothrops jararaca* venom. Circ Res 31:Suppl 2, 62–71

Haber, E. 1978. Renin inhibitors. N Engl J Med 298:1023–1025

Jaeger, P., Ferguson, R.K., Brunner, H.R., Kitchertz, E.J., and Gavras, H. 1978. Mechanism of blood pressure reduction by teprotide (SQ 20881) in rats. Kidney International 13:289–296

Kushiro, T., Lee, T.C., Gassia, J.P., Sabado, M., and Maxwell, M.H. 1978. The anti-hypertensive effect of the kallikrein-kinin system (KKS) in conscious rats with two kidney renovascular hypertension (RHT). Kidney International 14:697 (abstract)

Laragh, J.H., and Sealey, J.E. 1973. The renin-angiotensin-aldosterone hormonal system and regulation of sodium, potassium and blood pressure homeostasis. In *Handbook of Physiology: Renal Physiology,* ed. J. Orliff and R.W. Berliner, p 831. Washington, D.C.: American Physiological Society

Macdonald, G.J., Boyd, G.W., and Peart, W.S. 1975. Effect of the angiotensin II blocker 1-Sar-8-Ala angiotensin II on renal artery clip hypertension in the rat. Circ Res 37:640–646

Man in t'Veld, A.J., Schicht, I.M., Derkx, F.H.M., De Bruyn, J.H.B., and Schalekamp, M.A.D.H. 1980. Effects of an angiotensin-converting enzyme inhibitor (captopril) on blood pressure in anephric subjects. British Med J 1:288–290

Marks, E.S., Thurston, H., Bing, R.F., and Swales, J.D. 1979. Pressor responsiveness to angiotensin II in renovascular and steroid hypertension. Clin Sci 57:47s–50s

Marks, E.S., Bing, R.F., Thurston, H., Swales, J.D. 1980. Vasodepressor property of the converting enzyme inhibitor captopril (SQ 14225): the role of factors other than renin-angiotensin blockade in the rat. Clin Sci 58:1–6

Miller, E.D., Samuels, A.I., Haber, E., Barger, A.C. 1972. Inhibition of angiotensin conversion in experimental renovascular hypertension. Science 177:1108–1109

Muirhead, E.E., Prewitt, R.L., Brooks, B., Brosius, W.L. 1978. Antihypertensive action of the orally active converting enzyme inhibitor (SQ 14225) in spontaneously hypertensive rats. Circ Res 43:Suppl 1, 53–59

Murthy, V.S., Waldron, T.L., Goldberg, M.E., Vollmer, R.R. 1977. Inhibition of angiotensin converting enzyme by SQ 14225 in conscious rabbits. Eur J Pharmacol 46:207–212

Murthy, V.S., Waldron, T.L., Goldberg, M.E. 1978. The mechanism of bradykinin potentiation after inhibition of angiotensin converting enzyme by SQ 14225 in conscious rabbits. Circ Res 43:Suppl 1,41–45

Ng, K.K.F., Vane, J.R. 1967. Conversion of angiotensin I to angiotensin II. Nature (London) 216:762–766

Oparil, S., Koerner, T., O'Donoghue, J.K. 1979. Mechanism of angiotensin I converting enzyme inhibition by SQ 20881 (GLu-TRP-Pro-Arg-Pro-GLn-IIe-Pro-Pro) in vivo. Hypertension 1:13–22

Ondetti, M.A., Rubin, B., Cushman, D.W. 1977. Design of specific inhibitors of angiotensin-converting enzyme: new class of orally active antihypertensive agents. Science 196:441–444

Pals, D.T., Masucci, F.D., Denning, G.S., Sipos, F., and Fessler, D.C. 1971. Role of the pressor action of angiotensin II in experimental hypertension. Circ Res 29:673–681

Peart, W.S. 1965. The renin-angiotensin system. Pharmacol Rev 17:143–178

Rubin, B., Laffan, R.J., Kotler, D.G., O'Keefe, E.H., DeManio, D.A., and Goldberg, M.E. 1977. Pharmacologic properties of SQ 14225, an orally active specific inhibitor or angiotensin converting enzyme (ACE). Fed Proc:1049

Reiger, A.J.G., Millar, J.A., Lever, A.F., Morton, J.J., and Slack, B. 1977. Correction of renal hypertension in the rat by prolonged infusion of angiotensin inhibitors. Lancet 2:1317–1319

Sen, S., Smeby, R.R., Merlin Bumpus, F., and Turcotte, J.G. 1979. Role of renin-angiotensin system in chronic renal hypotensive rats. Hypertension 1:427–434

Streeton, D.H.P., Anderson, G.H.A., Freiberg, J.M., and Dalakos, T.G. 1975. Use of an angiotensin II antagonist (saralasin) in the recognition of 'angiotensinogenic' hypertension. N Engl J Med 292:657–662

Thurston, H. 1976. Vascular angiotensin receptors and their role in blood pressure control. Am J Med 61:768–778

Thurston, H., and Laragh, J.H. 1975. Prior receptor occupancy as a determinant of the pressor activity of infused angiotensin II in the rat. Circ Res 36:113–117

Thurston, H., and Swales, J.D. 1974a. Action of angiotensin II antagonist and antiserum upon the pressor response to renin: further evidence for the local generation of angiotensin II. Clin Sci Mol Med 46:273–276

Thurston, H., and Swales, J.D. 1974b.

Comparison of angiotensin II antagonist and antiserum infusions with nephrectomy in the rat with two-kidney Goldblatt hypertension. Circ Res 35:325–329

Thurston, H., and Swales, J.D. 1977. Blood pressure response of nephrectomized hypertensive rats to converting enzyme inhibition: evidence for persistent vascular renin activity. Clin Sci Mol Med 52:299–304

Thurston, H., and Swales, J.D. 1978. Converting enzyme inhibitor and saralasin infusion in rats: evidence for an additional vasodepressor property of converting enzyme inhibitor. Circ Res 42:588–592

Thurston, H., Hurst, B.C., Bing, R.F., and Swales, J.D. 1978. Role of persistent vascular renin after bilateral nephrectomy in Goldblatt two kidney hypertension. Clin Sci Mol Med 55:23s–26s

Thurston, H., Swales, J.D., Bing, R.F., Hurst, B.C., and Marks, E.S. 1979. Vascular renin like activity and blood pressure maintenance in the rat: studies of the effect of changes in sodium balance, hypertension and nephrectomy. Hypertension 1:643–649

Thurston, H., Bing, R.F., Marks, E.S., and Swales, J.D. 1980. Response of chronic renovascular hypertension to surgical correction or prolonged blockade of the renin-angiotensin system by two inhibitors. Clin Sci 58:15–20

Captopril: Factors Involved in its Mechanism of Action

Michael J. Antonaccio, Magdi Asaad, Bernard Rubin and Zola P. Horovitz
The Squibb Institute for Medical Research
Princeton, New Jersey 08540

INTRODUCTION

Elevated plasma renin activity (PRA) has been clearly shown to play a causal role in the early stages of certain forms of renal hypertension (Gavras et al., 1976; Davis, 1977; Ganten and Gross, 1979). Good evidence for such a role comes from studies using competitive receptor antagonists of angiotensin II (AII) and antibodies to renin and AII (Thurston and Swales, 1974; Swales and Thurston, 1973; Gavras et al., 1973; Carretero and Gulati; 1978; Dzau et al., 1980). In models of hypertension in which PRA is normal or low, these agents are not effective in reducing blood pressure, once again suggesting that "normal" PRA does not contribute to the maintenance of hypertension. However, the efficacy of captopril in clinical and experimental hypertension of diverse etiologies having low, normal or high PRA has questioned the apparent relationship between PRA and blood pressure and helped show it to be an untenably facile one.

Captopril is a highly specific and potent inhibitor of angiotensin converting enzyme (ACE). In reasonable concentrations that essentially abolish ACE activity, captopril has no reported inhibitory effects on receptor activation by any tested agonist, does not alter responses to sympathetic or parasympathetic nerve stimulation, nor does it enter the central nervous system (see Rubin et al., 1978a). Captopril is effective in virtually all forms of hypertension including human renal and essential hypertension with low, normal or high PRA (Rubin and Antonaccio,

1980), acute and chronic 2 kidney, 1 clip renal hypertension (Bengis et al., 1978; Rubin et al., 1978b; Antonaccio et al., 1979a; Sen et al., 1979), 2 kidney, 2 clip renal hypertension (R. Ferrone, personal communication), 1 kidney, 1 clip renal hypertension (Bengis et al., 1978; Sen et al., 1979; Rubin et al., 1980), 2 kidney, 2 wrapped renal hypertension (Rubin et al., 1978a), aortic coarctation hypertension (Laffan et al., 1978), spontaneously hypertensive rats (SHR) (Muirhead et al., 1978; Laffan et al., 1978; Antonaccio et al., 1979; Sen et al., 1979, 1980; Ferrone and Antonaccio, 1980) and stroke-prone SHR (Watanabe and Sokabe, 1979). Because there is no clear relationship between PRA before, and blood pressure reduction after, captopril, many authors have suggested that the drug may have activity independent of the renin-angiotensin system or even of ACE inhibition (Vollmer et al., 1978; Jandhyala et al., 1978; 1979; Larochelle et al., 1979; Marks et al., 1980; Thurston et al., 1980; Tree and Morton, 1980).

Captopril causes a marked reduction of blood pressure in both chronic renal hypertensive rats and SHR when PRA levels are not elevated, whereas AII antagonists are ineffective (see Rubin and Antonaccio, 1980). Thurston et al. (1980) and Marks et al. (1980) have suggested that this might be due to bradykinin potentiation by captopril. Since bradykinin releases prostaglandins, especially renal, and since this effect is also enhanced by captopril (Murthy et al., 1978; Moncada et al., 1979), it has been suggested that the antihypertensive action of captopril might be due, at least in part, to prostaglandin formation and release (Swartz et al., 1979; Crantz et al., 1979).

Recently, the vascular renin-angiotensin system has been implicated in the maintenance of hypertension (Garst et al., 1979; Asaad and Antonaccio, 1979; Thurston et al., 1980). Vascular renin levels are higher than respective normotensive controls in chronic 2 kidney, 1 clip renal hypertensive rats as well as SHR in which PRA is normal (Garst et al., 1979; Asaad and Antonaccio, 1979; Thurston et al., 1979; Antonaccio et al., 1980).

The purpose of this study was to examine the potential role of prostaglandin formation and release, and inhibition of the vascular renin-angiotensin system in the antihypertensive actions of captopril in various animal models.

METHODS

Blood Pressure Determinations

Spontaneously Hypertensive Rats

Twenty SHR and 20 WKY normotensive rats (NTR) (Charles River) were individually housed for at least one week with food and water ad lib. For the first part of the study, the animals were randomly divided into four groups. A group of SHR and WKY were treated with 100 mg/kg captopril by gavage daily for 8 days; other groups of SHR and WKY (controls) received 4 ml/kg water by gavage daily for 8 days. All the animals were weighed daily, and their systolic blood pressure was recorded indirectly using the tail cuff procedure as previously described (Antonaccio and Cavaliere, 1974). Immediately following the final blood pressure determination, the animals were anesthetized with sodium pentobarbital (35 mg/kg, i.p.). Approximately 5 ml blood was obtained without hemolysis by heart puncture in a syringe moistened with heparin (200 units/ml).

In another study, four groups of SHR were used. One group received captopril alone (30 mg/kg p.o. daily for 2 days). Another received MK–447 (2-aminoethyl-4-[1,1-dimethylethize]-6-iodophenol HC1) (2.5 mg/kg p.o. daily on 2 days), a third group received captopril plus MK–447, while a fourth group was pretreated with indomethacin (5 mg/kg p.o.) one hour prior to the combination of captopril and MK–447. All pressures were recorded directly in conscious rats according to the method of Weeks and Jones (1960) as modified by Laffan et al. (1972); see below).

In other studies, mean arterial blood pressure (MBP) of SHR was measured approximately 18 days before each of the dosage intervals described above. Indwelling abdominal aortic cannulas were implanted in rats anesthetized with pentobarbital sodium according to the method of Weeks and Jones (1960). After the 18-day recovery period, direct mean arterial blood pressures of conscious rats were recorded by the method of Laffan et al. (1972) modified as follows. The signal from the transducer was digitized in a 10 list A/D converter with input to PDP 11/05 computer. The program was designed to sense and store samples at a rate of 125/sec for each rat. These parameters were averaged and stored as the MBP. Data were obtained from each rat every 5 min for as long as described. Six such sets of data were averaged to give a mean value representing a 30 min sample and this 30 min figure was stored for subsequent analysis. The data were transferred serially to a host computer (PDP 11/40) for final analysis.

DOCA-Salt Hypertensive Rats

Male Sprague-Dawley rats, 200 to 250 gm, were fed a standard diet (Purina Laboratory Chow) and were allowed tap water ad lib. for 13 days following receipt. On day 14, a right unilateral nephrectomy was performed on all rats and, following recovery from pentobarbital anesthesia, 35 mg/kg i.p., the rats were placed on 1% NaCl solution as a replacement for drinking water until the end of the test period. Two days after nephrectomy and once a week thereafter for a total of 5 wk all rats received a subcutaneous injection of 50 mg desoxycorticosterone acetate (DOCA)/kg, in a 0.25% agar suspension. During the third week of dosage with DOCA, the rats were surgically prepared for blood pressure measurement according to the method of Weeks and Jones (1960), and approximately 2 wk later were dosed with drugs as described in the text.

Vascular Tissue Preparation and Renin Determinations

In the anesthetized rats, a midline incision was made, and the major vessels (abdominal aorta and vena cava) with their branches were exposed by careful, gentle dissection, trimmed of all connective tissue and then opened longitudinally, in situ. Immediately, the opened vessels were flushed with ice cold saline and traces of coagulated blood were removed manually. The vessels were then removed, washed in four successive ice cold saline solutions containing EDTA (1.5 mg/ml), and blotted over filter paper between each wash. The blot-dried tissues were then weighed, frozen using a dry ice–acetone mixture, and stored at −70°C until used.

For renin determinations, the vascular tissues were thawed, then cut into small pieces with a razor blade. The process of freezing and thawing was repeated several times. The tissues were then homogenized in ice cold normal saline containing EDTA (1.5 mg/ml) with a motor driven glass homogenizer fitted with a teflon pestle. During the homogenization, the tissues were kept over ice at all times. Aliquots of the homogenate were incubated with the renin substrate at 37°C; others were kept at 4°C.

Vascular renin concentrations as well as PRA were determined by radioimmunoassay utilizing the Squibb Angiotensin-I Immutope Assay Kit as previously described (Harris et al., 1978). Vascular tissue renin concentration was expressed as ng AI/g wet tissue/hr.

Preparation of Renin Substrate

Renin substrate was prepared according to a modification of the procedure by Boucher et al. (1967). Male Sprague-Dawley rats (250–300 gm) were nephrectomized by a bilateral flank incision. The kidney was decapsulated to avoid adrenalectomy and the renal pedicle was isolated. Twenty-four hours later, the rats were anesthetized with sodium pentobarbital (35 mg/kg i.p.), and blood was obtained without hemolysis by heart puncture in syringes moistened with 15% EDTA or heparin (200 units/ml). The blood was immediately placed over ice and then centrifuged at 0–5°C for 10 min at 10,000 rpm. To each 100 ml plasma, 4.7 g of solid EDTA was added and, while stirring slowly at room temperature, the pH of the plasma solution was adjusted to 8 by the dropwise addition of 5N NaOH. An hour later, the solution was cooled to 0–5°C and adjusted to pH 5.3 by the slow addition of 2.5N H_2SO_4. The substrate was salted out by the addition of solid $(NH4)_2SO_4$ (2.3 M) stirring for 10 min. The mixture was then centrifuged at 2000 x g for 1 hr and the precipitate was suspended in, and dialyzed against, cold

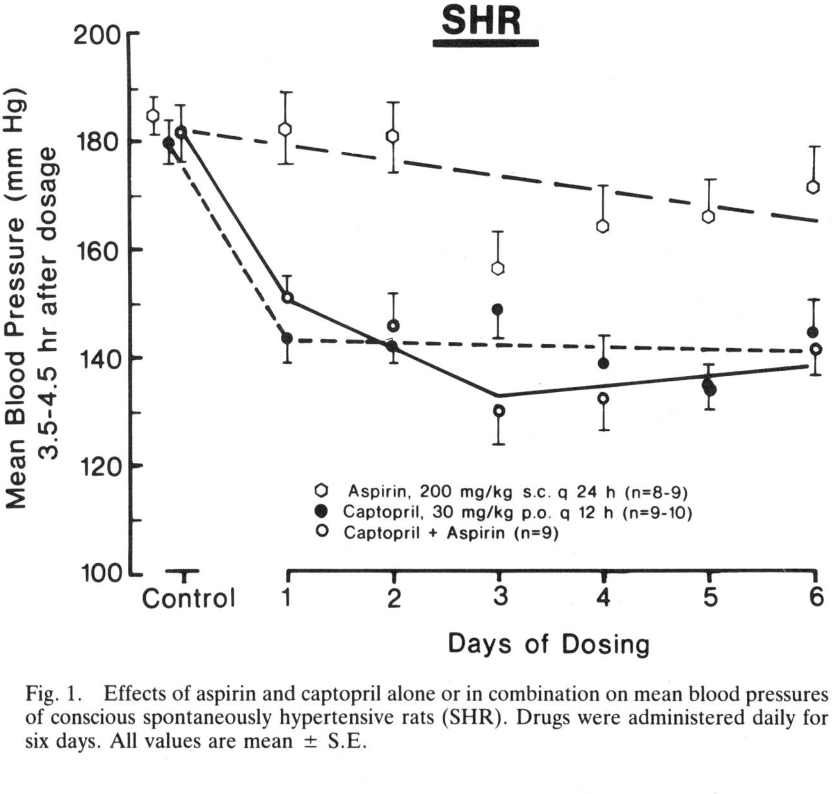

Fig. 1. Effects of aspirin and captopril alone or in combination on mean blood pressures of conscious spontaneously hypertensive rats (SHR). Drugs were administered daily for six days. All values are mean ± S.E.

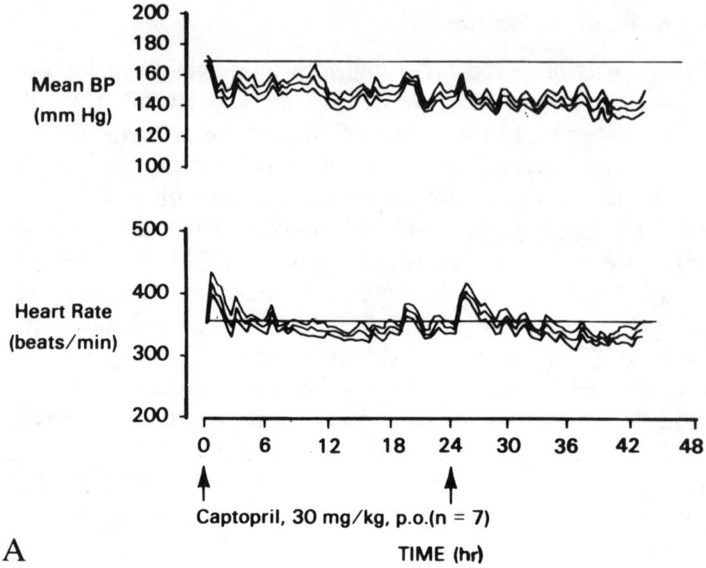

A

Fig. 2, A and B. Effects of captopril alone (*A*), and MK–447 alone (*B*), on mean blood pressures and heart rates of conscious SHR. All treatments were once daily for two consecutive days. The middle line is the mean bracketed by ± one S.E. of the mean.

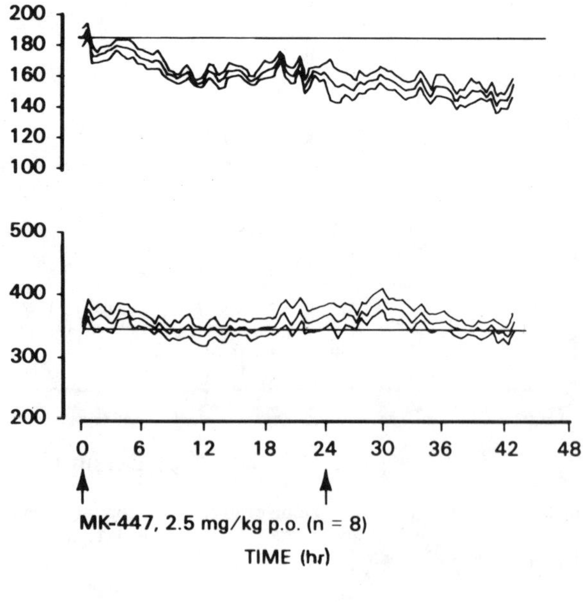

B

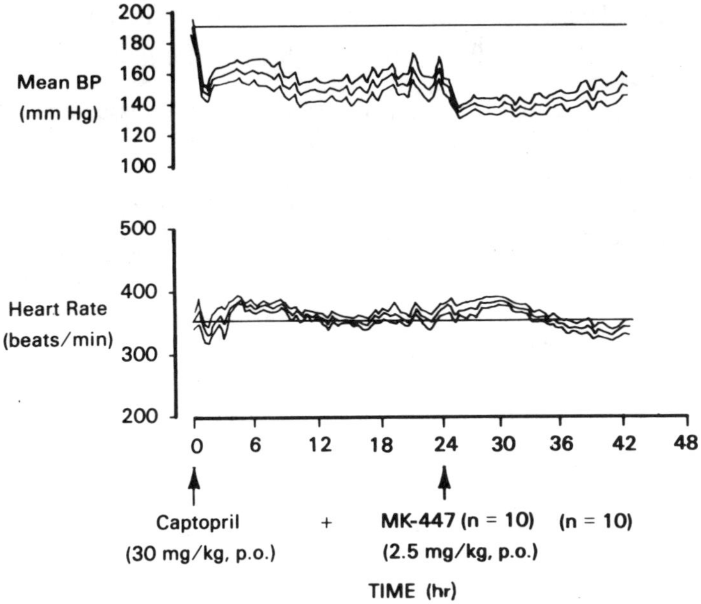

C

Fig. 2, C and D. Effects of the combination of captopril plus MK–447 (*C*), and preptreatment with indomethacin for one hour prior to the combination of captopril plus MK–447 (*D*) on mean blood pressure and heart rates of conscious SHR. All treatments were once daily for two consecutive days. The middle line is the mean bracketed by ± one S.E. of the means.

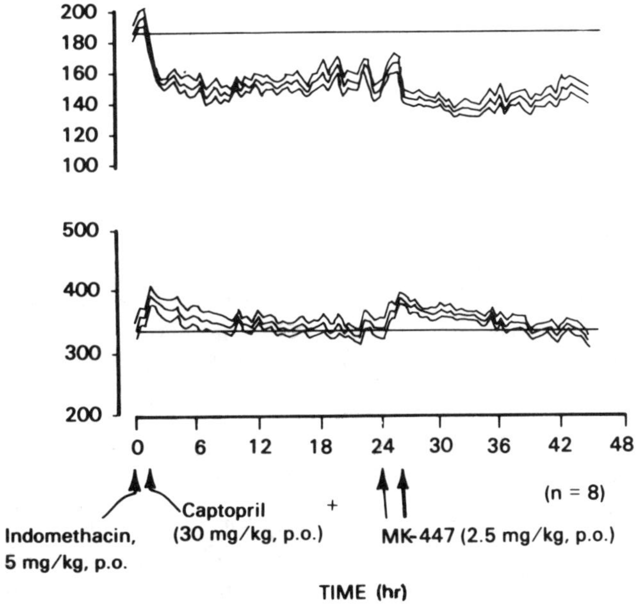

D

distilled water for 24 hr. The precipitate in the dialysis bag was removed by filtration and the solution was lyophilized. The dried substrate was stored at −20°C.

To ensure that the renin concentration in the tissue homogenates was the limiting factor in the assay, maximal substrate concentrations were used.

It was determined that 0.5 ml of the following renin substrate mixture was an adequate concentration when used with 1.0 ml of vascular tissue homogenate (10–50 mg/ml).

Renin Substrate Mixture	
Lyophilized substrate	0.750 g
Sodium azide (1%)	0.200 ml
HCl (1N)	0.025 ml
NaOH (1N)	0.025 ml
H_2O	14.500 ml

Plasma Na^+, K^+ and Protein Determination

Plasma Na^+ and K^+ concentrations were determined using the Flame Photometer 343 (Instrumentation Laboratory Inc.); plasma protein concentration was determined according to the procedure of Lowry et al. (1951).

RESULTS

Effects of Aspirin and Captopril Alone and in Combination on Mean Blood Pressure (MBP) of SHR

In SHR dosed daily with captopril (100 mg/kg p.o.) for 6 days, MBP was significantly reduced throughout the dosing period (Fig. 1). In contrast, treatment of SHR with aspirin alone (200 mg/kg s.c.) during this period had no significant effect on MBP (Fig. 1). Treatment of SHR with aspirin did not reduce the antihypertensive effects of captopril when both drugs were administered simultaneously to SHR (Fig. 1).

Effects of MK–447 and Captopril Alone and in Combination on MBP of SHR

Captropril (30 mg/kg p.o. daily for 2 days) caused a modest reduction of MBP in SHR (Fig. 2*A*). MK–447 (2.5 mg/kg p.o. daily for 2 days) had similar effects (Fig. 2*B*). The combination of captopril and MK–447

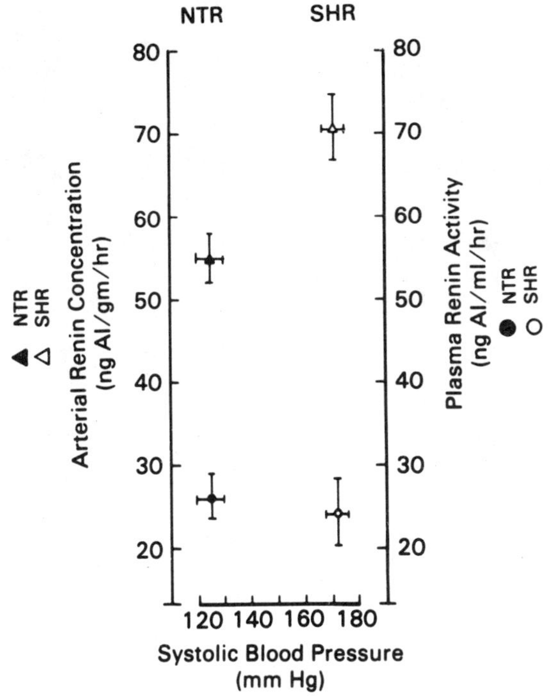

Fig. 3. The relationships among arterial renin concentration, plasma renin activity and systolic blood pressure of SHR and Wistar-Kyoto normotensive rats (NTR).

caused marked reductions in MBP which were substantially greater than either drug alone (Fig. 2*C*). Pretreatment of SHR with indomethacin (5 mg/kg p.o.) for 1 hr prior to the combination of MK–447 and captopril on two successive days caused only minor and insignificant reductions in MBP compared to the changes caused by the combination of MK–447 and captopril without indomethacin (Fig. 2*D*; *cf* Fig. 2*C*).

Vascular Renin Concentration and Plasma Renin Activity Before and After Captopril in SHR and DOCA-Salt Hypertensive Rats

SHR

The relationships among aortic renin concentration (ARC), plasma renin activity (PRA) and systolic blood pressure (SBP) of SHR and Wistar-Kyoto normotensive rats (NTR) are shown in Figure 3. PRA of SHR was slightly lower than NTR whereas ARC for SHR was significantly higher ($P<0.01$) than NTR and correlated well with SBP.

In SHR and NTR treated daily for 8 days with captopril (100 mg/kg p.o.), SBP was significantly reduced in comparison with water-vehicle

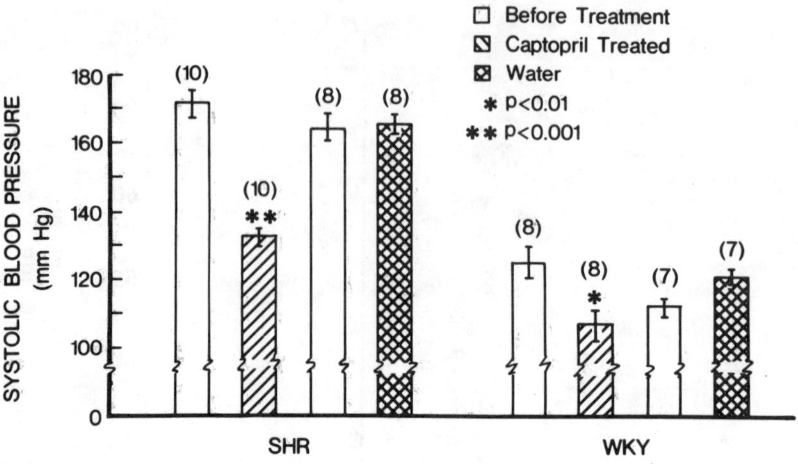

Fig. 4. Effects of captopril (100 mg/kg p.o. daily for 8 days) or water vehicle treatment on systolic blood pressures of conscious SHR and Wistar-Kyoto normotensive rats (WKY). Numbers in parentheses indicate number of rats per group.

treated rats (Fig. 4). ARC of untreated SHR was higher than in untreated NTR whereas venous renin concentration (VRC) was not significantly different between SHR and NTR (Fig. 5). After captopril, both ARC and VRC of SHR and NTR were significantly increased, the difference in ARC being significantly greater ($P<0.05$) in NTR than in SHR (Fig. 5).

Plasma concentrations of sodium, potassium and protein were not different in NTR and SHR either before or after captopril (Fig. 6).

DOCA-Salt Hypertensive Rats

In different groups of DOCA-salt hypertensive rats, neither captopril (100 mg/kg p.o.) nor hydrochlorothiazide (HCTZ; 10 mg/kg p.o.) alone given daily for 4 days had any substantial or consistent effect on mean arterial blood pressure (MBP) or heart rate (Fig. 7*A* and *B*). Only the results of the first day of dosing are shown for convenience since they are representative of the four-day results. In contrast to the lack of effect of either captopril or HCTZ alone on MBP of these rats, the combination of captopril plus HCTZ caused a significant and reproducible reduction in MBP (Fig. 7*C*).

Both PRA and ARC were markedly reduced in DOCA-salt hypertensive rats (Fig. 8, V) in comparison to their normotensive Sprague-Dawley controls (Fig. 8, NONE). Similarly, kidney renin concentration

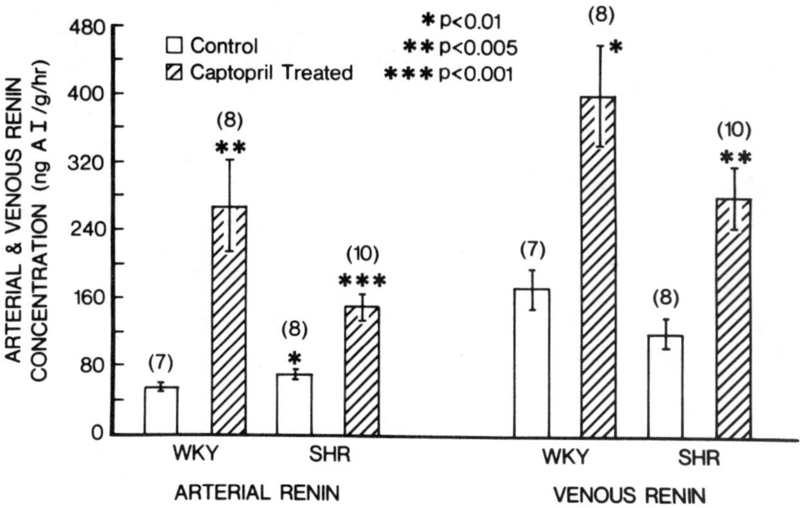

Fig. 5. Arterial and venous renin concentrations of SHR and WKY after either water (control) or captopril treatment for 8 days as described in Figure 4. Numbers in parentheses indicate number of rats per group.

was reduced from 666 ± 46 ng AI/mg kidney/hr in controls to 52 ± 6 in DOCA-salt rats ($P<0.001$). Neither captopril nor HCTZ alone or in combination had any significant effect on PRA of DOCA-salt rats (Fig. 8). In contrast, both captopril and HCTZ increased ARC in DOCA-salt rats, and the combination of captopril plus HCTZ caused the largest increase in ARC (Fig. 8). Obviously, the degree of ARC increase could not be accounted for by changes in PRA.

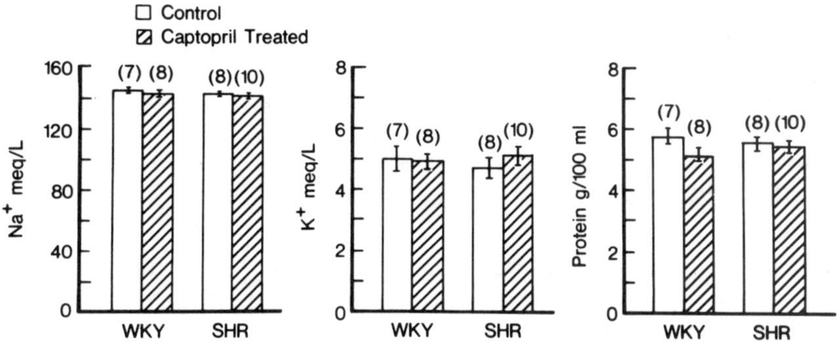

Fig. 6. Plasma sodium (Na^+), potassium (K^+) and protein concentrations in SHR and WKY rats after either water (control) or captopril treatment for 8 days as described in Figure 4. Numbers in parentheses indicate number of rats per group.

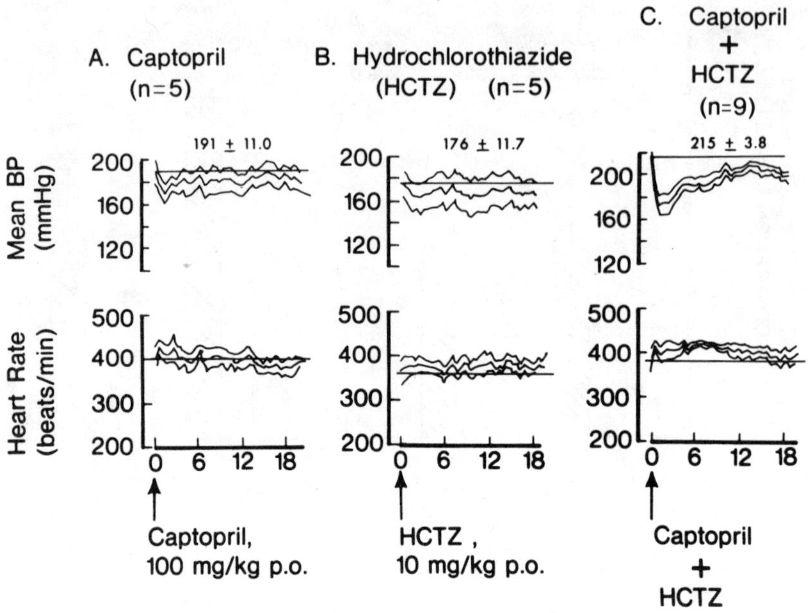

Fig. 7. Effects of captopril (A) or hydrochlorothiazide (B) alone or in combination (C) on mean blood pressures and heart rates of conscious DOCA-salt hypertensive rats. Numbers in parentheses indicate number of rats per group. Numbers in the middle of each panel indicate starting blood pressure.

Plasma sodium concentration was slightly elevated in DOCA-salt rats in comparison to controls, as well as in the DOCA-salt rats receiving captopril and HCTZ alone or in combination (Fig. 9). DOCA-salt rats had markedly lower plasma potassium concentrations than controls, which were not further altered by captopril or HCTZ alone or in combination (Fig. 9). Protein concentrations were similar in control and DOCA-salt rats, regardless of whether the latter were treated with captopril or HCTZ alone or in combination (Fig. 9).

DISCUSSION

Because of captopril's efficacy in most forms of hypertension regardless of etiology or PRA, several authors have suggested that captopril has an action independent of the renin-angiotensin system and/or ACE inhibition (*see* Introduction, this chapter, for references). Since ACE is the same enzyme as kininase II, which degrades kinins into inactive metabolites, there is an obvious possibility that inhibition of ACE could result in accumulation of kinins either circulating in plasma (McCaa et al., 1978) or locally in organs such as the kidneys (Olsen and Arrigoni-Martelli, 1979). Furthermore, bradykinin releases vasodilating and natriuretic prostaglandins, including PGE_2 and prostacyclin, from renal and other tissues (McGiff et al., 1975; Terragno and Terragno, 1980), an effect which is enhanced by captopril (Murthy et al., 1978; Moncada et al., 1979). Therefore, a potential role for at least the partial mediation

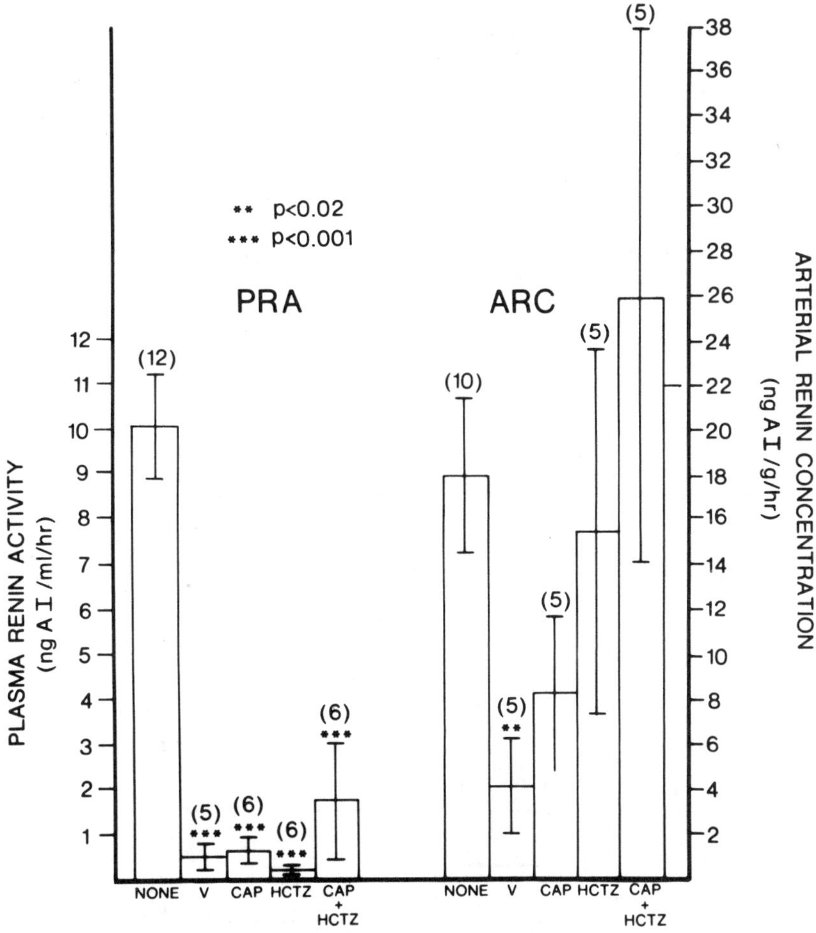

Fig. 8. Plasma renin activity (left side) and arterial renin concentration (right side) of normotensive, untreated Sprague-Dawley rats (NONE) and DOCA-salt hypertensive rats given either water vehicle (V), captropril (CAP), hydrochlorothiazide (HCTZ) or the combination of captropril and hydrochlorothiazide (CAP + HCTZ). Numbers in parentheses indicate the number of animals per group.

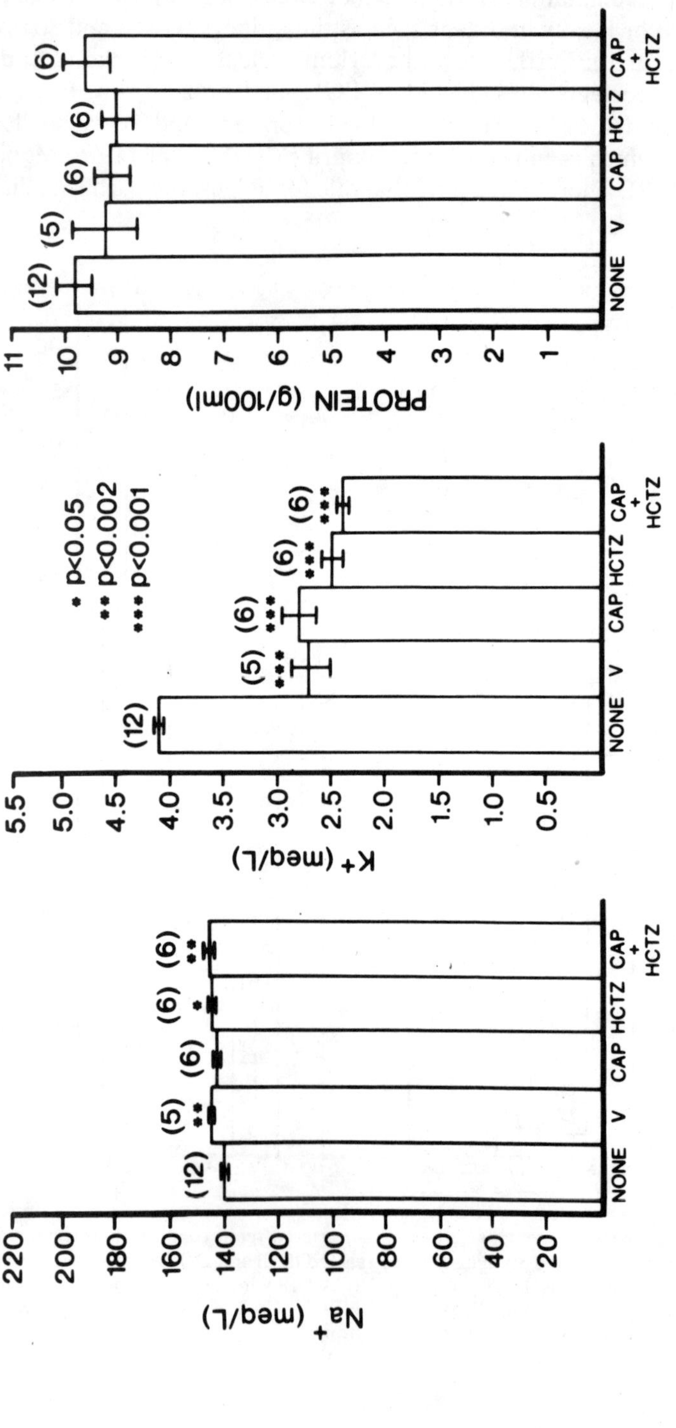

Fig. 9. Plasma concentrations of sodium (Na^+), potassium (K^+) and protein in DOCA-salt hypertensive rats. See Figure 8 for details.

of the hemodynamic actions of captopril exists and was examined in the present study. No evidence for a prostaglandin involvement in the antihypertensive effects of captopril in SHR was found. Inhibition of prostaglandin synthesis by aspirin was without effect on the reduction in blood pressure normally observed after captopril in SHR. These results are in good agreement with the authors' previous study in both SHR and NTR which additionally showed that indomethacin did not inhibit the increase in renin release caused by captopril (Antonaccio et al., 1979a). In other studies, indomethacin pretreatment actually enhanced the increase in renal blood flow caused by captopril in SHR and normotensive dogs (Antonaccio et al., 1980b; Tobia and Giardino, 1980).

MK–447 (2-aminoethy1-4-[1,1-dimethylethyl]-6-iodophenol HCl) is a diuretic and antihypertensive agent which stimulates the conversion of arachidonic acid to PGE_2 (Kuehl et al., 1978; Scriabine et al., 1979). In SHR, the antihypertensive action of MK–447 was partially antagonized by indomethacin, suggesting that prostaglandins play a role in the mediation of its effects (Scriabine et al., 1979). In the present study, MK–447 was also found to modestly decrease blood pressure of SHR and to enhance the antihypertensive actions of captopril, the latter effect probably a consequence of its diuretic action (Rubin et al., 1978b; Antonaccio et al., 1979). Prior treatment of SHR with indomethacin decreased the antihypertensive action of the combination of MK–447 plus captopril only to the extent that has previously been reported for the inhibition of the MK–447 response alone. These data, as well as the fact that urinary prostaglandin excretion after captopril in both animals and man is not altered by captopril (Bravo and Tarazi, 1979; Olsen and Arrigoni-Martelli, 1979), provide good evidence that prostaglandins play little or no role in the hemodynamic actions of captopril. Other indirect evidence supporting the lack of prostaglandin involvement derives from the effects of AI and AII on prostaglandin release. Injection of both AI and AII releases prostacyclin-like material into the blood of cats (Gryglewski et al., 1979). This effect of AI is prevented by captopril, suggesting that the prostacyclin release was a result of AII formation. Therefore, reduction of AII levels by captopril might be expected to *reduce* prostaglandin formation caused by this substance. The relevance of such an effect to any of captopril's actions is unknown but is certainly contrary to the suggestion that prostaglandin formation is enhanced after the drug.

In other studies, captopril decreased blood pressure in nephrectomized fluid or volume depleted hypertensive patients, and in animals with prior AII receptor blockade by saralasin (Vollmer et al., 1978; Jandhyala et al., 1978; Man In't Veld et al., 1979, 1980). Thus the suggestion has been made by several authors that captopril might have an antihypertensive action independent of ACE inhibition. Analogously, captopril decreases blood pressure in models of hypertension such as chronic renal hypertension and SHR in which PRA is normal and does not appear to play an obviously causal role in maintaining the hypertension (Rubin et al., 1978b; Antonaccio et al., 1979a,b). However, recent studies have shown that arterial renin concentration (ARC) in chronic 2 kidney, 1 clip renal hypertension and SHR are elevated relative to normotensive controls at a time when PRA is normal, suggesting a causal relationship between ARC and blood pressure (Thurston et al., 1979; Garst et al., 1979; Asaad and Antonaccio, 1979). The present study deomonstrates that arterial, but not venous, wall renin concentration is higher in SHR than in their genetically comparable normotensive counterpart, the Wistar-Kyoto rat, despite similar PRA in both groups. Furthermore, systolic blood pressure correlated with ARC but not PRA supporting a causal relationship. Captopril treatment of SHR for 8 days decreased SBP and elevated both PRA and ARC, as well as VRC, suggesting that vascular renin was under the same negative feedback inhibition by AII as plasma renin. The major portion of ARC in SHR apparently was not derived from renal renin since the levels did not change appreciably during a 24-hr period following bilateral nephrectomy (Antonaccio et al., 1980), results similar to those reported for 2 kidney, 1 clip renal hypertensive rats (Thurston et al., 1979).

In DOCA-salt hypertensive rats, both PRA and ARC were markedly depressed, as was kidney renin concentration, in comparison with normotensive controls. Neither captopril nor HCTZ alone had any pronounced effect on mean blood pressure in DOCA-salt rats whereas the combination of captopril with HCTZ was consistently antihypertensive. Gavras et al. (1975) have previously reported low PRA values in DOCA-salt hypertensive rats and also that the AII antagonist Sar[1], ala[8] actually increased blood pressure in these animals. In the present study, PRA was not significantly altered by treatment with either captopril or HCTZ alone or in combination. In contrast, both captopril and HCTZ caused elevations of ARC, the greatest increase occurring with the combination of both drugs. Based on these data, it is tempting to suggest that the elevation of ARC by HCTZ, and the consequent inhibition of local formation of AII by simultaneous captopril administration, was responsible for the antihypertensive action of the combination therapy.

Therefore, the authors feel that the antihypertensive action of captopril in normal or low plasma renin models is not inconsistent with its ability to inhibit AII formation. Rather, AII formation might be localized in the vasculature under certain circumstances and not necessarily be reflected in circulating levels or by PRA. This suggestion is supported by the finding that AII infusions over a period of days can slowly increase systemic arterial pressure without raising plasma levels of AII (Dickinson and Lawrence, 1963; Bean et al., 1979).

SUMMARY AND CONCLUSIONS

The authors' findings, as well as those of others, do not support the possibility that prostaglandins play any role in the hemodynamic effects of captopril in SHR. Furthermore, our research shows that elevated vascular renin activity may be important in the maintenance of normal or low reninemic hypertension. Finally, captopril may decrease blood pressure in normal or low reninemic hypertension by inhibiting the vascular renin-angiotensin system.

REFERENCES

Antonaccio, M.J., and Cavaliere, T. 1974. A comparison of the effects of some inotropic and chronotropic agents on isolated atria from normotensive (NTR) and spontaneously hypertensive (SHR) rats. Arch Int Pharmacodyn Ther 209:273–282

Antonaccio, M.J., Harris, D., Goldenberg, H., High, J.P., and Rubin, B. 1979a. The effects of captopril, propranolol and indomethacin on blood pressure and plasma renin activity in spontaneously hypertensive and normotensive rats. Proc Soc Exp Bio Med 162:429–433.

Antonaccio, M.J., Rubin, B., Horovitz, Z.P., Laffan, R.J., Goldberg, M.E., High, J.P., Harris, D.N., and Zaidi, I. 1979b. Effects of chronic treatment with captopril (SQ 14,-225), an orally-active inhibitor of angiotensin-I-convertiing enzyme, in spontaneously hypertensive rats. Jap J Pharmacol 29:285–294

Antonaccio, M.J., Rubin, B., Horovitz, Z.P., Mackaness, G., and Panasevich, R. 1979c. Long-term efficacy of captopril (SQ 14,225) in 2-kidney renal hypertensive rats. Clin Exp Hypertension 1:505–519

Antonaccio, M.J., Asaad, M.M., High, J., Schaeffer, T., and Rubin, B. 1980a. The role of vascular renin and kidneys in the antihypertensive action of captopril in spontaneously hypertensive rats (SHR). Fed Proc 34:813

Antonaccio, M.J., Rubin, B., and Horovitz, Z.P. 1980b. Captopril in animal models of hypertension. Clin Exp Hypertension 2:613–638

Asaad, M.M., and Antonaccio, M.J. 1979. The effects of captopril (SQ 14,225) on vascular tissue renin in spontaneously hypertensive (SHR) and WKY-normotensive (NTR) rats: relationship to systolic blood pressure (SBP) and plasma renin activity (PRA). Pharmacologist 21:212

Bean, B.L., Brown, J.L., Casals-Stenzel, J., Fraser, R., Lever, A.F., Millar, J.A., Morton, J.J., Petch, B., Riegger, A.J.G., Robertson, J.I.S., and Tree, M. 1979. The relation of arterial pressure and plasma angiotension II concentration. Circ Res 44:452–458

Bengis, R.G., Coleman, T.G., Young, D.B., and McCaa, R.E. 1978. Long-term blockade of angiotensin formation in various normotensive and hypertensive rat models using converting enzyme inhibitor (SQ 14,-225). Circ Res 43 (Suppl I):45–53

Boucher, R., Menrad, J., and Genest, J. 1967. A micromethod of measurement of renin in the plasma and kidney of rats. Canad J Physiol Pharmacol 45:881–890

Bravo, E.L., and Tarazi, R.C. 1979. Converting enzyme inhibition with an orally active compound in hypertensive man. Hypertension 1:39–46

Carretero, O.A., and Gulati, O.P. 1978. Effects of angiotensin antagonist in rats with acute, subacute, and chronic two-kidney renal hypertension. J Lab Clin Med 91:264–271

Crantz, F.R., Swartz, S.L., Hollenberg, N.K., Moore, T.J., Bluhy, R.G., Levine, L., and Williams, G.H. 1979. Role of prostaglandins in the hypotensive response to captopril in essential hypertension. Clin Res 27:592A

Davis, J.O. 1977. The pathogenesis of chronic renovascular hypertension. Circ Res 40:439–444

Dickinson, C.J., and Lawrence, J.R. 1963. A slowly developing pressor response to small concentrations of angiotensin. Lancet 1:1354–1356

Dzau, V.J., Kopelman, R.I., Barger, C.A., and Haber, E. 1980. Renin-specific antibody for study of cardiovascular homeostasis. Science 207:1091–1093

Ferrone, R.A., and Antonaccio, M.J. 1979. Prevention of the development of spontaneous hypertension in rats by captopril (SQ 14,225). Eur J Pharmacol 60:131–137

Ganten, D., and Gross, F. 1977. Interferences with the renin-angiotensin system and their effect on high blood pressure. In *Antihypertensive Agents*, ed. Franz Gross, pp. 518–546. New York Springer-Verlag

Garst, J.B., Koletsky, S., Wisenbaugh, P.E., Hadady, M., and Matthews, D. 1979. Arterial wall renin in the hypertensive rat. Clin Sci 56:41–46

Gavras, H., Brunner, H.R., Darracott-Vaughn, Jr., E., and Laragh, J.H. 1973. Angiotensin-sodium interaction in blood pressure maintenance of renal hypertensive and normotensive rats. Science 180:1369–1372

Gavras, H., Brunner, H.R., Laragh, H.J., Darracott-Vaughan, Jr., E., Koss, M., Cote, L.J., and Gavras, I. 1975. Malignant hypertension resulting from deoxycorticosterone acetate and salt excess. Circ Res 36:300–309

Gavras, H., Oliver, J.A., and Cannon, P.J. 1976. Interrelationships of renin, angiotensin II, and sodium in hypertension and renal failure. Ann Rev Med 485–521

Gryglewski, R.J., Korbut, R., and Splawinski, J. 1979. Endogenous mechanisms which regulate prostacyclin release. Haemostasis 8:294–299

Harris, D.N., Heran, C.L., Goldenberg, H.J., High, J.P., Laffan, R.J., Rubin B., Antonaccio, M.J., and Goldberg, M.E. 1978. Effects of SQ 14,225, an orally active inhibitor of angiotensin-converting enzyme on blood pressure, heart rate and plasma renin activity of conscious normotensive dogs. Eur J Pharmacol 51:345–349

Jandhyala, B.S., Washington, G.F., and Lokhandwala, M.F. 1978. Influence of inhibitor of angiotensin I converting enzyme with SQ 14,225 on thc arterial blood pressure of mongrel dogs. Res Comm Chem Pathol Pharmacol 22:257–265

Jandhyala, B.S., Nandiwada, P., Buckley, J.P., and Lokhandwala, M.F. 1979. Studies on the mechanism of the hypotensive action of

SQ 14,225 an angiotensin-converting enzyme inhibitor in anesthetized dogs. Res Comm Chem Pathol Pharmacol 25:429–446

Kuehl, F.A., Jr., Egan, R.W., Humes, J.L., Beveridge, G.C., and VanArman, C.G. 1978. Evidence for a pivotal role of the endoperoxide, PGG_2, in inflammatory processes. In *Biological Aspects of Prostaglandins and Thromboxanes*, eds. J. Fried and N. Kharasch, pp. 55–74. New York: Academic Press

Laffan, R.J., Peterson, A., Hitch, S.O., and Jeunelot, C. 1972. A technique for prolonged, cointinuous recording of blood pressure of unrestrained rats. Cardiovasc Res 6:319–324

Laffan. R.J., Goldberg, M.E., High J.P., Schaeffer, T.R., Waugh, M.H., and Rubin, B. 1978. Antihypertensive activity in rats of SQ 14,255, an orally active inhibitor of angiotensin converting enzyme. J Pharmacol Exp Ther 204:281–288

Larochelle, P., Genest, J., Kuchel, O., Boucher, R., Gutkowska, Y., and McKinstry, D. 1979. Effect of captopril (SQ 14,225) on blood pressure, plasma renin activity and angiotensin I converting enzyme activity. Can Med Assoc J 121:309–316

Lowry, O.H., Rosebrough, H.J., Farr, A.L., and Randall R.J. 1951. Protein measurement with the Folin phenol reagent. J Biol Chem 193:265–275

Man in't Veld, A.J., Wenting, G.J., and Schalekamp, M.A.D.H. 1979. Does captopril lower blood pressure in anephric patients? Br Med J 2:1110

Man in't Veld, A.J., Schicht, I.M., Derkx, F.H.M., de Bruyn, J.H.B., and Schalekamp, M.A.D.H. 1980. Effects of an angiotensin converting enzyme inhibitor (captopril) on blood pressure in anephric patients. Br Med J 1:288–290

Marks, E.S., Bing, R.F., Thurston, H., and Swales, J.D. 1980. Vasodepressor property of the converting enzyme inhibitor captopril (SQ 14,225): the role of factors other than renin-angiotensin blockade in the rat. Clin Sci 58:1–6

McCaa, R.E., Hall, J.E., and McCaa, C.S. 1978. The effects of angiotensin I-converting enzyme inhibitors on arterial blood pressure and urinary sodium excretion. Role of the renal renin-angiotensin and kallikrein-kinin systems. Circ Res 43(Suppl I):32–39

McGiff, J.C., Itskovitz, H.D., and Terragno, N.A. 1975. The actions of bradykinin and eledoisin in the canine isolated kidney: relationships to prostaglandins. Clin Sci Mol Med 49:125–131

Moncada, S., Mullane, K.M., and Vane, J.K. 1979. Prostacyclin-release by bradykinin in vivo. Br J Pharmacol 66:96P–97P

Muirhead, E.E., Prewitt, R.L., Jr., Brooks, B., and Brosius, W.L., Jr. 1978. Antihypertensive action of the orally active converting enzyme inhibitor (SQ 14,225) in spontaneously hypertensive rats. Circ Res 43(Suppl I):53–59

Murthy, V.S., Waldron, T.L., and Goldberg, M.E. 1978. The mechanism of bradykinin potentiation after inhibition of angiotensin-converting enzyme by SQ 14,225 in conscious rabbits. Circ Res 43(Suppl. I):40–45

Olsen, U.B., and Arrigoni-Martelli, E. 1979. The effects of kininase II inhibition by SQ 14,225 on kidney kallikrein-kinin and prostaglandin systems in conscious dogs. Eur J Pharmacol 54:229–234

Rubin, B., and Antonaccio, M.J. 1980. Captopril. In *The Pharmacology of Antihypertensive Drugs*, ed. Alexander Scriabine, pp. 21–42. New York: Raven Press

Rubin, B., Antonaccio, M.J., Goldberg, M.E., Harris, D.N., Itkin, A.G., Horovitz, Z.P., Pansevich, R.E., and Laffan, R.J. 1978a. Chronic antihypertensive effects of captopril (SQ 14,225), an orally active angiotensin I-converting enzyme inhibitor, in conscious 2-kidney renal hypertensive rats. Eur J Pharmacol 51:377–388

Rubin, B., Antonaccio, M.J., and Horovitz, Z.P. 1978b. Captopril (SQ 14,225) (D-3-mercapto-2-methylpropanoyl-L-proline): a novel orally active inhibitor of angiotensin-converting enzyme and antihypertensive agent. Progr Cardiovasc Dis 21:183–194

Sen, S., Smeby, R.R., Bumpus, F.M., and Turcotte, J.G. 1979. Role of renin-angiotensin system in chronic renal hypertensive rats. Hypertension 1:427–434

Sen, S., Tarazi, R.C., and Bumpus, F.M. 1980. Effect of converting enzyme inhibitor (SQ 14,225) on myocardial hypertrophy in spontaneously hypertensive rats. Hypertension 2:169–176

Scriabine, A., Watson, L.S., Russo, H.F., Ludden, C.T., Sweet, C.S., Fanelli, G.M., Jr., Bohidar, N.R., and Stone, C.A. 1979. Diuretic and antihypertensive effects of 2-aminoethyl-4-(1,1-dimethylethyl)-6-iodophenol hydrochloride (MK–447). J Pharmacol Exp Ther 208:148–154

Swales, J.D., and Thurston, H. 1978. Generation of angiotensin II at peripheral vascular levels: studies using angiotensin II antisera. Clin Sci Mol Med 45:691–700

Swartz, S.L., Crantz, F.R., Moore, T.J., Dluhy, R.G., Hollenberg, N.K., and Williams, G.H. 1979. Prostaglandin changes associated with the hypotensive response to captopril (SQ 14,225) in normotensive males. Clin Res 27:595H

Terragno, N.A., and Terragno, A. 1980. Release of vasoactive substances by kinins. In *Bradykinin, Kallidin and Kallikrein,* ed. Erwin G. Erdös, pp. 401–426. New York: Springer-Verlag

Thurston, H.,, and Swales J.D. 1974. Comparison of angiotensin II antagonist and antiserum infusion with nephrectomy in the rat with two-kidney Goldblatt hypertension. Circ Res 35:325–329

Thurston, H., Swales, J.D., Bind, R.F., Hurst, B.C., and Marks, E.S. 1979. Vascular renin-like activity and blood pressure maintenance in the rat. Hypertension 1:643–649

Thurston, H., Bing, R.F., Marks, E.S., and Swales, J.D. 1980. Response of chronic renovascular hypertension to surgical correction of prolonged blockade of the renin-angiotensin system by two inhibitors in the rat. Clin Sci 58:15–20

Tobia, A., and Giardino, E. 1980. Captopril-induced renal vasodilation in pentobarbital anesthetized dogs. Fed Proc 39:1189

Tree, M., and Morton, J.J. 1980. Captopril lowers blood pressure by an additional mechanism independent of angiotensin II. Clin Sci 58:2p

Vollmer, R.R., Boccagno, J.A., Harris, D.N., and Murthy, V.S. 1978. Hypotension induced by inhibition of angiotensin-converting enzyme in pentobarbital-anesthetized dogs. Eur J Pharmacol 51:39–45

Watanabe, T.X., and Sokabe, H. 1979. Acute vasodepressor effect of D-3-mercapto-2-methylpropanyol-L-proline (SQ 14,225) in the stroke-prone substrain of spontaneously hypertensive rats (SHRSP) Jap J Pharmacol 29:133–135

Weeks, J.R., and Jones, J.A. 1960. Routine direct measurement of arterial pressure in unanesthetized rats. Proc Soc Exp Biol Med 104:646–648

Discussion

H.R. BRUNNER: Oscar, I think last fall you showed some results in Germany where you already had a chronic experiment. Is my recollection right that after a few days or so of captopril administration there was no difference in blood pressure whether you administered bradykinin antibodies together with captopril or not?

O.A. CARRETERO: We have some few rats in which, with chronic administration of captopril, we get a significant decrease of the blood pressure but if we give kinin antibody to the rat we don't observe an increase in blood pressure. One of the problems with that type of experiment is that the effect of kinin's antibodies last a very short period of time, only 2 hrs, and the reason is that the kinin antibody cross-reacts with kininogen to form big complexes that are possibly picked up by the reticular endothelial system. But yes, we have some negative evidence.

H.R. BRUNNER: Did you look at renal blood flow also?

O.A. CARRETERO: No, we haven't done that.

F. GROSS: I have one question with respect to your double inhibition with saralasin and with captopril. Wouldn't it be possible that the competitive antagonism of saralasin is not complete, and that, if you, in addition, block the further production of angiotensin II, you get a more pronounced effect? I think this is another interpretation, and I don't need bradykinin to explain your effect.

C.I. JOHNSTON: One can demonstrate quite marked shifts, not only in the blood pressure dose responses to angiotensin II, but also in the renal vasoconstrictor dose responses to angiotensin II. With saralasin the renal blood flow dose responses to AII were shifted 100-fold to the right. Really quite massive doses were needed to have any effect, demonstrating that the receptors were completely blocked to exogenous angiotensin II. Now it may be that endogenously produced angiotensin II in the kidney might be a little bit different. But it has been shown by Mendelsohn that AII levels in the kidney fall after captopril.

O.A. CARRETERO: Colin, I would like to go back to your bradykinin measurements in blood. We have evidence that bradykinin does not circulate in nanogram amounts. You are using the method of Mashford and he reported much lower levels, like 200 pg. You report 1 or 2 ng, 10 times higher than those values obtained by Mashford and 100 times greater than the values we have obtained. Do you have any idea of what is causing this big discrepancy, particularly with the values of Mashford since you are using his same method in the same country?

C.I. JOHNSTON: True, and I have also used the same antibodies as Mashford, not in this study but previously. Yes, our values are about 5 times his. I think it does depend on the antibody and its cross-reactivity to fragments. As you know, one of the problems with bradykinin is that the fragments are not generally available. And, I know for example, in Mashford's work the levels he got depended very much on the particular antibodies he used. He had three different antibodies and they gave three different sets of results. We have only two antibodies that are sufficiently good to use for radioimmunoassay and both of those in our hands give the same result. Well, I think that is an open question on what is real and whether it actually does circulate. If in the rat, we give a big enough dose, we do measure an increase. With infusions in dogs we can measure linear increases and its extension across the lung using this radio-immune assay.

B.G. ZIMMERMAN: I would like to go back to the suggestions by Drs. Carretero and Johnston that intrarenally formed kinins may be responsible for the increase in renal blood flow seen with converting enzyme inhibitors. We have some data from our own laboratory in both anesthetized and conscious instrumented dogs that there is an increase in renal blood flow even in salt replete animals. We find that if we give the converting enzyme inhibitors intrarenally, that is, to the renal artery directly, we can show a very profound inhibition of the conversion of AI to AII, but no increase in renal blood flow. We subsequently give the converting enzyme inhibitors intravenously and then we do see the increase in renal blood flow. We have attributed that effect to a systemic action rather than an intrarenal action of converting enzyme inhibitors. Furthermore, we find that saralasin given intraarterially in nonhypotensive hemorrhaged dogs does block the increase in the renal blood flow, at least in our anesthetized dogs. So I think there is probably more to it than an influence of intrarenal bradykinin, when one asks what is responsible for the renal vasodilator effect of converting enzyme inhibitors.

O.A. CARRETERO: I think that you are right in relation to the total renal blood flow. What I was talking mainly about is distribution, and if you look at the slide that I showed you on the blood flow distribution in the kidney of the rat, the change in the total blood flow was really not very impressive. The blood flow that increased significantly was the inner cortex blood flow. I think if you take sodium repleted dog and measure total blood flow you will not find an increase and I agree with you.

B.G. ZIMMERMAN: We've never measured distribution of flow in these animals and agree there is a difference between a rat and a dog in that the dog does show the increase in total flow probably to a greater extent than the rat. Perhaps there is some kinin involvement in these dogs that we don't see unless we measure distribution flow. It's a possibility.

C.I. JOHNSON: I am aware of your studies and I cannot really explain the difference. I didn't want to intimate that all the renal blood flow changes may be due to kinin. We have used captopril intrarenally, as well as saralasin, and we get identical results. As you saw in our studies, with very adequte blockade with the competitive angiotensin inhibitor we could still not prevent the renal vasodilation after captopril. The interpretation of the results are a little bit difficult because in our hands saralasin, particularly given into the renal artery, has marked agonistic activities, and so the dog kidney starts in a very vasoconstricted state after the administration of saralasin. Lastly, we have also looked at the effect of captopril in normal sodium replete chronically instrumented conscious dogs and we find no effect of captopril on renal blood flow in such dogs.

M.J. ANTONACCIO: I would like to expand that question a little bit. It is shown in the literature that if you nephrectomize dogs and you still maintain saralasin infusion, captopril still lowers blood pressure. Does that rule out kinins or just perhaps renal kinins?

C.I. JOHNSTON: Probably, it would rule out the renal kinins but I have a view that kallikrein-kinins also exist in vessel walls something like the beautiful pictures you showed of angiotensin converting enzyme (see Fig. 1) and aortic renin.

R. GREENBERG: I would like to say something about the inactivation of bradykinin. Bakhle showed that the lung is the major site of inactivation of bradykinin. The isolated lungs of guinea pigs, rats and dogs inactivate 92 to 99% of bradykinin perfused through the pulmonary circulation. Inhibition of angiotensin converting enzyme with teprotide only reduced this inactivation to 64 to 68%. This illustrates that there are other enzymes capable of degrading bradykinin in the lung. Therefore, captopril might not have any effect on circulating bradykinin levels, while elevating bradykinin levels in the kidney.

I want to comment on the inhibitory effect of captopril on the pressor response to sympathetic nerve stimulation. I would like to point out that this might be a consequence of inhibition of angiotensin converting enzyme and not necessarily due to some other mechanism. Hughes and Roth showed that angiotensin increases the synthesis and release of norepinephrine in vascular smooth muscle in response to sympathetic nerve stimulation. It is, therefore, possible that angiotensin plays a modulatory role in sympathetic neurotransmission. Inhibition of the synthesis of angiotensin by captopril could therefore result in inhibition of the pressor response to sympathetic nerve stimulation.

D. GANTEN: I recommend a question to Dr. Antonaccio. I would like to confirm first what you said with respect to arterial wall renin in spontaneously hypertensive rats; we also found increased arterial wall renin in spontaneously hypertensive rats of the Strong strain. We did not find what you reported today, that following captopril treatment renin goes up in the arterial wall. That leads to my question. Our treatment was prolonged treatment for 6 months or longer. Have you done time course studies and seen whether the increase of arterial wall renin is correlated to plasma renin, or whether it decreases with time as we find?

M.J. ANTONACCIO: Dr. Asaad hasn't done both kinds of studies in detail but I think we can say that there is definitely a separation between plasma renin activity and arterial renin concentrations. For instance, if you give just a single dose of captopril, plasma renin activity rises very dramatically, but arterial renin concentration does not rise. So I think these kinds of studies must be done to determine the time course of these changes.

O.A. CARRETERO: I would like to ask a question of Dr. Antonaccio. It is in relation to the use of aprotinin (Trasylol®). First I think that you have to take into consideration that aprotinin is not a specific inhibitor of kallikrein. It is a polyvalent inhibitor. Second, when we use the doses that you have used, we don't inhibit the vasodepressor effect of exogenous injected kallikrein. I wonder if you are really inhibiting kallikrein. Did you inject any exogenous kallikrein to see that you block its vasodepressor effect?

M.J. ANTONACCIO: Let me first agree with you that it's a dirty drug to use. Unfortunately it is the only one available and I'm impressed by the fact that despite all the changes that we see with Trasylol itself we can still get a very nice fall in pressure which is not different from captopril alone. As to the second point, we have not injected kallikrein to see if the effect is blocked by Trasylol. However we used the same doses and route of administration as Nasjilleti used when he showed that he got rather dramatic inhibition of kallikrein, although he used indirect measurements.

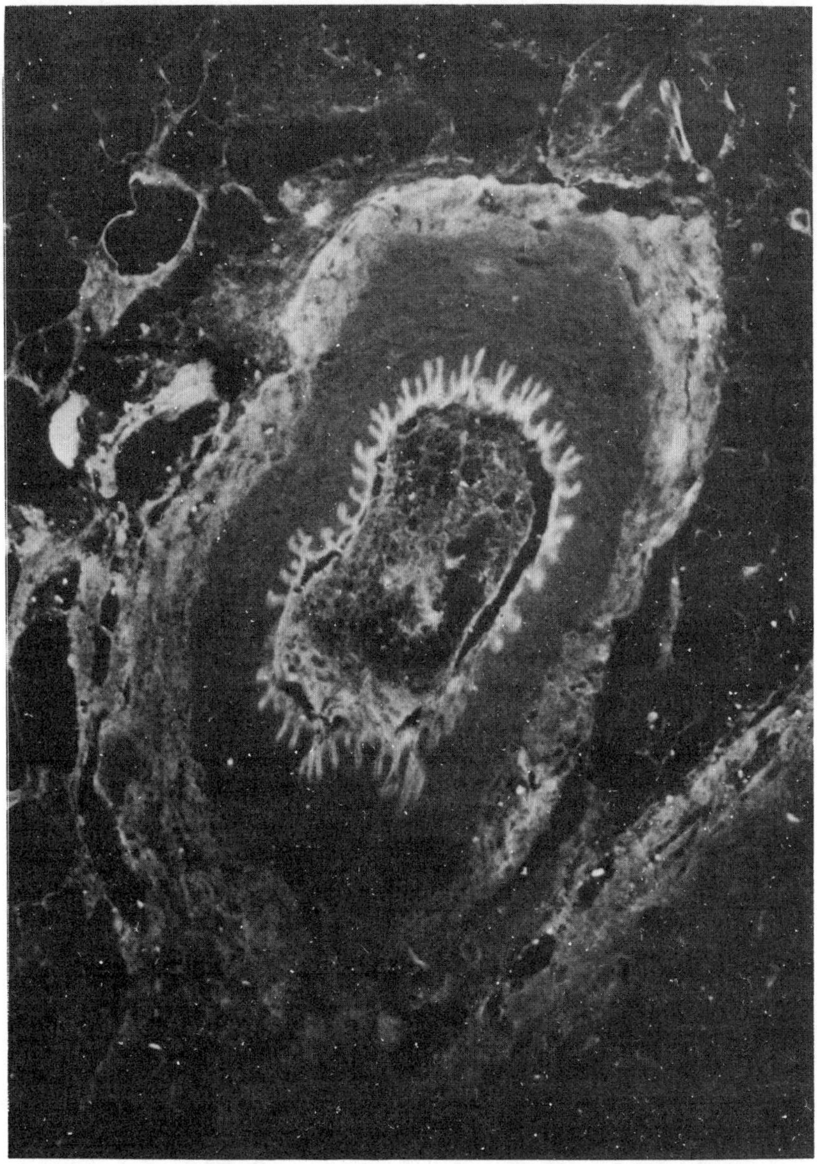

Fig. 1. Immunofluorescence of antibody to angiotensin converting enzyme in a small artery from guinea pig kidney. Note bright staining of the endothelium and adventitia with essentially nothing in the media. (Courtesy of Dr. George Zaki, Squibb Institute for Medical Research).

O.A. CARRETERO: Nasjilleti was not injecting exogenous kallikrein, he was looking in the kidney.

M.J. ANTONACCIO: Right. What I mean is, it is the same dose of Trasylol as he used when he showed he got inhibition. We didn't use exogenous kallikrein and we didn't monitor inhibition of kallikrein.

O.A. CARRETERO: May I make a comment about this? I think that every time you inject an inhibitor and claim that you are blocking an endogenous system, you have to test whether or not the inhibitor is doing the job. Nobody will accept it if we inject converting enzyme inhibitor and say it blocks angiotensin I, if we haven't tested it. So the same goes for the kallikrein-kinin system. We have to do it.

J.H. LARAGH: I wonder about the so-called arterial renin. I have two questions. Have you studied its distribution in arterial tree and, more importantly, how long after nephrectomy do you do your arterial renin measurements and is there vascular wall renin long-term after nephrectomy or is it of renal origin and just a hold-over from acute nephrectomy?

M.J. ANTONACCIO: To the first question: no, we have not looked at distribution. This is very high on the list. Number two, we have not looked at long-term nephrectomized animals. This happens to be a 24-hr study where there was no change. There are reports in the literature in dogs where it was reported that 11 days after bilateral nephrectomy, vascular renin-like activity was still seen. However, the problem with that study was that they assayed at pH 5.5, so it might not be renin at all, it might not be physiological. But I think that studies still have to be done.

C.I. JOHNSTON: I wonder if the people who measure aortic renin could tell us a bit more about its characteristics and specificity, and, whether or not it is the same enzyme as occurs in the kidney and the plasma? It seems to me that many people are mushing up the arteries and putting in the substrate and measuring the end product, but I'd like to hear some of the evidence that it is similar to or that it is really renin that we are measuring. Or if it is different, I imagine it is equally as important.

M.J. ANTONACCIO: I would like to get some comments from some people who measure it. First let me say that I think that it doesn't matter to me as a pharmacologist what the enzyme is that generates the angiotensin at these pHs. But I think I would like to know if it is in fact renin.

M. ASAAD: When we started working with vascular renin, one of our primary concerns was to ensure that what we are measuring is renin and not other acid proteases, such as cathepsin-D. Thus, we used homologous renin substrates and maintained our assay incubations at a physiological pH. The capability of vascular tissue homogenate, in our system, to generate angiotensin I from homologous renin substrate at physiological pH, clearly suggests that what we are dealing with is renin. We also found that when the pH of incubations was raised above 7.5, AI generation was drastically reduced. Such findings have also been reported with renal renin. Furthermore, Gold et al in the 1950s have reported that hog arterial renin is biochemically similar to kidney renin. We are in the process of the further characterization of rat vascular tissue renin. Actually, we are using the criteria proposed by Dr. Ganten regarding extrarenal renin. Again, I would like to call to your attention to two important points that our study raises. First, vascular tissue homogenates in our system can generate AI at a physiological pH from homologous renin substrate. Second, 8 days treatment with captopril resulted in increased renin levels in vascular tissues suggesting that vascular tissue renin was under typical negative feedback inhibition; a response seen with renal renin. However, we did not study the effect of more than 8 days captopril treatment on vascular tissue renin.

H. THURSTON: I would like to raise a question about aortic renin activity. We found measurements of aortic renin activity using different incubation pHs produced different results. In the dog experiments quoted, arterial renin activity persisted after bilateral nephrectomy but an incubation pH of less than 5 was used in the assay and the activity could be due to acid proteases.

We have measured aortic renin activity in the Goldblatt 2 kidney hypertensive rat using an incubation pH 5.3 and 6.5. The optimum pH for rat plasma and renal renin is 6.5 and the aortic renin activity measured at pH 5.3 was unchanged after bilateral nephrectomy. However, renin activity at pH 6.5 disappeared slowly with a much longer half-life, reaching normal levels after 24 hr. In addition, other workers have demonstrated the aortic renin activity measured at pH 6.5 is kinetically similar to rat renal renin. Now, since the aortic renin activity Dr. Antonaccio measured was not changed after nephrectomy in spontaneously hypertensive rats, how can he explain the failure of captopril to lower the blood pressure 24 hr after bilateral nephrectomy?

M.J. ANTONACCIO: I think that is something we are going to have to work on. I knew that as soon as I mentioned arterial renin, automatically this is the way I was proposing that captopril was working. There is a modest blood pressure lowering effect from captopril in salt repleted SHR 24 hr after nephrectomy. There is about a 15 to 20 mm reduction in pressure which might be related to the amount of renin that is in the vessels. When you salt deplete, you increase that amount and captopril can work a little better. But I don't know why you don't get a better effect. Maybe vascular renin is, in fact, not important.

H. THURSTON: The other point I wanted to make is that others have measured aortic renin activity in the spontaneously hypertensive rats. Barrett found normal levels and in addition, as the animals aged, the blood pressure rose but arterial renin levels fell. We have measured aortic renin in spontaneously hypertensive rats aged 10 to 11 months and the level at incubation pH 6.5 was quite low.

M.J. ANTONACCIO: Is it inhibited by anti–kidney renin antibodies?

H. THURSTON: I know of no experiments using rat antirenin antibodies.

O.A. CARRETERO: I would like to present another question to Dr. Antonaccio. This time it is in relation to prostaglandin and the use of indomethacin. Are you aware of a report that indomethacin increases the blood pressure in rabbits and tends to produce hypertension? The second question is if you are using indomethacin it would be nice to measure the prostaglandins to determine that their synthesis is inhibited. I would be very careful before completely eliminating prostaglandins. At least we should simultaneously measure different types to see that we really are inhibiting their synthesis.

M.J. ANTONACCIO: About the effects of indomethacin on pressure I think this is a very mixed bag. Some people report there is an increase, but most people report there is not. In fact, in rabbits under the very same conditions, Dr. McGiff reported an increase in blood pressure with indomethacin. At the last Federation meeting Dr. Muirhead reported there was no change in blood pressure under the same conditions. We've not seen any increase in blood pressure with long-term treatment of indomethacin on blood pressure. One of the problems is, as I showed, indomethacin is very toxic and it is very difficult to keep these animals going for very long periods of time. It makes me wonder about the specificity of indomethacin and is one of the reasons why we use aspirin. So, it is a problem and I don't know that the real story will come out for a long time yet about circulating levels of prostaglandins and the effects of inhibiting their synthesis on blood pressure. Concerning the amount of inhibition of prostaglandin synthetase by indomethacin in our studies, we didn't look at levels of prostaglandins but we did inject arachidonic acid. We know that arachidonic acid causes a very dramatic fall in blood pressure. Indomethacin, 5 mg/kg orally, inhibited that response some 92%. So, although that is not entirely adequate, I think if you have to go to those degrees of inhibition to show any effect whatsoever on captopril's antihypertensive actions, then you are searching for something which is not going to be a significant contribution to the drug's action.

Z.P. HOROVITZ: Let me make a statement that I think will be interesting and stimulate some controversy. I have been through these kinds of discussions on mechanisms of action before, and I have to take credit for even setting up the titles of the afternoon session, and I think what tends to happen is that we focus on site of action more than mechanism of action. Everything I have heard late this morning and this afternoon primarily asks, does the drug work in the central nervous system or does it work in the kidney or does it work on the arterial wall. Also we focused on the question: is the primary action through the renin-angiotensin system or bradykinin? But I haven't heard anything yet, in fact I haven't heard anything after about three years of research availability, that says that compounds like captopril or teprotide work or lower blood pressure by a means other than inhibiting the angiotensin converting enzyme or kininase II. If there is anybody in the audience who has any data that would support these drugs working by a completely different mechanism than by inhibiting this enzyme then I think it is important that we discuss those data. Recognizing that even though captopril may just inhibit the converting enzyme we are still not sure about site or indeed whether it is angiotensin or bradykinin that is important.

UNIDENTIFIED: Recently I surveyed the literature. There is an abstract just recently published by a group of people from I think South Africa. And in their clinical study they found out that the essential hypertensive patient, after receiving captopril had some kind of "kinin potentiating factor" which they say has increased after they received captopril. What they appear to be thinking might be that the bradykinin level increase is not really necessary, but bradykinin potentiating factor increase by captopril might be more important. Can anyone make any comment on this?

O.A. CARRETERO: Maybe I can answer the question. Converting enzyme inhibitors were originally called bradykinin potentiating factor, but really we didn't know their mechanism of action. When investigators talk about bradykinin potentiating factor, or BPF, they are referring to the converting enzyme inhibitor or kininase II inhibitor. But really they are the same thing.

D. GANTEN: I haven't been clear enough probably this morning. I thought I had given an indication of possible other mechanisms. And I'll be clearer now. One hypothesis, of course we don't have the proof but I've shown you the possibility, is that captopril modifies substance P, the vasodepressor substance, intravenously. Substance P is a competitive substrate for converting enzyme and the degradation is inhibited by converting enzyme inhibitors. If substance P is potentiated like bradykinin in the plasma, that could be another mechanism for the blood pressure lowering effect of captopril. Another possibility would be what I have indicated this morning, the effect on enkephalins and other neural peptides. The hemodynamic effects of these substances are rather complicated. The short chain peptides like leucine enkephalin and methionine enkephalin appear to be mainly pressor, whereas the long chain peptides like beta endorphin seem to be depressor. The second hypothesis which I can put forward is that the degradation of, say, beta endorphin in the plasma is inhibited by captopril and thereby another blood pressure decreasing mechanism is potentiated. A third possibility is that a medium chain peptide like gamma endorphin or alpha endorphin, which may be acting and has been found to be present in the medulla oblongata and has depressor effects, also might be potentiated in the brain in captopril's action. So there are another three peptides at least or groups of peptides which might be involved in the action of captopril and these have to be investigated.

M.J. ANTONACCIO: Let me just say quickly that naloxone does not touch the antihypertensive effects of captopril.

C.I. JOHNSTON: There was some Japanese work presented the 6th International Endocrine Congress in February and also published in Life Sciences, which I think has suggested that there may be another mechanism of action and that is that captopril attenuated the sympathetic vasoconstrictor response in the mesenteric vascular bed and also the noradrenaline outflow from that vessel, but teprotide did not. That either suggests that they have a different spectrum of enzymatic inhibition or that captopril has some other action beyond that of converting enzyme inhibition. I don't know if people have verified it. I believe Dr. Brunner has shown after prolonged captopril, responses to noradrenaline are reduced, and we in our laboratory have shown the vascular response to noradrenaline, after prolonged captopril, is reduced.

D.W. CUSHMAN: I just wanted a clarification from Dr. Ganten. Has anybody shown directly that substance P can be a substrate for angiotensin converting enzyme? This seems very unlikely, because substance P has a terminal amide group. Is there any evidence for this beyond the reported inhibition of substance P degradation by teprotide?

I.L. NATOFF: I am interested in the aspect of captopril having effects other than through ACE inhibition. As with antihypertensive beta blockers, beta blockade may be an indication of the biological activity of the compound, but it is not necessarily related to the antihypertensive effects that we see. And, by virtue of captopril working in low or normal renin hypertension, ACE inhibition could well be another instance of an indication of the compound's biological activity unrelated to its therapeutic effect. I've been interested in Jandhyala's initial observation that in the presence of saralasin blockade, captopril could still reduce the high perfusion pressure in the anesthetised dog. I have been taking this observation just one step further and have been using Gillespie and Muir preparations. These are pithed rats, in which the total sympathetic outflow is stimulated while recording blood pressure and heart rate. Now with each stimulation at submaximal frequency, we get a pressor response and tachycardia which records quite nicely. If we inject captopril in a dose of 1 mg/kg^{-1} i.v. (presumably lower doses also work, but certainly 1 mg/kg^{-1} i.v.) we get something in the order of 50 to 70% inhibition of the pressor response. We do not get any effect on the tachycardia response. This indicates therefore that it is not an inhibition of the sympathetic outflow which we are observing because the heart rate is unaffected. It must be an effect on the vascular smooth muscle. I suspect that in the initial stages of these observations we are looking at a direct dilatory effect of captopril, and whether there is any evidence for its effect on calcium flux, I don't know. But this is the stage of my observations now and I hope that it will contribute to the discussion that Dr. Horovitz has started.

M.J. ANTONACCIO: Thank you. Concerning your remarks about inhibiting specifically vascular responses to sympathetic stimulation in pithed rats, we have done the very same experiments and find exactly the same thing. Cardiac responses are not inhibited but vascular responses are. I think you would find that if you went back and reinfused angiotensin, you'd get your responses back. So, once again, it might be related to converting enzyme inhibition but at a different site than we normally think about, at the vasculature rather than in the kidneys.

I.L. NATOFF: I would just like to add that we have done this in the presence of saralasin infusion with the appropriate control, showing that AII inhibition is present. Both inhibiting AII with saralasin before administering the captopril and after establishing the reduction of the pressor response with captopril does not affect the inhibition due to the ACE inhibitor.

M.J. ANTONACCIO: I am getting a little concerned about the use of saralasin for blocking anything other than obvious endothelial or any membranous angiotensin receptors. Now that we know that in fact there are differences between using converting enzyme inhibitors and using polypeptides which are very large and do not distribute to the same areas as a small molecule like captopril, perhaps only the use of saralasin is not sufficient any longer. Perhaps we have to use compounds that we know distribute to the same areas as we are interested in. That might be one of the reasons saralasin doesn't work in SHR and chronic two kidney renal hypertension. The same is true for teprotide, whereas captopril is quite effective.

J.H. LARAGH: I was merely going to make a comment about the SHR because it has been extremely difficult to do renin measurements properly in rats, not only for technical reasons, but also because when you touch them, as you know, the renin secretion goes off the walls. Because of this we used to rely only on decapitation to collect samples. But the renin levels that are now being reported in SHR rats in the plasma are certainly adequate enough to contribute to the hypertension if the human experience is any guide. And while they are called "normal," and as you showed very nicely they are in the normal range, they are not "normal" for a blood pressure of 200. The physiological secretory response to a blood pressure of 200 results in a renin of zero. I would also remind you of some studies that Peter Chiu did with sodium depleted SHR rats that had quite high renin levels, in which he showed that the higher the renin the greater the response to either captopril or propranolol. With such evidence now emerging I think you don't need to invoke extra vascular renin to explain what you are seeing, especially when you take it in the context of the fact that the DOCA-salt animal gives much less response with a lower renin and the anephric with no plasma renin gives no response to captopril or beta blockers.

M.J. ANTONACCIO: Yes, I agree that that's a possibility, but then one would have to account for the lack of effect of angiotensin II antagonists in SHRs. This brings up another point for Dr. Thurston. I'm sorry that Dr. Micheal Peach isn't here. I'm going to say some things that he told me, which were not said in confidence. He found that in spontaneously hypertensive rats, if he infused high doses of about 100 μg/kg/min of antagonist, he could in fact get low pressures in these animals over a 12 hour period. I wonder, Dr. Thurston, if you tried to increase your dose of saralasin? You used a single dose in all your studies. Have you tried to push the doses to get a better effect?

H. THURSTON: We have not tried raising the dose above 10 μg/min, i.e., 40 μg/kg/min, but we have used lower doses. There is no difference in blood pressure reduction between the lower dose of 10 μg/kg/min and 40 μg/kg/min.

M.J. ANTONACCIO: Over long periods of time?

H. THURSTON: Over long periods of time, yes.

M.E. GOLDBERG: Why so much effort to show that there is only one action? Would you be unhappy to find several actions?

M.J. ANTONACCIO: Not at all. I would just like to have some evidence that there are in fact several actions.

R. GREENBERG: I just wanted to make a comment to both the effects of captopril on the sympathetic stimulation, and I don't want people to be left with the idea that if captopril has an effect on the blood pressure rise due to sympathetic stimulation that this may not have anything to do with angiotensin. Hughes and Roth, eight or nine years ago, reported that angiotensin would increase norepinephrine release from vascular smooth muscle. And therefore, if angiotensin is a control mechanism for sympathetics, then the inhibition of its synthesis may reduce blood pressure and may be responsible for inhibiting sympathetic stimulation.

F.W. WOLFF: I'd like to come back to Dr. Antonaccio's beautiful pictures of the concentrations of these materials in the walls of blood vessels (see Fig. 1). We've heard and learned clearly that there is no change in converting enzyme when one gives any other antihypertensive drug. However, could there be a change in the concentrations in the wall of blood vessels? Some 11 years ago we demonstrated that tolbutamide, when given to a dog in whom glucose was controlled, raised the blood pressure. We could never explain it in any way. These were dogs which were not hypoglycemic, but tolbutamide in dogs repeatedly raises the blood pressure. We got onto this because of Schering's work on the interaction of diazoxide and tolbutamide. Perhaps we are now getting close to an understanding of the vascular action of tolbutamide at least in the dog and the rat. Have any observations been made on local concentrations of these materials, the walls of blood vessels after various changes in blood pressure?

M.J. ANTONACCIO: You are anticipating all of our experiments. Again, Dr. F.G. Zaki, whose slides I used to show the localization of converting enzyme, has attempted to look at the distribution of captopril in various organs, and in particular in arterioles. So far I am afraid we have been unsuccessful in trying to show where it goes because, despite long autoradiographic times for development of these kinds of pictures, the captopril unfortunately seems to wash out of tissues and we can't, at this time at least, see anything. But he still has some much longer-term plates going. I don't know if that is going to show anything, but you are anticipating what we'd like to show also.

G.A. MACGREGOR: Dr. Antonaccio, the rats that you salt-depleted were nephrectomized after sodium depletion, whereas Dr. M. Schalekamp and co-workers' anephric patients were volume-depleted after nephrectomy. This may lead to differences. The fall in blood pressure with captopril, in volume-depleted anephric patients, was predominantly a postural one, something that we have not seen in normal subjects or patients with essential hypertension, even when captopril is combined with a diuretic. This suggests to me that the mechanism of the fall in blood pressure in anephric patients with captopril may be different from that we see in patients with kidneys. Volume-depleted, anephric patients might be very sensitive to very small amounts of any circulating vasodilator.

M.J. ANTONACCIO: I hate to beat a dead horse here, but it is possible—to make a long story very short—that these changes are in fact related to vascular angiotensin. We find that in situations where vascular angiotensin is elevated and you give captopril, only under those circumstances can you get inhibition of responses to sympathetic nerve stimulation. If in fact vascular levels are normal, relatively speaking, we cannot see any inhibition. For instance, in SHRs we see very nice inhibition. In 2 kidney, 2 clip renal hypertensive rats we see very nice inhibition; but if you go to a 1 kidney, 1 clip renal hypertensive rat you see nothing. Others have told me that in Sprague-Dawley rats, they see nothing. So I am beginning to believe that perhaps this might be really significant and might be the answer to some of the things that are happening that we don't understand. It's a possibility at this time. I don't want to be quoted as being incorrect next year so I'm just holding it up as a possibility.

M.E. GOLDBERG: I'd like to ask Dr. Antonaccio or Dr. Snyder whether they have studied captopril given both systemically and intraventricularly in conscious animals and whether they got an augmented antihypertensive effect or whether they studied newer converting enzyme inhibitors which have more facile entry into the central nervous system and whether they see any additional or augmented antihypertensive activity?

D. SNYDER: Other analogs of captopril which do penetrate the brain are not more effective than captopril. As far as the central actions of captopril in the SHR, we have not been able to show a hypotensive effect of captopril upon central administration until we see inhibition of the AI pressor response peripherally. We have to push the dose centrally to such a point that we now see the peripheral actions of the drug. In talking with Dr. Ganten before this session, he informed me that their data was obtained in the stroke-prone SHR whereas we have not used the stroke-prone SHR. A difference in the animal model could account for the different results between our findings and those of Dr. Ganten.

I would also like to make a comment on the effect of captopril on sympathetic nerve function. We have looked at the effects of captopril in a neurogenic model of hypertension which has been termed NTS hypertension. Dr. Reis's group has shown that electrolytic lesion of the nucleus tractus solitarii in rats results in an acute fulminating hypertension which is due to high sympathetic outflow as the result of central deafferentation of the baroreceptor reflex arch. Now in this acute form of hypertension there is no effect of captopril in a dose as high as 50 mg/kg. Also in the cat, evoking a pressor response through electrical stimulation of the diencephalic region of the brain, we have not been able to show an attenuation of the pressor response following captopril (50 mg/kg). Thus we have no evidence that captopril may be acting by altering sympathetic nervous function.

M.J. ANTONACCIO: Those were all normotensive?

D. SNYDER: The cats were normotensive. The rats in NTS hypertension were Sprague-Dawley.

M.J. ANTONACCIO: I think one of the best models to use to examine the vascular renin hypothesis is the DOCA-salt model. Here is an animal that we know not only has low plasma renin activity but very low vascular renin concentrations. These animals are currently being tested and we will have the answer at the end of the month. Did that answer your question? We've looked at other compounds that do get into the brain and we don't find them to be any more effective than captopril.

E. DIPASQUALE: Since you tried MK–447, have you superimposed treatment with prostacyclin or PGF or PGE?

M.J. ANTONACCIO: You mean have we infused?

E. DIPASQUALE: Well, instead of using the stimulators, have you used the actual end products of the pathway?

M.J. ANTONACCIO: No. Dr. Carretero, you showed the possible effect of bradykinin. With the first dose of captopril in SHR using antibodies, you found approximately 10 mm fall in blood pressure with captopril alone. Then you found if you added the antibody you saw no change in pressure but if you used the preimmune serum you found a 10mm rise in blood pressure. Is this physiological antagonism or is this a true antibody-antigen reaction?

O.A. CARRETERO: The controls injected with captopril were also injected with normal rabbit globulin. So those that had a 10 mm of mercury decrease in blood pressure were injected with the same amount of protein as those injected with the kinin antibody that did not have any change in blood pressure.

SECTION III

Hemodynamics, Fluid Balance and Clinical Implications

Robert E. McCaa, *Chairperson*

Aldosterone and Fluid Balance[1]

Robert E. McCaa
Department of Physiology and Biophysics, University of Mississippi School of Medicine, Jackson, Mississippi 39216

INTRODUCTION

The importance of the adrenal glands in the regulation of electrolyte and fluid metabolism was firmly established in 1926 when Banting and Gairns demonstrated that bilateral adrenalectomy resulted in a decrease in serum sodium concentration and an increase in serum potassium concentration in experimental animals. The symptoms of bilateral adrenalectomy could be delayed by the administration of sodium or accelerated by the administration of potassium (Marine and Baumann, 1927). In 1927 Hartman et al. demonstrated that an extract from the adrenal cortex contained a factor with high sodium retaining activity that prolonged the life of adrenalectomized animals. Luetscher and Deming (1950) found a high concentration of a "sodium retaining factor" in urine from patients with edema. In subsequent studies, Luetscher and his associates demonstrated that the "sodium retaining factor" was high in urine when sodium intake was restricted, normal in patients with hypopituitarism, and undetectable in patients with Addison's disease (Luetscher and Axelrad, 1954a,b; Luetscher et al., 1954; Luetscher and Curtis, 1955a,b,c; Luetscher et al., 1955).

In 1953 Simpson and Tait, using paper chromatographic techniques, reported the successful separation of the "sodium retaining factor" from cortisone. Since this steroid possessed a much greater effect on the renal handling of sodium and potassium than any other known substance, it was called "electrocortin." The isolation and purification of electrocortin from beef adrenal glands led to the characterization of its struc-

[1] These studies were supported by USPHS Grant HL 09921 from the National Institutes of Health.

ture. It was a naturally occurring biologically active steroid with an aldehyde group and accordingly given the definitive name "aldosterone" (Simpson et al., 1953, 1954a,b).

Since the identification of aldosterone 25 years ago, important contributions by a large number of scientists in the field have demonstrated that aldosterone plays a primary role in maintaining the overall adequacy of the circulatory system. A hypodynamic state of the circulation is usually followed within minutes by increased aldosterone secretion, and a hyperdynamic state of the circulation results in decreased aldosterone secretion. Despite the vast importance of aldosterone in overall circulatory control, the regulation of aldosterone biosynthesis and the role of aldosterone in the development and maintenance of hypertension have remained the subject of intense debate.

During the past fifteen years, we have collected considerable quantitative data in experimental animals and man in an effort to develop a composite quantitative control systems analysis for the regulation of aldosterone biosynthesis. This chapter presents a review of the experimental studies performed in our laboratory. Special emphasis is focused on several phases of our experimental studies: 1) the role of adrenocorticotropic hormone (ACTH), sodium, potassium, angiotensin II and angiotensin III (des-asp[1] angiotensin II) in the long-term regulation of aldosterone biosynthesis in intact animals; 2) the role of the pituitary and unidentified pituitary hormone(s) in the regulation of aldosterone biosynthesis; 3) the role of the renal renin-angiotensin system in mediating increased aldosterone secretion during sodium deficiency; 4) the role of the renal renin-angiotensin and kallikrein-kinin systems in mediating the hypotensive and natriuretic actions of angiotensin I converting enzyme (kininase II) inhibitors; 5) the role of the renin-angiotensin-aldosterone system in mediating elevated arterial blood pressure in experimental renovascular hypertension; and 6) the antihypertensive action of captopril in patients with severe, treatment-resistant accelerated or malignant hypertension.

Our studies differ from earlier investigations on aldosterone regulation in that they are long-term in nature, often lasting for several days or weeks. The animal experiments presented in this review were performed in intact unanesthetized mongrel dogs with chronic indwelling catheters in the femoral artery and vein. The venous catheter was used to maintain continuous infusion of various pharmacologic agents at a constant rate from a syringe infusion pump. Continuous 24 hr arterial blood pressure measurements were obtained from the arterial catheter by a pressure transducer and a recorder. The animals were housed in separate metabolic cages during the control, experimental and recovery stages of the study. The experimental animal preparation used in these studies and the procedure for collection and analysis of the blood samples for steroids and peptides by radioimmunoassay have been described in complete detail in previous publications (McCaa et al., 1972, 1973a, 1975, 1978a).

REGULATION OF ALDOSTERONE BIOSYNTHESIS

Though all the details of the regulatory mechanisms controlling aldosterone biosynthesis are not completely known, it is apparent that at least three different factors, and possibly several others, play important roles in the regulation of aldosterone secretion. These regulatory mechanisms include the pituitary secretion of adrenocorticotropic hormone (ACTH), alterations in sodium and potassium metabolism, and the renin-angiotensin system.

Role of ACTH

Much of our knowledge of the role of ACTH on aldosterone secretion has been obtained from studies in human beings maintained on long-term administration of ACTH or from studies of patients with Cushing's syndrome with chronic excessive secretion of endogenous ACTH. In both instances, aldosterone secretion remains at normal or even below normal levels. The normal or decreased production of aldosterone in the face of chronically high levels of ACTH is thought to result from suppression of aldosterone secretion by increased body fluid volume resulting from excessively high levels of circulating glucocorticoids, especially cortisol. It is generally believed by most investigators that ACTH, although important in maintaining the steroid precursors necessary for the synthesis of aldosterone, plays a minor role in the day-to-day control of aldosterone secretion except for the increases produced by acute stress such as surgical treatment, anxiety or physical trauma.

The response of plasma aldosterone concentration, plasma cortisol concentration, plasma renin activity and arterial blood pressure to continuous long-term infusion of adrenocorticotropic hormone in conscious dogs maintained on normal sodium intake is illustrated in Figure 1. Within 24 hr after continuous ACTH infusion at a rate of 500 μg/day, plasma aldosterone concentration increased from 6.7 ± 1.3 to 36.8 ± 5.4 ng/dl and remained significantly elevated above control levels for the next 3 days. However, by the fourth day plasma aldosterone concentration returned to the level that existed before ACTH infusion despite continuous ACTH infusion that maintained plasma cortisol concentration elevated sixteen times greater than control levels (0.73 ±

0.25 to 11.8 ± 2.9 μg/dl). A similar transient increase in aldosterone secretion in response to ACTH infusion was observed in sodium-deficient dogs. Although aldosterone secretion returned to control levels after 4 days of continuous ACTH infusion in sodium replete and sodium deplete dogs, arterial blood pressure increased by 15 to 20 mm Hg in both groups of dogs and the animals remained hypertensive until the ACTH infusion was discontinued (McCaa et al., 1978c). It is noteworthy that this increase in mean arterial blood pressure occurred while plasma aldosterone concentration returned to levels observed before ACTH

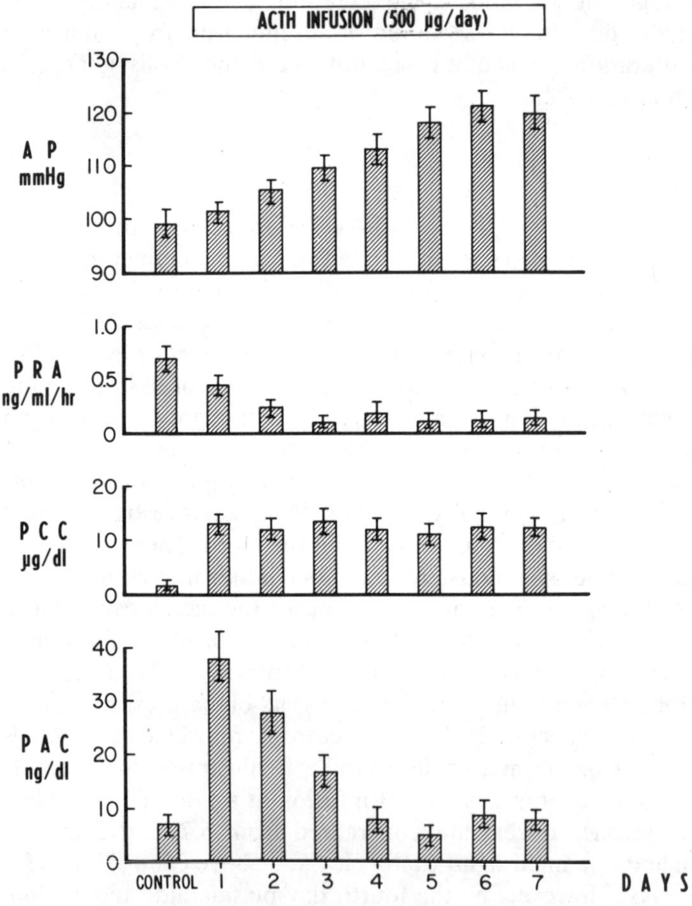

Fig. 1. The response of arterial blood pressure (AP), plasma renin activity (PRA), plasma cortisol concentration (PCC) and plasma aldosterone concentration (PAC) to continuous ACTH infusion (500 μg/day) in dogs on normal sodium intake (60 mEq/day). Vertical lines indicate ± S.E.

administration and while urinary sodium excretion remained elevated threefold. The increase in urinary sodium excretion was due to the increase in glomerular filtration rate that occurred with increased secretion of glucocorticoids. Since ACTH infusion, or infusion of high levels of aldosterone and cortisol, failed to increase arterial blood pressure in adrenalectomized dogs, ACTH-induced hypertension may be due to the production of an unidentified adrenocortical hormone(s) with little mineralocorticoid activity but with high pressor activity (Kastner and McCaa, 1978). Several other investigators have demonstrated ACTH-induced hypertension in experimental animals (Coghlan et al., 1977; Scoggins et al., 1978) and in man (New et al., 1975, 1976).

Role of Potassium

Increased serum potassium concentration induced either by increased dietary intake of potassium or by direct intravenous infusion of potassium ions results in marked increases in aldosterone secretion in experimental animals and man. Earlier studies indicated that large increases in serum potassium concentration were required to elicit increased aldosterone secretion. However, more recent studies have demonstrated marked increases in plasma aldosterone concentration in response to small increases in serum potassium concentration (Boyd et al., 1973; McCaa et al., 1974a), and clinical studies in nephrectomized human beings indicate that potassium may be the primary regulatory mechanism controlling aldosterone biosynthesis.

The aldosterone response to long-term infusion of potassium chloride in dogs maintained on normal sodium intake is illustrated in Figure 2. In response to continuous infusion of potassium chloride at the rate of 250 mEq/day, plasma aldosterone concentration increased from 6.2 ± 2.4 to 13.6 ± 3.4 ng/dl within 24 hr and averaged 24.4 ± 4.3 ng/dl by 48 hr after beginning potassium chloride infusion. Serum potassium concentration averaged 3.9 ± 0.5 mEq/L during the control period and increased to 4.7 ± 0.5 mEq/L within 48 hr after beginning potassium chloride infusion. In contrast to the transient increase in aldosterone secretion in response to continuous ACTH infusion, continuous potassium chloride infusion at a rate that increased serum potassium concentration only 20% produced a sustained increase in plasma aldosterone concentration of 400%. These studies demonstrate that slight increases in serum potassium concentration produce marked sustained increases in aldosterone secretion and indicate that alterations in potassium metabolism play a major role in both short-term and long-term regulation of aldosterone biosynthesis (McCaa et al., 1974a; Komor and Müller, 1979).

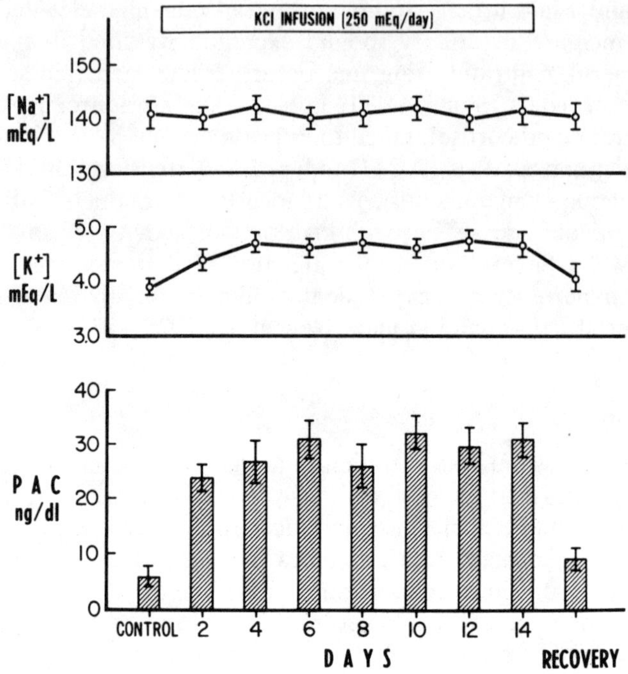

Fig. 2. The response of plasma aldosterone concentration, serum sodium concentration [Na^+], and serum potassium concentration [K^+] to long-term infusion of potassium chloride at the rate of 250 mEq/day in intact conscious dogs. Vertical lines indicate ± S.E.

Role of Sodium

Although slight changes in serum potassium concentration produce marked effects on aldosterone secretion, marked alterations in serum sodium concentration per se have little or no direct effect on adrenal aldosterone biosynthesis in experimental animals in vivo (McCaa et al., 1974c) or in vitro (Müller, 1965; Saruta et al., 1972). Yet, slight alterations in sodium balance have profound effects on aldosterone secretion in experimental animals and man. Since dietary sodium restriction or removal of sodium from the body by hemodialysis, diuresis or any of several other means is associated with parallel increases in plasma renin activity, plasma angiotensin II concentration and plasma aldosterone concentration, most investigators believe that the aldosterone response to alterations in sodium balance is mediated by increased activity of the renin-angiotensin system.

We have evaluated the direct effect of changes in serum sodium concentration on adrenal aldosterone production in an isolated adrenal preparation. We have utilized a technique for studying headless animals that has made it possible to study animals whose bodies have no secretory substances and no nervous control from the head. This procedure has been combined with total nephrectomy so that, in addition to the removal of the head factors, the renal renin-angiotensin system is also removed from the animals. Both of these procedures have been combined with hemodialysis so that the blood concentration of sodium and potassium and other substances, as well as the total body fluid volume, can be controlled (McCaa et al., 1973a). Figure 3 illustrates the aldosterone response to reduction in serum sodium concentration by hemodialysis in decapitated nephrectomized dogs. These studies demonstrate that acute reduction in serum sodium concentration by 25 mEq/

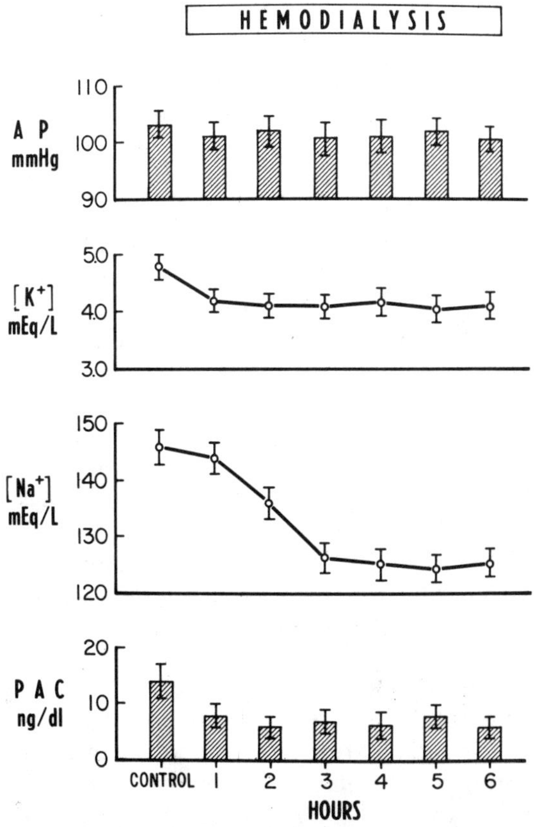

Fig. 3. The response of plasma aldosterone concentration (PAC) to a reduction in serum sodium concentration [Na^+] by hemodialysis in decapitated nephrectomized dogs. Vertical lines indicate ± S.E.

L failed to stimulate aldosterone secretion in decapitated nephrectomized animals. Similar results have been observed in hypophysectomized nephrectomized dogs (McCaa et al., 1974c). In addition to these studies, continuous infusion of antidiuretic hormone and hypotonic saline in conscious dogs for 14 days resulted in a decrease in plasma sodium concentration from 143.6 ± 0.5 to 125.0 ± 0.5 mEq/L, while plasma aldosterone concentration decreased from 9.1 ± 2.8 to 5.7 ± 1.6 ng/dl (Manning et al., 1979). Similar results have been observed in human beings maintained on antidiuretic hormone (Fichman et al., 1974), and in several clinical studies investigators have observed normal or decreased aldosterone secretion and plasma aldosterone concentration in patients with inappropriate ADH syndrome with plasma sodium concentration markedly reduced below normal levels. Collectively, these observations indicate that changes in sodium concentration per se have little or no direct effect on the zona glomerulosa to stimulate aldosterone secretion in experimental animals and man.

Role of Angiotensin II

In 1958, Gross demonstrated that sodium depletion resulted in increased renal renin content and increased renin secretion from the kidneys, followed by an increase in the plasma concentration of angiotensin II, and further that angiotensin II stimulated aldosterone secretion from the adrenal glands (Gross et al., 1958, 1960). These studies were supported by Davis et al. (1961) and others (Denton et al., 1959; Ganong and Mulrow, 1961; Mulrow et al., 1962) who demonstrated that the kidney released a humoral aldosterone-stimulating substance, and later by still other investigators who demonstrated that administration of exogenous angiotensin II increased aldosterone biosynthesis (Genest et al., 1960; Biron et al., 1961; Blair-West et al., 1962; Carpenter et al., 1961; Brown et al., 1972).

Although acute intravenous infusions of angiotensin II increase aldosterone secretion (Laragh et al., 1960), plasma concentration (Fraser et al., 1965; Horton, 1969; McCaa et al., 1973b) and urinary excretion (Genest et al., 1960; Biron et al., 1961), the physiological role of the renal renin-angiotensin system in the long-term regulation of aldosterone secretion has been less certain. In normal man, aldosterone secretion remained elevated during prolonged infusion of pressor doses of the vasopressor octapeptide, angiotensin II (Ames et al., 1965). In contrast, the adrenal glomerulosa cell in nephrectomized persons and in patients with hypopituitarism is almost totally insensitive to acute or prolonged infusion of angiotensin II (McCaa et al., 1973b; Williams et al., 1973; Weidmann et al., 1973). In conscious sodium replete sheep, long-term infusion of angiotensin II resulted in a transient increase in aldosterone secretion that lasted only 6 to 12 hr (Blair-West et al., 1973). Although acute intravenous infusion of angiotensin II in intact conscious rats resulted in a marked increase in aldosterone secretion (Coleman et al., 1974; Campbell et al., 1974), long-term infusion of angiotensin II in conscious rats failed to sustain aldosterone secretion or adrenal aldosterone production above control levels (Marieb and Mulrow, 1965; Bojesen, 1966; Komor and Müller, 1979; Henderson and McCaa, 1979).

In response to continuous angiotensin II infusion in intact conscious animals, we have observed marked increases in aldosterone secretion lasting 12 to 24 hr, followed by a sustained increase in aldosterone secretion above control levels but considerably below the levels observed during the acute period of angiotensin II infusion. The aldosterone response to the long-term infusion of angiotensin II in dogs on normal sodium intake is illustrated in Figure 4. During the acute phase of angiotensin II infusion at a rate of 15 ng/kg/min, plasma aldosterone concentration increased from 7.1 ± 2.9 to 48.3 ± 9.7 ng/dl, plasma cortisol concentration increased from 0.92 ± 0.35 to 3.85 ± 0.87 μg/dl and plasma renin activity decreased from 0.82 ± 0.23 to 0.39 ± 0.34 ng/ml/hr. Within 24 hr after beginning the angiotensin II infusion, plasma aldosterone concentration decreased from the high levels observed during the acute phase of angiotensin II infusion. During the continuous long-term angiotensin II infusion, plasma aldosterone concentration

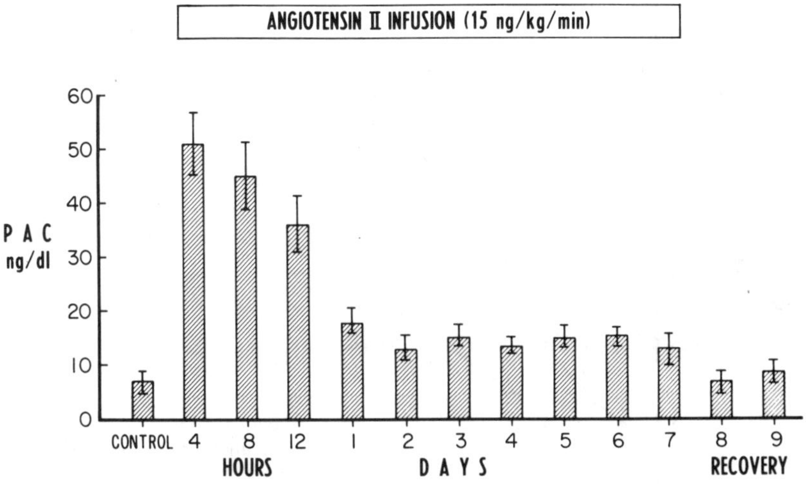

Fig. 4. The response of plasma aldosterone concentration to long-term infusion of angiotensin II at the rate of 5 ng/kg/min in intact conscious dogs on normal sodium intake (60 mEq/day). Vertical lines indicate ± S.E.

averaged 14.7 ± 4.6 ng/dl, plasma cortisol concentration averaged 1.03 ± 0.43 μg/dl, plasma renin activity was undetectable by radioimmunoassay, and arterial blood pressure remained elevated above control levels, averaging 142 ± 5 mm Hg, an increase of 42 mm Hg (McCaa, 1978). It is of interest to note that while a change in serum potassium concentration of only 20% above control level resulted in a sustained increase in plasma aldosterone concentration of 400%, an increase in blood angiotensin II concentration 400% above the levels observed during sodium deficiency was required for only a 100% sustained increase in plasma aldosterone concentration. Thus, our studies indicate that potassium is far more potent in stimulating aldosterone secretion than angiotensin II and may be the major regulatory mechanism controlling aldosterone secretion.

Role of Angiotensin III

The failure of long-term infusion of pressor doses of angiotensin II to sustain elevated aldosterone secretion to the same levels as those observed in untreated sodium deficient animals has led several investigators to suggest that the heptapeptide metabolite of angiotensin II,

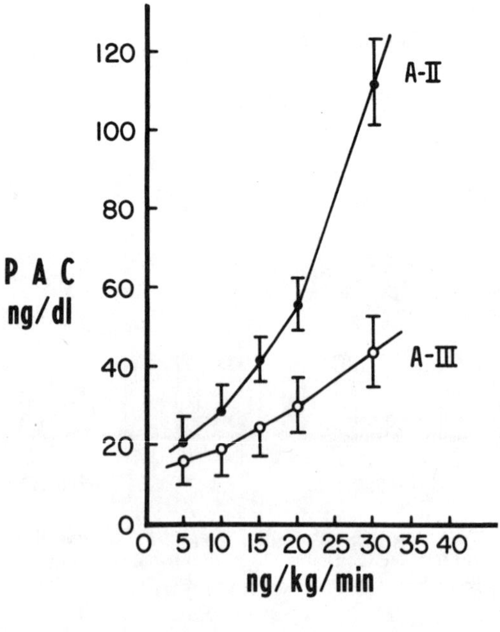

Fig. 5. Dose-response curve of angiotensin II/aldosterone and angiotensin III/aldosterone in dexamethasone treated, bilaterally nephrectomized dogs. Vertical lines indicate ± S.E.

des–asp[1] angiotensin II (angiotensin III), may be the active component of the renin-angiotensin system and play a physiological role in the regulation of aldosterone biosynthesis.

A comparison of the angiotensin II/aldosterone dose-response curve with the angiotensin III/aldosterone dose-response curve obtained in dexamethasone treated, bilaterally nephrectomized dogs is shown in Figure 5. These data demonstrate that plasma aldosterone concentration increased with each of the five rates of angiotensin II and angiotensin III infusion. However, at each of the five rates of infusion, angiotensin II was approximately twice as effective as angiotensin III in stimulating aldosterone secretion. In addition, an increase in arterial pressure was observed with each of the five rates of angiotensin II infusion, while angiotensin III failed to increase arterial pressure even at the higher rate of infusion. Also, long-term infusion of angiotensin II at the lowest rate (5 ng/kg/min) resulted in suppression of plasma renin activity to levels undetectable by radioimmunoassay, while long-term infusion of angiotensin III at the highest rate (30 ng/kg/min) had little effect on plasma renin activity.

Thus, studies from our laboratory (McCaa, 1978) and other laboratories (Carey et al., 1978; Aguilera et al., 1978) indicate that angiotensin III has less than one half the aldosterone stimulating effect of angiotensin II, only a slight effect on the feedback suppression of renin secretion and little or no effect on arterial blood pressure.

Role of Other Factors

It has become increasingly apparent that all aspects of the regulation of aldosterone secretion cannot be explained by the data presently available. For several years many investigators have postulated that other unidentified factors may be involved in the control of aldosterone biosynthesis. Recently, a new aldosterone stimulating hormone has been isolated from normal human urine (Saito et al., 1980; Bravo et al., 1980). This glycoprotein fraction is less potent than angiotensin II or ACTH in increasing aldosterone production on rabbit adrenal glomerulosa cells, and its effect is dependent on the presence of potassium. Also, studies from other laboratories have recently demonstrated that β-lipotropin has a potent effect on aldosterone production and may play a role in aldosterone regulation (Matsuoka et al., 1980).

Evidence for an Essential Role of the Pituitary

The regulation of aldosterone secretion has been studied extensively in intact animals and in man during sodium repletion and depletion. However, fewer investigations have been performed to evaluate the role of the pituitary in the control of aldosterone biosynthesis. Indeed, studies in patients with pituitary insufficiency (Lieberman and Luetscher, 1960; Williams et al., 1971) and in hypophysectomized animals (Ganong, Pemberton and Van Brunt, 1967b; Palmore and Mulrow, 1967; Lee et al., 1968) indicate that aldosterone secretion remains normal under resting conditions, but fails to respond with the usual increase after ACTH administration, angiotensin II infusion or sodium deprivation. These observations suggest that chronic deficiency of some pituitary secretory factor(s) may reduce or abolish the adrenal aldosterone response to sodium deprivation, ACTH and angiotensin II. Recently, we evaluated the response of plasma aldosterone concentration and plasma renin activity to dietary sodium restriction (10 mEq Na^+/day for 10 days) in normal intact persons and in 25 patients with pituitary insufficiency. In these studies, the response of plasma aldosterone concentration and plasma cortisol concentration to ACTH administration and to angiotensin II infusion was determined in each patient before and after dietary sodium restriction (McCaa and Langford, 1979).

In patients with pituitary insufficiency, plasma renin activity increased fivefold in response to dietary sodium restriction for 10 days. However, plasma aldosterone concentration failed to change from control levels in response to dietary sodium restriction. Despite the normal increase in plasma renin activity and blood angiotensin II concentration, and the failure of aldosterone secretion to increase during dietary sodium restriction, each patient came into sodium balance at 10 mEq of sodium per day without a change in arterial blood pressure.

While the patients with pituitary insufficiency were on normal sodium intake, angiotensin II and ACTH failed to stimulate aldosterone secretion. However, 10 days after dietary sodium restriction the aldosterone response to ACTH was markedly increased and quantitatively similar to the aldosterone response to ACTH observed in normal intact human subjects after dietary sodium restriction. In sharp contrast, during sodium deficiency, while the adrenal glomerulosa was markedly sensitive to ACTH, it remained almost totally refractory to angiotensin II, since plasma aldosterone concentration failed to increase during angiotensin II infusion for 60 min at a rate that increased arterial blood pressure.

Replacement therapy with ACTH gel for three months in patients with pituitary insufficiency failed to restore a normal aldosterone response to ACTH or angiotensin II. In previous studies, we have observed a normal aldosterone response to dietary sodium restriction, ACTH and angiotensin II in patients with isolated growth hormone deficiency (McCaa et al., 1978b). Thus, these data indicate that some pituitary factor(s), other than ACTH or growth hormone, is essential for a normal aldosterone response to sodium deficiency, ACTH or angiotensin II. These provocative studies indicate that the unknown pituitary hormone(s) may act as a conditioning or permissive factor.

REGULATION OF ALDOSTERONE BIOSYNTHESIS DURING SODIUM DEFICIENCY

Despite the controversy over the role of the renin-angiotensin system in the regulation of aldosterone biosynthesis and the control of arterial pressure during sodium deficiency, most investigators observe parallel increases in the activity of the renin-angiotensin system and aldosterone secretion in response to sodium depletion in experimental animals and man. Yet, some studies have indicated that changes in plasma angiotensin II concentration, the active component of the renin-angiotensin system, are not sufficient to explain completely the increased aldosterone secretion associated with sodium deficiency (Boyd et al., 1972; Mendelsohn et al., 1972; Blair-West et al., 1973). Also, several experimental studies, including studies from our laboratory, have demonstrated that long-term infusion of pressor doses of the octapeptide, angiotensin II, fails to sustain increased plasma aldosterone concentration to the same levels of those observed in untreated sodium deficient animals. Conversely, still other investigators have shown a marked potentiation in adrenal sensitivity to angiotensin II during sodium deficiency (Hollenberg et al., 1974; Oelkers et al., 1974), indicating that the aldosterone response to angiotensin II may be augmented with decreased sodium intake (Davis et al., 1969; Cowley and McCaa, 1976).

Alterations in Adrenal Sensitivity During Sodium Deficiency

Several investigators have suggested that the sensitivity of the adrenal glomerulosa cell to various stimuli (ACTH and angiotensin II) is markedly increased during sodium deficiency and that the increased sensitivity is mediated, at least in part, by the increase in blood angiotensin II concentration (Ganong and Boryczka, 1967; Ganong et al., 1967a). However, nearly all of the previous studies on the aldosterone response to ACTH and angiotensin II during alterations in sodium balance have been short-term in nature, lasting for only a few hours. Indeed, as shown in Figure 4, continuous angiotensin II infusion results in a marked increase in aldosterone secretion lasting for only a few hours, followed by a sustained increase in aldosterone secretion above preinfusion levels but well below the levels observed during the acute phase of angiotensin II infusion, and well below the aldosterone levels observed in untreated sodium deficient animals.

Aldosterone Response to Long-term Angiotensin II Infusion in Intact Dogs Before and After Dietary Sodium Restriction

Recently we determined the aldosterone response to long-term infusion of angiotensin II at four different rates in conscious dogs before (Fig. 6) and after (Fig. 7) dietary sodium restriction. Figure 6 and Figure 7 illustrate only the long-term steady-state levels of plasma aldosterone concentration in the same animals before and after dietary sodium re-

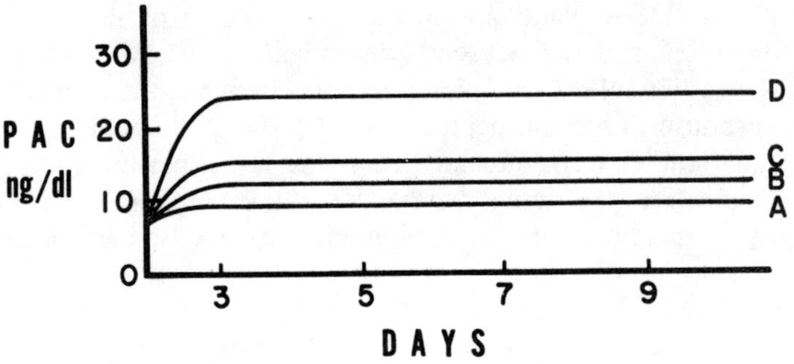

Fig. Fig. 6. The long-term steady state response of plasma aldosterone concentration (PAC) to angiotensin II infusion at four different rates (A = 5 ng/kg/min, B = 10 ng/kg/min, C = 15 ng/kg/min, D = 25 ng/kg/min) in sodium replete dogs.

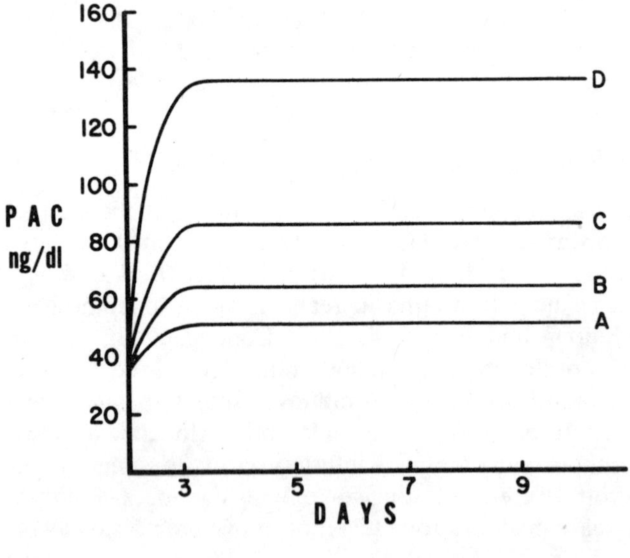

Fig. 7. The long-term steady state response of plasma aldosterone concentration (PAC) to angiotensin II infusion at four different rates (A = 5 ng/kg/min, B = 10 ng/kg/min, C = 15 ng/kg/min, D = 25 ng/kg/min) in sodium deplete dogs.

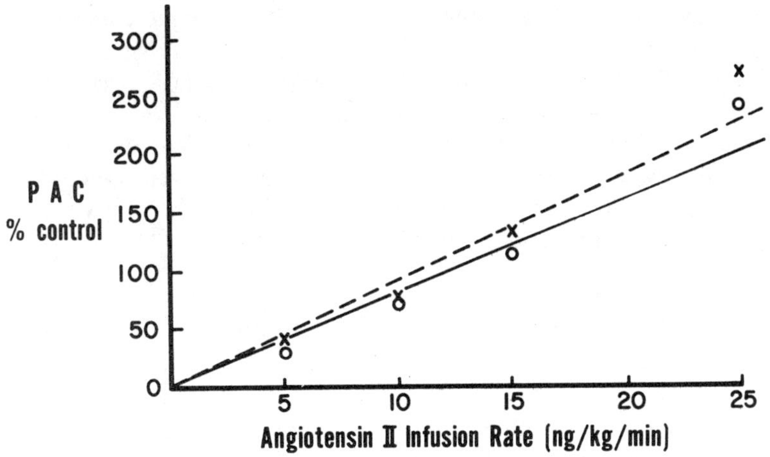

Fig. 8. The relationship between change in plasma aldosterone concentration and angiotensin II infusion at four different rates in dogs before and after dietary sodium restriction (5 mEq Na^+/day) for 21 days. ———— = sodium replete, — — — = sodium deplete.

striction. Plasma aldosterone concentration remained elevated above control levels at each of the four rates of angiotensin II infusion while the animals were sodium replete and sodium deplete. Figure 8 illustrates the relationship between the change in plasma aldosterone concentration and the rate of angiotensin II infusion in sodium replete and sodium deplete animals during long-term steady-state conditions. Thus, we have found very little difference in the percent change in plasma aldosterone concentration in response to angiotensin II at different rates of infusion in animals before and after sodium depletion. Therefore, our studies do not support the concept of increased adrenal sensitivity to angiotensin II during sodium deficiency.

Use of Pharmacologic Agents to Interrupt the Renin-Angiotensin System in Sodium Deficient Animals

In more recent studies, we have used several pharmacologic agents to interrupt the renal renin-angiotensin system in sodium deficient dogs in an effort to evaluate the role of the renin-angiotensin system in the maintenance of renal blood flow, urinary sodium excretion, plasma aldosterone concentration and arterial blood pressure during sodium deficiency. Our studies were designed to inhibit renin secretion with the beta-adrenergic blocking agent, propranolol, to block the action of angiotensin II at the vascular smooth muscle and adrenal glomerulosa receptor sites with angiotensin II inhibitory analogues, and to inhibit the conversion of angiotensin I to angiotensin II with the angiotensin I converting enzyme (kininase II) inhibitor, captopril.

All of these studies were long-term in nature and performed in dogs maintained on dietary sodium restriction (5 mEq Na^+/day) for at least 21 days. Since diuretics have profound effects on renin secretion and urinary sodium and potassium excretion, sodium deficiency was induced entirely by dietary sodium restriction.

Response of Arterial Blood Pressure, Urinary Sodium Excretion and Plasma Aldosterone Concentration to Suppression of Renin Release with Propranolol

Intact conscious sodium deficient dogs were maintained on long-term continuous propranolol infusion (200 mg/day) to determine the response of arterial blood pressure, urinary sodium excretion and plasma aldosterone concentration during suppression of renin secretion by beta-adrenergic blockade. Figure 9 illustrates the change in arterial pressure, urinary sodium excretion, and plasma aldosterone concentration after 10 days of continuous propranolol infusion. In response to beta-adre-

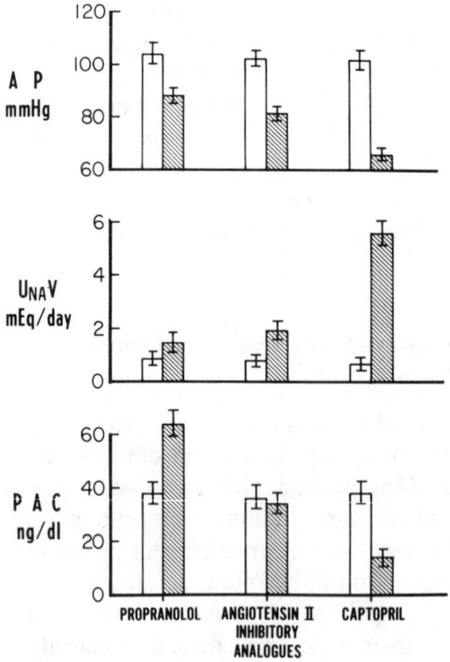

Fig. 9. The response of arterial blood pressure (AP), urinary sodium excretion ($U_{Na}V$) and plasma aldosterone concentration (PAC) in sodium deficient dogs after continuous infusion of propranolol, angiotensin II inhibitory analogs or captopril for 10 days. Vertical lines indicate ± S.E.

nergic blockade in sodium deficient dogs, plasma renin activity decreased from 3.63 ± 0.42 to 1.12 ± 0.26 ng/ml/hr ($P<0.001$), arterial blood pressure decreased from 100 ± 3 to 84 ± 2 mm Hg ($P<0.001$), plasma aldosterone concentration increased from 36.8 ± 6.9 to 73.6 ± 8.7 ng/dl ($P<0.001$), and urinary kallikrein activity increased from 19.7 ± 3.9 to 41.3 ± 4.8 EU/day ($P<0.001$). Also, during beta-adrenergic blockade with propanolol, serum sodium and potassium concentration increased from 143 ± 0.5 to 145 ± 0.5 (ns, $P<0.05$) and 4.3 ± 0.5 to 5.6 ± 0.5 ($P<0.001$) mEq/1, respectively, while urinary sodium and potassium excretion increased from 0.59 ± 0.25 to 1.12 ± 0.32 ($P<0.01$) and 34.7 ± 4.8 to 39.7 ± 5.1 ($P<0.01$) mEq/day, respectively.

In sodium deficient dogs, beta-adrenergic blockade with propranolol produced parallel decreases in plasma renin activity and arterial blood pressure, while urinary sodium excretion increased during propranolol administration. The increase in plasma aldosterone concentration during beta-adrenergic blockade was associated with a highly significant sustained increase in serum potassium concentration induced by propranolol administration. It is well established that propranolol, and other beta-adrenergic blocking agents, induce a massive efflux of intracellular potassium ions from red blood cells (Ekman et al., 1969), and from muscle cells in man (Mannien, 1970). The increase in urinary kallikrein activity observed during propranolol administration was probably due to the marked increase in aldosterone secretion, since considerable experimental evidence indicates a high positive correlation between kallikrein activity and changes in activity of sodium retaining steroids (Geller et al., 1972; Margolius et al., 1974, 1976; McCaa, 1978).

Response of Arterial Blood Pressure, Urinary Sodium Excretion and Plasma Aldosterone Concentration to Angiotensin II Receptor Blockade with Angiotensin II Inhibitory Analogs

Long-term infusion of two potent angiotensin II inhibitory analogs, Sar^1-Ala^8 angiotensin II and Sar^1-Ile^8 angiotensin II, was maintained in sodium deficient dogs to determine the response of arterial blood pressure, urinary sodium excretion and plasma aldosterone concentration during angiotensin II receptor blockade. Figure 9 illustrates the change in arterial pressure, urinary sodium excretion, and plasma aldosterone concentration after 10 days of Sar^1-Ala^8 angiotensin II infusion.

In response to continuous infusion of Sar^1-Ala^8 angiotensin II in sodium deficient dogs, arterial blood pressure decreased from 103 ± 4 to 78 ± 2 mm Hg ($P<0.001$) while urinary sodium excretion increased from 0.61 ± 0.23 to 1.84 ± 0.31 mEq/day ($P<0.001$). However, plasma aldosterone concentration failed to change significantly from control levels during long-term infusion of Sar^1-Ala^8 angiotensin II. In contrast, continuous long-term infusion of Sar[1]-Ile[8] angiotensin II resulted in a significant increase in plasma aldosterone concentration in sodium deficient dogs. The agonistic properties of the angiotensin II inhibitory analogs on the adrenal glomerulosa cells have been observed by several other investigators in experimental animals (Steele and Lowenstein, 1974; Williams et al., 1974; Beckerhoff et al., 1975; Bravo et al., 1975, 1976; McCaa, 1977a) and in man (Hollenberg et al., 1976; Sealey et al., 1976), and have led some investigators (Steele and Lowenstein, 1974; Williams et al., 1974; Bravo et al., 1975) to postulate that significant functional differences exist between the angiotensin II receptors in the vascular smooth muscle and the angiotensin II receptors in the adrenal glomerulosa cells.

Response of Arterial Blood Pressure, Urinary Sodium Excretion and Plasma Aldosterone Concentration to Inhibition of Angiotensin I Converting Enzyme (Kininase II) with Captopril

Although beta-adrenergic blockade with propranolol and angiotensin II receptor blockade with angiotensin II inhibitory analogues effectively decreased arterial blood pressure and increased urinary sodium excre-

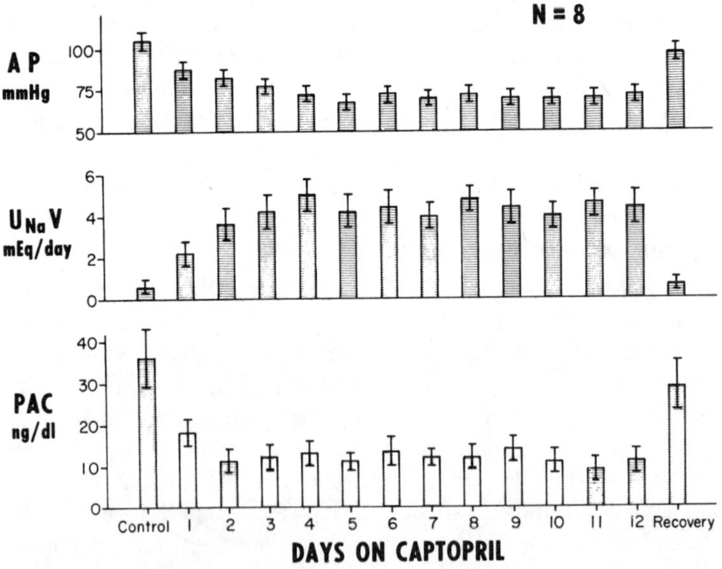

Fig. 10. The response of arterial blood pressure (AP), urinary sodium excretion ($U_{Na}V$) and plasma aldosterone concentration (PAC) to long-term captopril administration (400 mg/day) in sodium deficient dogs. Vertical lines indicate ± S.E. N indicates number of animals studied.

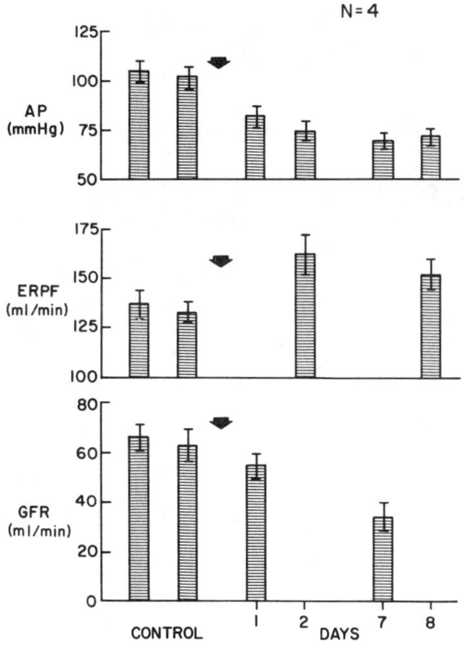

Fig. 11. The response of arterial blood pressure (AP), effective renal plasma flow (ERPF) and glomerular filtration rate (GFR) to long-term captopril administration (400 mg/day) in sodium deficient dogs. Vertical lines indicate ± S.E. N indicates number of dogs studied.

tion in sodium deficient dogs, the most effective pharmacologic agents used thus far for studying the role of the renin-angiotensin system in the regulation of aldosterone secretion and the control of arterial blood pressure during sodium deficiency have been the angiotensin I converting enzyme inhibitors, teprotide and captopril (Engel et al., 1972; Ondetti et al., 1977; McCaa, 1977a; McCaa et al., 1978a; Aguilera and Catt, 1978). Figure 9 demonstrates that the hypotensive and natriuretic actions of captopril are greater than that of propranolol and the angiotensin II inhibitory analogues. The long-term response of arterial blood pressure, urinary sodium excretion, and plasma aldosterone concentration to captopril is shown in Figure 10. During long-term captopril administration in sodium deficient dogs, arterial blood pressure decreased from 103 ± 5 to 67 ± 2 mm Hg ($P<0.001$), urinary sodium excretion increased from 0.65 ± 0.26 to 4.8 ± 1.2 mEq/day ($P<0.001$), and plasma aldosterone concentration decreased from 36.4 ± 6.8 to 14.2 ± 3.4 ng/dl ($P<0.001$). Also, Figure 11 illustrates that effective renal plasma flow increased from 136 ± 6 to 154 ± 8 ml/min ($P<0.01$) and remained elevated during long-term captopril administration while glomerular filtration rate decreased from 65 ± 9 to 36 ± 7 ml/min ($P<0.001$) and remained below control levels until arterial blood pressure returned to normal after captopril administration was discontinued.

ROLE OF THE RENAL KALLIKREIN-KININ SYSTEM

It is well established that the potent vasoactive octapeptide angiotensin II is produced by the enzymatic removal of a dipeptide from the biologically inactive decapeptide angiotensin I by angiotensin I converting enzyme. Two potent vasodepressor peptides, bradykinin and kallidin, are inactivated by the enzymatic removal of a dipeptide by kininase II. In 1970 Erdös and Yang showed that angiotensin I converting enzyme and kininase II are the same protein, peptidyl dipeptide hydrolase. In experimental animals (Nasjletti and Colina-Chourio, 1976; McCaa et al., 1978a) and in man (Swartz et al., 1979), inhibition of angiotensin I converting enzyme (kininase II) with captopril results in decreased concentrations of circulating and renal angiotensin II and increased concentrations of circulating and renal kinins. Inhibition of angiotensin II formation and increased concentration of circulating and renal kinins

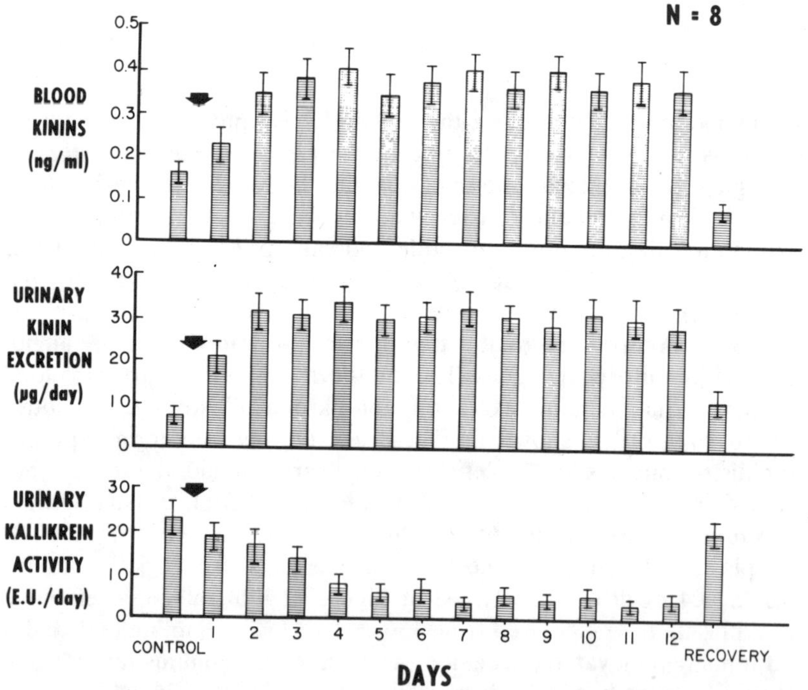

Fig. 12. The response of blood kinins, urinary kinin excretion and urinary kallikrein activity to long-term captopril administration in sodium deficient dogs. Vertical lines indicate ± S.E. N indicates number of animals studied.

both produce hypotensive and natriuretic actions in normal (Barraclough and Mills, 1965; Willis et al., 1969; Stein et al., 1972; Nasjletti et al., 1975), sodium deficient and experimental renovascular hypertensive animals (Bengis et al., 1978; McCaa et al., 1978a).

Response of Blood Kinins, Urinary Kinin Excretion and Urinary Kallikrein Activity to Long-term Captopril Administration

The response of blood kinin concentration, urinary kinin excretion and urinary kallikrein activity to long-term captopril administration in sodium deficient dogs is illustrated in Figure 12. During long-term captopril administration, the concentration of kinins in the blood increased from 0.17 ± 0.02 to 0.41 ± 0.05 ng/ml ($P<0.001$), urinary kinin excretion increased from 7.2 ± 1.5 to 31.4 ± 3.2 μg/day ($P<0.001$) and urinary kallikrein activity decreased from 23.6 ± 3.1 to 5.3 ± 1.2 EU/day ($P<0.001$).

The increase in blood kinins is due to a decrease in the rate of degradation of the kinins and not to an increase in kallikrein activity, since kallikrein activity decreased markedly in sodium deficient animals (McCaa et al., 1978a) and hypertensive man (Karlberg et al., 1980) during long-term captopril administration. It is noteworthy that urinary kallikrein excretion is a direct reflection of renal kallikrein release. The decrease in urinary kallikrein activity in the present study could have been caused by decreased plasma angiotensin II concentration (Nasjletti et al., 1975), decreased renal arterial perfusion pressure (Mills et al., 1976), decreased plasma aldosterone concentration (Margolius et al., 1976) or a combination of all of these factors. Indeed, it is also possible that a direct intrarenal kinin-kallikrein feedback mechanism may exist.

ROLE OF THE RENAL RENIN-ANGIOTENSIN AND KALLIKREIN-KININ SYSTEMS IN MEDIATING THE HYPOTENSIVE AND NATRIURETIC ACTIONS OF CAPTOPRIL

Although the precise role of the renal renin-angiotensin and the kallikrein-kinin systems in mediating the decrease in arterial blood pressure and increase in urinary sodium excretion during inhibition of angiotensin I converting enzyme (kininase II) still has not been evaluated quantitatively, the significant increase in circulating and renal concentration of kinins during long-term administration of captopril suggests that the hypotensive and natriuretic actions of the inhibitors of angiotensin I converting enzyme (kininase II) could be due in part to inhibition of angiotensin II formation and in part to the accumulation of kinins.

Response of Arterial Blood Pressure, Urinary Sodium Excretion and Plasma Aldosterone Concentration to Angiotensin II Infusion in Sodium Deficient Dogs Maintained on Captopril

To evaluate quantitatively the role of angiotensin II in mediating the hypotensive and natriuretic actions of captopril, we determined the response of arterial blood pressure, urinary sodium excretion and plasma aldosterone concentration to angiotensin II infusion in sodium deficient dogs maintained on captopril (Fig. 13) (McCaa, 1980). Sodium deficient dogs were maintained on continuous captopril infusion (400 mg/day) for 7 days. After seven days of captopril infusion, arterial blood pressure decreased from 102 ± 3 to 68 ± 4 mm Hg (P<0.001), urinary sodium excretion increased from 0.69 ± 0.24 to 7.8 ± 1.6 mEq/day (P<0.001)

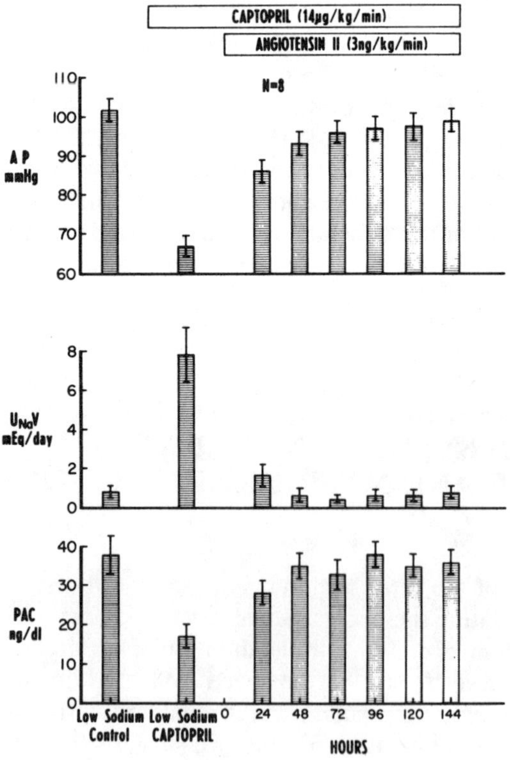

Fig. 13. The response of arterial blood pressure (AP), urinary sodium excretion ($U_{Na}V$) and plasma aldosterone concentration (PAC) to continuous angiotensin II infusion (3 ng/kg/min) in sodium deficient dogs maintained on continuous captopril infusion. Vertical lines indicate ± S.E. N indicates number of dogs studied.

and plasma aldosterone concentration decreased from 37.8 ± 6.5 to 16.3 ± 3.6 ng/dl (*P*<0.001). During the next 7 days the animals were maintained on captopril infusion at the rate of 400 mg/day and angiotensin II infusion at the rate of 3 ng/kg/min. In response to angiotensin II infusion (3 ng/kg/min), arterial blood pressure increased from 68 ± 4 to 86 ± 3 mm Hg (*P*<0.001), urinary sodium excretion decreased from 7.8 ± 1.6 to 1.7 ± 0.9 mEq/day (*P*<0.001) and plasma aldosterone concentration increased from 16.3 ± 3.6 to 27.5 ± 5.2 ng/dl (*P*<0.001) within 24 hr after beginning angiotensin II infusion. During the next 48 hr of continuous angiotensin II infusion, arterial blood pressure increased to 96 ± 3 mm Hg, urinary sodium excretion decreased to 0.53 ± 0.21 mEq/day and plasma aldosterone concentration increased to 34.3 ± 6.4 ng/dl. Continuous infusion of angiotensin II at the rate of 3 ng/kg/min restored arterial blood pressure, urinary sodium excretion and plasma aldosterone concentration to levels that were not significantly different from those observed in untreated sodium deficient dogs, indicating that the hypotensive and natriuretic actions of inhibitors of angiotensin I converting enzyme (kininase II) are due to the inhibition of angiotensin II formation and not to increased circulating and renal concentrations of kinins.

It is of interest to note that Gross et al. (1965) demonstrated over 20 years ago that the active component of the renin-angiotensin system, angiotensin II, has at least three major functions: a) it increases intrarenal vascular resistance; b) it acts on the renal tubules to conserve sodium; and c) it stimulates aldosterone secretion. Captopril administration inhibits angiotensin II formation in sodium deficient dogs, resulting in a marked decrease in intrarenal vascular resistance, an increase in renal blood flow, an increase in urinary sodium excretion, a decrease in plasma aldosterone concentration and a decrease in arterial blood pressure. Angiotensin II infusion at the rate of 3 ng/kg/min into sodium deficient dogs maintained on captopril increased intrarenal vascular resistance, induced sodium retention, stimulated aldosterone secretion and restored arterial blood pressure to levels that existed in untreated sodium deficient dogs.

As can be seen in Figure 14, when angiotensin II infusion was increased to higher infusion rates in sodium deficient dogs maintained on captopril, arterial blood pressure, urinary sodium excretion and plasma aldosterone concentration increased markedly. However, at the higher rates of angiotensin II infusion the increase in urinary sodium excretion was due to pressure induced diuresis and natriuresis.

Additional studies performed earlier have also suggested that the kallikrein-kinin system plays little or no role in mediating the hypotensive and natriuretic actions of captopril. First, it is well-established that plasma renin activity is markedly suppressed and urinary kallikrein activity is markedly elevated in experimental animals in which hypertension is induced by administration of salt-retaining steroid hormones and sodium chloride. Studies by Douglas et al. (1979) have demonstrated that captopril does not exert its antihypertensive action on either the development of or on established desoxycorticosterone-salt hypertension in rats. Second, in conscious sodium deficient animals maintained on continuous Trasylol administration to inhibit blood and renal kallikrein activity, the hypotensive and natriuretic response to captopril

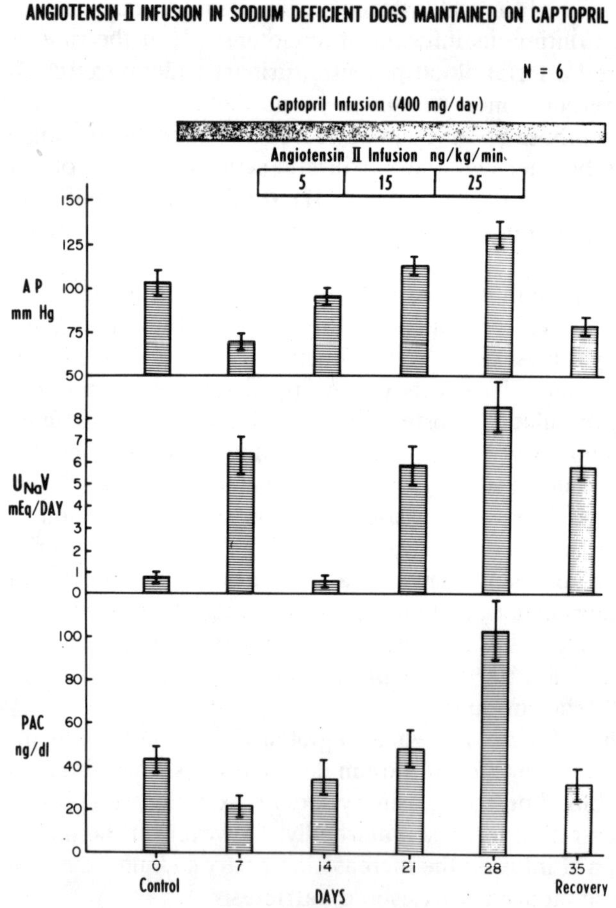

Fig. 14. The response of arterial blood pressure (AP), urinary sodium excretion ($U_{Na}V$) and plasma aldosterone concentration (PAC) to angiotensin II infusion at three different rates in sodium deficient dogs maintained on long-term captopril administration. Vertical lines indicate ± S.E. N indicates number of animals studied.

administration was quantitatively similar to the response observed to captopril in untreated control animals. Antonaccio et al. (1980) have made similar observations in the spontaneously hypertensive rat. Finally, continuous infusion of bradykinin into sodium deficient animals at rates that produce blood kinin concentrations comparable to those observed during long-term administration of captopril failed to induce a sustained decrease in arterial blood pressure and a sustained increase in urinary sodium excretion.

Evidence that the kallikrein-kinin system plays little or no role in the antihypertensive response to captopril has been provided by recent research studies in experimental animals using renin inhibitors and angiotensin II antagonists to interrupt the renin-angiotensin system. Sen et al. (1979) demonstrated that the phospholipid renin-inhibitor, URI–73A, and the angiotensin II inhibitory analogue, Sar^1-Thr^8 angiotensin II, were equally as effective as captopril in lowering arterial blood pressure in high renin models of renovascular hypertensive rats. Additional studies by Antonaccio et al. (1980) have demonstrated that captopril produced no further hypotensive response in spontaneously hypertensive rats after URI–73A administration. Since URI–73A and Sar^1-Thr^8 angiotensin II have no apparent bradykinin potentiating effects, these investigators concluded that all three agents were working by interfering only with the renin-angiotensin system.

Thus, we have concluded from all of these experimental studies that the hypotensive and natriuretic actions of inhibitors of angiotensin I converting enzyme (kininase II) are due to the decrease in circulating and renal concentration of angiotensin II and not to the accumulation of circulating and renal kinins.

ROLE OF THE RENIN-ANGIOTENSIN-ALDOSTERONE SYSTEM IN EXPERIMENTAL RENOVASCULAR HYPERTENSION

The role of the renin-angiotensin-aldosterone system in the development and maintenance of experimental renovascular hypertension has been studied extensively in experimental animal models (Carpenter et al., 1961; Singer et al., 1963; Gross et al., 1965; McCaa et al., 1965; Blair-West et al., 1968; Liard et al., 1974). In experimental 1 kidney benign renovascular hypertension, plasma renin activity, blood angiotensin II concentration and plasma aldosterone concentration are elevated during the early developmental phase of hypertension. However, plasma renin activity, blood angiotensin II concentration and plasma aldosterone concentration return to near normal levels after 3 to 5 days, while arterial blood pressure remains elevated. In contrast, in experimental renovascular malignant hypertension, plasma renin activity, blood angiotensin II concentration and plasma aldosterone concentration remain elevated throughout the development and maintenance of hypertension. Most investigators agree that the increased arterial blood pressure in malignant hypertensive animals and human beings is mediated by the powerful vasoconstrictor action of the octapeptide, angiotensin II, on the vascular smooth muscle while the increased aldosterone secretory rate is produced by the action of the octapeptide on the adrenal glomerulosa cells.

Captopril, the potent, orally active inhibitor of angiotensin I converting enzyme (kininase II) is the most effective pharmacologic agent known to interrupt the renin-angiotensin system, and has provided a means by which the importance of the system can be evaluated quantitatively in animals with experimental hypertension and in hypertensive human beings. Recently, we evaluated the response of arterial blood pressure, plasma renin activity and plasma aldosterone concentration to inhibition of angiotensin II formation in rats with established 1 and 2 kidney Goldblatt benign and 1 and 2 kidney Goldblatt malignant hypertension (McCaa et al., 1979). Our studies were performed in intact unanesthetized male Sprague-Dawley rats with chronic indwelling catheters in the aorta. Benign Goldblatt hypertensive rats were produced by placing a 0.25 mm silver clip on the left renal artery while the contralateral kidney was either removed (1 kidney) or remained intact (2 kidney). Malignant Goldblatt hypertensive rats were produced by placing a 0.20 mm silver clip on the left renal artery while the contralateral kidney was either removed (1 kidney) or remained intact (2 kidney). Captopril was administered to the experimental animals by adding the powder to the drinking water (0.5 mg/ml) (Bengis et al., 1978).

The Response of Arterial Blood Pressure, Plasma Renin Activity and Plasma Aldosterone Concentration to Long-term Captopril Administration in Normal, Sodium Deficient and Experimental Renovascular Hypertensive Rats

Normal Rats

The response of arterial blood pressure, plasma renin activity and plasma aldosterone concentration to captopril administration was studied in normal and sodium depleted rats. During a 2 wk period of captopril administration, arterial blood pressure decreased progressively from 103 ± 3 to 89 ± 2 mm Hg, plasma renin activity increased from 1.0 ± 0.18

to 12.3 ± 2.7 ng/ml/hr, while plasma aldosterone concentration failed to change significantly during captopril treatment averaging 18.3 ± 4.2 ng/dl in normal rats.

Sodium Deficient Rats

During a 2 wk period of captopril administration in sodium depleted rats, arterial blood pressure decreased markedly within 24 hr after treatment from 103 ± 2 to 68 ± 2 mm Hg, plasma renin activity increased from 7.8 ± 2.3 to 36.3 ± 5.6 ng/ml/hr, while plasma aldosterone concentration decreased to near sodium replete levels, averaging 20.7 ± 4.3 ng/dl ($P<0.001$) 3 days after beginning captopril administration.

The Relationship Between Initial Plasma Renin Activity and Change in Mean Arterial Blood Pressure in Response to Captopril Administration in Four Groups of Renovascular Hypertensive Rats

The relationship between the initial plasma renin activity and the decrease in mean arterial blood pressure in response to captopril administration for 7 days in 1 and 2 kidney benign and 1 and 2 kidney malignant hypertensive rats is illustrated in Figure 15.

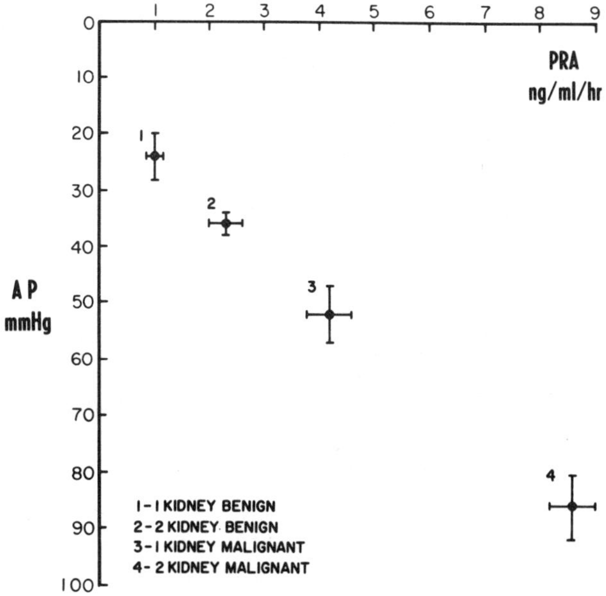

Fig. 15. The relationship between the initial plasma renin activity (PRA) and the change in arterial blood pressure (AP) in 1 and 2 kidney benign and 1 and 2 kidney malignant renovascular hypertensive rats. Vertical and horizontal lines indicate ± S.E.

1 Kidney Goldblatt Benign Hypertensive Rats

During a 2 wk control period before captopril administration, arterial blood pressure averaged 153 ± 5 mm Hg, plasma renin activity averaged 0.7 ± 0.2 ng/ml/hr and plasma aldosterone concentration averaged 16.7 ± 3.8 ng/dl in 1 kidney Goldblatt benign hypertensive rats. In response to captopril administration, arterial blood pressure decreased progressively, averaging 139 ± 5 mm Hg ($P<0.001$) after seven days; plasma renin activity increased markedly, averaging 13.7 ± 3.4 ng/ml/hr ($P<0.001$) after 7 days, while plasma aldosterone concentration failed to change significantly from control levels, averaging 17.3 ± 4.1 ng/dl after 7 days.

2 Kidney Goldblatt Benign Hypertensive Rats

In 2 kidney Goldblatt benign hypertensive rats during a 2 wk control period before captopril administration, arterial blood pressure averaged 133 ± 5 mm Hg, plasma renin activity averaged 2.8 ± 0.5 ng/ml/hr and plasma aldosterone concentration averaged 31.2 ± 5.3 ng/dl. In response to captopril administration, arterial blood pressure decreased progressively for 7 days, averaging 98 ± 3 mm Hg ($P<0.001$); plasma renin activity increased markedly averaging 19.3 ± 3.3 ng/ml/hr ($P<0.001$) after 7 days; while plasma aldosterone concentration decreased from 31.2 ± 5.3 to 22.7 ± 4.3 ng/dl, and remained near normal levels, averaging 19.0 ± 4.2 ng/dl ($P<0.01$).

1 Kidney Goldblatt Malignant Hypertensive Rats

In 1 kidney Goldblatt malignant hypertensive rats during a 2 wk control period before captopril administration, arterial blood pressure averaged 192 ± 5 mm Hg, plasma renin activity averaged 5.1 ± 0.4 ng/ml/hr and plasma aldosterone concentration averaged 43.3 ± 6.7 ng/dl. In response to captopril administration, arterial blood pressure decreased progressively, averaging 136 ± 5 mm Hg ($P<0.001$) after 7 days; plasma renin activity increased to 22.8 ± 2.3 ng/ml/hr; and plasma aldosterone concentration decreased from 43.3 ± 6.7 to 21.2 ± 4.3 ng/dl ($P<0.01$), and remained near normal levels during the next 4 days of captopril treatment, averaging 19.3 ± 4.5 ng/dl ($P<0.01$).

2 Kidney Goldblatt Malignant Hypertensive Rats

In 2 kidney Goldblatt malignant hypertensive rats during a 2 wk control period before captopril administration, arterial blood pressure averaged 192 ± 5 mm Hg, plasma renin activity averaged 9.8 ± 1.6 ng/ml/hr and plasma aldosterone concentration averaged 83.8 ± 12.4 ng/dl. In response to captopril administration, arterial pressure decreased from 192 ± 5 to 108 ± 3 mm Hg ($P<0.001$) within 24 hr and averaged 96 ± 3 mm Hg after 7 days. Plasma renin activity increased markedly and remained elevated during captopril administration, averaging 29.8 ± 4.3 ng/ml/hr ($P<0.001$) after 7 days. Plasma aldosterone concentration decreased from 83.8 ± 12.4 to 25.6 ± 6.9 ng/dl ($P<0.001$) within 48 hr after captopril treatment, and remained near normal levels during captopril administration. During a recovery period of 1 wk, arterial blood pressure, plasma renin activity and plasma aldosterone concentration returned toward the levels that existed during the control period in all four groups of renovascular hypertensive rats.

Relationship Between Plasma Renin Activity and Plasma Aldosterone Concentration in Experimental Renovascular Hypertensive Rats

The relationship between plasma renin activity and plasma aldosterone concentration in normal, sodium depleted and experimental renovascular hypertensive rats is illustrated in Figure 16. In normal rats, plasma renin activity averaged 1.0 ± 0.18 ng/ml/hr and plasma aldosterone concentration averaged 18.3 ± 4.2 ng/dl. Plasma renin activity and plasma aldosterone concentration were below normal in 1 kidney benign Goldblatt hypertensive rats averaging 0.7 ± 0.2 ng/ml/hr and 16.7 ±

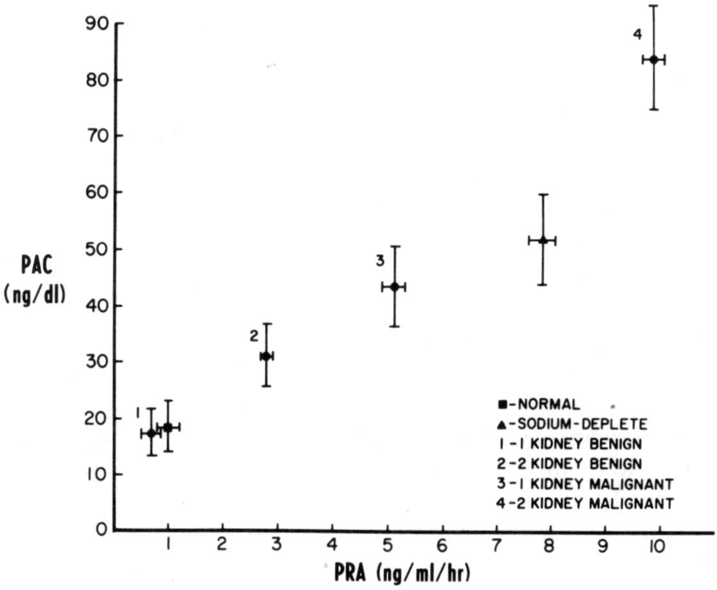

Fig. 16. The relationship between plasma renin activity (PRA) and plasma aldosterone concentration (PAC) in normal rats (square), sodium-depleted rats (triangle) and in four groups of rats with experimental renovascular hypertension (circles). Horizontal and vertical lines indicate ± S.E.

3.8 ng/dl, respectively. In sodium depleted rats and the other three groups of experimental renovascular hypertensive rats used in these investigations, plasma renin activity and plasma aldosterone concentration were markedly raised above the levels observed in normal rats. The highest values for plasma renin activity and plasma aldosterone concentration were observed in the 2 kidney Goldblatt malignant hypertensive rats. However, during long-term blockade of angiotensin II formation with captopril, plasma aldosterone concentration returned to near normal levels in the sodium depleted rats, the 2 kidney Goldblatt benign hypertensive rats, and in the 1 and 2 kidney Goldblatt malignant hypertensive rats, while plasma renin activity increased markedly during inhibition of angiotensin II formation (McCaa, R.E. et al., 1979).

ANTIHYPERTENSIVE ACTION OF CAPTOPRIL IN PATIENTS WITH SEVERE, TREATMENT-RESISTANT MALIGNANT HYPERTENSION

The knowledge acquired from the research studies in experimental renovascular hypertensive animals, using angiotensin inhibitors to delineate the role of the renin-angiotensin-aldosterone system, led us logically to extend our investigations to human beings with hypertension (McCaa, C.S., et al., 1979). Investigative studies were performed in human beings with a history of severe hypertension with supine diastolic blood pressure > 120 mm Hg before conventional antihypertensive therapy. The patients had either severe essential hypertension uncontrolled by conventional antihypertensive therapy, or malignant hypertension and supine diastolic blood pressure >100 mm Hg with or without antihypertensive therapy. This research investigation in human beings was approved by the Institutional Human Investigation Committee, performed in the Clinical Research Center at the University of Mississippi School of Medicine and informed consent was obtained from each patient before enrollment in the study.

Upon entry into the study, all prestudy antihypertensive drugs were replaced with "standard triple therapy" medication. The optimal standard triple therapy was hydralazine (50 mg, 4 times daily), propranolol (80 mg, 4 times daily) and hydrochlorothiazide (25 mg, 4 times daily). Patients were maintained on standard triple therapy for 2 wk as outpatients. If diastolic blood pressure remained above 100 mm Hg, the patient was randomly assigned to captopril or continued on standard triple therapy. The patients were admitted to the Clinical Research Center during the initiation of captopril and underwent single dose titration to obtain an effective dose, i.e., lowering of supine diastolic blood pressure by at least 10 mm Hg. Such an effective dose was given every 8 hr. If the maximal dose, 150 mg of captopril 3 times daily, did not reduce the diastolic blood pressure at least 10 mm Hg, hydrochlorothiazide (50 mg twice daily) was added.

The Response of Arterial Blood Pressure, Plasma Renin Activity and Plasma Aldosterone Concentration to Long-term Administration of Captopril in Patients with Severe, Treatment-Resistant Malignant Hypertension

The response of arterial blood pressure, plasma renin activity and plasma aldosterone concentration to inhibition of angiotensin I converting enzyme with captopril in patients with severe, treatment-resistant accelerated hypertension is illustrated in Figure 17. During the period on standard triple therapy, arterial blood pressure averaged 192 ± 8 mm Hg systolic and 137 ± 6 mm Hg diastolic, plasma renin activity averaged 3.37 ± 0.96 ng/ml/hr and plasma aldosterone concentration averaged 43.7 ± 6.8 ng/dl. Within 2 hr after captopril administration, there was a precipitous fall in arterial blood pressure and plasma aldosterone concentration, and a marked increase in plasma renin activity. Twenty-four hours after initiation of captopril medication, arterial blood pressure averaged 148 ± 6 mm Hg systolic ($P<0.001$)and 93 ± 5 mm Hg diastolic ($P<0.001$), plasma aldosterone concentration averaged 21.3 ± 4.2 ng/dl ($P<0.01$) and plasma renin activity averaged 22.7 ± 1.8 ng/ml/hr ($P<0.001$). Although arterial blood pressure remained significantly below the values observed during the control period, the pressure did tend to increase after 3 days, and on the seventh day hydrochlorothiazide was given in combination with captopril. During long-term administration of the combination of captopril and hydrochlorothiazide, arterial blood pressure remained significantly below pretreatment control values, averaging 141 ± 6 mm Hg systolic and 91 ± 4 mm Hg diastolic; plasma renin activity remained above control values, averaging 12.6 ± 1.4 ng/ml/hr, and plasma aldosterone concentration remained below control values, averaging 23.5 ± 4.3 ng/dl.

Our results in human beings demonstrate the dramatic effectiveness of captopril in lowering elevated arterial blood pressure in severe, treatment resistant accelerated or malignant hypertension. The pragmatic usefulness of captopril, the lack of tolerance to its antihypertensive action, and freedom from side effects are impressive. Although arterial blood pressure remained significantly below the values observed during the control period, blood pressure did tend to increase after 7 days of treatment. However, the combination of captopril and hydrochlorothiazide therapy maintained arterial blood pressure significantly below

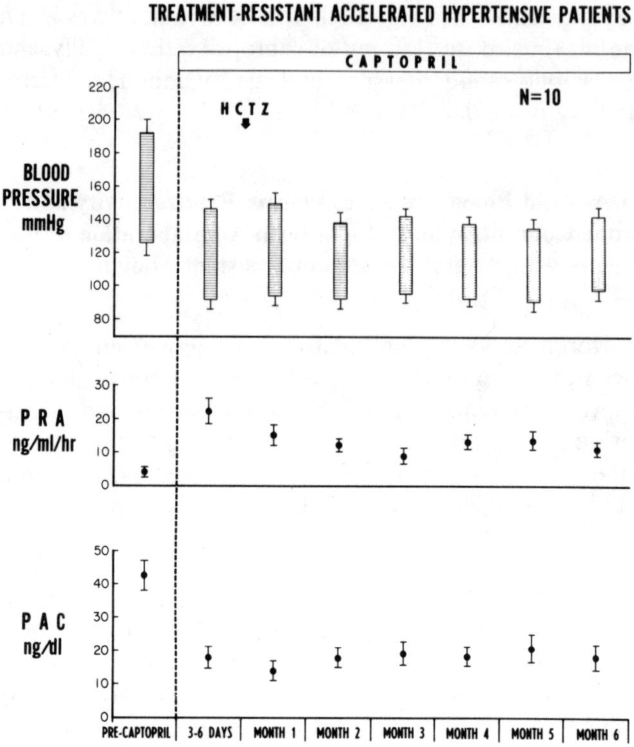

Fig. 17. The response of arterial blood pressure, plasma renin activity (PRA) and plasma aldosterone concentration (PAC) to long-term administration of captopril and combined captopril and hydrochlorothiazide therapy in patients with severe, treatment resistant accelerated or malignant hypertension. Vertical lines indicate ± S.E.

pretreatment control levels. Thus, our data demonstrate that the renin-angiotensin-aldosterone system actively participates in the maintenance of blood pressure in the vast majority of patients with hypertension and that blockade of this system with the potent, orally active inhibitor of angiotensin I converting enzyme (kininase II), captopril, represents an effective advance in the treatment of this disease.

CONCLUSIONS

The role of the renal renin-angiotensin system in the regulation of aldosterone biosynthesis and the development and maintenance of experimental renovascular hypertension has been the subject of intense debate for many years. In this chapter I have presented our studies on the long-term effects of various stimuli on aldosterone secretion in intact conscious experimental animals. I have also presented evidence for an important role of the renal renin-angiotensin system, aldosterone, and volume factors in blood pressure control in experimental animals and in human beings. When possible I have collated our data obtained in conscious animals under long-term steady-state conditions with observations of other investigators who have used different experimental approaches to study the renin-angiotensin-aldosterone system in other animal species and in man.

Our studies demonstrate several important facts that we believe will be significant in finally understanding aldosterone regulation and its role in fluid volume control. First, continuous ACTH infusion induces a transient increase in aldosterone secretion and a sustained increase in arterial blood pressure. ACTH induced hypertension is neither observed in adrenalectomized animals nor produced by infusion of high levels of aldosterone and cortisol in intact animals. It is of interest to note that captopril administration lowers arterial blood pressure in ACTH induced hypertension. Second, the marked sustained increase in aldosterone secretion (400%) induced by infusion of potassium chloride at a rate that produced a slight increase in serum potassium concentration (20%) demonstrates the important role of potassium in aldosterone regulation. Third, although sodium concentration per se has little or no effect on aldosterone secretion, alterations in sodium balance have marked effects on aldosterone secretion. When sodium deficiency is induced by dietary sodium restriction, increased aldosterone secretion is maintained by both increased activity of the renin-angiotensin system and positive potassium balance. On the other hand, increased aldosterone secretion during sodium deficiency induced by diuretics is caused by increased activity of the renin-angiotensin system. However, it is important to remember that chronic deficiency of some unidentified pituitary factor(s) prevents an increase in aldosterone secretion in response to ACTH or angiotensin II.

Our studies in several models of experimental renovascular hypertensive animals and in human beings with severe, treatment resistant malignant hypertension demonstrate the highly positive correlation between the activity of the renin-angiotensin system and the elevated arterial blood pressure. In all high renin models of experimental hypertension we have observed a dramatic antihypertensive action in response to captopril administration. Although the mechanism of action of angiotensin I converting enzyme inhibitors will be debated for some time, captopril opens a new era of therapeutics that will be both exciting and rewarding for researchers and clinicians.

REFERENCES

Aguilera, G., and Catt, K.J. 1978. Regulation of aldosterone secretion by the renin-angiotensin system during sodium restriction in rats. Physiol Sci 75:4057–4061

Aguilera, G., Baukal, A., Fujita, K., Hauger, R., and Catt, K.J. 1978. Metabolism and biological activities of angiotensin peptides in isolated rat glomerulosa cells. The Endocrine Soc(Abs):80

Ames, R.P., Borkowski, A.J., Sicinski, A.M., and Laragh, J.H. 1965. Prolonged infusions of angiotensin II and norepinephrine on blood pressure, electrolyte balance, and aldosterone and cortisol secretion in normal man and in cirrhosis with ascites. J Clin Invest 44:1171–1186

Antonaccio, M.J., Horovitz, Z.P., and Rubin, B. 1980. Angiotensin converting enzyme inhibition in experimental models of hypertension. Clin Exp Hypertension 2:613–638

Banting, F.G., and Gairns, S. 1926. Suprarenal insufficiency. Am J Physiol 44:100–113

Barraclough, M.A., and Mills, I.H. 1965. Effect of bradykinin on renal function. Clin Sci 28:69–74

Bayard, R., Cooke, C.R., Tiller, D.J., Beitins, I.Z., Kowarski, A., Walker, W.G., and Migeon, C.J. 1971. Regulation of aldosterone secretion in anephric man. J Clin Invest 50:1585–1595

Beckerhoff, R., Uhlschmid, G., Furrer, J., Nussberger, J., Schmied, U., Vetter, W.J., and Siegenthaler, W. 1975. In vivo effects of angiotensin antagonists on plasma aldosterone in the dog. Eur J Pharmacol 34:363–367

Bengis, R.G., Coleman, T.G., Young, D.B., and McCaa, R.E. 1978. Long-term blockade of angiotensin formation in various normotensive and hypertensive rat models using converting enzyme inhibitor (SQ 14,-225). Circ Res 43:I-45–I-53

Biron, P., Koiw, E., Nowaczynski, J., Brouillet, J., and Genest, J. 1961. The effects of intravenous infusions of valine-5 angiotensin II and other pressor agents on the urinary electrolytes and corticosteroids, including aldosterone. J Clin Invest 40:338–347

Blair-West, J.R., Coghlan, J.P., Denton, D.A., Goding, J.R., Munro, J.A., Peterson, R.E., and Wintour, M. 1962. Humoral stimulation of adrenal cortical secretion. J Clin Invest 41:1606–1627

Blair-West, J.R., Coghlan, J.P., Denton, D.A., Orchard, E., Scoggins, B.A., and Wright, R.D. 1968. Renin-angiotensin-aldosterone system and sodium balance in experimental renal hypertension. Endocrinology 83:1199–1209

Blair-West, J.R., Coghlan, J.P., Denton, D.A., Funder, J.W., and Scoggins, B.A. 1972. Role of the renin-angiotensin system in control of aldosterone secretion. Adv Exp Med Biol 17:167–181

Blair-West, J.R., Coghlan, J.P., Cran, E., Denton, D.A., Funder, J.W., and Scoggins, B.A. 1973. Increased aldosterone secretion during sodium depletion with inhibition of renin release. Am J Physiol 224:1409–1414

Bojesen, E. 1966. Concentration of aldosterone and corticosterone in peripheral plasma of rats. The effects of salt depletion, salt repletion, potassium loading and intravenous injections of renin and angiotensin II. Eur J Steroids 1:145–169

Boyd, G.W., Adamson, A.R., Arnold, M., James, V.H.T., and Peart, W.S. 1972. Role of angiotensin II in the control of aldosterone in man. Clin Sci 42:91–104

Boyd, J., Mulrow, P.J., Palmore, W.P., and Silvo, P. 1973. Importance of potassium in the regulation of aldosterone production. Circ Res 32–33(Suppl. 1):39–45

Bravo, E.L., Khosla, M.C., and Bumpus, F.M. 1975. Vascular and adrenocortical responses to a specific antagonist of angiotensin II. Am J Physiol 228:110–114

Bravo, E.L., Khosla, M.C., and Bumpus, F.M. 1976. Comparative studies of the humoral and arterial pressure responses to Sar^1-Ala^8, Sar^1-Ile^8 and Sar^1-Thr^8-angiotensin II in the trained, unanesthetized dog. Prog Biochem Pharmacol 12:33–40

Bravo, E.L., Saito, I., Zanella, T., Sen, S., and Bumpus, F.M. 1980. *In vitro* steroidogenic properties of a new hypertension-producing compound isolated from normal human urine. J Clin Endocrin Metab 51:176–178

Brown, R.D., Strott, C.A., and Liddle, G.W. 1972. Site of stimulation of aldosterone biosynthesis by angiotensin and potassium. J Clin Invest 51:1413–1418

Campbell, W.B., Brooks, S.N., and Pettinger, W.A. 1974. Angiotensin II- and angiotensin III-induced aldosterone release *in vivo* in the rat. Science 184:994–996

Carey, R.M., Vaughan, D., Jr., Peach, M.J., and Ayers, C. 1978. Activity of [des-$Aspartyl^1$]-angiotensin II and angiotensin II in man—Differences in blood pressure and adrenocortical responses during normal and low sodium intake. J Clin Invest 61:20–31

Carpenter, C.C.J., Davis, J.O., and Ayers, C.R. 1961. Relation of renin angiotensin II and experimental renal hypertension to aldosterone secretion. J Clin Invest 40:2026–2042

Coghlan, J.P., Denton, D.A., Fan, J.S.K., McDougall, J.G., and Scoggins, B.A. 1977. Further studies on the ACTH induced hypertension in sheep—Involvement of new hypertensinogenic steroid hormones. In *Juvenile Hypertension,* eds. M.I. New and L.S. Levine, pp. 69–77. New York: Raven Press

Coleman, T.G., McCaa, R.E., McCaa, C.S. 1974. Effect of angiotensin II on aldosterone secretion in the conscious rat. J Endocrinol 60:421–427.

Cowley, A.W., Jr., and McCaa, R.E. 1976. Acute and chronic dose response relationships of angiotensin, aldosterone and arterial pressure at varying levels of sodium intake. Circ Res 39:788–797

Davis, J.O. 1961. Mechanisms regulating the secretion and metabolism of aldosterone in experimental secondary hyper-aldosteronism. Recent Prog Horm Res 17:293

Davis, W.W., Burwell, L., and Bartter, F.C. 1969. Inhibition of the effects of angiotensin II on adrenal steroid production by dietary sodium. Proc Natl Acad Sci USA 63:718–723

Denton, D.A., Goding, J.R., and Wright, R.D. 1959. Control of adrenal secretion of electrolyte-active steroids. Br Med J 2:447–522

Douglas, B.H., Langford, H.G., and McCaa, R.E. 1979. Response of mineralocorticoid hypertensive animals to an angiotensin I converting enzyme inhibitor (40496). Proc Soc Exp Biol Med 161:86–87

Ekman, A., Manninen, V., and Salminin, S. 1969. Ion movements in red cells treated with propranolol. Acta Physiol Scand 74:333–334

Engel, S.L., Schaeffer, T.R., Gold, B.I., and Rubin, B. 1972. Inhibition of pressor effects of angiotensin I and augmentation of depressor effects of bradykinin by synthetic peptides. Proc Soc Exp Biol Med 140:240–244

Erdös, E.G. 1979. Kininases. In *Bradykinin, Kallidin and Kallikrein. Handbook of Experimental Pharmacology,* Suppl to Vol. 25, ed.

E.G. Erdös, pp. 427–481. Heidelberg: Springer-Verlag

Erdös, E.G., and Yang, H.Y.T. 1970. Kininases. In *Handbook of Experimental Pharmacology,* ed. E.G. Erdös, pp. 298–323. New York: Springer-Verlag

Fichman, M.P., Michelakis, A.M., and Horton, R. 1974. Regulation of aldosterone in the syndrome of inappropriate antidiuretic hormone secretion (SIADH). J Clin Endocrinol Metab 39:136–144

Fraser, R., James, V.H., Brown, J.J., Isaac, P., Lever, A.F., and Robertson, J.I.S. 1965. Effects of angiotensin and of furosemide on plasma aldosterone, corticosterone, cortisol and renin in man. Lancet 2:989–991

Ganong, W.F., and Boryczka, A.T. 1967. Effect of a low sodium diet on aldosterone-stimulating activity of angiotensin II in dogs. Proc Soc Exp Biol Med 124:1230

Ganong, W.F., and Mulrow, P.J. 1961. Evidence of secretion of an aldosterone-stimulating substance by the kidney. Nature 190:1115

Ganong, W.F., Boryczka, A.T., and Shackelford, R. 1967a. Effect of renin on adrenocortical sensitivity to ACTH and angiotensin II in dogs. Endocrinology 80:703

Ganong, W.F., Pemberton, D.L., and Van Brunt, E.E. 1967b. Adrenocortical responsiveness to ACTH and angiotensin II in hypophysectomized dogs and dogs treated with large doses of glucocorticoids. Endocrinology 81:1147–1150

Geller, R.G., Margolius, H.S., Pisano, J.J., and Keiser, H.R. 1972. Effects of mineralocorticoids, altered sodium intake and adrenalectomy on urinary kallikrein in rats. Circ Res 31:857–861

Genest, J., Nowaczynski, W., Koiw, E., Sandor, T., and Biron, P. 1960. Adrenocortical function in essential hypertension. In *Essential Hypertension,* eds. K.D. Bock and P.T. Cottier, pp. 126–146. Berlin: Springer-Verlag

Gross, F. 1958. Renin and Hypertensin: Physiologische oder pathologische Wirkstoff? Klin Wochenschr 36:693

Gross, F. 1960. Adrenocortical function and renal pressor mechanisms in experimental hypertension. In *Essential Hypertension,* eds. K.D. Bock and P.T. Cottier, p. 92. Berlin: Springer-Verlag

Gross, F.H., Brunner, H.R., and Zeigler, M. 1965. Renin-angiotensin system, aldosterone, and sodium balance. Rec Prog Horm Res 21:119–177

Hartman, F.A., MacArthur, C.G., and Hartman, W.E. 1927. A substance which prolongs the life of adrenalectomized cats. Proc Soc Exp Biol Med 25:69–70

Henderson, B.S., and McCaa, R.E. 1979. Aldosterone response to prolonged infusion of angiotensin II, angiotensin III, adrenocorticotropic hormone, and potassium chloride in conscious rats. Fed Proc(Abs) 38:1302

Hollenberg, N.K., Chenitz, W.R., Adams, D.F., and Williams, G.H. 1974. Reciprocal influence of salt intake on adrenal glomerulosa and renal vascular responses to angiotensin II in normal man. J Clin Invest 54:34–42

Hollenberg, N.K., Williams, G.H., Burger, B., Ishikawa, I., and Adams, D.F. 1976. Blockade and stimulation of renal, adrenal and vascular angiotensin II receptors with 1-Sar, 8-Ala angiotensin II in normal man. J Clin Invest 57:39–46

Horton, R., 1969. Stimulation and suppression of aldosterone in plasma of normal man and primary aldosteronism. J Clin Invest 48:1230–1236

Karlberg, B.E., Ohman, K.P., Nilsson, O.R., and Wettre, S. 1980. Captopril lowers urinary kallikrein in hypertensive patients. Lancet 1:150–151

Kastner, P.R., and McCaa, R.E. 1978. Blood pressure response to chronic infusion of ACTH, cortisol and aldosterone in adrenalectomized dogs. Physiologist 21(Abs):62

Komor, J., and Müller, J. 1979. Effects of prolonged infusions of potassium chloride, adrenocorticotrophin or angiotensin II upon serum aldosterone concentration and the conversion of corticosterone to aldosterone in rats. Acta Endocrinol 90:680–691

Laragh, J.H., Angers, M., Kelly, W.G., and Lieberman, S. 1960. Hypotensive agents and pressor substances: Effects of epinephrine, norepinephrine, angiotensin II, and others on the secretory rate of aldosterone in man. J Am Med Assoc 174:234–240

Lee, R.C., Van Der Wal, B., and DeWied, D. 1968. Influence of the anterior pituitary on the aldosterone secretory response to dietary sodium restriction in the rat. J Endocrinol 42:465–475

Liard, J.F., Cowley, A.W., Jr., McCaa, R.E., McCaa, C.S., and Guyton, A.C. 1974. Renin, aldosterone, body fluid volumes, and the baroreceptor reflex in the development and reversal of Goldblatt hypertension in conscious dogs. Circ Res 34:549–559

Lieberman, A.H., and Luetscher, J.A., Jr. 1960. Some effects of abnormalities of pituitary, adrenal or thyroid function on excretion of aldosterone and the response to corticotropin or sodium deprivation. J Clin Endocrinol Metab 20:1004–1016

Luetscher, J.A., Jr., and Axelrad, B.J. 1954a. Increased aldosterone output during sodium deprivation in normal man. Proc Soc Exp Biol Med 87:650–653

Luetscher, J.A., Jr., and Axelrad, B.J. 1954b. Sodium-retaining corticoid in the urine of normal children and adults and of patients with hypoadrenalism or hypopituitarism. J Clin Endocrin Metab 14:1086–1089

Luetscher, J. A., Jr., and Curtis, R. H. 1955a. Aldosterone: observations on the regulation of sodium and potassium balance. Ann Intern Med 43:658–666

Luetscher, J. A., Jr., and Curtis, R. H. 1955b. Observations on aldosterone in human urine. Fed Proc 14:746–751

Luetscher, J. A., Jr., and Curtis, R. H. 1955c. Relationship of aldosterone in urine to sodium balance and to some other endocrine functions. J Clin Invest 34:951

Luetscher, J. A., Jr., and Deming, Q. B. 1950. Treatment of nephrosis with cortisone. J Clin Invest 29:1576–1587

Luetscher, J. A., Jr., Neher, R., and Wettstein, A. 1954. Isolation of crystalline aldosterone from the urine of a nephrotic patient. Experientia 10:456–458

Luetscher, J. A., Jr., Dowdy, A., Harvey, J., Neher, R., and Wettstein, A. 1955. Isolation of crystalline aldosterone from the urine of a child with the nephrotic syndrome. J Biol Chem 217:505–512

Manninen, V. 1970. Movements of sodium and potassium ions and their tracers in propranolol-treated red cells and diaphragm muscle. Acta Physiol Scand 355(suppl. I):1–76

Manning, R. D., Jr., Guyton, A. C., Coleman, T. G., and McCaa, R. E. 1979. Hypertension in dogs during antidiuretic hormone and hypotonic saline infusion. Am J Physiol 236:H314–H322

Margolius, H. S., Horwitz, D., Geller, R. G., Alexander, R. W., Gill, J. R., Jr., Pisano, J. J., and Keiser, H. R. 1974. Urinary kallikrein in normal subjects: relationships to sodium intake and to sodium retaining steroids. Circ Res 35:812–819

Margolius, H. S., Horwitz, D., Pisano, J. J., and Keiser, H. R. 1976. Relationships among urinary kallikrein, mineralocorticoids and human hypertensive disease. Fed Proc 35:203–206

Marieb, N. J., and Mulrow, P. J. 1965. Role of renin-angiotensin system in the regulation of aldosterone secretion in the rat. Endocrinology 76:657–664

Marine, D., and Baumann, E. J. 1927. Duration of life after suprarenalectomy in cats and attempts to prolong it by injections of solutions containing sodium salts, glucose and glycerol. Am J Physiol 81:86–100

Matsuoka, H., Mulrow, P. J., and Li, C.H. 1980. β-Lipotropin: a new aldosterone stimulating factor. Science 209:307–309

McCaa, C. S., Richardson, T.Q., McCaa, R. E., Sulya, L. L., and Guyton, A. C. 1965. Aldosterone secretion by dogs during the developmental phase of Goldblatt hypertension. J Endocrinol 33:97–102

McCaa, C. S., Langford, H. G., Cushman, W. C., and McCaa, R. E. 1979. Response of arterial blood pressure, plasma renin activity and plasma aldosterone concentration to long-term administration of captopril in patients with severe, treatment-resistant malignant hypertension. Clin Sci 57:371s–373s

McCaa, R. E. 1977a. Role of the renin-angiotensin system in the regulation of aldosterone biosynthesis and arterial pressure during sodium deficiency. Circ Res 40:I–157–I–162

McCaa, R. E. 1977b. Response of arterial pressure and aldosterone biosynthesis to continuous infusion of propranolol in intact conscious sodium deficient dogs. Physiologist 20(Abs):62

McCaa, R. E. 1978. Aldosterone response to long-term infusion of angiotensin II and angiotensin III in conscious dogs before and after dietary sodium restriction. Endocrinology 103:458–464

McCaa, R. E. 1980. Specificity of converting enzyme inhibition. In *Enzymatic Release of Vasoactive Peptides,* eds. F. Gross and G. Vogel, pp. 383–399. New York: Raven Press

McCaa, R. E., and Langford, H. G. 1979. Aldosterone response to adrenocorticotropic hormone and angiotensin II in patients with panhypopituitarism before and after dietary sodium restriction. Fed Proc 38:1203

McCaa, R. E., McCaa, C. S., Read, D. G., Bower, J. D., and Guyton, A. C. 1972. Increased plasma aldosterone concentration in response to hemodialysis in nephrectomized man. Circ Res 31:473–480

McCaa, R. E., McCaa, C. S., Cowley, A. W., Jr., Ott, C. E., and Guyton, A. C. 1973a. Stimulation of aldosterone secretion by hemorrhage in dogs after nephrectomy and decapitation. Circ Res 22:356–362

McCaa, R. E., Read, V. H., Cowley, A. W., Jr., Bower, J. D., Smith, G. V., and McCaa, C. S. 1973b. Influence of acute stimuli on plasma aldosterone concentration in anephric man and kidney allograft recipients. Circ Res 33:313–322

McCaa, R. E., McCaa, C. S., and Guyton, A. C. 1974a. Role of angiotensin II and potassium in the long-term regulation of aldosterone secretion in intact conscious dogs. Circ Res 36/37(Suppl. 1):57–67

McCaa, R. E., Ott, C. W., and McCaa, C. S. 1974b. Relation between plasma potassium concentration and aldosterone secretion in nephrectomized dogs. IRCS 2:1263

McCaa, R. E., Young, D. B., Guyton, A. C., and McCaa, C. S. 1974c. Evidence for a role of an unidentified pituitary factor in regulating aldosterone secretion during altered sodium balance. Circ Res 34–35(Suppl. 1):15–25

McCaa, R. E., Hall, J. E., and McCaa, C. S. 1978a. The effects of angiotensin I-converting enzyme inhibitors on arterial blood pressure and urinary sodium excretion—Role of the renal renin-angiotensin and kallikrein-kinin systems. Circ Res 43(Suppl. 1):32–39

McCaa, R. E., Montalvo, J. M., and McCaa, C. S. 1978b. Role of Growth Hormone in the Regulation of Aldosterone Biosynthesis. J Clin Endocrinol Metab 46:247–253

McCaa, R. E., Muirhead, E. E., Pitcock, J. A., Kastner, P. R., and McCaa, C. S. 1978c. The response of plasma aldosterone concentration and arterial blood pressure to long-term infusion of adrenocorticotropic hormone (ACTH) in conscious dogs during sodium repletion and sodium depletion. The Endocrine Society(Abs):401

McCaa, R. E., McCaa, C. S., Bengis, R. G., and Guyton, A. C. 1979. Role of aldosterone in experimental hypertension. J Endocrinol 81:69P–78P

Mendelsohn, F. A. O., Johnston, C. I., Doyle, A. E., Scoggins, B. A., Denton, D. A., and Coghlan, J. P. 1972. Renin, angiotensin II, and adrenal corticosteroid relationships during sodium deprivation and angiotensin infusion in normotensive and hypertensive man. Circ Res 31:728–739

Mills, I. H., MacFarlane N. A. A., Ward P. E., and Obika L. F. O. 1976. The renal kallikrein-kinin system and the regulation of salt and water excretion. Fed Proc 35:181–188

Muller, J., 1965. Aldosterone stimulation *in vitro*. 1. Evaluation of assay procedure and determination of aldosterone-stimulating activity in a human urine extract. Acta Endocrinol (Kbh)48:283–296

Mulrow, P. J., Ganong, W. F., Cera, G., Kuljian, A. 1962. The nature of the aldosterone-stimulating factor in dog kidneys. J Clin Invest 41:505

Nasjletti, A., and Colina-Chourio, J. 1976. Interaction of mineralocorticoids, renal prostaglandins and the renal kallikrein-kinin system. Fed Proc 35:189–193

Nasjletti, A., Colina-Chourio, J. and McGiff, J. C. 1975. Disappearance of bradykinin in the renal circulation of dogs: effects of kininase inhibition. Circ Res 37:59–65

New, M. I., Saenger, P. H., Peterson, R. E., and Ulick, S. 1975. Evidence for an ACTH stimulable hormone causing hypertension. Bull N Y Acad Med 41:1179

New, M. I., Peterson, R. E., Saenger, P., and Levine, L. S. 1976. Evidence for an unidentified ACTH-induced steroid hormone causing hypertension. J Clin Endocrinol Metab 43:1283–1293

Oelkers, W., Brown, J. J., Fraser, R., Lever, A. R., Morton, J. J., and Robertson, J. I. S. 1974. Sensitization of the adrenal cortex to angiotensin II in sodium-deplete man. Circ Res 34:69–77

Ondetti, M. A., Rubin, B., and Cushman, D. W. 1977. Design of specific inhibitors of angiotensin converting enzyme: new class of orally active antihypertensive agents. Science 196:441–444

Palmore, W. P., and Mulrow, P. J. 1967. Control of aldosterone secretion by the pituitary gland. Science 158:1482–1484

Saito, I., Zanella, T., Bravo, E. L., Sen, S., and Bumpus, F. M. 1980. *In vitro* steroidogenic properties of a new aldosterone stimulating hormone: interaction with potassium. Fed Proc 39(Abs):1073

Saruta, T., Cook, R., and Kaplan, N.M. 1972. Adrenocortical steroidogenesis. Studies on the mechanisms of action of angiotensin and electrolytes. J Clin Invest 51:2239–2245

Scoggins, B. A., Butkus, A., Coghlan, J. P., Denton, D. A., Fan, J. S., Humphrey, T. J., and Whitworth, J. A. 1978. Adrenocorticotropic hormone-induced hypertension in sheep. A model for the study of the effect of steroid hormones on blood pressure. Circ Res 43(Suppl 1):76–81

Sealey, J. E., Wallace, J. M., Case,

D. B., and Laragh, J. H. 1976. Inhibition of aldosterone secretion by converting enzyme blockade contrasted with the agonist/antagonist effect of an angiotensin II analogue. Endocrinology 98(Abs):160

Sen, S., Smeby, R.R., Bumpus, F.M., and Turcotte, J.G. 1979. Role of renin-angiotensin system in chronic renal hypertensive rats. Hypertension 4:427–434

Simpson, S.A., and Tait, J.F. 1953. Physico-chemical methods of detection of a previously unidentified adrenal hormone. Mem Soc Endocrin 2:9–24

Simpson, S.A., Tait, J.F., Wettstein, A., Neher, R., von Euw, J., and Reichstein, T. 1953. Isolierung eines neuen kristallisierten Hormons aus Nebennieren mit besonders hoher Wirksamkeit auf den Mineralstoffwechsel. Experientia 9:333–335

Simpson, S.A., Tait, J.F., Wettstein, A., Neher, R., von Euw, J., Schindler, O., and Reichstein, T. 1954a. Aldosteron, Isolierung und Eigenschaften: Über Bestandteile der Nebennierenrinde und verwandte Stoffe. Helv Chim Acta 37:1163–1200

Simpson, S.A., Tait, J.F., Wettstein, A., Neher, R., von Euw, J., Schindler, O., and Reichstein, T. 1954b. Die Konstitution des Aldosterons. Über Bestandteile der Nebennierenrinde und verwandte Stoffe. Helv Chim Acta 37:1200–1223

Singer, B., Losito, C., and Salmon, S. 1963. Aldosterone and corticosterone secretion rates in rats with experimental renal hypertension. Acta Endocrinol (Kbh)44:505–518

Steele, J.M., Jr., and Lowenstein, J. 1974. Differential effects of an angiotensin II analogue on pressor and adrenal receptors in the rabbit. Circ Res 35:592–600

Stein, J.H., Congbalay, R.C., Karsh, D.L., Osgood, R.W., and Ferris, T.F. 1972. The effect of bradykinin on proximal tubular sodium reabsorption in the dog: evidence for functional nephron heterogeneity. J Clin Invest 51:1709–1721

Swartz, S.L., Williams, G.H., Hollenberg, N.K., Moore, T.J., and Dluhy, R.G. 1979. Converting enzyme inhibition in essential hypertension: the hypotensive response does not reflect only reduced angiotensin II formation. Hypertension 1:106–111

Weidmann, P., Horton, R., Maxwell, M.H., Franklin, S.S., and Fichman, M. 1973. Dynamic studies of aldosterone in anephric man. Kidney Internat 4:280–288

Williams, G.H., and Hollenberg, N.K. 1977. Accentuated vascular and endocrine response to SQ 20,881 in hypertension. N Engl J Med 297:184–188

Williams, G.H., Rose, L.I., Dluhy, R.G., Dingman, J.F., and Lauler, D.P. 1971. Aldosterone response to sodium restriction and ACTH stimulation in panhypopituitarism. J Clin Endocrinol Metab 32:27–34

Williams, G.H., Bailey, G.L., Hampers, C.L., Lauler, D.P., Merrill, J.P., Underwood, R.H., Blair-West, J.R., Coghlan, J.P., Denton, D.A., Scoggins, B.A., and Wright, D.R. 1973. Studies on the metabolism of aldosterone in chronic renal failure and anephric man. Kidney Internat 4:280–288

Williams, G.H., McConnell, L.M., Raux, M.C., and Hollenberg, N.K. 1974. Evidence for different angiotensin II receptors in the rat adrenal glomerulosa and rabbit vascular smooth muscle cells. Circ Res 34:384–390

Willis, L.R., Ludens, J.H., Hook, J.B., and Williamson, H.E. 1969. Mechanism of natriuretic action of bradykinin. Am J Physiol 217:1–5

Alterations in Renal Function Produced by Angiotensin Converting Enzyme Inhibitors[1]

Ben G. Zimmerman and Pancras C. Wong
Department of Pharmacology, University of Minnesota
Minneapolis, Minnesota 55455

INTRODUCTION

Teprotide (SQ 20,881), a nonapeptide, and captopril (SQ 14,225), an orally effective propanoyl proline analogue, are angiotensin converting enzyme inhibitors (CEIs) which have prominent cardiovascular effects. From both a therapeutic standpoint in the treatment of hypertension and for physiologic applications in studying the renin-angiotensin system, their effect on renal function is of great interest. This chapter deals first with observations of effects of the CEIs on renal blood flow, sodium excretion and glomerular filtration rate in the experimental animal and man. We try to relate these changes in renal function to the influence of the renin-angiotensin and the kallikrein-kinin systems on the kidney, and discuss whether the CEIs act by affecting these systems. In addition, we consider whether the renal actions of the CEIs are in part mediated by renal prostaglandins.

EFFECTS ON RENAL BLOOD FLOW, SODIUM EXCRETION AND GLOMERULAR FILTRATION RATE

Based on early studies carried out in anesthetized animals, it was found that the CEIs increase renal blood flow (RBF) and sodium excretion

[1] Portions of this research were supported by USPHS grants HL–08570 and HL–17871.

(Bailie and Barbour, 1975; Nasjletti et al., 1975). Distribution of RBF also seems to be affected. Captopril (SQ 14,225) increased outer cortical more than inner cortical flow in the rat kidney (Jones et al., 1979), whereas teprotide (SQ 20,881) shifted flow to the inner cortex in the dog (Bailie and Barbour, 1975).Since both studies were carried out in anesthetized animals, the discrepant results may be due to a difference in renal hemodynamic response to the anesthetics employed. Conditions which result in higher PRA bring about greater effects of the CEIs. The influence of the renin-angiotensin system on the kidney is greater, and so too is the effect of the CEI. The RBF increase caused by teprotide and captopril is especially prominent in these conditions, e.g. in the salt deplete state (Kimbrough et al., 1977; Hollenberg et al., 1977; Arendshorst and Finn, 1977; Gavras ert al.,1978), renovascular hypertension (Ayers et al., 1974) and heart failure (Freeman et al., 1979). When considering the effect of the CEIs on RBF, the fall in blood pressure (BP) produced by these agents should be taken into account since changes in BP alone affect blood flow. This is usually done by calculating renal vascular resistance and making comparisons on that basis. Because renal autoregulation allows renal vascular resistance to compensate for changes in BP in order to maintain RBF constant, we feel that a better gauge for comparison of renal vasodilator effects of the CEIs is the change in RBF per se rather than renal vascular resistance. Generally speaking, the increase in RBF caused by the CEIs is moderate to quite large depending on the experimental conditions. In the conscious salt deplete dog or human, teprotide administered intrarenally (Kimbrough et al., 1977) or intravenously (Hollenberg et al., 1977) increased RBF by approximately 29%. In situations where PRA is even higher, namely, in the anesthetized salt deprived dog (Burger et al., 1976) and salt deprived essential hypertensive human (Williams and Hollenberg, 1977), a greater percentage increase in RBF was obtained, 40 and 55%, respectively. Results obtained in anesthetized salt replete animals were more variable, which is probably due to different prevailing PRA levels. Murthy et al. (1978), Nasjletti et al. (1975) and Britton et al. (1979) obtained mean increases in RBF of 11.1, 16.5 and 45.1%, respectively, in the anesthetized dog. In our laboratory no significant increase in RBF was found after captopril (0.5 mg/kg i.v.), but an increase of approximately 24% was obtained after the dogs had their PRA raised by non-hypotensive hemorrhage (Wong and Zimmerman, 1980).

The increase in sodium excretion elicited by the CEIs appears to be similar (approximately twofold) regardless of whether the agents are administered to the anesthetized salt replete dog (Nasjletti et al., 1975), conscious salt deplete dog (McCaa et al., 1978), salt replete human

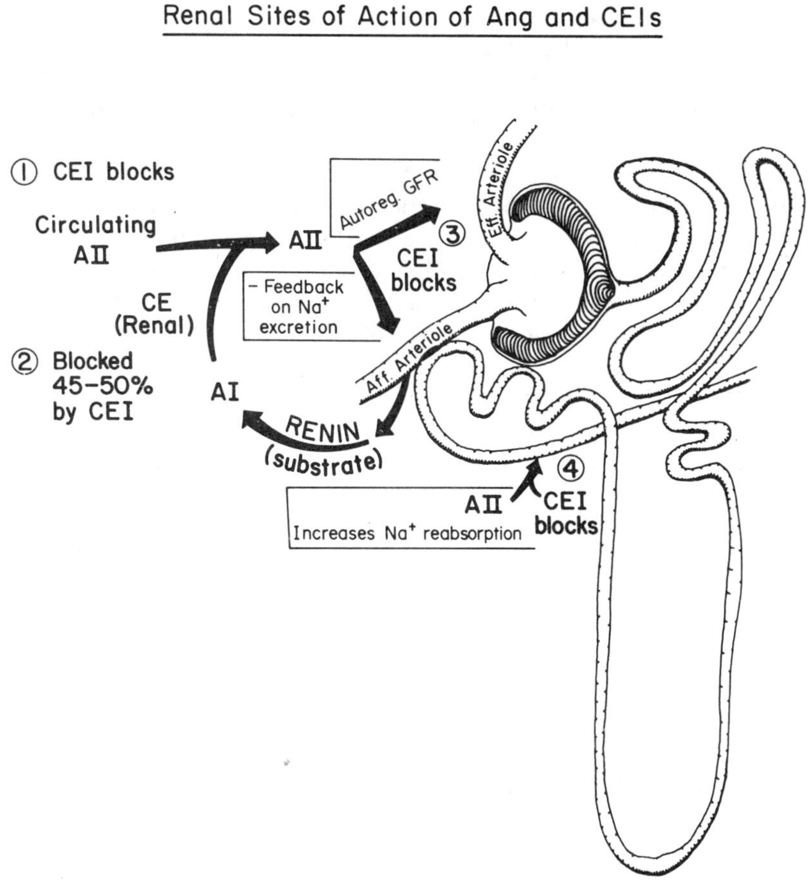

Fig. 1. Angiotensin II formed locally in the kidney exerts a negative feedback effect on sodium excretion. Macula densa senses increase in tubular flow rate (or sodium or chloride delivery), and causes renin release from juxtaglomerular cells in afferent arteriole which induces angiotensin II formation. Angiotensin II acts to constrict afferent arteriole, thus decreasing glomerular filtration rate (Thurau and Schnermann, 1965). Angiotensin II also mediates autoregulation of glomerular filtration rate. Fall in renal perfusion pressure elicits renin release and local formation of angiotensin II. Constriction of efferent arteriole maintains glomerular capillary pressure, and thus glomerular filtration rate in face of decrease in systemic arterial pressure (Hall et al., 1977a). Although the emphasis has been on locally formed angiotensin II as being mediator of these events, there may also be a role for circulating angiotensin II. CEIs block the renal influence of circulating angiotensin II (1) much more effectively than that of locally formed angiotensin II (2) on renal vasculature (see this review), and this partially accounts for the renal vasodilator effect of CEIs. Bradykinin potentiation may also contribute to renal vasodilator effect of CEIs. Since CEIs interfere with the conversion of angiotensin I to angiotensin II by 45–50%, it is possible that the suppression of feedback regulation of glomerular filtration rate is due to partial inhibition of renal CE (3). Natriuretic effect of CEIs is brought about by inhibition of facilitating action of angiotensin II on reabsorption in distal tubule (4).

(Millar and Johnston, 1979) or salt deplete hypertensive human (Hollenberg et al., 1979). In the canine model of low output heart failure, sodium excretion was more markedly increased by captopril (Freeman et al., 1979). Augmented sodium excretion by converting enzyme inhibition may relate to interference with the effect of endogenous angiotensin on the renal tubule (*see* 4 in Fig. 1). It has been demonstrated that low (physiologic) concentrations of angiotensin facilitate, while higher doses decrease, sodium reabsorption in the dog (Fagard et al., 1976; Johnson and Malvin, 1977; Vander, 1963), possibly by an effect of the peptide on the distal tubule (Vander, 1963). The proposal has been put forth that endogenous angiotensin can regulate sodium excretion by a direct effect on the renal tubule and that agents which interfere with the function of renin-angiotensin system abolish this effect (Hall et al., 1977b). Potentiation of the renal tubular effect of kinins cannot be ruled out in the increase in sodium excretion produced by the CEIs.

Since changes in glomerular filtration rate induced by the CEIs are variable, it is not certain whether their effect is exerted at a pre- or postglomerular site. Both McCaa et al. (1978) and Hall et al. (1977a) reported a decrease in glomerular filtration rate in conscious sodium deplete dogs given captopril over a period of days, whereas Navar et al. (1979) showed no change in this parameter after teprotide in salt replete dogs with the renal artery constricted. If, as in the former studies, glomerular filtration rate falls in the face of an increase in renal plasma flow, it can be concluded that efferent arteriolar resistance decreased. Navar et al. (1979) demonstrated, however, that both afferent and efferent arteriolar resistance were decreased by teprotide and that this explained why, in their study, RBF increased and GFR remained the same after converting enzyme inhibition. Based on the results to date, we would suggest that circulating, and possibly intrarenally formed, angiotensin act on both pre- and postglomerular vessels (Figure 1).

The mechanism(s) of these effects of the CEIs on renal function have been mainly attributed to two causes. Blockade of the renal influence of angiotensin II due to inhibition of its formation, and potentiation of the renal actions of kinins by inactivation of kininase II, are most commonly proposed to explain the effects of the CEIs on renal function. More is added below about the mechanisms of these actions.

Effect on Normal Renal Function

Previous studies failed to reveal any effect of teprotide on RBF when given i.v., or a small decrease in RBF when given i.a. to the kidney of conscious sodium replete dogs (Gavras et al., 1978; Kimbrough et al., 1977; Burger et al., 1976). More recent work has uncovered renal vasodilator and natriuretic effects of captopril in sodium replete human subjects (Hollenberg, personal communication; Millar et al., 1979). We found that captopril decreased BP slightly and increased RBF moderately in conscious normotensive dogs with a low PRA (Zimmerman et al., 1980). These were unexpected findings since it was believed that there was no influence of the renin-angiotensin system on BP or RBF under basal conditions. More recent work has indicated, however, that the prevailing PRA of normotensive humans may be sufficient to provide a small degree of angiotensinogenic vascular tone (Conway et al., 1979; Niarchos et al., 1979). Because captopril may cause renal vasodilatation in normotensive animals or humans by an action unrelated to converting enzyme inhibition, as has been postulated for the hypotensive effect when high doses are given (Vollmer et al., 1978), we sought a similar vasodilator action of teprotide. We administered either teprotide or captopril in separate sessions to 14 conscious instrumented normotensive dogs in which we recorded BP and RBF and determined PRA. The majority of animals were initially given 0.1 mg/kg of captopril or 0.25

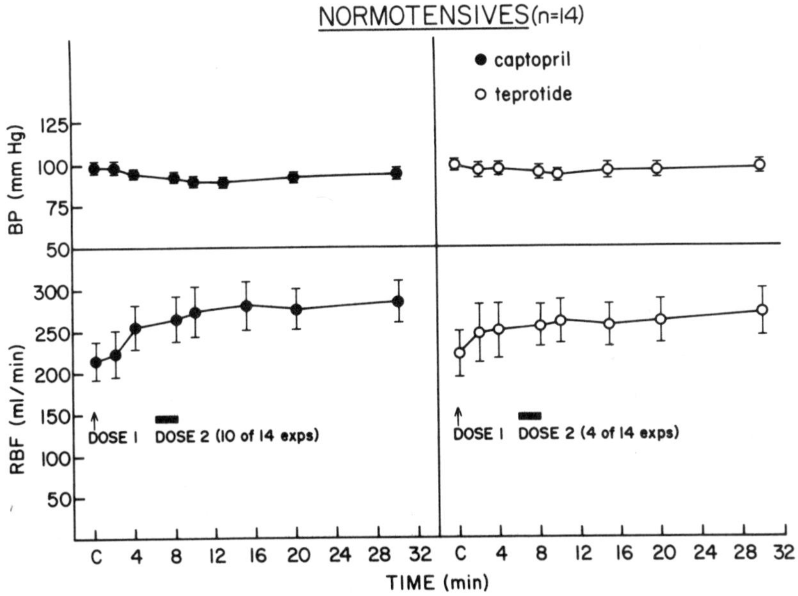

Fig. 2. Blood pressure and renal blood flow in control period and after captopril or teprotide in 14 normotensive dogs. Dose 1 of captopril was 0.1 mg/kg in 10 and 0.2 mg/kg in 4 experiments, and Dose 1 of teprotide was 0.25 mg/kg in 12 and 0.5 mg/kg in 2 experiments. Dose 2, 0.2 mg/kg of captopril in 10 experiments, and 0.5 mg/kg of teprotide in 4 experiments, was administered during the time intervals designated by solid bar. Decrease in BP at 8 min was significant at $P < 0.05$ (paired t-test) and increase in RBF was significant at $P < 0.01$ (analysis of variance).

mg/kg of teprotide (Dose 1). Ten of 14 dogs were also administered 0.2 mg/kg of captopril (Dose 2), and 4 of 14 dogs received 0.5 mg/kg of teprotide (Dose 2) after peak effect of the first dose had been attained. Both agents caused a small decrease in BP and a slowly developing rise in RBF (Fig. 2). It appeared that the second dose of captopril may have contributed additionally to the effect of the first dose on RBF, whereas with teprotide this was not apparent because of the small number of experiments in which a second dose was given. These effects on BP and RBF of both CEIs were statistically significant (*see* legend, Fig. 2). The peak increase in RBF caused by Dose 1 of captopril and Dose 1 of teprotide was 24.1 ± 4.2% and 12.8 ± 2.7%, respectively, and these values differed statistically ($P < 0.02$). Thus, captopril appears to be more effective in increasing RBF than teprotide under basal conditions. It is possible that the doses of these agents that we used represent different points on their respective dose-response curves, and this may account for the apparent difference in potency. The lesser ability of teprotide than captopril to cause renal vasodilatation in the salt replete

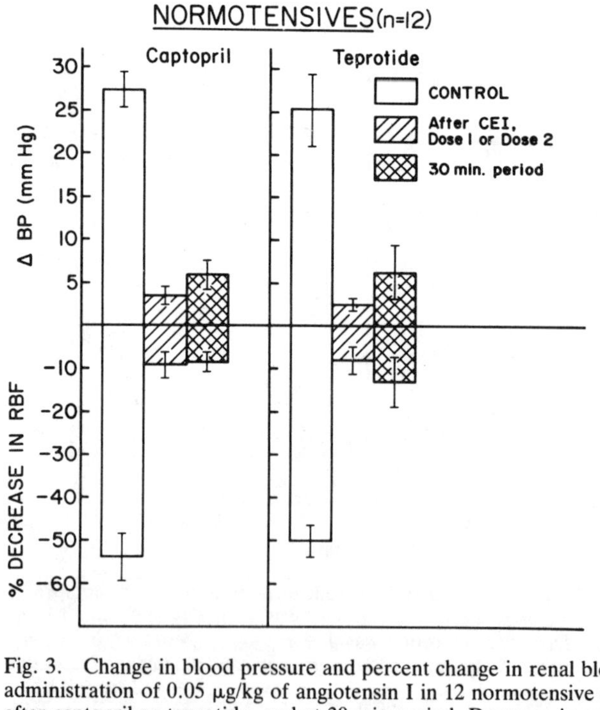

Fig. 3. Change in blood pressure and percent change in renal blood flow caused by i.v. administration of 0.05 μg/kg of angiotensin I in 12 normotensive dogs in control period, after captopril or teprotide, and at 30 min period. Decrease in responses was significant at $P < 0.01$ by analysis of variance.

dog may account for the failure of others to demonstrate this effect of teprotide previously.

Control PRA levels were 1.20 ± 0.23 and 0.71 ± 0.12 ng angiotensin I/ml/hr, respectively, for the captopril and teprotide experiments and were increased to 6.10 ± 3.24 and 1.47 ± 0.50 ng angiotensin I/ml/hr after Dose 1 (or Dose 2, if a second dose had been administered). Thus, a renal vasodilator effect of CEIs occurs in spite of a low control PRA.

Responses to angiotensin I injected i.v. were obtained in the control period, after the peak effect of Dose 1 or Dose 2, and at the 30 min period. As can be seen in Figure 3, the BP increase and RBF decrease were nearly abolished by the CEIs.

Effect on Renal Function in Hypertensive Dog and Human

Relatively few studies have defined the renal response to CEIs in the experimental hypertensive animal or human. Williams and Hollenberg (1977) demonstrated that teprotide caused an increase in RBF, decrease in BP, increase in PRA and decrease in plasma angiotensin II in salt deplete normotensive and essential hypertensive subjects. In another study, Hollenberg and collaborators (1979) found that teprotide given i.v. to salt deplete renovascular and essential hypertensive humans increased glomerular filtration rate only in those subjects with a depressed control glomerular filtration rate. These results suggest that in the salt deprived human with hypertension the renin-angiotensin system influences renal vascular tone. This influence appears greater in the hypertensive than in the normotensive humans, as was reported above (Williams and Hollenberg, 1977). Intrarenal infusion of teprotide decreased renal artery, but not aortic, pressure in the 1 kidney, 1 clip Goldblatt hypertensive dog (Ayers et al., 1974). It was concluded from this study that angiotensin II, presumably formed in the kidney, mediated renal vasoconstriction in these animals.

We carried out experiments to determine the renal response to captopril and teprotide in conscious salt replete 2 kidney, 1 clip Goldblatt hypertensive dogs. As in the normotensive animals referred to above, most dogs received, as a first dose, 0.1 mg/kg of captopril or 0.25 mg/kg of teprotide. Both CEIs caused a fall in BP and increase in RBF which was of a similar magnitude in the hypertensive animals (Fig. 4), in contrast to the greater effect of captopril on RBF in the normotensive dogs. When the maximum increments in RBF were compared, teprotide tended to increase RBF more than captopril (39 ± 8.9% vs. 26 ± 3.8%). The difference was not, however, statistically significant. A second dose of captopril (0.2 mg/kg) was administered in 7 of 9 of the hypertensives, and a second dose of teprotide (0.5 mg/kg) was given in

3 of 9 experiments. There was an additional effect of both agents on RBF with the higher doses. The control PRA levels were 11.1 ± 2.6 and 8.1 ± 2.3 ng angiotensin I/ml/hr, for the captopril and teprotide experiments, respectively, and were increased to 37 ± 7.9 and 31.0 ± 9.8 ng angiotensin I/ml/hr after Dose 1 (or Dose 2, if a second dose had been administered).

Responses to angiotensin I injected i.v. were obtained as in the experiments on the normotensive dogs. As can be seen in Figure 5, the RBF responses to angiotensin I were smaller in the hypertensive animals than in the normotensives, due to higher PRA. The CEIs markedly depressed the BP increase and RBF decrease elicited by angiotensin I.

The results of the experiments in the hypertensive dogs suggest that renal vascular tone is influenced by the renin-angiotensin system as it is in the normotensive dogs, and, based on the relative increments in RBF obtained with the CEIs, the tone is not much greater than that in the normotensive animals. Since BP is decreased more by the CEI in

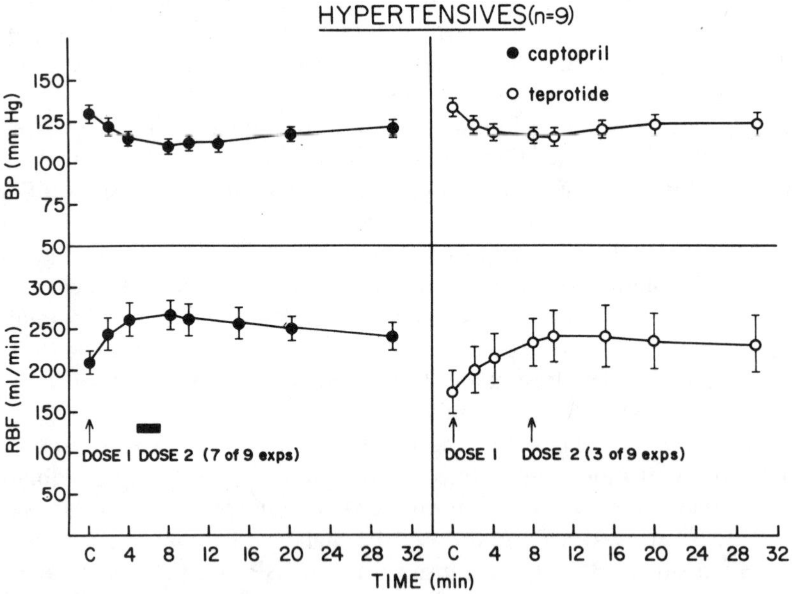

Fig. 4. Blood pressure and renal blood flow in control period and after captopril or teprotide in 9 hypertensive dogs. Dose 1 of captopril was 0.1 mg/kg in 7 and 0.2 mg/kg in 2 experiments, and Dose 1 of teprotide was 0.25 mg/kg in 7 experiments and 0.5 mg/kg in 2 experiments. Dose 2, 0.2 mg/kg of captopril in 7 experiments and 0.5 mg/kg of teprotide in 3 experiments, was administered during the time intervals designated by solid bar or arrow. Decrease in BP at 8 min was significant at $P < 0.01$ (paired t-test) and increase in RBF was significant at $P < 0.05$ for captopril and 0.01 for teprotide (analysis of variance).

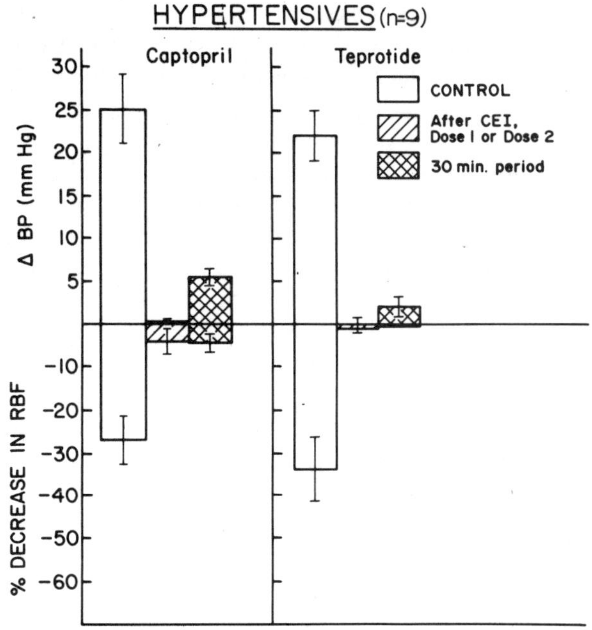

Fig. 5. Changes in blood pressure and percent change in renal blood flow caused by i.v. administration of 0.05 μg/kg of angiotensin I in 9 hypertensive dogs in control period, after captopril or teprotide, and at 30 min period. Decreases in responses were significant at $P < 0.01$ (analysis of variance).

the hypertensives, there is a greater decrease in renal vascular resistance in this group than in the normotensive animals. Teprotide caused a smaller increase in RBF in the normotensives and approximately equivalent increase in the hypertensives compared to captopril. At present there is no explanation for this apparent difference between these agents.

Mechanism of Renal Vasodilator Response

As proposed above, two mechanisms are likely to explain the increase in RBF seen after converting enzyme inhibition in the experimental animal or human. Inhibition of angiotensin II formation and/or a decreased rate of kinin metabolism could bring about renal vasodilatation. A decrease in the circulating angiotensin II level or in angiotensin II formed locally in the kidney are both possible. Similarly, the prevailing circulating kinin level or intrarenal kinin concentration could be augmented by the CEIs. There are few data available which would indicate whether an angiotensin or kinin mechanism, or possibly both, are involved in this renal vasodilator effect. In fact controversy exists as to whether the circulating level of kinins actually increases after a CEI. Nasjletti et al. (1975) detected a twofold increase in circulating kinins and an even greater increase in urinary kinin excretion after giving teprotide to the anesthetized dog. Removal of 50 ml blood samples which was necessary to carry out the bioassay of kinins may, however, have influenced the results of this study. McCaa and co-workers (1978), using a similar bioassay procedure, also found a sustained twofold increase in blood kinins in conscious salt deplete dogs orally dosed with captopril. Urinary kinins were increased 1 hr following captopril administration in conscious salt replete dogs (Olsen and Arrigoni-Martelli, 1979). In the rat, high (greater than 3 mg/kg), but not low, oral doses of captopril increased blood bradykinin levels (Matthews and Johnston, 1979). Teprotide was shown to increase bradykinin levels in salt deplete hypertensive patients, but not in normotensives (Millar and Johnston, 1979). Another group of investigators found no alteration in the circulating bradykinin level in hypertensive patients, but did detect an increase in urinary kinin which correlated with the depressor effect of teprotide (Vinci et al., 1979). Disagreement between investigations concerning kinin levels may represent differing experimental conditions and/or methods of kinin assay.

Some evidence has appeared which suggests that decreased formation of angiotensin II is involved in the renal vasodilator action of captopril. In essential or renovascular hypertensive patients, there appears to be a correlation between the PRA prior to drug administration and the increase in RBF seen after the drug (Murao et al., 1979). The angiotensin antagonist saralasin has been found to block completely or to reduce the renal vasodilator response to captopril in the anesthetized dog that had undergone nonhypotensive hemorrhage (Wong and Zimmerman, 1980) and also in the anesthetized salt loaded rabbit (Hollenberg, personal communication). Administration of saralasin would not be expected to block renal vasodilatation due to bradykinin.

Prostaglandin Mediation

Because of the relationship between the kallikrein-kinin system and renal prostaglandins, it has been suggested that the CEIs may act in part through prostaglandins. Bradykinin increases prostaglandin E synthesis by renal medullary interstitial cells in culture (Zusman et al., 1977), and also releases prostaglandin E from the canine kidney in situ (McGiff et al., 1972). Several studies have addressed the question of whether prostaglandins are released by the CEIs and mediate their renal vasodilator effect. Johns et al. (1977) failed to find an increase in prostaglandin E in rabbit renal venous blood after teprotide, and we similarly were unable to detect an increase in prostaglandin E concentration or secretion rate after captopril in the dog (Wong and Zimmerman, 1980). An increase in arterial prostaglandin E was observed by Vinci and co-workers (1979) in human essential hypertensives after captopril administration. Because of the almost complete removal of prostaglandin E by the lungs after a single pass, it is puzzling as to what this finding represents. In addition, recent work has demonstrated consistently that indomethacin does not block or even reduce the renal vasodilator effect of captopril or teprotide (Lokhandwala and Jandhyala, 1979; Wong and Zimmerman, 1980). Taken together, these results would appear to rule out involvement of renal prostaglandins in the response of the renal vasculature to the CEIs.

Intrarenal and Extrarenal Converting Enzyme Blockade

Thurau and Schnermann (1965) conceived the idea that an intrarenal renin-angiotensin system mediates control of afferent arteriolar resistance (Fig. 1). This was looked upon as a feedback system for regulation of sodium excretion, and the sodium or chloride sensing apparatus was located in the macula densa. Hall et al. (1977a) have proposed an intrarenal angiotensin mechanism which they believe adjusts efferent arteriolar tone (Fig. 1). A detailed discussion of these mechanisms is, however, beyond the scope of this review. Several investigations (Ayers et al., 1974; Hall et al., 1979a; Ploth et al., 1979) have suggested that the intrarenal administration of CEIs interrupts the function of the renal renin-angiotensin system. It is conceivable, therefore, that the intravenous administration of CEIs may cause renal vasodilatation by inhibition of the renal converting enzyme, besides the pulmonary converting enzyme. It has not been established whether renal and/or extrarenal converting enzyme inhibition is involved in the renal response to either teprotide or captopril. Investigations were conducted in our laboratory to correlate the ability of these CEIs to block the renal vasoconstrictor response to angiotensin I with their renal vasodilator effect when given i.a. to the kidney. Interestingly, the vasoconstrictor response to angiotensin I was maximally inhibited with the CEIs by approximately 50% (Fig. 1), and yet RBF did not change. Subsequent administration of either CEI intravenously caused no further decrease in the renal vasoconstrictor response to angiotensin I, but did increase RBF and decrease BP. The results of these investigations pointed out two important aspects of the renal actions of the CEIs. First, these agents have only a limited capacity to block formation of angiotensin

II from angiotensin I in the kidney, and, second, a change in concentration of circulating rather than intrarenal peptide(s) is responsible for the RBF effect of these drugs (Fig. 1).

CONCLUSIONS

1. Angiotensin converting enzyme inhibitors (CEIs) cause renal vasodilatation and natriuresis in both the sodium replete and deplete states, as well as in Goldblatt hypertension. Glomerular filtration rate is usually depressed, but may not change or may even increase depending upon experimental conditions. This suggests an angiotensinogenic influence on the kidney which affects renal vascular resistance, glomerular filtration and sodium reabsorption in both the normal and hypertensive state.
2. The mechanism of renal vasodilatation caused by the CEIs appears primarily due to decreased formation of angiotensin II, rather than to an increase in bradykinin. Contribution to the vasodilator effect by bradykinin has, however, not been ruled out. Prostaglandins do not play a role in the increase in renal blood flow caused by the CEIs.
3. Renal vasodilatation caused by the CEIs does not depend on a depressed formation of intrarenal angiotensin II, but is directly related to a systemic effect, most likely a change in the concentration of circulating angiotensin II or bradykinin.

ACKNOWLEDGMENTS

The authors wish to thank Randolph Marchessault, Edward Kraft, Patricia Friedman and Brenda Zimmerman for their excellent technical assistance, Theodora Danielson for artwork and Marilyn Mettenbrink for secretarial assistance.

REFERENCES

Arendshorst, W.J., and Finn, W.F. 1977. Renal hemodynamics in the rat before and during inhibition of angiotensin II. Am J Physiol 233:F290–F297

Ayers, C.R., Vaughan, E.D., Jr., Yancey, M.R., Bing, K.T., Johnson, C.C., and Morton, C. 1974. Effect of 1-sarcosine-8-alanine angiotensin II and converting enzyme inhibitor on renin release in dog acute renovascular hypertension. Circ Res 34/35(Suppl I):I–27–I–33

Bailie, M.D., and Barbour, J.A. 1975. Effect of inhibition of peptidase activity on distribution of intrarenal blood flow. Am J Physiol 228:850–853

Britton, S.L., Beierwaltes, W.H., Fiksen-Olsen, M.J., and Romero, J.C. 1979. Intrarenal vascular effects of (Des-Asp[1]) angiotensin I and angiotensin III in the dog. Circ Res 44:666–671

Burger, B.M., Hopkins, T., Tulloch, A., and Hollenberg, N.K. 1976. The role of angiotensin in the canine renal vascular response to barbiturate anesthesia. Circ Res 35:196–202

Conway, J., Hatton, R., Keddie, J., and Dawes, P. 1979. The role of angiotensin in the control of blood pressure during sodium depletion. Hypertension 1:402–409

Fagard, R.H., Cowley, A.W., Jr., Navar, G., Langford, H.G., and Guyton, A.C. 1976. Renal responses to slight elevations of renal arterial plasma angiotensin II concentration in dogs. Clin Exp Pharmacol Physiol 3:531–538

Freeman, R.H., Davis, J.O., Williams, G.M., DeForrest, J.M., Seymour, A.A., and Rowe, B.P. 1979. Effects of the oral converting enzyme inhibitor, SQ 14225, in a model of low cardiac output in dogs. Circ Res 45:540–545

Gavras, H., Liang, C., and Brunner, H.R. 1978. Redistribution of regional blood flow after inhibition of the angiotensin-converting enzyme. Circ Res 43(Suppl I):I–59–I–63

Hall, J.E., Guyton, A.C., Jackson, T.E., Coleman, T.G., Lohmeier, T.E., and Trippodo, N.C. 1977a. Control of glomerular filtration rate by renin-angiotensin system. Am J Physiol 233:F366–F372

Hall, J.E., Guyton, A.C., Trippodo, N.C., Lohmeier, T.E., McCaa, R.E., and Cowley, A.W., Jr. 1977b. Intrarenal control of electrolyte excretion by angiotensin II. Am J Physiol 232:F538–F544

Hall, J.E., Coleman, T.G., Guyton, A.C., Balfe, J.W., and Salgado, H.C. 1979a. Intrarenal role of angiotensin II and (des-Asp[1]) angiotensin II. Am J Physiol F252–F259

Hall, J.E., Guyton, A.C., Smith, M.J., Jr., and Coleman, T.G. 1979b. Chronic blockade of angiotensin II formation during sodium deprivation. Am J Physiol 237:F424–F432

Hollenberg, N.K., Williams, G.H., Taub, K.J., Ishikawa, I., Brown, C., and Adams, D.F. 1977. Renal vascular response to interruption of the renin-angiotensin system in normal man. Kidney International 12:285–293

Hollenberg, N.K., Swartz, S.L., Passan, D.R., and Williams, G.H. 1979. Increased glomerular filtration rate after converting-enzyme inhibition in essential hypertension. N Engl J Med 301:9–12

Johns, E.J., Murdock, R., and Singer, B. 1977. The effect of converting enzyme inhibitor (SQ 20881) on the release of prostaglandins by rabbit kidney, in vivo. Br J Pharmacol 60:573–581

Johnson, M.D., and Malvin, R.L. 1977. Stimulation of renal sodium reabsorption by angiotensin II. Am J Physiol 232:F298–F306

Jones, D.R., Butt, T.J., and Wallis, A.T. 1979. The effects of captopril on the renal circulation of the rat. Med J Aust Specl Supp 2:XI–XII

Kimbrough, H.M., Jr., Vaughan, D.E., Jr., Carey, R.M., and Ayers, C.R. 1977. Effect of intrarenal angiotensin II blockade on renal function in conscious dogs. Circ Res 40:174–178

Lokhandwala, M.R., and Jandhyala, B.S. 1979. Sympathetic nerve function and renal hemodynamics during hemorrhage: Role of pre- and postsynaptic interactions between angiotensin II and prostaglandins. J Pharmacol Exp Ther 211:539–545

Matthews, P.G., and Johnston, C.I. 1979. Changes in endogenous circulating angiotensin and bradykinin after inhibition of converting enzyme (kininase II). Med J Aust Specl Suppl 2:XII–XV

McCaa, R.E., Hall, J.E., and McCaa, C.S. 1978. The effects of angiotensin I converting enzyme inhibitors on arterial blood pressure and urinary sodium excretion. Role of the renal renin-angiotensin and kallikrein-kinin system. Circ Res 43(Part II):I–32–I–39

McGiff, J.C., Terragno, N.R., Malik, K.U., and Lonigro, A.J. 1972. Release of prostaglandin E-like substance from canine kidney by bradykinin. Circ Res 31:36–43

Millar, J.A., and Johnston, C.I. 1979. Sequential changes in circulating levels of angiotensin I and II, renin and bradykinin after captopril. Med J Aust Specl Suppl 2:XV–XVII

Murao, S., Takeda, T., et al. 1979. Action of the angiotensin converting enzyme inhibitor on hemodynamics and renal functions in hypertensive patients. Abstract. The 22nd General Assembly of the Japanese Society of Nephrology, p. 1

Murthy, V.S., Waldron, T.L., and Goldberg, M.E. 1978. Inhibition of angiotensin converting enzyme by SQ 14,225 in the anesthetized dog, hemodynamic and renal vascular effects. Proc Soc Exp Biol Med 157:121–124

Nasjletti, A., Colina-Chourio, J., and McGiff, J.C. 1975. Disappearance of bradykinin in the renal circulation of dogs. Effects of kininase inhibition. Circ Res 37:59–65

Navar, L.G., LaGrange, R.A., Bell, P.D., Thomas, C.E., and Ploth, D.W. 1979. Glomerular and renal hemodynamics during converting enzyme inhibition (SQ 20,881) in the dog. Hypertension 1:371–377

Niarchos, A.P., Pickering, T.G., Case, D.B., Sullivan, P., and Laragh, J.H. 1979. Role of the renin-angiotensin system in blood pressure regulation. The cardiovascular effects of converting enzyme inhibition in normotensive subjects. Circ Res 45:829–837

Olsen, U.B., and Arrigoni-Martelli, E. 1979. The effects of kininase II inhibition by SQ 14225 on kidney kallikrein-kinin and prostaglandin systems in conscious dogs. Eur J Pharmacol 54:229–234

Ploth, D.W., Rudolph, J., LaGrange, R., and Navar, L.G. 1979. Tubuloglomerular feedback and single nephron function after converting enzyme inhibition in the rat. J. Clin Invest 64:1325–1335

Thurau, K., and Schnermann, J. 1965. Die Natrium Konzentration an den Macula densazellen als regulierender Kaktor dur das Glomerulumfiltrat (Mikropunktions-versuche). Klin Wochenschr 43:410–413

Vander, A.J. 1963. Inhibition of distal tubular sodium reabsorption by angiotensin II. Am J Physiol 205:133–138

Vinci, J.M., Horwitz, D., Zusman, R.M., Pisano, J.J., Catt, K.J., and Keiser, H.R. 1979. The effect of converting enzyme inhibition with SQ 20,881 on plasma and urinary kinins, prostaglandin E and angiotensin II in hypertensive man. Hypertension 1:416–426

Vollmer, R.R., Boccagno, J.A., Harris, D.N., and Murthy, V.S. 1978. Hypotension induced by inhibition of angiotensin-converting enzyme in pentobarbital-anesthetized dogs. Eur J Pharmacol 51:39–45

Williams, G.H., and Hollenberg, N.K. 1977. Accentuated vascular and endocrine response to SQ 20881 in hypertension. N Eng J Med 297:184–188

Wong, P.C., and Zimmerman, B.G. 1980. Mechanism of captopril-induced renal vasodilation in anesthetized dogs after nonhypotensive hemorrhage. J Pharmacol Exp Ther 215:104–109

Zimmerman, B.G., Mommsen, C., and Kraft, E. 1980. Renal vasodilatation caused by captopril in conscious normotensive and Goldblatt hypertensive dogs. Proc Soc Exp Biol Med 164:459–465

Zusman, R.M., and Keiser, H.R. 1977. Prostaglandin biosynthesis by rabbit renomedullary interstitial cells in tissue culture. J. Clin Invest 60:215–223

Hemodynamic Response to Acute and Chronic Angiotensin Converting Enzyme Inhibition (Captopril) in Hypertensive Patients[1]

Robert Fagard, Paul Lijnen and Antoon Amery
Hypertension and Cardiovascular Rehabilitation Unit
Department of Pathophysiology
University of Leuven (K.U.Leuven)
Leuven, Belgium

INTRODUCTION

Ondetti et al. (1977) have described 2D-methyl-3-mercaptopropanoyl-L-proline (SQ 14,225; captopril, Squibb Institute for Medical Research) which inhibits the conversion of angiotensin I to angiotensin II and has been effective in reducing arterial pressure (Case et al., 1978; Brunner et al., 1979; Fagard et al., 1980).

In the present study we report the hemodynamic changes in the systemic and pulmonary circulation observed during acute (75 min) and chronic (2 mo) treatment with captopril of patients with moderate to severe essential or renovascular hypertension.

METHODS

Fourteen patients (7 male) with hypertension (5 essential, 9 renovascular) were studied; their ages averaged 46±2 years and their weight 71±4 kg. Informed consent was obtained before admission into the

[1] The research was supported by the Belgian IWONL.

study. The diagnosis was based on history, physical examination, appropriate laboratory tests and an intravenous pyelogram. A renal arteriogram and renal vein blood sampling for measurement of plasma renin activity were performed when indicated.

After the described investigations, all patients were started on 3 placebo tablets a day and were followed on an outpatient basis for about 1 mo except for 2 patients in whom captopril was started earlier because of the severity of hypertension. Their 24 hr urinary sodium excretion averaged 43.8 ± 7.8 mmole; there was no significant difference in urinary sodium excretion between the acute and the chronic study.

Acute Studies

The first dose of captopril was given to the patient in the morning, while recumbent in the laboratory where room temperature was 18 to 22° C and humidity 40 to 60%. Before its administration, a small catheter (Vygon, 115.09) was introduced in the brachial artery of all patients for sampling of blood and for recording the intraarterial pressure using the Elema-Schonander EMT pressure transducer; pressure was continuously recorded on a Mingograph 81 recorder and mean brachial artery pressure was obtained by electrical damping every 5 min. Heart rate was calculated from the ECG. Cardiac output was measured using the direct O_2-Fick method, as previously described (Fagard et al., 1978). A balloon-tipped Swan-Ganz 93.110.5F floating catheter was introduced in the anticubital vein. Systemic vascular resistance was calculated from mean arterial pressure and cardiac output; pulmonary vascular resistance from mean pulmonary artery and capillary wedge pressures and cardiac output; resistances were expressed as mm Hg/l/min.

After the introduction of the catheters, a control period of about 60 min was observed, and two determinations of cardiac output were performed (and averaged for analysis) during the last 15 min. Then the patients received one 25 mg tablet of captopril and pressures and heart rate were monitored for another 75 min; cardiac output was determined twice between 60 and 75 min. The catheters were removed and the patients remained in the surroundings of the laboratory for about 3 hr before returning home.

Plasma angiotensin II was determined in arterial blood as described by Lijnen et al. (1978); normal values are 9 to 59 pg/ml.

Chronic Studies

Patients were followed on an outpatient basis at biweekly intervals, and the initial dose of captopril (25 mg, 3 times daily) was increased at each visit until control of blood pressure was obtained or maximal dosage reached. Eleven patients could be restudied in the laboratory after 2 mo when on their final dose, which was 150 to 600 mg/day. They took their morning dose of captopril (50 to 200 mg) in the laboratory and the hemodynamic data were obtained about 60 to 75 min later (cardiac output could not be measured in 1 patient).

Statistical methods are regression analysis and two-tailed Student's t-test for paired data. The dispersion of the data is given by standard error (SE).

RESULTS

Systemic Hemodynamics

Mean brachial artery pressure decreased from a control value of 136±6 to 120±4 mm Hg (P<0.001) 75 min after acute administration of captopril; mean brachial artery pressure reached a plateau after 55 min. After 2 mo mean brachial artery pressure had dropped by a further 7.6±2.4 mm Hg (P<0.01).

Table 1 gives the percentage changes from control of mean brachial artery pressure, systemic vascular resistance (control: 22±2 mm Hg/l/min), cardiac output (6.5±0.4 l/min), heart rate (74±4 beats/min),

Table 1. Haemodynamic data during acute and chronic treatment with captopril expressed as change from control. Values are means ± SE.

	Acute[a]	Chronic[a]	P (chronic vs acute)
ΔMBAP (%)	−11.5±1.9***	−18.4± 3.1***	<0.05
ΔSVR (%)	− 9.6±2.5**	−29.4± 3.5***	<0.001
ΔCO (%)	− 1.3±2.3[NS]	+16.0± 4.4**	<0.02
ΔHR (%)	+ 2.3±1.6[NS]	+ 0.5± 2.9[NS]	NS
ΔSV (%)	− 2.9±1.8[NS]	+15.8± 4.0**	<0.01
ΔVO_2 (%)	− 1.4±1.1[NS]	+14.7± 4.9*	<0.01
Δ(a–v)O_2 (%)	+ 1.3±2.1[NS]	− 2.2± 2.2[NS]	NS
ΔMRAP (mm Hg)	—	− 1.4± 0.5*	—
ΔMPCWP (mm Hg)	− 2.3±0.3***	− 2.8± 0.7*	NS
ΔMPAP (mm Hg)	− 1.9±0.5**	− 0.7± 0.7[NS]	NS
ΔPVR (%)	+ 8.6±4.5[NS]	+16.4±13.0[NS]	NS

Abbreviations: MBAP: mean brachial artery pressure; SVR: systemic vascular resistance; CO: cardiac output; HR: heart rate; SV: stroke volume; VO_2: oxygen consumption; (a–v)O_2: arteriovenous oxygen difference; MRAP: mean right atrial pressure; MPCWP: mean pulmonary capillary wedge pressure; MPAP: mean pulmonary artery pressure; PVR: pulmonary vascular resistance.

[a] P vs control, *P <0.05, **P <0.01, ***P <0.001;

stroke volume (92 ± 8 ml), oxygen consumption (270 ± 13 ml/min)and systemic arteriovenous difference in oxygen content (41 ± 2 ml/l). The acute dose of captopril produced a significant decrease of mean brachial artery pressure (−12%) and of systemic vascular resistance (−10%), with unchanged cardiac output, heart rate, stroke volume, oxygen consumption and arteriovenous oxygen content difference. During chronic treatment, mean brachial artery pressure dropped further, and the decrease of systemic vascular resistance reached 29%. Oxygen consumption significantly increased from 260 ± 9 to 302 ± 11 ml/min ($P<0.01$), with unchanged arteriovenous oxygen content difference, resulting in a significant increase of cardiac output characterized by an increased stroke volume with unchanged heart rate.

Pulmonary Hemodynamics

Mean pulmonary capillary wedge pressure decreased by 2.3 ± 0.3 mm Hg ($P<0.001$) from a control level of 5.6 ± 0.6 mm Hg, in response to the acute dose; the decrease was maintained during chronic treatment. Also, mean right atrial pressure, only measured at introduction of the Swan-Ganz catheter, had decreased significantly by 1.4 ± 0.5 mm Hg ($P<0.05$) from its control level of 0.9 ± 0.7 mm Hg when the patients were restudied after 2 mo. Pulmonary vascular resistance remained essentially unchanged during the acute and chronic treatment with captopril.

Relationship Between Hemodynamic and Biochemical Data

The changes of mean brachial artery pressure ($r = -0.80$; $P<0.001$) and of systemic vascular resistance ($r = -0.75$; $P<0.005$) in response to the acute dose of captopril were significantly related to the arterial log plasma angiotensin II obtained prior to the captopril administration.

During chronic treatment the changes of mean brachial artery pressure ($r = -0.56$; $P = 0.05$) and of systemic vascular resistance ($r = -0.69$; $P<0.05$) remained significantly related to the control log plasma angiotensin II.

DISCUSSION

Acute Hemodynamic Effect of Captopril

The angiotensin converting enzyme inhibitor captopril effectively decreased systemic arterial pressure in hypertensive patients. The acute hypotensive effect of the drug was based on a decrease of systemic vascular resistance, suggesting arteriolar dilatation, with essentially unchanged heart rate and cardiac output. A similar hemodynamic response to captopril has been observed by Fouad et al. (1980). Converting enzyme inhibition with teprotide increased cardiac output (with unchanged heart rate) when it reduced arterial pressure in patients with acute hypertension after coronary bypass surgery (Niarchos et al., 1979a). Furthermore captopril, like saralasin (Fagard et al., 1978) and teprotide (Niarchos et al., 1979a) acutely decreased pulmonary capillary wedge pressure; this may be due to unloading of the left ventricle by the decrease of arterial pressure and/or to pooling of blood by arteriolar and/or venous dilatation, though the effect of angiotensin on the capacitance vessels remains controversial (Borucki et al., 1978; Coleman et al., 1975). If venous dilatation occurred, the consequent reduction of venous return may help to explain the unchanged cardiac output but not the absence of tachycardia; increases of cardiac output and heart rate normally occur with potent arteriolar dilators such as diazoxide, while combined arteriolar and venous dilatation (nitroprusside) only produces tachycardia (Koch-Weser, 1974). These hemodynamic effects of captopril are similar to those which characterize the acute hypotensive effect of the angiotensin II antagonist saralasin. Indeed, when the latter decreases blood pressure, it reduces systemic vascular resistance and left ventricular filling pressure with unchanged cardiac output (Fagard et al., 1978; Borucki et al., 1978; Mookherjee et al., 1978), though some claim a drop of cardiac output in some of the studied subjects (de Carvalho et al., 1978).

Chronic Hemodynamic and Metabolic Effects of Captopril

The hemodynamic findings during chronic treatment with captopril were different from the acute observations. However we cannot ascertain from our study if the differences were due to the duration of treatment or to the higher dose of captopril. Systemic vascular resistance, which was reduced by 10% in the acute phase, dropped by 29% after 2 mo. Cardiac output rose by 15%, related to an increased oxygen consumption. The rise of cardiac output was also characterized by an increased stroke volume with unchanged heart rate. The overall hemodynamic response to chronic captopril is unlike the usual response to arteriolar dilatation with drugs such as minoxidil, where the rise in cardiac output is associated with a decreased arteriovenous difference in oxygen content and unchanged oxygen consumption with elevated heart rate and stroke volume (Reybrouck et al., 1976), or to combined arteriolar and venous dilatation with nitroprusside, where cardiac output even falls with a decreased stroke volume and elevated heart rate (Koch-Weser, 1974).

The decrease of pulmonary capillary wedge pressure was maintained during chronic treatment with captopril, and also right atrial pressure had dropped, indicating decreased filling of the circulation and possible dilatation of the venous capacitance vessels. Unloading of the left ventricle may have contributed to the drop of the wedge pressure, but, in view of the unchanged pulmonary artery pressure and pulmonary vascular resistance, there was no unloading of the right ventricle. The fact that the acute decrease of the pulmonary artery pressure was not maintained chronically in the presence of a reduced wedge pressure can be explained by the increased cardiac output. It is clear from the increases of stroke volume and of cardiac output that venous return was increased during chronic captopril treatment, so that any dilating effect of captopril on the venous capacitance vessels (which would reduce venous return) seems to have been overcome by the mechanisms responsible for the elevated cardiac output.

Effects of Captopril on Pulmonary Vascular Resistance

Captopril did not affect pulmonary vascular resistance, which is in accordance with the absence of any effect on the pulmonary vessels of the angiotensin II antagonist saralasin in hypertensive man (Fagard et al., 1978; Mookherjee et al., 1978), and on hypoxic pulmonary vasoconstriction in dog (Fanburg et al., 1976; Hales et al., 1977) and rat (McMurthy et al., 1976). Also converting enzyme inhibition did not affect hypoxic vasoconstriction in the dog (Fanburg et al., 1976), although indirect evidence suggested that angiotensin II plays an integral role in the hypoxic pressor response (Berkov, 1974). Niarchos et al. (1979b), however, did find a reduction of pulmonary vascular resistance when teprotide was administered to patients with acute hypertension after coronary bypass surgery.

Mechanisms of Hypotensive Effect

In the present study, the changes of mean brachial artery pressure and of systemic vascular resistance were to some extent related to the pretreatment plasma angiotensin II level, both for the acute and for the chronic observations. This suggests that the antihypertensive effect of captopril depends at least partly on its interference with the renin-angiotensin system. Other suggested mechanisms, such as bradykinin accumulation (Williams & Hollenberg, 1977), cannot, however, be excluded from the present study.

SUMMARY

The effects of the angiotensin converting enzyme inhibitor captopril were studied in 14 patients with hypertension 75 min after ingestion of 25 mg of the drug, and after 2 mo when on 150 to 600 mg/day. Captopril acutely decreased mean brachial artery pressure by 16.3 mm Hg ($P<0.001$) while the chronic reduction on the higher dose was significantly greater by 7.6 mm Hg ($P<0.01$). The changes of mean brachial artery pressure were significantly related to control plasma levels of angiotensin II. The acute hypotensive effect of the drug was based on a 10% ($P<0.01$) decrease of systemic vascular resistance with essentially unchanged cardiac output and heart rate. Pulmonary capillary wedge pressure decreased by 2.3 mm Hg ($P<0.001$) with unchanged pulmonary vascular resistance. During chronic treatment systemic vascular resistance decreased by 29% ($P<0.001$). Oxygen consumption increased by 15% ($P<0.01$), with a concomitant 16% rise in cardiac output ($P<0.01$), characterized by an increased stroke volume and unchanged heart rate. The drop of pulmonary capillary wedge pressure was maintained, and also right atrial pressure decreased (-1.4 mm Hg; $P<0.05$). The data indicate that the action of captopril is characterized by arteriolar and possibly venous dilatation.

ACKNOWLEDGMENTS

We gratefully acknowledge the assistance of Miss S. Taelemans, Miss L. Lommelen, Miss G. Huybens, Miss R. Nuyts, Mrs L. Jans, Mrs G. Melotte, Mrs R. Deruyck, Mrs C. Willems, Mr L. Cockx, Mr J. Delsupehe, Mr J. Huysecom and Mr J. Romont.

We thank the Squibb Institute for Medical Research for the generous supply of drugs.

REFERENCES

Berkov, S. 1974. Hypoxic pulmonary vasoconstriction in the rat. The necessary role of angiotensin II. Circ Res 35:256–61.

Borucki, L.J., Levenson, D., and Hollenberg, N.K. 1978. Cardiovascular responses to blockade of angiotensin and alpha-adrenergic receptors. Am J Physiol 235:F199–F202.

Brunner, H.R., Gavras, H., Waeber, B., Kershaw, G.R., Turini, G.A., Vukovich, R.A., McKinstry, I., and Gavras, I. 1979. Oral angiotensin-converting enzyme inhibitor in long-term treatment of hypertensive patients. Ann Intern Med 90:19–23

Case, D.B., Atlas, S.A., Laragh, J.H., Sealy, J.E., Sullivan, P.A., and

McKinstry, D.N. 1978. Clinical experience with blockade of the renin-angiotensin-aldosterone system by an oral converting-enzyme inhibitor (SQ 14225, captopril) in hypertensive patients. Progr Cardiovasc Dis 21:195–206.

Coleman, T.G., Cowley, A.W., and Guyton, A.C. 1975. Angiotensin and the hemodynamics of chronic salt deprivation. Am J Physiol 229:167–71

de Carvalho, J.G.R., Dunn, F.G., Kem, D.C., Chrysant, S.G., and Frohlich, E.D. 1978. Hemodynamic correlates of saralasin-induced arterial pressure changes. Circulation 57:373–78

Fagard, R., Amery, A., Lijnen, P., Reybrouck, T., Joossens, J.V., Billiet, L., Moerman, E., and De Schaepdryver, A. 1978. Effects of 1-sar-8-ala-angiotensin II on systemic and pulmonary haemodynamics in hypertensive patients. Clin Exp Pharmacol Physiol 5:457–64

Fagard, R.H., Amery, A.K., Lijnen, P.J., and Reybrouck, T. 1980. Comparative study of an angiotensin II analog and a converting enzyme inhibitor. Kid Int 17:647–653

Fanburg, B.L., Mieszala, J.R., and Levine, H.J. 1976. Absence of role for angiotensin II in the pulmonary vascular response to hypoxia in the intact dog. Clin Res 24:613A

Fouad, F.M., Ceimo, J.M.K., Tarazi, R.C., and Bravo, E.L. 1980. Contrasts and similarities of acute hemodynamic responses to specific antagonism of angiotensin II (sar^1thr^8A II) and to inhibition of converting enzyme (captopril). Circulation 61:163–69

Hales, C.A., Rouse, E.T., and Kazemi, H. 1977. Failure of saralasin acetate, a competitive inhibitor of angiotensin II, to diminish alveolar hypoxic vasoconstriction in the dog. Cardiovasc Res II:541–46

Koch-Weser, J. 1974. Vasodilator drugs in the treatment of hypertension. Arch Intern Med 133:1017–27

Lijnen, P., Amery, A., Fagard, R., and Katz, F. 1978. Radioimmunoassay of angiotensin II in unextracted plasma. Clin Chim Acta 88:403–12

McMurthy, I.F., Davidson, A.B., Reeves, J.T., and Grover, R.F. 1976. Inhibition of hypoxic pulmonary vasoconstriction by calcium antagonists in isolated rat lungs. Circ Res 38:99–104

Mookherjee, S., Obeid, A., Warner, R., Anderson, G., Eich, R., and Smulyan, H. 1978. Systemic and pulmonary haemodynamic effects of saralasin infusion in hypertension. Am J Cardiol 42:987–92

Niarchos, A.P., Roberts, A.J., Case, D.B., Gay, W.A., and Laragh, J.H. 1979a. Hemodynamic characteristics of hypertension after coronary bypass surgery and effects of converting enzyme inhibitor. Am J Cardiol 43:586–93.

Niarchos, A.P., Roberts, J.A., and Laragh, J.H. 1979b. Effects of the converting enzyme inhibitor (SQ 20881, Teprotide) on the pulmonary circulation in man. Am J Med 67:785–91.

Ondetti, M.A., Rubin, B., and Cushman, D.W. 1977. Design of specific inhibitors of angiotensin converting enzyme: New class of orally active antihypertensive agents. Science 196:441–44.

Reybrouck, T., Amery, A., Billiet, L., and Fagard, R. 1976. Triple antihypertensive therapy: a hemodynamic approach. In *The Arterial Hypertensive Disease,* ed. G. Rorive and H. Van Cauwenberghe, pp. 255. New York: Masson.

Williams, G.H., and Hollenberg, N.K. 1977. Accentuated vascular and endocrine response to SQ 20881 in hypertension. N Engl J Med 297:184–88.

Hemodynamic Effects of Long-Term Captopril Therapy in Hypertensive Man

Emmanuel L. Bravo, Robert C. Tarazi and Fetnat M. Fouad
The Cleveland Clinic Foundation, Research Division
Cleveland, Ohio 44106

INTRODUCTION

Captopril has proved to be an effective antihypertensive agent in certain subsets of the human hypertensive population (Gavras et al., 1978; Brunner et al., 1979; Bravo and Tarazi, 1979). It remains controversial whether its blood pressure lowering effect is wholly dependent on interference of the renin-angiotensin system. The immediate (acute) and short-term (5 to 7 days) changes associated with blood pressure reduction appear to be related to decreases in total peripheral resistance without any significant changes in cardiac output or heart rate (Cody et al., 1979).

The present report summarizes our experience with hemodynamic changes associated with the use of captopril over a more prolonged period. The data revealed a complex pattern that could not be predicted from the initial response or from our current knowledge of the pharmacology of converting enzyme inhibitors, suggesting the contribution of other variables in its long-term antihypertensive effect.

MATERIALS AND METHODS

Patient Population

The patients investigated have been derived from the patient population followed by the Research Division of the Cleveland Clinic who are on long-term (6 to 28 months) captopril therapy for their hypertension.

There were 21 patients, ranging in age from 32 to 68 years, 5 of whom were females and 16 males. Of these, 16 have essential hypertension; 5 have arteriographic and biochemical evidence of significant renal arterial occlusive disease. Six patients were on captopril alone (Group I); in 9, hydrochlorothiazide (25 to 50 mg BID) was added after an initial trial on captopril alone did not produce satisfactory blood pressure control (Group II); in 6, captopril was added to their diuretic regimen (Group III). Therefore, at the time of this assessment, 6 patients were on captopril alone and 15 were on combined captopril-diuretic regimen. The dosage of captopril varied from 200 to 400 mg/day.

Hemodynamic Studies

Prior to treatment, cardiac output was determined by thermal dilution; during this assessment, it was determined by a noninvasive isotopic method utilizing 99^m TC–HSA. Correlation of both methods as performed in our laboratory is very close (r = 0.97; $P < 0.001$).

All patients were studied fasting, in the morning, after at least 30 to 45 min of supine rest. Plasma volume was first measured, followed thereafter by cardiac output measurements.

Volume and Cardiac Output Measurements

Plasma volume was measured by radioiodinated (^{125}I) serum albumin with 10 min equilibration period. Total blood volume was calculated from the plasma volume and the simultaneously corrected packed cell volume. Results are calculated in ml/cm height in order to minimize the effect of age and overweight; normal value (± 1SD) for plasma volume and total blood volume average 15.3±1.7 and 24.1±2.4, respectively for women and 18.4±1.9 and 30.9±3.0, respectively for men. Values are expressed as percent of normal (% N) to permit inclusion of both men and women in subsequent analyses (Tarazi et al., 1970a). The variability of the method is 8%.

Thermodilution curves were obtained from injections of 10 ml of 5% dextrose. Cardiac output was calculated as the average of three determinations that did not differ by more than 5% (Cody et al., 1979).

Cardiac output using 99^m Tc–HSA was obtained in duplicate from the first pass radionuclide study. The method for cardiac output calculation and its validation was previously reported (Fouad et al., 1979).

Blood pressure was measured by sphygmomanometer. Mean arterial pressure was calculated as diastolic blood pressure plus 1/3 pulse pressure, total peripheral resistance as the ratio of mean arterial pressure to cardiac index and expressed in arbitrary unit.

Cardiovascular Response to Head-up Tilt

Studies were usually performed in the morning after the patient had been supine for at least 30 min. Arterial blood pressure and heart rate were measured for 15 min at 5-min intervals, then the patient was tilted head-up at 60° for 20 min. Blood pressure and heart rate were measured at 5-min intervals as before.

Analysis of Results

The hemodynamic correlates of blood pressure response were first sought. Then, using cardiac output as the independent variable, changes in blood pressure, heart rate, blood volume and total peripheral resistance were assessed in relationship to cardiac output values. This allowed a study of the correlates of cardiac output responses as a continuum without arbitrarily subdividing the patients into different treatment groups.

Statistical analysis and significance of results were calculated using standard methods for calculation of standard and paired t-tests and of correlation coefficients (Croxton and Cowden, 1941). Values are reported as mean ± SEM.

RESULTS

Hemodynamic Effects

Blood pressure fell significantly in all three groups (Δ39 ± 5 (SE) mm Hg in Group I, Δ40 ± 6 in Group II, and Δ34 ± 7 in Group III). The depressor response was equivalent between those who were maintained on captopril alone and those on combined diuretic–captopril therapy. The reduction of arterial pressure was primarily achieved through decreased total peripheral resistance (Table 1). Supine resting heart rate was essentially unchanged. As a group, cardiac output remained unchanged, whereas total peripheral resistance was significantly reduced in close correlation with the change in mean arterial pressure (Fig. 1).

Although group averages did not reveal significant alterations in cardiac output, a closer analysis demonstrated marked individual variations in cardiac output values (Fig. 2). Thus, we found that cardiac output values in 6 patients were significantly increased, unchanged in 8, and significantly decreased in 7. These changes were independent of duration and type of treatment, and of the hemodynamic findings prior to therapy.

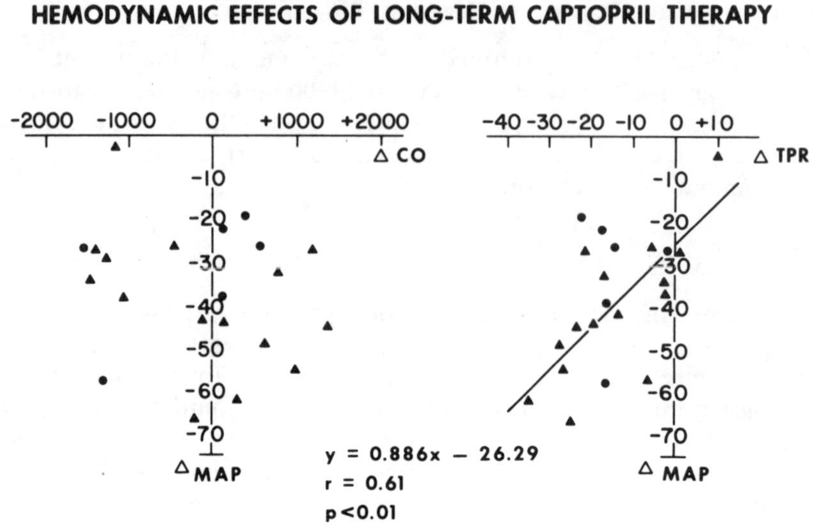

Fig. 1. Hemodynamic correlates of the depressor response to captopril. Solid circles represent patients on captopril alone; solid triangles, patients on captopril-diuretic regimen. MAP = mean arterial pressure, CO = cardiac output, TPR = total peripheral resistance.

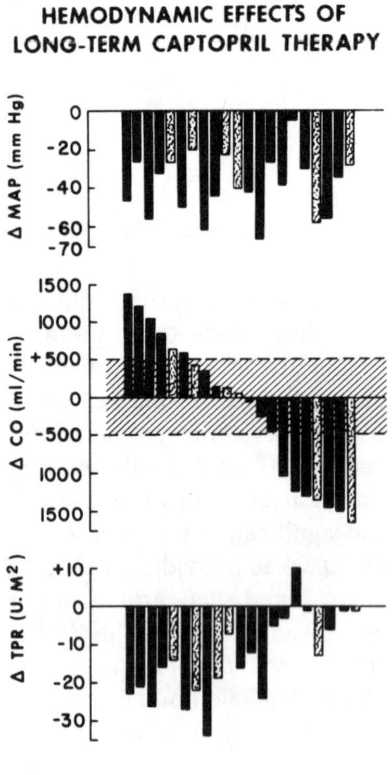

Fig. 2. Relationship among ΔMAP (mean arterial pressure), ΔCO (cardiac output) and ΔTPR (total peripheral resistance) during long-term captopril administration. CO is the independent variable and arranged from the biggest increase to the biggest decrease. Solid bars represent patients on captopril-diuretic and stippled bars, patients on captopril alone. The interrupted lines indicate the variability of CO measurements.

Prolonged treatment with captopril did not lead to adverse cardiovascular responses (Fig. 3). Head-up tilting at 60° for 20 min produced normal blood pressure and heart rate responses regardless of the treatment regimen utilized.

Table 1. Systemic hemodynamic findings with long-term captopril administration.

Patient Groups	Mean Arterial Pressure (mmHg)	Cardiac Output (ml/min)	Total Peripheral Resistance (U.M.²)	Heart Rate (beats/min)
Group I (n = 6)				
Control	132±7	5566±507	48±5	74±6
Captopril	100±4*	5284±305	36±3*	70±5
Group II (n = 9)				
Control	142±3	5227±211	52±3	70±5
Captopril	129±6	5248±264	47±4	70±5
+ Diuretic	98±2*	5135±238	37±2*	73±3
Group III (n = 6)				
Diuretic	137±3	5457±527	51±5	73±4
+Captopril	103±4*	5346±549	39±3*	72±4

All values are expressed as mean ± SEM.
*Significantly different from Control

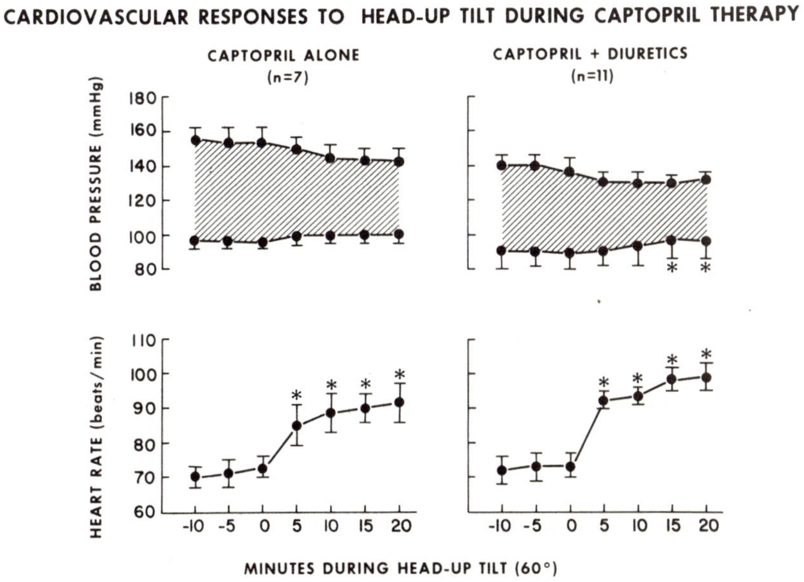

Fig. 3. Blood pressure and heart rate responses to head up tilt in patients on long-term captopril therapy.

Volume Effects

Patients maintained on captopril alone had a small and insignificant fall in plasma volume (from 96 ± 6 to 93 ± 6% of normal). Patients who were given diuretics during maintenance captopril therapy had the greatest fall in plasma volume (from 97 ± 5 to 90 ± 4% of normal); however, this was not a statistically significant change. The addition of captopril to those on maintained diuretic therapy produced no additional fall in plasma volume (from 98 ± 5 to 97 ± 5% of normal).

Relationship Between Cardiac Output and Other Hemodynamic Correlates

Cardiac output values correlated directly with changes in blood volume (Fig. 4)—the greater the fall in blood volume, the lower the cardiac output values. This weak but significant correlation ($r = 0.445$, $P < 0.05$) obtained whether or not patients were maintained on diuretics. Changes in cardiac output did not correlate with change in heart rate ($r = 0.413$, $P > 0.05$). Multiple regression analysis among changes in cardiac output, heart rate and blood volume revealed a correlation coefficient of 0.589 ($P < 0.05$).

The changes in cardiac output were not related to pretreatment systemic vascular resistance and mean arterial pressure, or to the degree of blood pressure fall.

Relationship of Change in Total Peripheral Resistance and Blood Volume

Changes in total peripheral resistance correlated inversely with blood volume ($r = 0.523$, $P < 0.01$)—the greater the fall in total peripheral resistance, the greater the tendency for volume retention (Fig. 5).

DISCUSSION

The dominant hemodynamic mechanism for the antihypertensive effect of captopril seems established—i.e., maintained reduction of systemic vascular resistance. This was a common feature in all our previous studies (Cody et al., 1979), whether measurements were made 30 to 90 min following a single oral dose or after 3 to 5 days of maintained administration, and whether patients were on normal or low dietary sodium during therapy. Similar results were obtained by Fagard et al. (1980) and Muirhead et al. (1978) in spontaneously hypertensive rats.

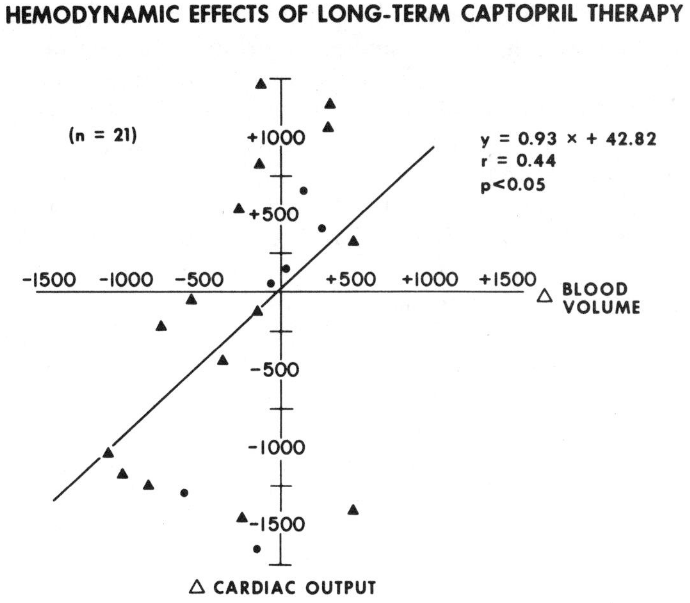

Fig. 4. Correlation between changes in blood volume and changes in cardiac output. Other details as in Figure 1.

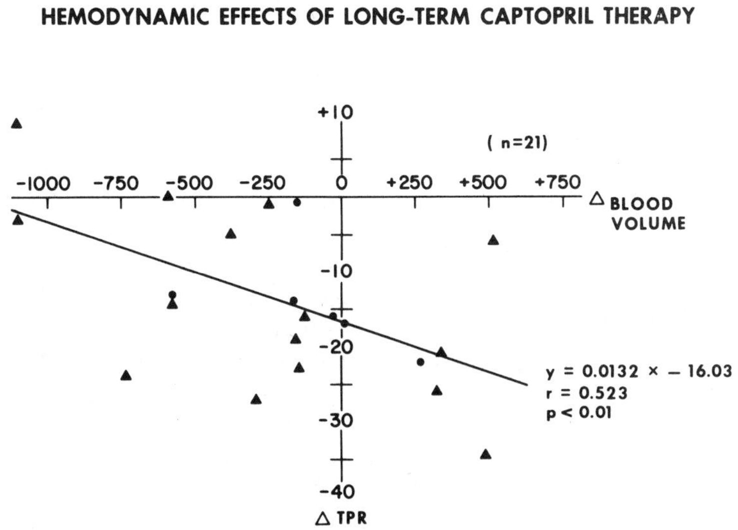

Fig. 5. Correlation between changes in blood volume and changes in total peripheral resistance. Other details as in Figure 1.

However, in contrast with the uniform lack of cardiac output changes in acute and short-term studies, surprising variations in cardiac output were found during long-term therapy. The differences in cardiac output among patients were not related to the height of pretreatment blood pressure, to the degree of blood pressure control or to alterations in cardiac performance. Similarly, variations in cardiac output showed no significant correlation either on simple or multiple regression analysis with changes in heart rate. However, there was a modest but definite correlation with alterations in blood volume. Blood volume as an average for the group remained remarkably constant; however, as is often found, this constancy covered a spectrum of change from unaltered to grossly diminished in some patients. It was in patients with diminished blood volume that all instances of reduced cardiac output were seen.

The correlation between total blood volume and cardiac output was significant, but modest, indicating the interplay of factors in this relationship other than the magnitude of changes in intravascular volume alone. There are two possible explanations for this correlation. The first is a speculation that a reduction in output was secondary to hypovolemia; although attractive, this hypothesis seemed unlikely, because in the presence of intact sympathetic activity, hypovolemia could readily be compensated for by venous constriction, maintaining essentially a normal venous return. On the other hand, the positive correlation between cardiac output and blood volume might only be a reflection of the inverse relationship between blood volume and total peripheral resistance that has been documented in many studies of hypertension (Tarazi et al., 1970a; Tarazi et al., 1973). Those with the lowest systemic vascular resistance would be expected to show some increase in total blood volume. Such an explanation will bring captopril in line with other antihypertensive medication, except for the fact that the changes in total blood volume were relatively minor and were not associated with increased weight or expansion of extracellular fluid volume (Tarazi et al., 1976).

The hemodynamic evolution during captopril therapy can be contrasted with two other types of therapeutic modalities. The first are the variations in hemodynamics during β-adrenergic blockade. In contrast with captopril, the cardiac output variations with propranolol therapy showed first a diminution in output followed by a normalization of that index as total peripheral resistance dropped towards or even below normal (Tarazi et al., 1980). Diuretics also showed the same general pattern of evolution (Conway and Palmero, 1963; Tarazi et al., 1970b; Dustan et al., 1974). These three classes of drugs had a common feature of lack of volume expansion despite blood pressure reduction. This could be attributed to diuretic effect of one, but probably to the reduction or normalization of aldosterone levels by β-adrenergic blockade and by converting enzyme inhibition. Although perhaps not very important in the immediate hemodynamic effect of captopril, maintained normalization of aldosterone production probably plays a very important role in the maintenance of responsiveness and the avoidance of plasma volume expansion and, therefore, in the modulation of the final pattern.

SUMMARY AND CONCLUSIONS

This review of our experience with captopril has outlined a complexity of effects, the pattern of which changed with maintenance of therapy. The basic mechanism of its antihypertensive action seems well established—i.e., a lowering of total peripheral resistance. However, in contrast to previous reports demonstrating no alterations in cardiac output, the present findings in longer-term studies revealed marked variability in cardiac output values. To the extent that hemodynamic changes reflect the factors involved in the action of a drug, these results suggest the interference of variables other than angiotensin activity in the antihypertensive action of captopril.

ACKNOWLEDGMENT

The authors would like to thank E. R. Squibb & Sons, Inc. for the gift of captopril.

REFERENCES

Bravo E.L., and Tarazi R.C. 1979. Converting enzyme inhibition with an orally-active compound in hypertensive man. Hypertension *1*:39–46

Brunner, H.R., Gavras H., Waeber B., Kershaw G.R., Turini G.A., Vukovich R.A., and McKinstry D.N. 1979. Oral angiotensin-converting enzyme inhibitor in long-term treatment of hypertensive patients. Ann Intern Med *90*:19–23

Cody R.J., Jr., Tarazi R.C., Bravo E.L., and Fouad F.M. 1979. Haemodynamics of an orally-active converting enzyme inhibitor (SQ 14,-225) in hypertensive patients. Clin Sci Mol Med *55*:453–459

Conway, J., and Palmero H. 1963. The vascular effects of the thiazide diuretics. Arch Intern Med *111*:203–207

Croxton, F.E., and Cowden D.J. 1941. *Applied General Statistics*. New York:Prentice-Hall

Dustan, H.P., Tarazi R.C., and Bravo E.L. 1974. Diuretic and diet treatment of hypertension. Arch Intern Med *133*:1007–1013

Fagard R., Amery A., Reybrouck T., Lijnen P., and Leon B. 1980. Acute and chronic systemic and pulmonary haemodynamic effects of angiotensin converting enzyme inhibition (captopril) in hypertensive patients. Am J Cardiol, in press

Fouad F.M., Tarazi R.C., MacIntyre W.J., and Durant D. 1979. Venous delay a major source of error in isotopic cardiac output determination. Am Heart J *97*:477–484

Gavras H., Brunner H.R., Turini G.A., Kershaw, G.R., Tifft C.P., Cuttelod S., Gavras I., Vukovich R.A., and McKinstry D.N. 1978. Antihypertensive effect of the oral angiotensin converting-enzyme inhibitor SQ 14,225 in man. N Engl J Med *298*:991–995

Muirhead E.E., Prewitt R.L., Jr., Brooks B., and Brosius W.L., Jr. 1978. Anti-hypertensive action of the orally-active converting enzyme inhibitor (SQ 14,225) in spontaneously hypertensive rats. Circ Res *43*(Suppl I):I–53–I–59

Tarazi R.C., Dustan H.P., Frohlich E.D., Gifford R.W., Jr., and Hoffman G.C. 1970a. Plasma volume and chronic hypertension: relationship to arterial pressure levels in different hypertensive diseases. Arch Intern Med *125*:835–842

Tarazi R.C., Dustan H.P., and Frohlich E.D. 1970b. Long-term thiazide therapy in essential hypertension. Circulation *41*:709–717

Tarazi R.C., Ibrahim M.H., Bravo E.L., and Dustan H.P. 1973. Hemodynamic characteristics of primary aldosteronism. New Engl J Med *289*:1330–1335

Tarazi, R.C., Dustan H.P., and Bravo E.L. 1976. Hemodynamic effects of propranolol in hypertension. Postgrad Med J *52*:92–100

Tarazi, R.C., Bravo E.L., Fouad F.M., Omvik P., and Cody R.J., Jr. 1980. Hemodynamic and volume changes associated with captopril. Hypertension, 2:576–585

Effects of Converting Enzyme Inhibition in Circulatory Shock[1]

Allan M. Lefer and George J. Trachte[2]
Department of Physiology, Jefferson Medical College
Thomas Jefferson University
Philadelphia, Pennsylvania 19107

INTRODUCTION

The discovery of angiotensin converting enzyme inhibitors (CEI) in the venom of the South American pit viper (*Bothrops jararaca*) by Ferriera (1965) has provided a convenient method for the evaluation of renin-angiotensin actions in a variety of conditions. Prior to this development, renin-angiotensin contributions were tested by nephrectomy, a crude and nonspecific method. Early trials utilizing nephrectomy revealed an increased sensitivity to hemorrhage in these animals (Bahnson, 1943; Hamilton and Collins, 1942). However, renal actions independent of the renin-angiotensin system could have also been involved in the enhanced susceptibility to hemorrhage.

The advent of converting enzyme inhibitors provided a far more specific means of examining the role of the renin-angiotensin system. The first use of converting enzyme inhibitors in shock utilized hemorrhaged dogs anesthetized with pentobarbital sodium (Errington and Rocha e Silva, 1974). These investigators noted enhanced cardiac output and decreased total peripheral resistance in shocked dogs treated with a nonapeptide inhibitor of converting enzyme (teprotide). This effect was observed immediately upon administration of the converting enzyme inhibitors even during oligemia while mean arterial blood pressure (MABP) was maintained at 40 mm Hg. No effect on mean bleedout

[1] Contribution of the Ischemia-Shock Research Center

[2] Present address: Department of Pharmacology, University of Virginia School of Medicine, Charlottesville, Virginia

volume was observed. Morton et al. (1977) later also found increased survival in hemorrhaged dogs following treatment with teprotide.

Following these encouraging results, we attempted to reexamine the effects of the renin-angiotensin system in hemorrhagic shock, including indicators of tissue damage (i.e., lysosomal enzyme and proteolytic activity), mesenteric perfusion (i.e., superior mesenteric artery blood flow [SMAF]) and formation of the endogenous negative inotropic humoral agent, myocardial depressant factor (MDF). With this multifaceted approach, we attempted to elucidate the mechanism of the effect of converting enzyme inhibitors in hemorrhagic shock. The newly developed converting enzyme inhibitor, captopril, was employed in these studies.

CAPTOPRIL IN HEMORRHAGIC SHOCK

The initial step in examining the effects of captopril in hemorrhagic shock was to ensure that the dose given provided an adequate blockade of converting enzyme. Representative tracings of pressor and splanchnic

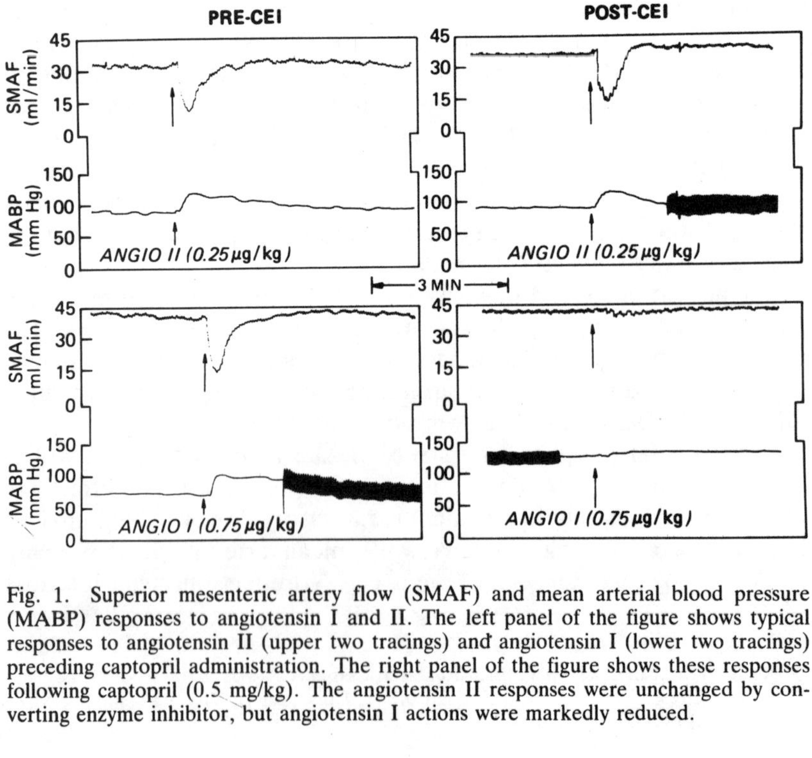

Fig. 1. Superior mesenteric artery flow (SMAF) and mean arterial blood pressure (MABP) responses to angiotensin I and II. The left panel of the figure shows typical responses to angiotensin II (upper two tracings) and angiotensin I (lower two tracings) preceding captopril administration. The right panel of the figure shows these responses following captopril (0.5 mg/kg). The angiotensin II responses were unchanged by converting enzyme inhibitor, but angiotensin I actions were markedly reduced.

vasocontrictor actions of angiotensin I and angiotensin II are shown in Figure 1. The upper tracings show an intravenous injection of angiotensin II (250 ng/kg) to decrease mesenteric blood flow and elevate arterial blood pressure to an equal extent prior to and following administration of captopril at 0.5 mg/kg. The increase in MABP was 29 ± 2 mm Hg (mean ± SEM) before and after captopril, and SMAF decreased 58 ± 10% before and 55 ± 5% after converting enzyme inhibitor in response to angiotensin II. In contrast, angiotensin I (750 ng/kg) produced a 52 ± 12% decrease in SMAF before and a 12 ± 4% decrease after converting enzyme inhibitor. The pressor response to angiotensin I was attenuated from 33 ± 3 mm Hg to 6 ± 1 mm Hg after captopril. Thus, virtually complete inhibition of angiotensin converting enzyme was achieved at doses of captopril that did not directly alter the hemodynamic status of the anesthetized non–salt depleted cat.

As a result of these and other studies, a regimen of captopril infusion of 0.5 mg/kg/hr was adopted, after a priming dose of 0.5 mg/kg, starting 30 min after the onset of hemorrhage and continuing for the remainder of the 270 min experimental period. This regimen did not alter baseline hemodynamics with regard to MABP, heart rate or central venous pressure in nonhemorrhaged control cats. Moreover, this regimen of captopril did not alter the bleedout or reuptake process of the hemorrhage protocol in shock cats. In this regard, the total bleedout volumes of the hemorrhaged cats receiving the vehicle was 31.0 ± 2.3 ml/kg, and for those receiving the captopril the volumes were 29.8 ± 2.6 ml/kg, a trivial and nonsignificant difference.

Table 1 summarizes the initial and final MABP of the four groups of cats studied. All groups exhibited comparable initial MABP. Both groups of sham shock cats demonstrated a slight decline of about 20 mm Hg over the 270 min experimental period. At the end of the experiment, the final MABP was 83 to 87% of initial values. In contrast, MABP of untreated hemorrhaged cats declined after reinfusion to 51 mm Hg, a value only 34% of initial MABP. This drop is consistent with an early mortality occurring approximately 90 to 120 min after reinfusion. However, hemorrhaged cats infused with captopril did not develop a postoligemic decline in MABP. In these treated cats, MABP declined to only 75% of initial pressures, a value just slightly below that of nonshocked control cats. This MABP profile is consistent with significantly improved survival (Lefer, 1974).

Plasma angiotensin concentrations were determined by radioimmunoassay (RIA) and are shown in Figure 2. All groups showed elevations in these values with time, probably resulting from dehydration caused by insensible water loss and blood sampling. However, very large elevations in circulating angiotensin II concentrations were noted in bled cats receiving vehicle, concentrations rising from 0.10 ± 0.03 to 1.3 ± 0.17 ng/ml during oligemia and declining after reinfusion to 1.08 ± 0.21 ng/ml. Captopril blunted these elevations to 0.50 ± 0.18 ng/ml prior to, and 0.36 ± 0.08 ng/ml following, reinfusion. Both of these

Table 1. Initial and final blood pressures of anesthetized cats.

Group	Mean Arterial Blood Pressure (mm Hg) 0 Min	Mean Arterial Blood Pressure (mm Hg) 270 Min	Final as % of Initial
Sham shock + 0.9% NaCl	147±9	122±10	83%
Sham shock + Captopril	142±8	124±7	87%
Hemorrhagic shock + 0.9% NaCl	149±10	51±8*	34%*
Hemorrhagic shock + Captopril	128±8	96±9	75%

*All values are means ± SEM for 6–7 cats in each group.
P <0.01 compared to sham shock control

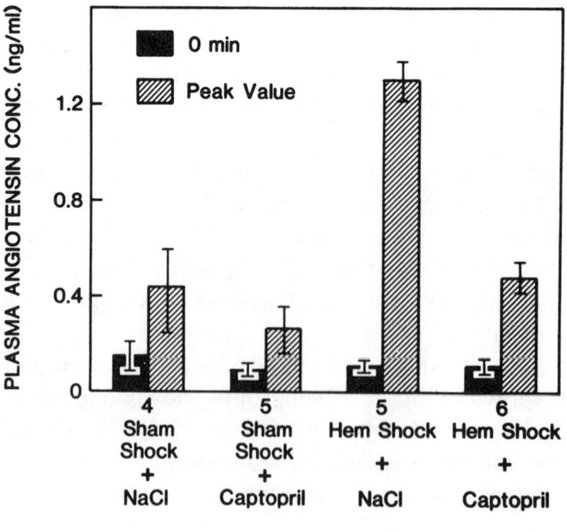

Fig. 2. Plasma angiotensin concentrations expressed as ng/ml, assessed by specific radioimmunoassay for angiotensin II. Values are shown at 0 time (black bars) and at peak value, 150 to 210 min, (hatched bars). All values are means ± SEM. Numbers at the bottom of the bars are numbers of cats studied in each group.

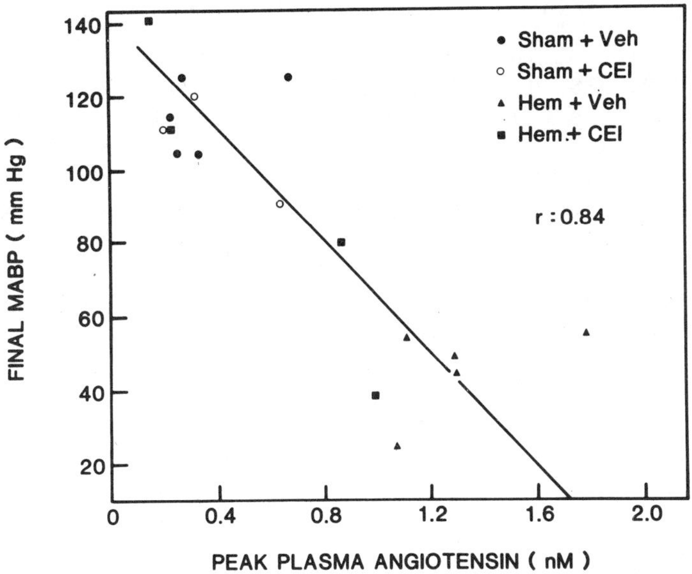

Fig. 3. Relationship between peak angiotensin plasma levels and final MABP in the four groups. The data indicate an inverse relationship between maximal angiotensin plasma concentrations and final arterial pressure with an equation of $y = -1.72x + 11.7$ and $r = 0.84$. Each point represents data from one cat.

increases were significantly less than the corresponding values observed in shocked cats given saline ($P < 0.02$). Interestingly, angiotensin concentrations continued to decline in cats receiving the vehicle despite the rapidly developing hypotension at 210 min, and exhibited only a small rebound at 270 min (0.79 ± 0.22 ng/ml), when pressures were again at shock levels (51 mm Hg).

Figure 3 shows the relationship between peak plasma angiotensin concentrations and final MABP in the four groups of cats. Vehicle treated shocked cats exhibited high peak angiotensin concentrations (i.e., about 1.0 to 1.4 ng/ml), and none of these cats had a final blood pressure greater than 55 mm Hg. Captopril treated shock animals and sham shock cats developed lower peak angiotensin II concentrations in general, and also had higher blood pressures at the termination of the experiment. The correlation coefficient for these data was 0.84, the relationship having a significance level of $P < 0.001$. These results indicate an inverse relationship between peak angiotensin concentrations and MABP, supporting the captopril data and indicating a role for angiotensin in exacerbating the pathophysiology of hemorrhagic shock.

CATHEPSIN D AND MDF

Table 2 summarizes plasma cathepsin D and myocardial depressant factor (MDF) activities of plasma samples obtained at the termination of the experiments. Sham shock cats exhibited typically low activities of both substances. Shock cats receiving the vehicle developed elevated activities of the toxic peptide to 63 ± 6 units, but captopril significantly limited this accumulation to 27 ± 4 units. Similar findings were obtained with respect to the lysosomal protease, cathepsin D. The prevention of MDF formation would be expected from the effect of captopril in stabilizing lysosomal membranes, since MDF is formed by the hydrolytic action of cathepsins and other proteases on endogenous pancreatic proteins (Glenn and Lefer, 1970; Lefer, 1974).

REPLACEMENT OF ANGIOTENSIN II

Results from this carefully controlled study with captopril indicate a significant protective action of this converting enzyme inhibitor in shock, but left some doubt whether the mechanism of the protective action

Table 2. Plasma cathepsin D and MDF activities in cats.

	Final Values (270 min)	
Group	Cathepsin D Activity[a] (Units/mg protein)	MDF Activity[b] (Units/10 ml)
Sham shock + 0.9% NaCl	2.4±0.5	17±3
Sham shock + Captopril	2.7±0.6	19±3
Hemorrhagic shock + 0.9% NaCl	16.4±2.9	63±6
Hemorrhagic shock + Captopril	6.9±1.4	27±4

All values are means ± SEM for 6 to 7 cats.

[a] Units expressed nmol tyrosine released at 37°C/hr.

[b] Units expressed as 1% decrease in developed tension of isolated cat papillary muscles at 37°C and 1 Hz.

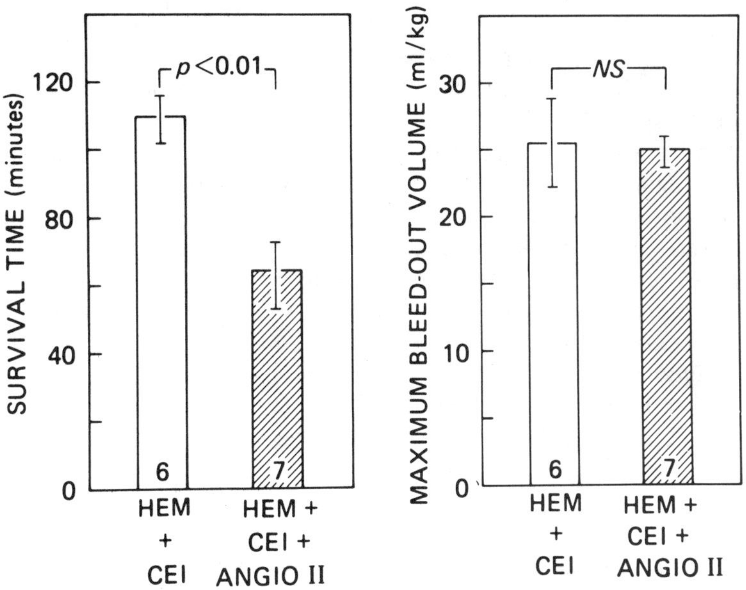

Fig. 4. Average survival times and bleedout volumes in hemorrhaged cats given captopril or captopril with angiotensin II (10 μg/kg/hr). Angiotensin II reduced postreinfusion survival times from 110 ± 5 min to 64 ± 11 min. The angiotensin II infusion did not increase the amount of blood necessary to withdraw to reduce MABP to 40 mmHg and maintain it there for 120 min.

occurred via blockade of angiotensin II or III formation or by some alternate action. For this reason, a replacement dose of angiotensin II was infused in conjunction with captopril in shocked animals. In this manner, we hoped to delineate angiotensin dependent actions of captopril and to further understand the role of angiotensin in hemorrhagic shock. The dose of angiotensin employed was calculated to reproduce maximal plasma angiotensin concentrations occurring in shock (Gavras et al., 1975).

Figure 4 summarizes the effect of angiotensin II (10 μg/kg/hr) on survival in captopril treated cats. The angiotensin infusion reduced survival time from 110 ± 5 min to 64 ± 11 min postreinfusion. The survival time was significantly lower in angiotensin infused cats ($P < 0.025$), despite the finding that there was no difference in maximal bleedout volume between the two groups.

Additional studies showed that not only was angiotensin II infusion able to offset the protective effect of captopril in hemorrhagic shock, but captopril was unable to potentiate the effects of bradykinin on blood pressure and blood flow during hypotension. Thus, bradykinin depressed blood pressure 20 ± 2 mm Hg at normotensive levels (i.e., 140 mm Hg). Captopril potentiated this depressor effect to 54 ± 14 mm Hg ($P < 0.01$). However, at a blood pressure of 75 mm Hg, bradykinin decreased MABP by 18 ± 2 mm Hg before captopril, and by 19 ± 2 mm Hg after captopril. This difference was not significant. In contrast, this regimen of captopril (0.5 mg/kg) significantly attenuated the pressor response to angiotensin I (1 μg/kg) from 82 ± 10 to 10 ± 3 mm Hg at normotensive pressures as well as at hypotensive levels from 105 ± 14 to 16 ± 8. Thus, the ability of captopril to inhibit the conversion of angiotensin I to angiotensin II is not diminished by hypotensive levels, whereas captopril's ability to potentiate bradykinin is severely curtailed in hypotension (Fig. 5). These results emphasize inhibition of angiotensin I to II conversion as the primary mechanism of the protective action of converting enzyme inhibitors in hemorrhagic shock, and minimize any role of kinin potentiation in this biological setting.

ANTIBODIES TO ANGIOTENSIN II

Another method of reducing the circulatory effects of angiotensin is development of antibodies to angiotensin II. We successfully developed high angiotensin II antibody titers over a 6 mo period in two rabbits. Both rabbits showed severely curtailed pressor responses to high doses of exogenously administered angiotensin II. These results may be of interest in attempting to develop new approaches to the study of the role of angiotensin in pathophysiologic situations like hemorrhagic shock.

DISCUSSION

The results of our studies show that inhibition of angiotension converting enzyme reduces mortality during severe hemorrhagic shock. Captopril was shown to improve postreinfusion arterial pressure, and to reduce cellular damage as measured by lysosomal enzyme release, plasma proteolysis and myocardial depressant factor (MDF) formation. Additionally, angiotensin II infusion was shown to reverse captopril enhancement of survival and to reduce the postreinfusion splanchnic blood flow. These data are in agreement with studies by Errington and Rocha e Silva (1974) and Morton et al. (1978), who found another converting enzyme inhibitor to improve cardiac output and survival in dogs. In all of these investigations, shed blood was replaced after a specified period of oligemia. In contrast, Yamashita et al. (1977) and Erdos et al. (1974) found

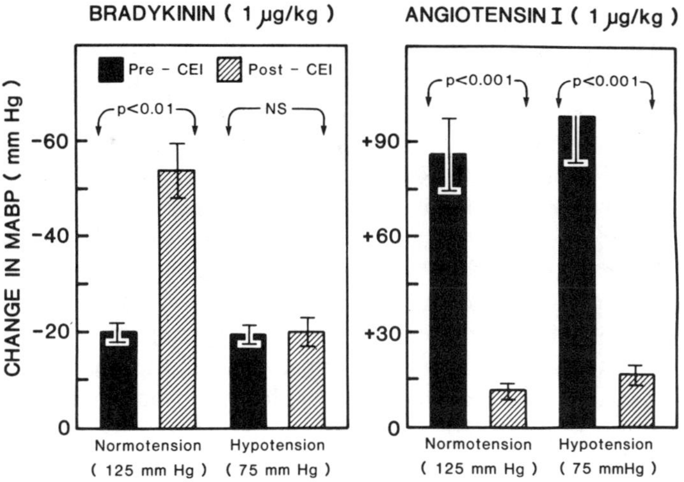

Fig. 5. Decreases in blood pressure in response to bradykinin (left panel) and increases in blood presure response to angiotensin I (right panel) before intravenous injection of 500 μg/kg of the converting enzyme inhibitor (CEI) captopril (black bars) and after captopril (hatched bars). Height of bars indicate means, brackets are ± SEM for 6 measurements. The data in the left pair of bars in each panel was obtained at 125 mm Hg and the right pair at 75 mm Hg. Note that at 75 mm Hg the potentiation of the bradykinin effect was abolished after captopril, whereas the degree of inhibition of the angiotensin I effect was still comparable to the pre-CEI value.

the converting enzyme inhibitor, teprotide, to impair the response to hemorrhage as assessed by recovery of blood pressure after rapid bleeding not replaced by infusion. Thus replacement of fluid volume may be essential to beneficial actions of converting enzyme inhibitors.

Our results do not indicate a significant difference in bleedout volume with captopril administration. Feuerstein and Cohen (1979) also found no initial difference in the amount of blood removed from pentobarbital anesthetized cats treated with captopril or its vehicle in their experimental protocol. Additionally, Samar and Coleman (1979) have recently shown captopril to have no effect on the recovery of mean circulatory pressure in hemorrhaged dogs. These data, together with our own, may indicate that captopril has less of an effect on the vasoconstriction in response to hemorrhage than does teprotide.

The hypothesis that a converting enzyme inhibitor exerts its beneficial effect via an improvement in splanchnic perfusion during hemorrhage was not substantiated by these experiments. No enhancement of SMAF was observed in captopril treated cats during oligemia. Moreover, replacement of angiotensin II in captopril treated cats failed to decrease SMAF during oligemia, although an increased splanchnic perfusion was noted following reinfusion in those cats receiving captopril when compared to animals treated with captopril and given angiotensin. These data are in agreement with earlier results of McNeil et al. (1970), who showed that either a renal (presumably renin-angiotensin) or a hypothalamic (presumably ADH) agent could mediate the splanchnic vasoconstriction occurring during hemorrhage. Thus, other humoral agents are capable of constricting the mesenteric vasculature in the absence of angiotensin, and captopril does not modulate shock severity by increasing blood flow to splanchnic organs during oligemia. However, maintenance of postreinfusion splanchnic blood flow may be involved in prevention of the shock process in agreement with a recent report utilizing an angiotensin antagonist in shock (Trachte and Lefer, 1979).

An alternative explanation for the improved hemodynamic status of shock cats given captopril could involve improved blood flow in another vascular bed. Recent reports by Lokhandwala and Jandhyala (1979) and Hock et al. (1979) indicate that the converting enzyme inhibitor, teprotide, and the angiotensin antagonist, saralasin, prevent the renal constriction which normally occurs in hemorrhage. Maintenance of blood flow to this critical excretory and gluconeogenic organ could be of vital importance during the development of circulatory shock.

Our investigations indicated a direct membrane-stabilizing action of captopril to retard lysosomal enzyme release. Captopril also inhibited proteolysis by a direct mechanism and probably by a secondary mechanism related to inhibition of angiotensin II. This was observed in the study of angiotensin replacement after captopril in which angiotensin treated cats had higher plasma amino-nitrogen values. Additionally, saralasin, an angiotensin antagonist, has been shown to retard proteolytic activity in hemorrhagic shock (Trachte and Lefer, 1979a). These data strongly suggest that angiotensin II acts to either increase proteolysis in shock or to reduce the removal of the proteolytic products (e.g., amino acids) possibly by renal artery constriction or by reducing blood flow to the liver in the postoligemic state.

Myocardial depressant factor (MDF) formation was also reduced by captopril in these studies. This effect is probably a result of the reduction of cathepsin D and proteolytic activity attributable to captopril, because both of these processes are intimately involved in MDF formation (Lefer, 1974; Lefer and Spath, 1974).

SUMMARY

The results of these investigations indicate that converting enzyme inhibitors enhance survival in hemorrhagic shock (Morton et al., 1978; Trachte and Lefer, 1978; Trachte and Lefer, 1979a), improve renal perfusion in hemorrhage (Lokandwala and Jandhyala, 1979; Hock et al., 1979; Jakschick et al., 1974) and postreinfusion splanchnic perfusion (Trachte and Lefer, 1979a). Captopril was also shown to reduce lysosomal enzyme release and subsequent plasma accumulation of these hydrolases, probably contributing to the reduced proteolysis. This combination of actions resulted in lowered circulating MDF activities. These data indicate a positive effect of converting enzyme inhibition which appears to be independent of bradykinin potentiation. Thus the potent vasoconstrictor, angiotensin II, is hypersecreted in response to hemorrhage, and this substance participates in the pathogenesis of the shock state. The other major humoral vasoconstrictors, including ADH (Errington and Rocha e Silva, 1974) and norepinephrine (Lillehei, 1964), have already been implicated as agents contributing to the production of hemorrhagic shock. The current evidence therefore indicates that vasopressors in general that function to elevate peripheral resistance tend to reduce tissue perfusion and exacerbate hemorrhagic shock with the consequent tissue ischemia, particularly in the splanchnic region.

REFERENCES

Arregui, A., Lee, C., Emson, P.C., and Iverson, L.L. 1979. Separation of human brain converting enzyme from enkephalin degrading activity. Eur J Pharmacol 59:141–144

Bahnson, H.T. 1943. Role of the kidneys in the resistance of rats to hemorrhage. Amer J Physiol 140:416–419

Boyd, G.W., Landau, J., and Peart, W.S. 1967. Radioimmunoassay for determining plasma levels of angiotensin II in man. Lancet 2:1002–1005

Brand, E.D., and Lefer, A.M. 1966. Myocardial depressant factor in plasma from cats in irreversible postoligemic shock. Proc Soc Exptl Biol Med 122:200–203

Erdös, E.G., Massion, W.H., Downs, D.R., and Gecse, A. 1974. Effect of the inhibition of angiotensin I converting enzyme in endotoxin and hemorrhagic shock. Proc Soc Exptl Biol Med 145:948–951

Errington, M.L., and Rocha e Silva, M. 1974. On the role of vasopressin and angiotensin in the development of irreversible haemorrhagic shock. J Physiol 242:119–141

Ferriera, S.H. 1965. A bradykinin potentiating factor (BPF) present in the venom of *Bothrops jararaca*. Br J Pharmacol Chemother 24:163–169

Feuerstein, G., and Cohen, S. 1979. Modification by SQ 14,225 of blood pressure and adrenal catecholamine response to hemorrhage. Eur J Pharmacol 55:203–206

Gavras, H., Kremer, D., Brown, J.J., Gray, B., Lever, A.F., MacAdam, R.F., Medina, A., Morton, J.J., and Robertson, J.I.S. 1975. Angiotensin and norepinephrine-induced myocardial lesions: experimental and clinical studies in rabbits and man. Amer Heart J 89:321–332

Glenn, T.M., and Lefer, A.M. 1970. Role of lysosomes in the pathogenesis of splanchnic ischemia shock in cats. Circ Res 27:783–797

Hamilton, A.S., and Collins, D.A. 1942. The homeostatic role of a renal humoral mechanism in hemorrhage and shock. Amer J Physiol 136:275–284

Hock, C.E., Passmore, J.C., Levin, J.I., and Neiberger, R.E. 1979. Effects of alpha blockade and angiotensin antagonist on intrarenal blood flow distribution during hemorrhage in dogs. Physiologist 22:57

Jakshik, B.A., Marshall, G.R., Kourik, J.L., and Needleman, P. 1974. Profile of circulating vasoactive substances in hemorrhagic shock and their pharmacological manipulation. J Clin Invest 54:842–852

Lefer, A.M. 1974. Myocardial depressant factor and circulatory shock. Klin Wschr 52:358–370

Lefer, A.M., and Spath, J.A. 1974. Pancreatic hypoperfusion and the production of a myocardial depressant factor in hemorrhagic shock. Ann Surg 179:868–876

Lillehei, R.C., Bloch, J.H., Longerbeam, J.K. and Manax, W.G. 1964. The nature of irreversible shock: experimental and clinical observations. Ann Surg 160:682–710

Lijnen, P.J., Amery, A.K., Fagard, R.H., and Katz, F.H. 1978. Radioimmunoassay of angiotensin II in unextracted plasma. Clin Chim Acta 88:403–412 (1978)

Lokhandwala, M.F., and Jandhyala, B.S. 1979. The role of angiotensin II and prostaglandins in control of renal hemodynamics during hemorrhage. Physiologist 22:79

McNeill, J.R., Stark, R.D., and Greenway, C.V. 1970. Intestinal vasoconstriction after hemorrhage: roles of vasopressin and angiotensin. Amer J Physiol 219:1342–1347

Morton, J.J., Semple, P.F., Ledingham, I.M., Stuart, B., Tehrani, M.A., Garcia, A.R., and McGarrity, G. 1977. Effect of angiotensin-converting enzyme inhibitor (SQ 20,881) on the plasma concentration of angiotensin I, angiotensin II and arginine vasopressin in the dog during hemorrhagic shock. Circ Res 41:301–308

Murthy, V.S., Waldron, T.L., Goldberg, M.E., and Vollmer, P.Q. 1977. Inhibition of angiotensin converting enzyme by SQ 14,225 in conscious rabbits. Eur J Pharmacol 46:207–212

Samar, R.E., and Coleman, T.G. 1979. Whole body response of the peripheral circulation following hemorrhage in the rat. Amer J. Physiol 236:H206–H210

Trachte, G.J., and Lefer, A.M. 1978. Beneficial action of a new converting enzyme inhibitor (SQ 14,225) in hemorrhagic shock in cats. Circ Res 43:576-582

Trachte, G.J., and Lefer, A.M. 1979. Effect of angiotensin II receptor blockade by saralasin in hemorrhagic shock. Amer J Physiol 236:H280–H285

Trachte, G.J., and Lefer, A.M. 1979. Mechanism of the protective effect of angiotensin converting enzyme inhibition in hemorrhagic shock. Proc Soc Exptl Biol Med 162:54–57

Yamashita, M., Oyama, T., and Kudo, T. 1977. Effect of the inhibitor of angiotensin I converting enzyme on endocrine function and renal perfusion in hemorrhagic shock. Can Anesth Soc J 24:695–701

Angiotensin Converting Enzyme Inhibitors in Congestive Heart Failure[1]

Kanu Chatterjee, Richard Ader and
William W. Parmley
University of California, San Francisco
Division of Cardiology and Cardiovascular Research Institute
San Francisco, California 94143

INTRODUCTION

In patients with acute or chronic heart failure associated with low cardiac output, systemic vascular resistance tends to increase as a compensatory mechanism to maintain arterial pressure. This increase in systemic vascular resistance, however, also increases the resistance to ejection or arterial impedance which can further reduce cardiac output (Cohn, J., 1973; Chatterjee and Parmley, 1977). Thus, a vicious cycle may develop that will ultimately lead to a lower steady state level of cardiac output and an inappropriately high systemic vascular resistance. Interruption of this cycle by decreasing systemic vascular resistance is the basis for the use of vasodilator therapy, which has been shown to improve cardiac performance in patients with heart failure. The mechanism by which systemic vascular resistance is elevated in patients with heart failure associated with low cardiac output is incompletely understood. It is likely that several neural, humoral or neurohumoral factors might be operative (Zelis et al., 1979). Enhanced activity of the sympathetic nervous system and increased circulating catecholamines have been thought to contribute to peripheral arteriolar vasoconstriction in these patients. In the presence of heart failure, there may also be increased stiffness of the arteriolar wall due to local retention of sodium and water. Recent studies have also suggested that in patients with heart failure associated with low cardiac output the renin-angiotensin system

[1] This work was supported in part by NIH Program Project Grant HL06285.

may be stimulated, and increased levels of angiotensin might contribute to an increased systemic vascular resistance (Merrill et al., 1946; Genest et al., 1968; Brown et al., 1970; Johnson et al., 1968). Renin released by the kidneys catalyzes the formation of angiotensin I, which is converted to angiotensin II, the active vasoconstrictor, by an angiotensin converting enzyme found widely in the vascular bed. The mechanism of the stimulation of renin in the presence of heart failure, however, is not entirely known. Alteration in the sodium load presented to the macula densa, a reduction in renal perfusion and stimulation of renal sympathetic nerves or increased circulating catecholamines may all be involved in the stimulation of renin production. Whatever the mechanism might be, if the renin-angiotensin system is stimulated, it is likely to cause peripheral vasoconstriction and an increase in systemic vascular resistance. In such circumstances, attenuation or inhibition of the effects of angiotensin would be expected to reduce systemic vascular resistance and thereby improve cardiac performance. Indeed, with the use of saralasin, a specific competitive angiotensin antagonist, improvement in left ventricular function has been demonstrated in patients with chronic congestive heart failure (Gavras et al., 1977). Recently, the hemodynamic effects of angiotensin converting enzyme inhibitors have been evaluated in patients with chronic heart failure and a beneficial response has been reported (Curtis et al., 1979; Gavras et al., 1979; Ader et al., 1979; Turini et al., 1979).

HEMODYNAMIC EFFECTS OF THE CONVERTING ENZYME INHIBITOR TEPROTIDE IN PATIENTS WITH HEART FAILURE (CURTIS ET AL., 1979; GAVRAS ET AL., 1979).

In several recent investigations, the hemodynamic effects of teprotide have been evaluated in patients with normotensive and hypotensive heart failure. In most patients there was a modest reduction of arterial pressure following intravenous administration of teprotide. The left ventricular filling pressure and right atrial pressure decreased significantly in the majority of patients. There was also a marked decrease in the calculated systemic vascular resistance and pulmonary vascular resistance. The fall in arterial pressure and systemic vascular resistance in response to teprotide were significantly correlated with control plasma renin activity. Along with a decrease in systemic vascular resistance, cardiac output, stroke volume and stroke work increased in these patients. Cardiac output increased by an average of 20 to 25%. Increased cardiac output, stroke volume and stroke work, along with a fall in left ventricular filling pressure, indicated improved cardiac performance.

Changes in coronary hemodynamics and myocardial metabolic function have also been evaluated in patients with heart failure following the administration of intravenous teprotide. In general, teprotide tends to decrease the determinants of myocardial oxygen demand. Arterial pressure falls, and heart rate either remains unchanged or may even decrease. The rate pressure product, an index of myocardial oxygen demand, tends to decrease. Furthermore, as there is a decrease in left ventricular filling pressure, it is likely there is also a concomitant decrease in left ventricular diastolic volume, which will tend to reduce left ventricular wall stress. The net effect is usually a reduction in the determinants of myocardial oxygen demand. Along with a reduction in myocardial oxygen demand, myocardial oxygen consumption also tends to decrease. Left ventricular mechanical performance, however, improves as suggested by an increase in left ventricular stroke work and minute work and a concomitant fall in left ventricular filling pressure. These findings suggest that myocardial efficiency may improve, because improved mechanical performance is observed without any significant increase in the metabolic cost.

HEMODYNAMIC EFFECTS OF ORAL ANGIOTENSIN CONVERTING ENZYME INHIBITOR, CAPTOPRIL, IN CHRONIC HEART FAILURE (ADER ET AL., 1979; TURINI ET AL., 1979; DAVIS ET AL., 1979; LEVINE ET AL., 1979; MASON ET AL., 1980).

The evaluation of hemodynamic and clinical responses to the oral converting enzyme inhibitor, captopril, in patients with chronic heart failure, has been the subject of many recent investigations. The hemodynamic effects of oral captopril (25 mg) in a group of patients with chronic refractory heart failure are summarized in Table 1. Within ½ hr, a beneficial hemodynamic response was observed. Cardiac output increased significantly and remained elevated for approximatley 7 hrs following a single oral dose of captopril (Fig. 1). Similarly, pulmonary capillary wedge pressure fell and remained at a lower level compared to control for approximately 7 hrs. An increase in cardiac output along with a fall in pulmonary capillary wedge pressure indicated improved cardiac performance in these patients. There was also a significant reduction in systemic vascular resistance (Fig. 2) and pulmonary vascular

Table 1. Initial hemodynamic effects of oral captopril in ten patients with chronic heart failure.

	$Control_1$	25 mg	$Control_2$	50 mg	$Control_3$	100 mg
HR (beats/min)	78 ± 15	67 ± 15	77 ± 9	67 ± 13	77 ± 13	65 ± 13
MAP (mm Hg)	92 ± 19	71 ± 20	88 ± 16	72 ± 20	90 ± 21	70 ± 20
PCW (mm Hg)	29 ± 8	15 ± 6	27 ± 6	15 ± 5	26 ± 5	15 ± 5
MPAP (mm Hg)	41 ± 9	28 ± 6	40 ± 10	28 ± 5	38 ± 7	27 ± 4
RAP (mm Hg)	11 ± 4	8 ± 4	12 ± 6	6 ± 4	11 ± 7	6 ± 5
CO (L/min)	3.96 ± 1.68	5.05 ± 1.78	4.57 ± 1.89	5.34 ± 1.95	4.03 ± 2.02	5.53 ± 1.95
SVR (dynes·sec·cm^{-5})	1798 ± 653	1066 ± 362	1554 ± 633	1089 ± 309	1812 ± 728	1066 ± 366
SWI (gm-m/m^2)	31 ± 11	39 ± 8	33 ± 12	41 ± 11	32 ± 10	38 ± 10

CO = cardiac output; PCW = pulmonary capillary wedge; HR = heart rate; MAP = mean arterial pressure; SVR = systemic vascular resistance; SWI = stroke work index; MPAP = mean pulmonary artery pressure; RAP = right atrial pressure.

Statistical Analysis

Variable	CO	PCW	HR	$\overline{MAP}$	SVR	SWI	$\overline{PAP}$	RAP
Controls 1 vs 2 vs 3	NS	NS	NS	NS	NS	NS	NS	NS
$Control_1$ vs 25 mg	$P<0.05$	$P<0.05$	$P<0.05$	$P<0.05$	$P<0.05$	$P<0.05$	$P<0.05$	$P<0.05$
$Control_2$ vs 50 mg	$P<0.05$	$P<0.05$	$P<0.05$	$P<0.05$	$P<0.05$	$P<0.05$	$P<0.05$	$P<0.05$
$Control_3$ vs 100 mg	$P<0.05$	$P<0.05$	$P<0.05$	$P<0.05$	$P<0.05$	NS	$P<0.05$	$P<0.05$
25 mg vs 50 mg vs 100 mg	NS	NS	NS	NS	NS	NS	NS	NS

Statistical analysis was performed using two-way analysis of variance with mutliple range test and mixed effects model. A Student-Newman-Keuls Multiple Comparison Test to compare the three drug dose responses and the three controls was used when the analysis of variance was significant ($p<.05$).

(Reproduced with permission from Circulation 61:931–937, 1980.)

Table 2. Seven patients.

	Control	Immediate Response	8 wk	Control vs. Immediate Response	Control vs 8 wk
HR	78±16	67±15	73±17	*P*<0.05	*P*<0.05
MAP	89±16	70±15	68±9	*P*<0.05	*P*<0.05
PCW	29±9	16±4	13±7	*P*<0.05	*P*<0.05
PAP	40±10	28±7	27±14	*P*<0.05	*P*<0.05
RAP	12±3	9±3	5±2	*P*<0.05	*P*<0.05
CO	3.54±1.66	4.38±1.30	5.09±1.54	*P*<0.05	*P*<0.05
SV	46±20	68±22	72±26	*P*<0.05	*P*<0.05
SWI	28±12	36±8	35±7	NS	NS
SVR	1933±753	1175±384	1084±327	*P*<0.05	*P*<0.05

SV = stroke volume. Other abbreviations as in Table 1.
(Reproduced with permission from *Circulation* 61:931–937, 1980.)

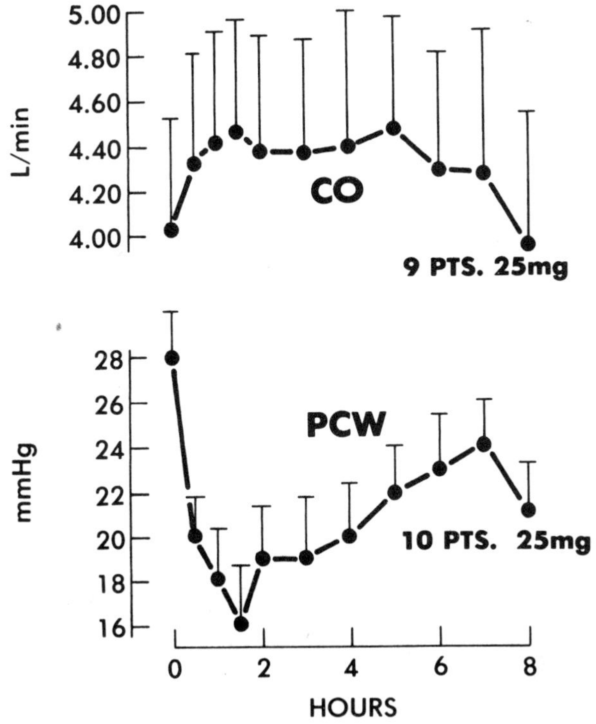

Fig. 1. Changes in cardiac output (CO) and pulmonary wedge pressure (PCW) following a single oral dose (25 mg) of captopril are illustrated. Peak effects were observed at 1½ hr and the duration of action exceeded 6 hr. Increase in cardiac output along with a decrease in pulmonary capillary wedge pressure indicated improved left ventricular performance in these patients.

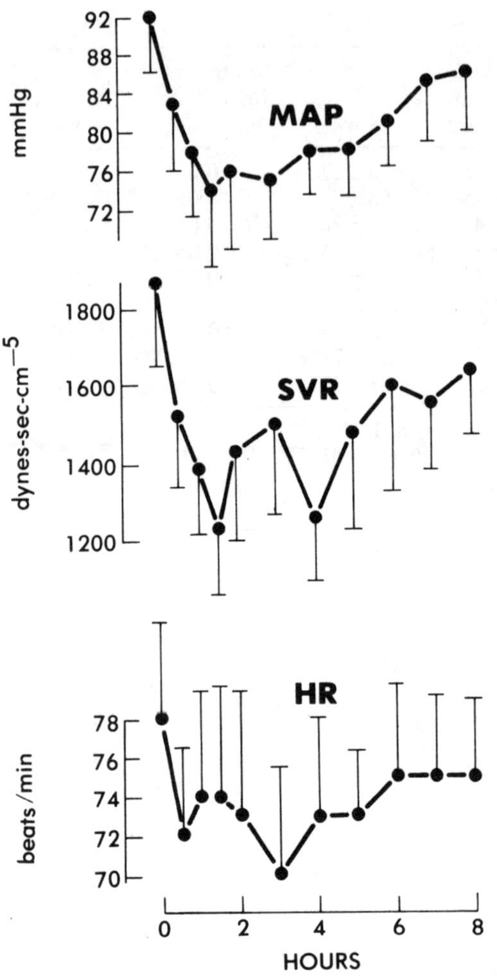

Fig. 2. Changes in mean arterial pressure, systemic vascular resistance, and heart rate following 25 mg oral dose of captopril are illustrated. Mean arterial pressure and systemic vascular resistance decreased within half an hour and remained lower compared to the control values for approximately 7 hr.

resistance and an increase in stroke volume and stroke work indices. The changes in heart rate and mean arterial pressure in individual patients following 25 mg of oral captopril are illustrated in Figure 3. Heart rate and mean arterial pressure decreased in all patients. Only a modest decrease in arterial pressure was observed in most patients; in two patients, however, arterial pressure fell markedly. At the time of maximum decrease in arterial pressure, heart rate also decreased. In most patients, the magnitude of decrease in heart rate was small; however, in occasional patients marked bradycardia was observed, which, however, did not persist with continued therapy. Although, in this study, some decrease in heart rate was observed in almost all patients, in other studies no change in heart rate has been reported during captopril therapy (Turini et al., 1979; Davis et al, 1979).

The hemodynamic responses to 50 and 100 mg doses of captopril was compared in the same patients to those after a 25 mg oral dose. The control cardiac output, heart rate, intravascular and intracardiac pressures, systemic and pulmonary vascular resistances, and stroke volume and stroke work indices before each dose of captopril were similar. The maximum changes in these hemodynamic variables following 25, 50, or 100 mg of oral captopril are summarized in Table 1. Cardiac output increased by similar magnitude following each dose of captopril. Left ventricular filling pressure fell by an average of 46% following the 25 mg dose; the fall in left ventricular filling pressure after 50 or 100 mg was similar. The magnitude of decrease in systemic vascular resistance (−41%), mean arterial pressure (−23%), heart rate (−14%), and right

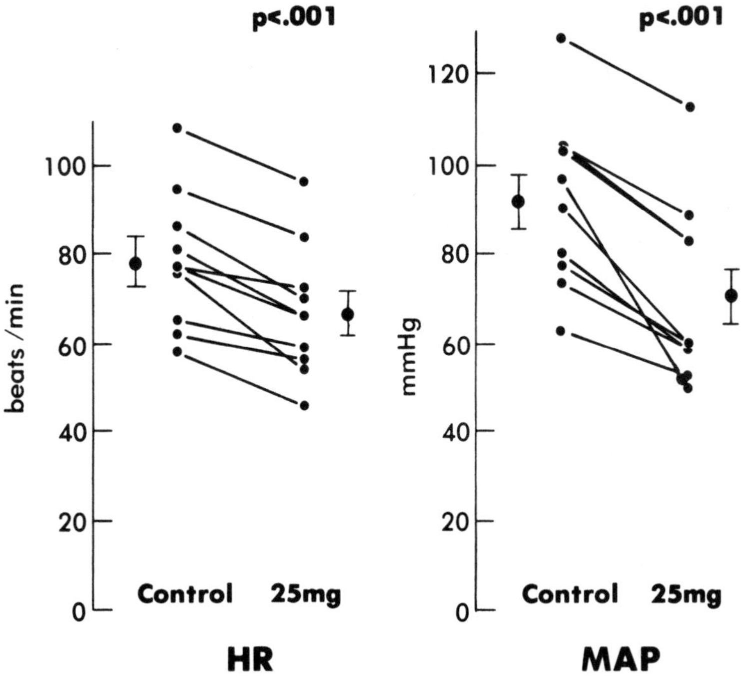

Fig. 3. Changes in heart rate (HR) and mean arterial pressure (MAP) following a single oral dose of captopril (25 mg) in individual patients are illustrated. Mean arterial pressure decreased in all patients. In two patients significant hypotension occurred. Heart rate decreased in all patients, but marked bradycardia was observed only in one patient. (Reproduced by permission of *Circulation* 61:934, 1980.)

atrial pressure (−27%) following the 25 mg dose was almost identical to that following the 50 mg or 100 mg doses. Although in this study the hemodynamic responses after 50 or 100 mg doses of captopril were no better than those after 25 mg, other investigators have reported the necessity of using a larger dose to produce optimal hemodynamic responses (Davis et al., 1979; Levine et al., 1979; Mason et al., 1980). It appears, therefore, that further studies will be needed to determine the optimal dose range of captopril for treatment of patients with chronic heart failure.

That captopril can produce beneficial hemodynamic and clinical responses in patients with chronic congestive heart failure has been reported by other investigators. Turini et al. (1979) reported a 38% in-

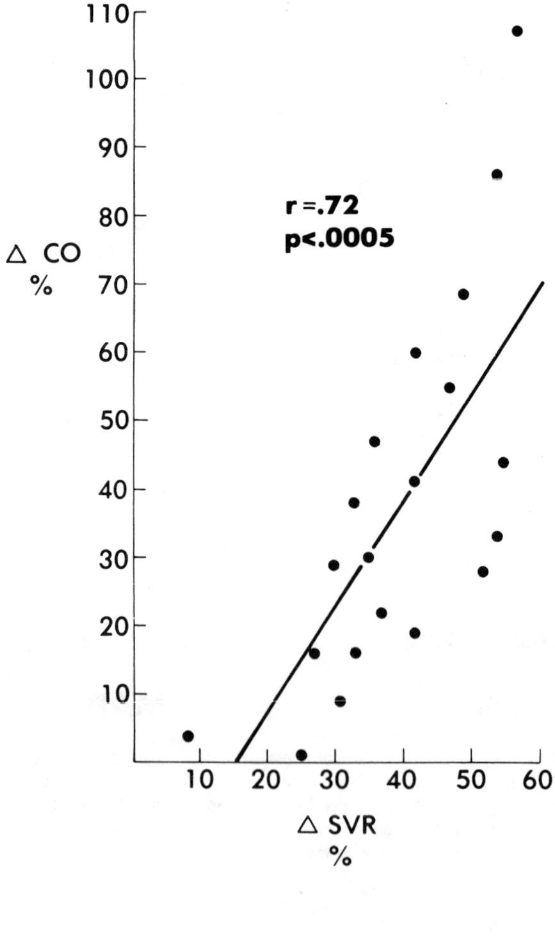

Fig. 4. Relation between the changes in cardiac output on vertical axis (% CO) and changes in systemic vascular resistance on horizontal axis (% SVR) following captopril therapy is illustrated. A significant correlation was found between the relative changes in systemic vascular resistance and cardiac output.

crease in cardiac index and a 25% decrease in left ventricular filling pressure after administration of 25 mg of captopril in a group of patients with chronic refractory heart failure. Davis et al. (1979) reported a 30% increase in cardiac index and a 37% decrease in pulmonary capillary wedge pressure after a single daily dosage of 25 to 150 mg of oral captopril. In the study reported by Levine et al. (1979) there was a 19% increase in cardiac index and a 46% decrease in pulmonary capillary wedge pressure. A 16% increase in cardiac output and a 38% decrease in pulmonary capillary wedge pressure were also reported by Mason et al (1980). It seems, therefore, that a 15 to 30% increase in cardiac output is expected following captopril therapy, but a more pronounced fall in left ventricular filling pressure may occur. Although captopril produces beneficial hemodynamic responses in patients with chronic heart failure, the mechanism is not entirely clear. The increase in cardiac output and stroke volume is associated with a significant reduction in systemic vascular resistance. Furthermore, a significant correlation is usually found between the relative changes in systemic vascular resistance and cardiac output (Fig. 4). A reduction in left ventricular outflow resistance, therefore, must play a role in improving cardiac performance in these patients. The reduction of systemic vascular resistance during captopril therapy, however, may not be entirely due to the attenuation of the effects of angiotensin II. In most studies only a general correlation has been found between the magnitude of fall in systemic vascular resistance and the initial level of plasma renin activity (Curtis et al., 1979; Gavras et al., 1979; Davis et al., 1979; Ader et al., 1980). Furthermore, no correlation is found between the initial level of systemic vascular resistance and the control plasma renin activity. This poor correlation between systemic vascular resistance and plasma renin activity suggests that the reduction of systemic vascular resistance by captopril may not be entirely due to a decrease in the levels of circulating angiotensin II. It has been demonstrated that captopril inhibits the degradation of bradykinin, and therefore kinin-induced vasodilatation with reduction of systemic vascular resistance remains a possibility (Ferreira, 1965; Engel et al., 1972). The mechanism for an elevated systemic vascular resistance in congestive heart failure may be multifactorial (Zelis et al., 1979). Augmented sympathetic activity with consequence arteriolar constriction and increased stiffness of the vascular walls due to sodium and water retention may be important contributing factors. Decreased sympathetic activity following captopril therapy may also be contributory in reducing systemic vascular resistance. Indeed, recent studies have indicated that, following administration of angiotensin converting enzyme inhibitors, there is a fall in the level of circulating catecholamines (Turini et al., 1979).

The mechanism for the decrease in systemic and pulmonary venous pressures with captopril that has been observed in almost all studies also remains unclear. It is generally accepted that angiotensin causes constriction of resistance of precapillary vessels while the veins or capacitance vessels are relatively insensitive to the direct constricting effect of angiotensin (Gross and Bock, 1962; Rose et al., 1962; Cohn, 1966; Hadduchal, 1962). During the systemic administration of angiotensin, however, peripheral venous tone and central venous pressures consistently increase (DePasquale and Burch, 1963a, 1963b; Emerson, 1966; Scroop et al., 1965). It is conceivable, therefore, that captopril may cause venodilatation by inhibiting this indirect venoconstricting effect of angiotensin. Recent studies with the use of plethysmography have indicated that captopril may significantly decrease the peripheral venous tone in patients with chronic congestive heart failure (Mason et al., 1979). Furthermore, a reduction in aldosterone level that has been observed following captopril therapy may also be contributory in reducing peripheral venous tone and, therefore, reduction of systemic and pulmonary venous pressures (Fouad et al., 1979). The mechanism for the decrease in heart rate that is observed in some patients receiving captopril therapy also remains unclear. Cardioacceleration following systemic administration of angiotensin has been reported previously (Nishith et al., 1962). It has been postulated that the chronotropic response to angiotensin may be related to the release of endogenous catecholamines from adrenergic nerve endings in the myocardium or from stellate ganglion stimulation (Krashey et al., 1967). Irrespective of the underlying mechanisms for angiotensin induced cardioacceleration, its inhibition by captopril might explain the decrease in heart rate observed in some patients. A reduction in heart rate with a concomitment decrease in arterial pressure and pulmonary capillary wedge pressure may, however, be beneficial in terms of metabolic cost, particularly in patients with ischemic heart disease. It is very likely that myocardial oxygen consumption will decrease because of the reduction in the determinants of myocardial oxygen demand. However, further clinical studies are indicated to evaluate the effect of captopril on coronary hemodynamics and myocardial metabolism.

RESULTS OF LONG-TERM THERAPY WITH CAPTOPRIL

Sustained improvement in the clinical status of patients with chronic heart failure have been reported with continued long-term captopril therapy. Three of the 10 patients reported by Davis et al. (1979) who accepted long-term treatment noted marked improvement in their clinical status and effort tolerance. Levine et al (1979) reported that, of 8 patients treated chronically with captopril, 6 improved by 1 or 2 New York Heart Association classes after 2 to 5 mo. of therapy. The exercise time in these patients increased an average of 88% on the graded treadmill test. A 47% increase in treadmill exercise duration was also reported by Mason et al (1980). Fouad et al. (1979) also observed clinical improvement in their patients during long-term captopril therapy.

Changes in the clinical status in 7 patients with severe refractory heart failure at the end of 8 wk of maintenance captopril therapy are illustrated in Figure 5. Five of these 7 patients were in Class III and the remaining two in Class IV before the initiation of therapy. At the end of 8 wk, 4 of the 5 Class III patients were in Class II and both Class IV patients were in Class III. Thus, 6 of the 7 patients on maintenance therapy had improvement by one clinical class; only one patient showed no clinical improvement. Exercise duration in these patients also increased by an average of 48%. These clinical observations indicate that sustained improvement in the symptoms of congestive heart failure and increased effort tolerance can occur in patients with chronic heart failure receiving long-term captopril therapy. The changes in cardiac performance and ventricular dynamics were investigated following long-term captopril therapy in a limited number of patients. The hemodynamic changes after initiation of captopril therapy and at 8 wk of maintenance therapy

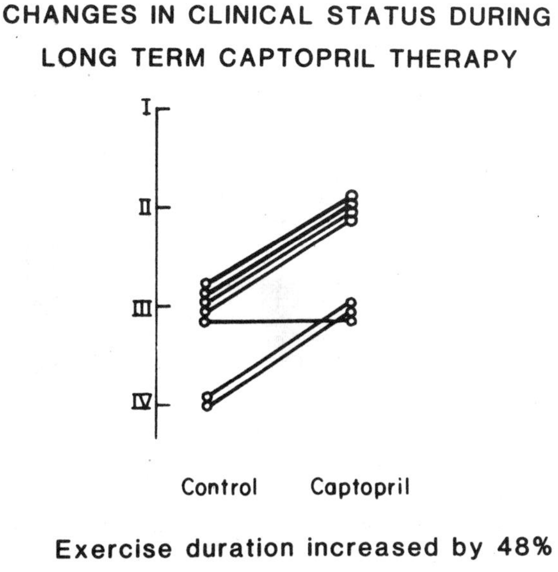

Fig. 5. Changes in the clinical status of 7 patients with severe chronic congestive heart failure after 8 wk of maintenance captopril therapy. Six of the 7 patients had improvement by one clinical class.

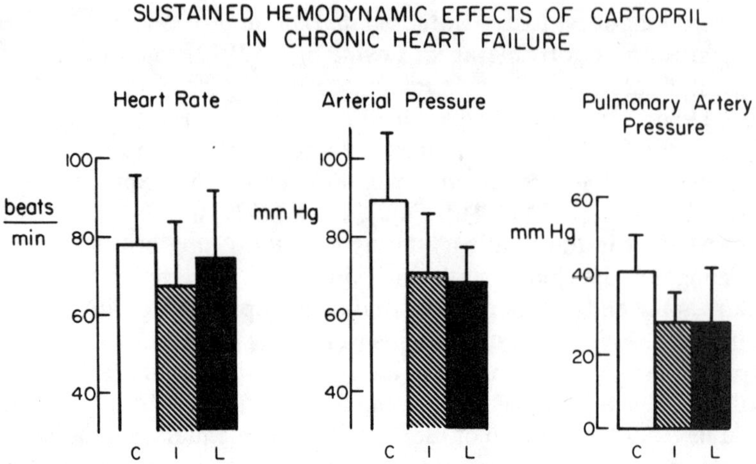

Fig. 6. Changes in heart rate, mean arterial pressure and pulmonary artery pressure following initiation (I) and after maintenance (L) captopril therapy in patients with chronic heart failure. Heart rate decreased initially and remained lower compared to the control values during maintenance therapy. Similarly, mean arterial pressure and pulmonary artery pressure decreased at the beginning of the therapy and remained lower compared to the control values (C) during maintenance therapy.

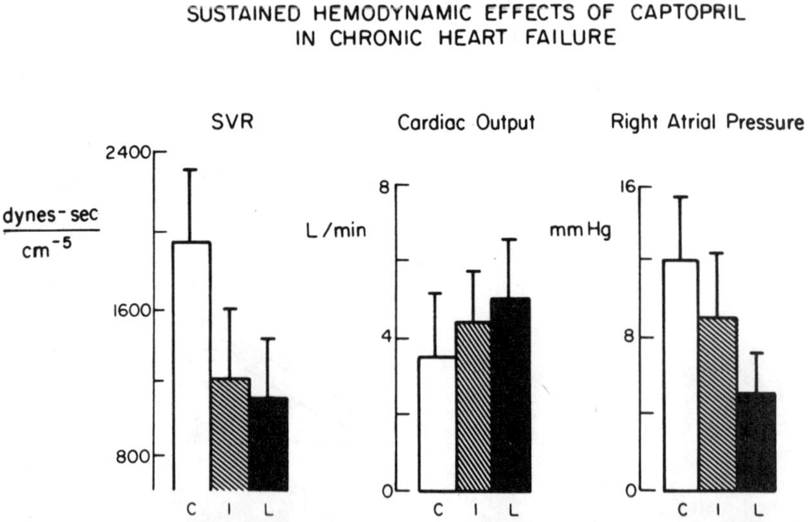

Fig. 7. Changes in systemic vascular resistance (SVR), cardiac output (CO) and right atrial pressure (RAP) at the initiation (I) and during maintenance (L) captopril therapy in patients with chronic heart failure. System vascular resistance and RAP decreased significantly after the initiation of therapy and remained lower compared to the control (C) during maintenance therapy. CO, however, increased initially and remained elevated compared to the control during maintenance therapy.

in 7 patients with chronic congestive heart failure are summarized in table 2 and illustrated in Figures 6 to 8. After the initiation of captopril therapy, mean arterial pressure decreased by 21%, and at 8 wk of maintenance therapy a 23% decrease compared to control was observed in these patients. Initially, captopril produced an increase of cardiac output by 24%, but at 8 wk of maintenance therapy a 41% increase was observed. The magnitude of decrease in systemic vascular resistance at the initiation of treatment and during maintenance therapy was similar. Mean right atrial pressure and pulmonary arterial pressure decreased significantly initially and remained lower at the time of restudy. The changes in stroke volume and pulmonary capillary wedge pressure at the beginning of the treatment and after 8 wk of captopril therapy are illustrated in Figure 6. In all 7 patients, stroke volume increased initially and this increase was maintained at 8 wk. Similarly, pulmonary capillary wedge pressure decreased in all patients initially and remained low at the time of restudy. The persistent increase in stroke volume along with a sustained decrease in pulmonary capillary wedge pressure indicated a sustained improvement in left ventricular function in these patients. Mason et al. (1980) have also reported a sustained beneficial hemodynamic response to long-term captopril therapy in patients with chronic congestive heart failure. An increase in left ventricular ejection fraction,

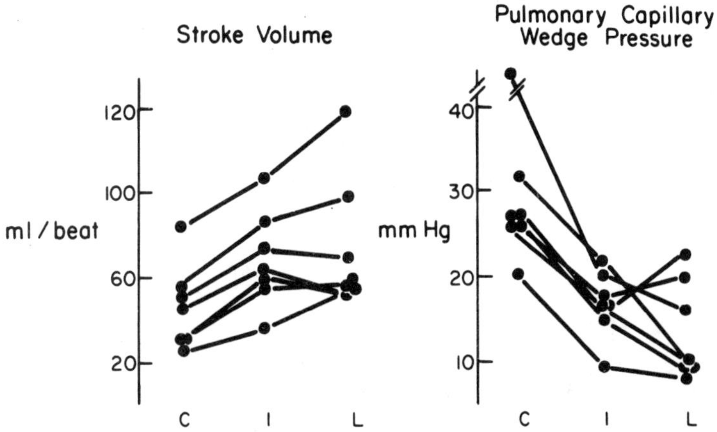

Fig. 8. Changes in stroke volume and pulmonary capillary wedge pressure (PCWP) at the initiation (I) and during maintenance (L) captopril therapy in patients with chronic heart failure. Following the initiation of therapy, stroke volume increased and pulmonary capillary wedge pressure decreased. During maintenance therapy stroke volume remained elevated and PCWP remained lower compared to the control value (C). Persistent increase in stroke volume along with a decrease in PCWP indicated sustained improvement in left ventricular performance in these patients. (Reproduced by permission of *Circulation* 61:935, 1980.)

determined by gated blood pool scintigraphy, was observed following 6 mo of captopril therapy. A reduction in left ventricular end-diastolic diameter (by echocardiography) may occur in some patients (Mason et al., 1980). These preliminary investigations suggest that chronic oral captopril therapy may provide sustained beneficial effects on cardiac dynamics in patients with chronic refractory heart failure. It needs to be emphasized, however, that the improvement in cardiac performance that is observed during long-term therapy may not be entirely due to the hemodynamic effects of captopril. The improvement in ventricular dynamics that occurs initially may be contributory in maintaining improved cardiac performance.

ADVERSE EFFECTS

Most studies indicate that captopril is well tolerated by the patients with chronic heart failure. In some patients, a marked hypotension can occur after the first dose of captopril. In the majority of patients, however, the magnitude of decrease in arterial pressure is only modest. Occasional patients may experience significant bradycardia at the initiation of treatment. Skin rash may develop in occasional patients, and discontinuation of thcrapy may be required.

In summary, angiotensin converting enzyme inhibitors, whether given intravenously or orally, can produce beneficial hemodynamic and clinical responses in patients with chronic refractory heart failure. Most patients tolerate these agents without experiencing significant side effects. Long-term therapy with the use of oral angiotensin converting enzyme inhibitor is also feasible, and it appears that sustained beneficial hemodynamic and clinical responses are observed in some patients with chronic congestive heart failure.

REFERENCES

Ader, R., Chatterjee, J., and Ports, T., et al. 1979. Beneficial hemodynamic effects of angiotensin converting enzyme inhibitor in chronic refractory heart failure. Am J Cardiol 43:404.

Ader, R., Chatterjee, K., Ports, T., Brundage, B., Hiramatsu, B., and Parmley, W. 1980. Immediate and sustained clinical improvement in chronic heart failure by an oral angiotensin converting enzyme inhibitor. Circulation, 61:931–937.

Brown, J.J., Davis, D. L., Johnson, V.W., et al. 1970. Renin relationship in congestive cardiac failure, treated and untreated. Am Heart J 80:329–342.

Chatterjee, K., and Parmley, W.W. 1977. The role of vasodilator therapy in heart failure. Prog Cardiovasc Dis 19:301–325

Cohn, J.N. 1966. Relationship of plasma volume changes to resistance and capacitance vessel effects of sympathomimetic amines and angiotensin in man. Clin Sci 30:267–278

Cohn, J.N. 1973. Vasodilator therapy for heart failure: the influence of impedance on left ventricular performance. Circulation 48:5–8

Curtiss, C., Cohn, J.N., Vrobel, T., et al. 1979. Role of the reninangiotensin system in the systemic vasoconstriction of chronic congestive heart failure. Circulation 58:763–770

Davis, R., Ribner, H.S., Keung, E., Sonnenblick, E.H., and Lejemtel, T.H. 1979. Effect of Captopril in heart failure. N Engl J Med 301:117–121

DePasquale, N.P., and Burch, G.E. 1963a. Effect of angiotensin II on the intact forearm veins of man. Circ Res 13:239–245

DePasquale, N.P., and Burch, G.E. 1963b. Angiotensin II, digital blood flow, and the precapillary and postcapillary blood vessels of man. Ann Intern Med 58:278–292

Emerson, T.E., Jr. 1966. Effects of angiotensin, epinephrine, norepinephrine and vasopressin on venous return. Am J Physiol 210:933–942

Engel, S.L., Schaeffer, T.R. Gold, B.I., et al. 1972. Inhibition of pressor effects of angiotensin I and augmentation of depressor effects of bradykinin by synthetic peptides. Proc Soc Exp Biol Med 140:240–244

Ferreira, S.H. 1965. A bradykinin-potentiating factor present in the venom of *Bothrops Jararaca*. Br J Pharmacol & Chemother 24:163–169

Fouad, F. M., Celmo, J.M., Tarazi, R.C., and Bravo, E.L. 1979. Converting enzyme inhibition (CEI) in normotensive heart failure (HF): Long-term hemodynamic and humoral effects. Circulation 59 and 60:II–130

Gavras, H., Flessas, A., Ryan, T.J., et al. 1977. Angiotensin II inhibition: treatment of congestive cardiac failure in a high renin hypertension. JAMA 238:880–882

Gavras, H., Faxon, D.P., Berkoberr, J., et al. 1979. Angiotensin converting enzyme inhibition in patients with congestive heart failure. Circulation 58:770–776

Genest, J., Granger, P., de-Champlain, J., et al. 1968. Endocrine factors in congestive heart failure. Am J Cardiol 22:35–42

Gross, F., and Bock, K.D. 1962. Some contributions to the pharmacology of synthetic angiotensin. Circulation 25:193–199

Haddy, F.J., Molnar, J.I., Borden, C.W., et al. 1962. Comparison of direct effects of angiotensin and other vasoactive agents on small and large blood vessels in several vascular beds. Circulation 25:239–246

Johnson, C.I., David, J.O., Robb, C.A., et al. 1968. Plasma renin in chronic experimental heart failure and during renal sodium "escape" from mineralocorticoids. Circ Res 22:113–125

Krasney, J.A., Thompson, J.L., and Lowe, R.F. 1967. Cardiac effects of angiotensin injections into perfused right coronary artery. Amer J Physiol 213:134–138

Levine, T.B., Carlyle, P.F., Gron, K.A., Franciosa, J.A., and Cohn, J.N. 1979. Hemodynamic and clinical response to Captopril in congestive heart failure. Circulation 59 and 60:II–39

Mason, D.T., Awan, N.A., Hermanovich, J., and Joye, J.A. 1979. Oral Captopril in congestive heart failure: Efficacy by cardiac catheterization, forearm plethysmography, nuclear scintigraphy, echocardiography, treadmill exercise and symptoms. Circulation 59 & 60:II–230

Mason, D.T., Hermanovich, J., Evensen, M., and Awan, N.A. 1980. Oral Captopril in ambulatory management of severe congestive heart failure; sustained beneficial effects of venticular function with 6 months therapy shown by cardiac catheterization, nuclear scintigraphy, echocardiography, treadmill exercise and symptomatology. Am J Cardiol 45:411

Merrill, A.J., Morrison, J.L., and Brannon, E.S. 1946. Concentration of renin in renal venous blood in patients with chronic heart failure. Am J Med 1:468–472

Nishith, S.D., Davis, L.D., and Youmans, W.B. 1962. Cardio-accelerator action of angiotensin. Amer J Physiol 202:237–240

Rose, J.C., Kot, P.A., Cohn, J.N., et al. 1962. Comparison of effects of angiotensin and norepinephrine on pulmonary circulation, systemic arteries, and veins and systemic vascular capacity in the dog. Circulation 25:247–252

Scroop, G.C., Walsh, J.A., and Whelan, R.F. 1965. A comparison of the effects of interaarterial and intravenous infusions of angiotensin and noradrenaline on the circulation in man. Clin Sci 29:315–326

Turini, G.A., Brunner, H.R., Gribic, M., Waeber, B., and Gavras, H. 1979. Improvement of chronic congestive heart failure by oral Captopril. Lancet 1:1213–1215

Zelis, R., Flain, S.F., Moskowitz, R.M., et al. 1979. How much can we expect from vasodilator therapy in congestive heart failure? Circulation 59:1092–1097

Captopril—Other Indications

Jeffrey A. Stritar
The Squibb Institute for Medical Research
Princeton, New Jersey 08540

INTRODUCTION

The possibility has been suggested that angiotensin converting enzyme (ACE) inhibition might clarify the pathogenesis of or provide specific treatment for essentially every clinical state with abnormalities of the renin-angiotensin-aldosterone system or any of its components. With the exception of scleroderma renal crisis, which might be considered a subset of malignant hypertension, there are either only scattered case reports or limited preclinical data available. Thus, the therapeutic use of catopril for indications other than congestive heart failure and hypertension is speculative at this time.

Plasma renin activity is elevated in cirrhosis and the hepatorenal syndrome, acute renal failure and scleroderma renal crisis (Wilkinson et al., 1979; Brown et al., 1970; Gavras et al., 1977). An inappropriate response of the renin-angiotensin system to upright posture may play a part in the pathogenesis of idiopathic edema (Kuchel et al., 1970). Patients with Bartter's Syndrome are normotensive, despite activation of the renin-angiotensin system, perhaps due to hyposensitivity to angiotensin II (Bartter et al., 1962) and serum levels of angiotensin converting enzyme are elevated in about half of the patients with sarcoidosis (DeRemme and Rohrbach, 1980). Serum ACE is also elevated in patients with viral hepatitis and in a smaller percentage of patients with a variety of nonsarcoid pulmonary diseases (Schweisfurth and Wernze, 1979).

The fact that the highest concentrations of ACE are in the endothelial cells of the pulmonary vasculature suggests that angiotensin II and perhaps bradykinin may be important in regulating pulmonary vascular resistance and be involved in the pathogenesis of primary and hypoxic pulmonary hypertension. Since the net effect of converting enzyme inhibitor is vasodilation, captopril may also be useful in the management of coronary artery disease and peripheral vascular disease.

PULMONARY HYPERTENSION

Pulmonary hypertension, in the absence of left-sided congestive heart failure or disease of the pulmonary veins, may be secondary to recurrent pulmonary embolism, intrinsic disease of the lungs such as chronic obstructive lung disease, or a primary process of unknown etiology. The latter primary process affects mainly young and middle-aged women and is a progressive disease that results in death in an average of seven years. Pulmonary hypertension secondary to parenchymal pulmonary disease may result from chronic hypoxia or from obliteration of pulmonary vessels.

In the early stages of both primary pulmonary hypertension and hypoxic pulmonary hypertension, the administration of oxygen or vasodilators may reduce pulmonary vascular resistance. In addition, correction of hypoxemia by the administration of oxygen for 15 hr or more daily has significantly reduced pulmonary artery pressure and pulmonary vascular resistance in patients with chronic bronchitis (Stark et al., 1972). No specific treatment has been found for primary pulmonary hypertension, and vasodilators such as isoproterenol (Daoud et al., 1978), regitine (Ruskin and Hutter, 1979), diazoxide (Klinke and Gilbert, 1980) and hydralazine (Rubin and Peter, 1980) have been effective in reducing pulmonary vascular resistance and improving exercise tolerance only in selected patients.

Inhalation of gas mixtures containing low concentrations of oxygen results in pulmonary vasoconstriction (Hauge, 1968); however, there is evidence that the direct effect of hypoxia on pulmonary smooth muscle is vasodilatation (Lloyd, 1968). This discrepancy suggests that there may be a chemical mediator of the pulmonary vascular response to hypoxia, and hypersensitivity to or abnormal levels of such a mediator could also play a part in the pathogenesis of primary pulmonary hypertension. Because the endothelial cells of the pulmonary vasculature contain the highest concentrations of angiotensin converting enzyme, both angiotensin II and bradykinin have been suggested to be important in maintaining pulmonary vascular tone (Berkov, 1974; Zakheim et al., 1975).

At subpressor doses, angiotensin II has been shown to augment the vasoconstrictor response to hypoxia in the isolated perfused rat lung (Berkov, 1974). However, the effects of converting enzyme inhibition on the development of hypoxic pulmonary hypertension in animal

Table 1. Effects of teprotide in man.

13 Patients with Hypertension after Coronary Bypass Surgery		
ΔPVR*	−33 ± 7%	($P < 0.005$)
Δ $\overline{PA}$**	−23 ± 3%	($P < 0.005$)
Δ CO	+16 ± 6%	($P < 0.01$)

* correlated with PRA ($r = 0.74$, $P < 0.01$)
**correlated with PRA ($r = 0.82$, $P < 0.01$)
Niarchos et al. 1979. Am J Med 67:785

models have been somewhat inconsistent. In rats exposed to chronic hypoxia for 21 days, teprotide (SQ 20,881) prevented right ventricular hypertrophy and the vascular changes of chronic alveolar hypoxia (Zakheim et al., 1975). On the other hand, treatment of rats with captopril during 2 to 3 wk of hypoxia reduced pulmonary artery pressure, but did not influence the degree of right ventricular hypertrophy and vascular changes (Rabinovitch, 1980).

There are no data on the effects of captopril on the pulmonary circulation, except in severe congestive heart failure. However, the effects of converting enzyme inhibition by teprotide on the pulmonary circulation in man were studied by Niarchos et al. (1979). Table 1 shows that in 13 patients who developed systemic hypertension following coronary bypass surgery, they observed a 30% decrease in pulmonary vascular resistance and 23% decrease in mean pulmonary artery pressure, despite a mean increase of 16% in cardiac output. The changes in pulmonary vascular resistance and pulmonary artery pressures were significantly correlated with plasma renin activity, but were independent of alterations in left ventricular function (left ventricular filling pressure, cardiac output and mean arterial pressure). It should be noted that the mean pulmonary artery pressure for these 13 patients was only 13 ± 1 mm Hg, which is considerably below that of most patients with symptomatic pulmonary hypertension. Although these results are preliminary, they indicate the need for further investigation of the role of angiotensin II (and bradykinin) in regulating pulmonary vascular resistance, and justify the study of the therapeutic effects of captopril in patients with pulmonary hypertension.

ACUTE RENAL FAILURE, HEPATORENAL SYNDROME

Renal ischemia is the commonest cause of acute renal failure (ARF, acute tubular necrosis) and clinical states associated with reduced renal perfusion are frequent antecedents of ARF. Nephrotoxins such as heavy metals, aminoglyoside antibiotics and contrast agents produce a similar if not identical clinical syndrome. Although the histological findings are variable in ARF of different causes, severe reduction of glomerular filtration rate and elevation of plasma renin activity with only moderate reduction of renal blood flow are generally present. The pathogenesis of ARF may involve both vascular events—e.g., afferent arteriole spasm due to renal nerve stimulation or angiotensin II—and tubular events—e.g., obstruction by casts or debris (Anderson and Schrier, 1980).

Elevated plasma renin activity, secondary aldosteronism and reduction in renal blood flow are also commonly present in liver disorders. Secondary aldosteronism results from impaired hepatic metabolism of aldosterone as well as increased secretion, and serum angiotensin converting enzyme is also elevated in patients with cirrhosis or viral hepatitis.

Functional renal failure may occur in association with either chronic or acute hepatic failure. The "hepatorenal syndrome" occurs most frequently in patients with ascites and cirrhosis but has been reported in fulminant hepatic failure, viral hepatitis, following hemihepatectomy, in primary tumors of the liver and in biliary obstruction. Excessive diuresis or gastrointestinal bleeding are frequent causes of the hepatrorenal syndrome in cirrhotic patients. The biochemical abnormalities have the features of a prerenal disorder (low urine sodium concentration, hyperosmolar urine), but the renal failure rarely responds to intravascular volume expansion, and in the presence of irreversible hepatic failure, the syndrome has a mortality approaching 100%.

The marked elevation of plasma renin activity in both the hepatorenal syndrome and ARF, at least initially, suggests that intense renal vasospasm caused by high levels of angiotensin II may be important in the pathogenesis of these disorders. Alternatively, the activation of the renin-angiotensin system may be a consequence, rather than the cause, of the decrease in renal perfusion.

The effects of converting enzyme inhibition on renal function were reviewed earlier (see B.G. Zimmerman, this volume). Captopril increases renal blood flow, but reduces filtration fraction in several animal models.

The effects of converting enzyme inhibtion on renal blood flow in man appear to depend on the state of sodium balance. Williams and Hollenberg (1977) studied the effects of teprotide (SQ 20,881) in normal subjects and patients with essential hypertension on 10 mEq sodium diets. In the normal subjects, modest decreases in mean blood pressure (−5 mm Hg) and modest increases in renal blood flow (+50 ml/100g/min) were observed. The same dose of teprotide (30 μg/kg) reduced

Table 2. Captopril in acute renal failure.

*Did not prevent or modify course of $HgCl_2$, glycerol or ischemia induced renal failure in rats
**"Reversible renal failure" in man

*(Mimran, A. et al. 1979. Eur J Clin Inves 9:24; Mimran, A. et al. 1979. Kidney Int 15:451; Mason, J. et al. 1980. Drug Devel Eval 4:30–41)

**(Farrow, P., et al. 1979. Brit Med J 1:1680; Woodhouse, K., et al. 1979. Brit Med J 2:1146–1147; Collste, P. et al. 1979. Brit Med J 2:621–613)

mean blood pressure by 11 mm Hg and increased renal blood flow by 137 ml/100g/min in hypertensive patients. Teprotide also increased endogenous creatinine clearance in patients with essential hypertension on sodium restricted diets (Hollenberg et al., 1979). The increases in creatinine clearance were greater in patients with more severe hypertension and greater renal impairment prior to administration of teprotide. In two patients with advanced bilateral renal artery stenosis, however, no enhancement in creatinine clearance was observed.

The effects of captopril on effective renal flow (EFPR) and glomerular filtration rate in patients with essential hypertension were assessed by Mimran et al. (1979). EFPR rose and GFR was unchanged in eight patients on unrestricted sodium intake. EFPR and GFR increased in a single patient on a low sodium diet.

Captopril did not prevent or modify the course of $HgCl_2$, glycerol- or ischemia-induced acute renal failure in rats (Table 2). These experiments appear to eliminate a part for either angiotensin II or bradykinin in the pathogenesis of ARF. However, although high doses were administered directly into the kidney in the ischemic model, the amounts that reached the intrarenal site of converting enzyme may have been insufficient to block the intrarenal formation of angiotensin II.

The effects of captropril on RBF and GFR in ARF and the hepatorenal syndrome in man have not been studied. The results of the animal studies and some clinical observations suggest that such studies should be approached with caution, particularly in disorders associated with a decrease in renal perfusion (e.g., hemorrhage, shock, surgery). Thus, the reduction of blood pressure to normal levels by captopril in several patients with compromised renal perfusion due to tight renal artery stenosis has been associated with reversible impairment of renal function (Table 2). Captopril has, however, been successful in modifying the acute renal failure of scleroderma renal crisis.

SCLERODERMA

Diffuse scleroderma (progressive systemic sclerosis) is a chronic progressive multisystem disorder characterized by the abnormal proliferation of fibrous connective tissue. Vasomotor phenomena such as Raynaud's phenomenon may precede other manifestations by years. In the skin, there is swelling, atrophy and thickening. Hypomotility of the gastrointestinal tract, particularly the distal third of the esophagus, occurs in about 90% of patients. In the cardiomyopathy of scleroderma, cardiac muscle is replaced by connective tissue and pulmonary fibrosis is usually observed in involvement of the lungs, but pulmonary hypertension may occur in the absence of pulmonary fibrosis.

Renal involvement occurs in about half the patients with scleroderma and is the most dramatic if not most frequent cause of death. Renal failure, usually associated with malignant hypertension, can occur with explosive onset in an otherwise stable patient without previous hypertension or obvious renal involvement. As in acute renal failure or malignant hypertension without scleroderma, plasma renin activity is markedly elevated in most cases (Gavras et al., 1977; Cannon et al., 1974).

The characteristic lesion in the kidney of scleroderma patients with renal failure is adventitial fibrosis and concentric intimal proliferation of the smaller arcuate and interlobular arteries. Fibronoid necrosis of smaller arteries and afferent arterioles indistinguishable from that seen in malignant hypertension in any other setting is also common (Cannon et al., 1974).

It has been proposed that vascular changes and the renin-angiotensin system are involved in the pathogenesis of scleroderma. Vascular lesions in the kidney occur in the absence of renal impairment and hypertension, and similar lesions are seen in other tissues. Mild vascular changes in the kidneys of patients with normal renal function appear to be associated with elevated plasma renin activity or substantial rise in response to cold pressor testing (Kovalchik et al., 1978). Elevated plasma renin activity in an otherwise stable patient may also be predictive of the imminent onset of renal crisis (Gavras et al., 1977). Finally, induction of Raynaud's phenomenon by cutaneous cooling results in renal vasospasm and a significant decrease in renal cortical blood flow in patients with scleroderma, but not in normal subjects (Cannon et al., 1974).

Prior to the report by Richardson (1973) of the prolonged survival of a patient following bilateral nephrectomy and transplantation, scleroderma renal crisis was generally considered to be rapidly and uniformly fatal. The first successful medical management was reported five years later (Moorthy et al., 1978; Mitnick and Feig, 1978; Wasner et al., 1978; Lam et al., 1978), and survival of scleroderma renal crisis without progression to end-stage renal failure continues to be unusual even with available antihypertensive drugs. Lopez-Ovejero et al. (1979) reported the reversal of renal crisis in 2 patients treated with captopril. Since then, more than 30 patients with scleroderma renal crisis have been treated with captopril, most of them under an emergency use protocol. Completed case reports and follow-up data through April 1, 1980 have been received for 18 patients, treated by 14 different investigators.

To qualify for emergency use of captopril, each patient had to have a severe hypertension resistance to multi-drug therapy consisting of a vasodilator, a sympatholytic and a diuretic. The etiology of the hypertension was not a criterion for selection; therefore, the 18 patients with scleroderma renal crisis are only a small fraction of the patients treated under the protocol. The diagnosis of scleroderma renal crises was made independently by each investigator. No attempt was made to apply uniform diagnostic criteria retroactively.

Including the 2 patients reported by Lopez-Ovejero, data are available for 20 patients with scleroderma renal crisis treated with captopril. There were 13 females and 7 males. The mean age was 44 yr (range 24 to 63) and the mean duration from the time of diagnosis of scleroderma to the onset of renal crisis was 2.5 yr (range 0.5 to 8). In addition to severe hypertension and progressive renal impairment over a period of several weeks, the majority had either retinal hemorrhages or papilledema. Nine had digital ulcers before the onset of renal crisis and most had evidence of diminished distal perfusion. Only 2 patients had renal biopsies: both were characteristic of scleroderma renal disease. Plasma renin activity measured before the administration of captopril in 9 patients was at least twice the upper limit of normal in 8. Microangiopathic changes on the peripheral smear were noted for 2 patients.

The diagnosis of renal crisis was made within 2 wk or less of starting treatment for 11 of the patients. For 6, the onset of renal crisis as suggested by the patients' history was uncertain but probably less than one month. For only 2 was the onset of renal deterioration more than 1 mo before treatment with captopril began (2 and 4 mo). The latter 2 patients were already on maintenance hemodialysis when they first received captopril.

The mean duration of treatment was 6.6 mo (range 2 wk to 23 mo). Six of the 20 had captopril withdrawn after 2 mo or less of treatment: 3 of the 6 were normotensive without any antihypertensive therapy, one refused all therapy and died, and two had captopril withdrawn because of lack of response. The two who had no response were on chronic dialysis and were fluid overloaded.

Two of the 3 patients who required no antihypertensive therapy after captopril was withdrawn died within 1 mo: both had significant myocardial involvement before the onset of renal crisis. The third patient who required no further antihypertensive therapy after captopril was stopped had chest pain and congestive heart failure prior to treatment with captopril and had a hypotensive episode and myocardial infarction during dialysis after several days of captopril treatment. Thus, the fact that all antihypertensive therapy could be withdrawn reflected the severity of disease in other organ systems and a poor prognosis rather than a "cure" of hypertension by captopril.

Of the 14 patients treated for more than 2 mo, 13 continued to be treated with captopril through April 1, 1980. One patient, who had developed renal crisis 1 mo before and required maintenance dialysis when captopril was started, died after 3 mo of treatment. This patient was an insulin-dependent diabetic of 9 yr duration, with recurrent congestive heart failure and electrolyte problems (Table 3).

For the 13 patients still being treated as of April 1, 1980, the mean duration of treatment was 9.6 mo (range 3 to 23); 2 required dialysis before captopril was started. The mean serum creatinine for the others before captopril was 4.0 mg% (range 1.4 to 9.4 mg%), and as of April 1, 1980, had decreased to 3.0 mg%. One of the 2 patients initially on dialysis continued to require dialysis after 7 mo of treatment; the serum creatinine of the other decreased to 1.9 mg% after 2 mo and was 1.4 mg% after 4 mo of treatment.

Renal function was stable in 6 and improved in 6 patients. Substantial improvement was noted in 3 of 7 whose serum creatinine levels were 4 mg% or greater when captopril was started. One patient has had a modest decrease in renal function, the serum creatinine increasing from 4.2 to 6.2 mg%, corresponding to a decrease in creatinine clearance from ~15 to ~10 cc/min. This patient has also required intermittent dialysis because of fluid overload.

Table 3. Patients treated for more than two months.

	Number of Patients
Died (Cardiac Arrest)	1
Renal Function	
Improved	6
Stable	6
Worse	1
Total	14

Duration of treatment: average 9.6 mo (range, 3–23)
Pretreatment serum creatinine: 4.0 mg% (1.4–9.4)
April, 1980 serum creatinine: 3.0 mg%

Blood pressure was controlled by captopril in all except 3 patients on dialysis who have had recurrent fluid overload. Most had an excellent response to the first dose, and more than half were controlled with captopril alone. Only 3 required an additional antihypertensive drug other than a diuretic (or dialysis): prazosin, minoxdil and clonidine were each required in one patient.

As noted previously, patients with scleroderma renal crisis were only a fraction of those receiving captopril for emergency use. Since follow-up data requested on case report forms mainly concerned blood pressure and adverse drug reactions rather than the underlying disease, information about other manifestations of scleroderma during treatment is incomplete. However, dramatic improvement in blood flow in the capillaries of the fingernail beds was noted in several patients within minutes of taking the first dose. The microangiopathic changes noted on the peripheral smears of two patients resolved within 2 wk. Digital ulcers, present before the onset of renal crisis, healed in 4 of 7 patients after 5 to 8 wk treatment, but were noted to recur in one patient. Finally, Raynaud's phenomena and subjective skin changes were reported for several patients. It cannot be certain that the improvement in other disease manifestations was related to captopril per se since similar improvement has been noted in other long-term survivors of scleroderma renal crisis receiving a variety of antihypertensive drugs.

In summary, captopril was remarkably effective in controlling blood pressure and stabilizing renal function in patients with scleroderma renal crisis. As of April 1, 1980, there have been 13 long-term survivors out of 20 for whom follow-up data are available. Renal function stabilized or improved in patients with all degrees of pretreatment renal impairment. A poorer prognosis was associated, however, with patients who had progressed to complete renal failure before starting on captopril and in those with significant cardiac involvement.

IDIOPATHIC EDEMA

Idiopathic edema, a syndrome of uncertain etiology characterized by periods of edema, most commonly occurs in women and appears to correlate with the menstrual cycle. Although the "idiopathic" nature of the syndrome in otherwise healthy women has been questioned and the edema attributed to sodium and water retention when their customary diuretic intake is stopped (MacGregor et al., 1979), others have identified an abnormal capillary leak (Edwards and Bayliss, 1976) in the upright posture and excessive morning-to-evening weight gain (Streeten et al., 1973). It has also been suggested the diurnal weight

Table 4. 32-year-old woman with idiopathic edema.

	Captopril		Placebo		Captopril			
	Week		Week			Week		
WK	1	2	3	4	5	6	7	8
MEWD (kg)	0.5	0.5	1.3	1.7	0.8	0.6	0.5	0.4

MEWD = mean morning-to-evening weight gain
(Mimran, A. et al. 1979. N Engl J Med 301:1289–1290)

gain may be due to an inappropriate response of the renin-angiotensin-aldosterone system to upright posture (Streeten et al., 1973).

Mimran and Targhetta (1979) compared the effects of captopril and placebo in a 32-yr-old woman with idiopathic edema (Table 4). Six weeks after spironolactone was withdrawn the patient was treated with 50 mg of captopril T.I.D. for 2 wk, followed by 2 wk of placebo treatment and an additional 4 wk on captopril. Diurnal weight fluctuations were sharply reduced by captopril without a significant effect on blood pressure. Daily weight gain during treatment with captopril averaged from 0.4 to 0.8 kg, while on placebo the patient gained from 1.1 to 1.7 kg per day. Further study of the effects of captopril in idiopathic edema would appear worthwhile.

SARCOIDOSIS

Sarcoidosis is a multisystem disease of unknown etiology that most frequently effects young adults. The lungs, hilar and peripheral lymph nodes, the skin, liver, spleen, eyes and phalangeal bones are commonly involved. The characteristic pathology is noncaseating granulomatous inflammation. Pulmonary involvement is the most frequent indication for treatment and the inflammation readily responds to corticosteroids.

Active disease is associated with elevated serum angiotensin converting enzyme activity in up to 75% of patients (DeRemme and Rohrbach, 1980). Levels of converting enzyme activity parallel disease activity and decrease in patients undergoing spontaneous remission or responding to steroid therapy. Although ACE levels are generally believed to reflect disease activity rather than participate in the inflammatory process, it is conceivable that inhibitors of ACE could have anti-inflammatory effects in sarcoidosis.

Granulomatous inflammation in one animal model is reduced by captopril. The administration of captopril reduced the granulomatous response to schistosome eggs in the lungs, liver, colon and ileum in mice.

However, withdrawal of captopril from mice previously unsensitized to schistosome eggs induced a flare in the granulomatous response (Weinstock et al., 1980).

SUMMARY

In conclusion, captopril is likely to be effective in the treatment of a variety of disorders in addition to hypertension and congestive heart failure. Treatment with captopril is effective in controlling the malignant hypertension and limiting renal damage in scleroderma renal crisis, and might be useful in ARF of other causes. ACE inhibition by teprotide reduces pulmonary artery pressure and pulmonary vascular resistance, suggesting that captopril would be useful pulmonary hypertension. The healing of digital ulcers in scleroderma patients suggests that captopril may be of benefit in occlusive vascular disease of other causes. Finally, there is a variety of diseases—such as idiopathic edema and sarcoidosis—in which captopril could help to unravel the pathogenesis as well as provide therapy.

REFERENCES

Anderson, R.J., and Schrier, R.W. 1980. Acute Renal Failure. In *Harrison's Principles of Internal Medicine*, Ninth Edition, eds. K.J. Isselbacher, R.D. Adams, E. Braunwald, R.G. Petersdorf, and J.D. Wilson, pp. 1293–1299. New York: McGraw-Hill Book Company

Bartter, F.C., Pronove, P., Gill, J.R., et al. 1962. Hyperplasia of the juxtaglomerular complex with hyperaldosteronism and hypokalemic alkalosis: a new syndrome. Am J Med 33:811–828

Berkov, S. 1974. Hypoxic pulmonary vasoconstriction in the rat. Circ Res 35:256–261

Brown, J.J., Gleadle, R.I., Lawson, D.H., et al. 1970. Renin and acute renal failure: studies in man. Brit Med J 1:253–258

Cannon, P.J., Hassar, M., Case, D.B., Casarella, W.J., Sommers, S.C., and Leroy, E.C. 1974. The relationship of hypertension and renal failure in scleroderma (progressive systemic sclerosis) to structural and functional abnormalities of the renal cortical circulation. *Medicine* 53:1–46

Daoud, F.S., Reeves, J.T., and Kelly, D.B. 1978. Isoproterenol as a potential pulmonary vasodilator in primary pulmonary hypertension. Am J Cardiol 42:817–822

DeRemme, R.A., and Rohrback, M.S. 1980. Serum angiotensin-converting enzyme activity in evaluating the clinical course of sarcoidosis. Ann Intern Med 92:361–365

Edwards, O.M., and Bayliss, R.I.S. 1976. Q J Med 45:125

Gavras, H.P., Gavras, I., Cannon, P.J., Brunner, H.R., and Laragh, J.H. 1977. Is elevated plasma renin activity of prognostic importance in progressive systemic sclerosis? Arch Intern Med 137:1554–1558

Hauge, A. 1968. Conditions governing the pressor response to ventilation hypoxia in isolated perfused rat lungs. Acta Physiol Scan 72:33–44

Hollenberg, N.K., Swartz, S.L., Passan, D.R., and Williams, G.H. 1979. Increased glomerular filtration rate after converting-enzyme inhibition in essential hypertension. N Engl J Med 301:9–12

Klinke, W.P., and Gilbert, J.A. 1980. Diazoxide in primary pulmonary hypertension. N Engl J Med 302:91–92

Kovalchik, M.T., Guggenheim, S.J., Silverman, M.H., Robertson, J.S., and Steigerwald, J.C. 1978. The kidney in progressive systemic sclerosis. Ann Intern Med 89:881–887

Kuchel, O., Horky, K., Gregorova, I. *et al.*, 1970. Inappropriate response to upright posture: a precipitating factor in the pathogenesis of idiopathic edema. Ann Intern Med 73:245–252

Lam, M., Ricanati, E.S., Kahn, M.A. and Kushner, I., 1978. Reversal of severe renal failure in systemic sclerosis. Ann Intern Med 89:642–643

Lloyd, T.C. 1968. Hypoxic pulmonary vasoconstriction: role of perivascular tissue. J Appl Physiol 25:560–565

Lopez-Ovejero, J.A., Saal, S.D., D'Angelo, W.A., Cheigh, J.S., Stenzel, K.H., and Laragh, J.H. 1979. Reversal of vascular and renal crises of scleroderma by oral angiotensin-converting enzyme blockade. N Engl Med J 300:1417–1419

MacGregor, G.A., Roulston, J.E., Markandu, N.D., Jones, J.C., and De Wardener, H.E. 1979. Is "idiopathic" edema idiopathic? Lancet 1:397–400

Mimran, A., and Targhetta, R. 1979. Captopril treatment of idiopathic edema. N Engl J Med 301:1289–1290

Mimran, A., Brunner, H.R., Turini, G.A., Waeber, B., and Brunner, D. 1979. Effect of captopril on renal vascular tone in patients with essential hypertension. Clinical Science 57:421s–432s

Mitnick, P.D., and Feig, P.U. 1978. Control of hypertension and reversal of renal failure in scleroderma. N Engl J Med 299:871–872

Moorthy, A.V., Wu, M.J., Beirne, G.J. and Sundstrom, W.S., 1978. Control of hypertension with acute renal failure in scleroderma without nephrectomy. Lancet 1:563–564

Niarchos, A.P., Roberts, A.J., and Laragh, J.H. 1979. Effects of the converting enzyme inhibitor (SQ 20,881) on the pulmonary circulation in man. Am J Med 67:785–791

Rabinovitch, M., Gamble, W.J., Williams, G., and Reid, L. 1980. SQ 14,225 converting enzyme inhibitor diminishes pulmonary artery hypertension secondary to chronic hypoxia in rat. Fed Proc 39:765

Richardson, J.A. 1973. Hemodialysis and kidney transplantation for renal failure from scleroderma. Arthritis and Rheum 16:265–271

Rubin, L.J., and Peter, R.H. 1980. Oral hydralazine therapy for primary pulmonary hypertension. N Engl J Med 302:69–73

Ruskin, J.M., and Hutter, A.M., Jr. 1979. Primary pulmonary hypertension treated with oral phentolamine. Ann Intern Med 90:772–774

Schweisfurth, H., and Wernze, H. 1979. Changes of serum angiotensin I converting enzyme in patients with viral hepatitis and liver cirrhosis. Acta Hepato-Gastroenterol 26:207–210

Stark, R.D., Finnegan, P., and Bishop, J.M. 1972. Daily requirement of oxygen to reverse pulmonary hypertension in patients with chronic bronchitis. Brit Med J 3:724–728

Streeten, D.H.P., Dalakos, T.G., Souma, M., Fellerman, H., Clift, G.V., Schletter, F.E., Stevenson, C.T., and Speller, P.J. 1973. Studies of the pathogenesis of idiopathic oedema: the roles of postural changes in plasma volume, plasma renin activity, aldosterone secretion rate and glomerular filtration rate in the retention of sodium and water. Clin Sci Mol Med 45:347–373

Wasner, C., Cooke, C.R., and Fries, J.F. 1978. Successful medical treatment of scleroderma renal crisis. N Engl J Med 299:873–875

Weinstock, J.V., Boros, D.L., Ehrinpreis, M.N., and Gee, J.B. 1980. Effect of angiotensin I converting enzyme (ACE) inhibitor on the granulomatous response to *Schistoma Mansoni* eggs in mice. Gastroenterology 78:1289

Wilkinson, S.P., Smith, I.K., and Williams, R. 1979. Changes in plasma renin activity in cirrhosis: a reappraisal based on studies in 67 patients and "low renin" cirrhosis. Hypertension 1:125–129

Williams, G.H., and Hollenberg, N.K. 1977. Accentuated vascular and endocrine response to SQ 20,881 in hypertension. N Engl J Med 297:184–188

Zakheim, R.M., Mattioli, L., Molteni, A., Mullis, K.B., and Bartley, J. 1975. Prevention of pulmonary vascular changes of chronic alveolar hypoxia by inhibition of angiotensin I-converting enzyme in the rat. Lab Invest 33:57–61

Discussion

R.K. FERGUSON: Dr. Chatterjee, in your patients with congestive heart failure you are saying that when you start captopril the blood pressure does or does not fall significantly when you get the increase in cardiac function and performance?

K. CHATTERJEE: It does fall. The arterial pressure fell in all patients and cardiac output increased significantly. It suggests therefore, that in patients with chronic heart failure with markedly elevated systemic vascular resistance the blood pressure fell despite an increase in cardiac output. That means the resistance must have fallen much more.

R.K. FERGUSON: But having heard you say it before I thought that the idea was to unload the heart without decreasing blood pressure to any great degree.

K. CHATTERJEE: No question. I think that if we are going to treat patients with chronic heart failure or acute heart failure who are normotensive or hypotensive at the time of treatment, particularly in patients with coronary artery disease, I think significant hypotension may be a problem, because the reduction of perfusion pressure may actually decrease the perfusion and therefore enhance ischemia. But I think what happens in these patients overall, metabolic function remains unchanged because of simultaneous reduction in pulmonary capillary wedge pressure and presumably left ventriclar diastolic volume and reduction of heart rate and arterial pressure decrease the overall myocardial oxygen consumption. But I shall be still worried at the time of treatment with captopril in patients with heart failure who are normotensive or hypotensive, to make sure that they don't develop severe hypotension at the time of treatment.

R.K. FERGUSON: And I take it that just like with other vasodilators in congestive heart failure, as blood pressure drops, you do not get the compensatory increase in heart rate even in the standing position.

K. CHATTERJEE: That is right. In other words, as you know, almost all vasodilators except phenotolamine tend to cause no change in heart rate or even slight decrease in heart rate in patients with chronic heart failure even with some degree of fall in arterial pressure. With captopril we saw more than what we usually see with other vasodilators. And this lack of increase in heart rate really has not been clearly explained. Even with hydralazine for example, which tends to increase heart rate uniformly in patients with hypertension, and without heart failure, with the same drug even when used in large doses (800 or 1200 mg a day) it does not increase heart rate in patients with chronic heart failure. That is a uniform response and why it happens has not been clearly defined.

R.K. FERGUSON: Did you imply that with captopril you actually got a decrease in heart rate, that is, a bradycardia?

K. CHATTERJEE: Yes, we did. We saw, with the first dose of captopril, whether it was 25 or 50 or 100 mg, a significant decrease in heart rate. Some patients' drop in heart rate was over 10 to 15 beats. But when you look at the change in stroke volume, it markedly increased because the heart rate decreased. In terms of changes in cardiac performance there was dramatic improvement. However, each patient had a decrease in heart rate initially. With continued therapy at 8 wk this bradycardic response was not as great as what was seen initially. So somethings happens during continued therapy. It might be that during continued therapy with persistent improvement in cardiac performance, catecholamine drive probably decreases and therefore no persistent bradycardia is seen.

C.I. JOHNSTON: I wonder if I just ask Dr. Chatterjee to tell us a little bit about the long term effects on the patients' saline balance and weight. How much of the improvement does he think may have been caused by a natriuresis and a decrease in weight?

K. CHATTERJEE: Of the 7 patients on long term therapy up to eight wk, we did not change any dose of diuretic because that was the protocol. In one patient we had to reduce the dose of captopril because of hypotension. After 8 wk of captopril therapy, in about 3 patients we had to reduce the dose of diuretics. But there was no significant change in weight. I think this is interesting in our experience. We have a large group of patients with severe chronic heart failure on long term vasodilator therapy with other vasodilator agents. We had seen that there might be increases in weight with continued therapy, but we did not see weight gain with captopril, although I must admit the number of patients that are on captopril therapy now is too small, only about 5 and 7 patients.

C.I. JOHNSTON: We have used captopril very much in congestive cardiac failure but in one or two of our patients, we have been a little bit concerned in the decrease in creatinine clearance, acutely. The hypotension doesn't seem to produce many symptoms but it does seem to elevate the plasma creatinine acutely.

K. CHATTERJEE: We looked at that. We did not see anything. We looked at the creatinine clearance acutely and at late follow-up. There was no change. In fact we didn't see any of the proteinuria that has been described. But again I must admit that the number of patients on long-term therapy are too small to derive any conclusion about the associated side effects or changes in renal function.

C.I. JOHNSTON: Could I just ask Dr. Bravo one short question? I wonder if he could tell us whether he thinks some of the hemodynamic changes might be due to changes in venous compliance. Angiotensin does constrict veins particularly in vitro. Have you measured venous compliance by plethysmography?

E.L. BRAVO: We have not measured venous compliance directly, Dr. Johnston, but others have done so. Collier and Robinson reported blocking the effects of angiotensin I with converting enzyme inhibitor (teprotide) on hand vein and forearm arteries in man. Recently Johns et al. reported that converting enzyme inhibition dilates forearm vasculature in severe hypertensive individuals. Indirect data also sugest that converting enzyme inhibition increases venous compliance, because if it does not then cardiac output should increase. This does not occure as blood pressure is falling.

K. CHATTERJEE: There is some data about changes in venous capacitance. Dr. Mason's group demonstrated a marked increase in venous capacitance and decreases in venous tone. In the May issue of *Circulation* there was also a paper looking at changes in venous tone after teprotide, and they did not find any significant change. So I think there is a variable response. But I agree there has to be some redistribution of blood volume, otherwise we can't explain why there will be such a dramatic decrease in pulmonary capillary wedge pressure and right atrial pressure. We also looked at the changes in left ventricular echocardiographic dimension after acute therapy. We did not see any significant change in end-diastolic volume of the ventricle. At the same time, however, there was dramatic decrease in pulmonary capillary wedge pressure. That raises the question whether there is any effect of captopril on pulmonary venous compliance.

G.M HOCKEL: Two questions for Dr. Zimmerman. In the acute anesthetized study that was presented, did you measure sodium excretion? And if so, did sodium excretion change during your intrarenal infusion of converting enzyme inhibitor when there was no change in renal blood flow? Secondly, in your conscious dog model, you stated that the observed changes in renal blood flow with teprotide had not been extended past 30 min, and I wondered if you saw any changes in flow longer than 30 min in time with captopril.

B.G. ZIMMERMAN: As far as sodium excretion is concerned, it was not measured. It is possible, though, that there is an effect of converting enzyme inhibition on electrolyte excretion separate from an effect on renal hemodynamics. This is a possibility which probably should be explored. As far as the extension of the study beyond 30 min with teprotide, as I said before we really haven't systematically studied that. I'm sure the effect lasts for longer than 30 min in terms of angiotensin I blockade, and at least in the normal dogs when there is an increase in blood flow it extends beyond that time. In the hypertensives there tends to be a waning of the renal vasodilatation with time but it is still present at 30 min even in the hypertensive.

G.M. HOCKEL: How about with captopril?

B.G. ZIMMERMAN: It is pretty much the same story with captopril, though again we really haven't looked at those time periods systematically beyond the 30 min period.

E.H. BLAINE: I direct this question to Dr. McCaa. Bob, you showed the increase in sodium excretion with captopril in sodium deficient dogs of approximately twofold, if I recall the data correctly. Obviously that sodium loss cannot go on for an indefinite period. I wonder if you have any feeling for the mechanism that comes into play to prevent the continued loss of sodium. Is it an increase in aldosterone or a decrease in blood pressure or perhaps a decrease in GFR? A second question is, do you have any feeling for the magnitude of the natriuresis that occurs with chronic captopril treatment in non–sodium depleted dogs?

R.E. MCCAA: Dr. Blaine, we have maintained intact conscious sodium replete dogs on captopril administration for approximately 2 wk. We observed an increase in urinary sodium excretion in these animals with only a slight decrease in arterial blood pressure (12 mm Hg). In addition to these studies, Dr. Roy Bengis maintained intact rats on a normal sodium intake and captopril administration for 1 wk. Dr. Bengis made similar observations in his studies; that is, an increase in urinary sodium excretion and a decrease in arterial blood pressure (14 mm Hg). In conscious dogs maintained on dietary sodium restriction (5 mEq Na^+/day for 21 days), urinary sodium excretion averaged 0.6 – 0.8 mEq/day. In response to captopril administration in sodium deficient animals, urinary sodium excretion increased five- to tenfold, averaging almost 5 mEq/day. Often during these studies we observed animals in which urinary sodium excretion increased to 9 mEq/day. However, these animals lost weight over the 3 wk period during administration of the converting enzyme inhibitor. As you recall from our earlier studies, this marked increase in urinary sodium excretion was associated with an increase in effective renal plasma flow and a sharp decrease in arterial blood pressure, glomerular filtration rate, and plasma aldosterone concentration. We are convinced that the increase in urinary sodium excretion in these sodium deficient dogs maintained on captopril administration was not due to the decrease in aldosterone concentration, since urinary sodium excretion increased to approximately the same level in sodium deficient dogs maintained on high rates of aldosterone infusion to prevent the decrease in plasma aldosterone concentration in response to captopril administration. In more recent studies, we have maintained sodium deficient dogs on captopril administration and an infusion of angiotensin II at very low rates (3 ng/kg/min). The addition of angiotensin II to the infusate at low infusion rates in sodium deficient dogs maintained on captopril restored sodium excretion to the levels observed in untreated sodium deficient animals (0.7 mEq/day). We have concluded from these studies that under long-term conditions angiotensin II plays a much more potent sodium controlling role than aldosterone.

H.R. BRUNNER: I would like to ask a question of Dr. Lefer. If I understand right, your beautiful experiments were done with pretreatment. Have you tried to treat already existent hemorrhagic shock? And if yes, at what point can you still get an effect with captopril?

A.M. LEFER: Yes, the experiments I described today were not all pretreatment. Some of them had captopril given at 1 hr after the onset of hemorrhage, and within that framework the same results were obtained. In other words, the protection occurs anytime during that one hour of treatment. We have not tried any later times because in the 2½ hr hemorrage model you didn't really have much more time to delay.

B. MCGRATH: My question to Dr. Lefer. Yesterday we heard information on possible modulation of sympathetic activity by captopril. In your hemorrhagic studies did you look at changes in catecholamines at all?

A.M. LEFER: No we didn't. It is certainly possible that there could be alterations in sympathetic tone. In fact there could be interaction between catecholamines and the prostaglandin release as well. As you know, some of the E-type prostaglandins actively diminish sympathetic tone and it is possible, as Dr. Zimmerman showed, that there are interactions between angiotensin and the prostaglandins. So there certainly could be a variety of types of interactions such as you suggest, but I don't know of anyone who has studied these precisely in any type of shock.

G.A. MACGREGOR: Dr. McCaa, the increase in sodium excretion with captopril in your experiment was entirely due to the fall in angiotensin II and nothing to do with the fall in aldosterone? Is that correct?

R.E. MCCAA: That is correct, Dr. MacGregor. These conclusions are based on the observation that urinary sodium excretion increased in sodium deficient dogs maintained on captopril, whether or not plasma aldosterone concentration remained elevated or decreased.

G.A. MACGREGOR: If that is so, it seens to contradict a lot of work about aldosterone. One vivid example is patients with Addison's disease who, if not given mineralocorticoids, get grossly volume depleted and have very high renins. In spite of the very high level of angiotensin II they remain in negative sodium balance until they are given mineralocorticoids. It seems, therefore, that in man aldosterone is essential to maintain normal sodium balance, and that high levels of angiotensin II are not sufficient in themselves to do this.

R.E. MCCAA: I agree with you entirely, Graham. There certainly is very good evidence in experimental animals and in human beings that aldosterone plays a very important role in electrolyte and fluid metabolism. Yet, I would like to comment further on your question. In our laboratory, we have used both teprotide and captopril to inhibit angiotensin II formation in animal models with experimental hypertension. Both teprotide and captopril effectively lowered arterial blood pressure and increased urinary sodium excretion in all high renin models of experimental renovascular hypertension we studied. Is the decrease in arterial blood pressure and the increase in urinary sodium excretion in these experimental hypertensive animals due to the increase in both angiotensin II and aldosterone concentrations or only to angiotensin II concentration? We believe it is only due to the inhibition of angiotensin II formation. Let me present another exciting model we have used recently in our experimental studies. Aldosterone secretion and plasma aldosterone concentration remain normal in patients with pituitary insufficiency or in the complete absence of the pituitary gland. When these patients are maintained on a dietary sodium restricted diet (10 mEq Na^+/day) for 2 wk, they come into balance (10 mEq/day) in the absence of an increase in aldosterone secretion or a decrease in arterial blood pressure. With our current knowledge of the physiology of aldosterone, this is difficult to understand. Could these patients be maintained in sodium balance at 10 mEq/day by the increase in plasma angiotensin II concentration? Plasma angiotensin II concentration increased between three- to five-fold. We are currently testing this hypothesis in sodium deficient hypophysectomized animals maintained on captopril.

G.A. MACGREGOR: I accept you experiment, but it does raise the paradox of why does aldosterone in man cause sodium retention. I wonder whether you had considered doing different experiments along the same lines to see if you get the same results.

R.E. McCaa: I would agree, and that is why I presented these data to make that point. We have some of these studies currently in progress in our laboratory.

F.W. Wolf: When Dr. McCaa mentioned hypopituitarism and was surprised about the lack of response of aldosterone as I understood it, well, I asked myself a question immediately. Are you dealing with chronic cases of hypopituitarism? If you are you've got something approximating a Sheehan's syndrome. I happen to have written up some of these cases many years ago. These patients don't have normal adrenals. They have hypoadrenocorticism and I am not at all surprised that there is no other steroid secretion. I do not know whether you have seen the autopsies of these patients. They have got minute little adrenals. You can hardly see them. They have got minute little ovaries. They have minute little everything. That's a classical Sheehan's syndrome. I was going to ask Dr. Chatterjee about his beautiful clinical talk. I see quite clearly that on chronic therapy of congestive heart failure with converting enzyme inhibitors there is improvement in exercise tolerance. And the heart rate does not increase. I may be anticipating this afternoon's session, but it's an important point. I'm concerned if I see an effect of a drug, let's say the blood pressure is lowered, and there is no physiologic increase of pulse rate. What happens to exercise tolerance, if you have congestive heart failure and severe hypertension, if you take converting enzyme inhibitor? Does your pulse rate go down if you are put on the treadmill? How do you respond? Does your exercise tolerance diminish? Does it remain the same?

K. Chatterjee: It depends on the magnitude of the response in terms of increase in stroke volume. Patients with chronic complete heart block, with normal left ventriclar function, maintain normal cardiac output at rest and can increase cardiac output to some extent by compensatory increase in stroke volume. These patients with chronic heart failure, even though their heart rate response is reduced at rest, do increase their heart rate during exercise. The magnitude of increase in heart rate response with conventional therapy and adding vasodilator appears to be identical. So there is an increase in heart rate during exercise. The cardiac output and stroke volume are still increased in the majority of patients. So the increase in cardiac output during exercise under chronic therapy is both due to an increase in heart rate and increase in stroke volume. The patients do take the advantage of increases in heart rate, although, resting heart rates appear to be lower.

R.E. MCCAA: I would like to respond to the first question that Dr. Wolf asked. In 1950, John Luetscher and his associates at Stanford found a high concentration of the "sodium retaining factor" in urine from patients with edema, and in 1954, Dr. Luetscher demonstrated that the concentration to this "sodium retaining factor" in urine was high when sodium intake was restricted in patients, while normal values were found in patients with hypopituitarism and no detectable level was observed in patients with Addison's disease. During the past 20 years, several investigators have observed normal or slightly elevated aldosterone secretion and plasma aldosterone concentration in patients with chronic pituitary insufficiency. There are several significant facts that should be mentioned about aldosterone metabolism in patients with pituitary insufficiency. First, despite the normal increase in plasma renin activity and blood angiotensin II concentration in response to dietary sodium restriction in patients with pituitary insufficiency, plasma aldosterone concentration fails to increase significantly. Although aldosterone secretion fails to increase during dietary sodium restriction, these patients come into sodium balance at 10 mEq without a decrease in arterial blood pressure. It is of interest to note that ACTH and angiotensin II fail to stimulate aldosterone secretion in patients with pituitary insufficiency while they are maintained on normal sodium intake. However, after dietary sodium restriction the aldosterone response to ACTH is quantitatively similar to the response observed in normal intact subjects. In striking contrast, while the adrenal is supersensitive to ACTH, it remains almost totally refractory to angiotensin II, since aldosterone secretion fails to increase in response to angiotensin II infusion for one hour at a rate that increases arterial blood pressure in patients with pituitary insufficiency. I would also like to add that observations have been made in chronic hypophysectomized animals. We are currently evaluating aldosterone metabolism in patients with pituitary insufficiency and in chronically hypophysectomized animals.

F.J. CONWAY: I'd like to try and get Dr. Lefer to expand a little bit on his experiments. It really is fascinating to see the converting enzyme inhibitor prolonging the life of these animals in hemorrhagic shock. Could you tell us, particularly when you administer the converting enzyme inhibitor during the hemorrhagic state, whether there is a difference in the hemodynamic response to the converting enzyme inhibitor as, say, compared with an alpha blocker, which also prolongs life. I am thinking of the cardiac response and the distribution of blood flow particularly to the splanchnic area.

A.M. LEFER: We see no gross obvious hemodynamic effects when we give captopril in hemorrhagic shock. It's remarkable in its absence of any overt effect on arterial pressure, venous pressure or on blood flow. Initially, we were controlling systemic pressure, so we didn't see an effect, but there was no change in bleedout volume so there seems to be no overt effect. We have not measured distribution of blood flow. Dr. H. Gavras has shown this very nicely in normotension, and found changes in the splanchnic region. We suspect that there might be changes although we did not see any significant effects on superior mesenteric artery flow. However, it is quite possible that there is a redistribution of flow within the splanchnic region. If there were changes to preserve the perfusion of the pancreas, the intestine and perhaps of the liver, this could partially help explain some of the beneficial effect of captopril in preventing the release of lysosomal hydrolases and preventing the production of the toxic peptide MDF. This is quite a reasonable possibility. There would have to be significant change to do this but it is quite possible, but we don't have any direct microsphere data at the moment to answer that question.

E.L. BRAVO: May I pose a question to Dr. Chatterjee. Assuming for a moment that bradykinin does not play an important role in the antihypertensive effect of converting enzyme inhibition, and also assuming that the bulk of the evidence indicates that the activity of the renin-angiotensin system is not increased in some patients with congestive heart failure, how do you explain the mechanism by which captopril lowers systemic vascular resistance?

K. CHATTERJEE: I wish I knew. I agree that we did not find a good correlation with initial renin level and the magnitude of reduction in systemic resistance. And we know the systemic vascular resistance decreases markedly in these patients. I really don't know the prime mechanism of reduction by captopril. There is some evidence that sodium is retained in the arterial wall, and reduced sodium content and reduction of wall stiffness may be contributing. But we thought that reduction of systemic resistance is too dramatic and rapid to be explained by this mechanism. I think that we have to think of an initiating mechanism. If we reduce systemic resistance and arterial pressure, it will reduce afterload wall stress and will reduce the left ventricular afterload further.

A.M. LEFER: I would like to ask Dr. Stritar if he knows of any data in which captopril has been tried even on an emergency basis or a small case study in myocardial ischemia or myocardiac infarction? It would seem to me that with the data that is known now in congestive failure, captopril might work in acute myocardial infarction.

J.A. STRITAR: Nobody is working on it right now. There is a study that will be starting soon. There is however some adverse information in the literature. There is a report on two cases that supposedly had aggravation of angina and I think one had a myocardial infarction while on captopril. One of the patients with scleroderma developed a hypotension on dialysis and a myocardial infarction, but she had just been in heart failure before she started captopril. I don't think there is really any good information.

K. CHATTERJEE: We are doing a study on coronary dynamics and myocardial oxygen consumption and it is very preliminary. It seems that oxygen consumption decreases proportional to the decrease in oxygen demand. If the heart rate falls, the blood pressure falls and oxygen consumption decreases. But as I said, the net effect will be probably related to the degree of reduction of coronary flow due to reduction in diastolic pressure—i.e., the perfusion pressure—and decrease in myocardial oxygen demand. I am a little worried about giving captopril in a patient with acute coronary artery disease because we saw, at least in some patients, the dramatic decrease in arterial pressure.

E.L. BRAVO: Maybe I can answer the question partially. Dr. Braunwald recently reported some evidence that converting enzyme inhibition can limit infarct size. For these studies he utilized an experimental model of selective coronary occlusion in the dog.

B.G. ZIMMERMAN: I just wanted to get back to the question about the lack of correlation between plasma renin activity and the possible effect of the captopril on systemic vascular resistance. I don't believe that there has to be an excellent correlation between the absolute value of PRA and the response you get with captopril. It is nice when you get that, but even in the lower renin states there still could be a relationship and though there may not be a good correlation coefficient, it is still a possible relationship.

K. CHATTERJEE: We know that norepinephrine levels are markedly elevated in these patients with chronic heart failure. In fact all of the determinations that we did were both acute and chronic, so it is possible that withdrawal of the norepinephrine effect to some extent may also account for the change in the systemic vascular resistance. Also, the decrease in afterload, whatever way we do it, by either measuring initial fall in arterial pressure, fall in systemic resistance, and therefore reduction of diastolic volume and increase in afterload, increase in cardiac output may actually give us a further decrease in systemic resistance, because after all we are calculating the resistance from the change in cardiac output.

F. FYHRQUIST: I would like to ask Dr. Lefer about shock prevention. We yesterday learned from Dr. Ganten that captopril is effective also in preventing degradation of enkephalins and opiates. On the other hand, there have been two reports, one in *Nature* and another in

Science, on the possible role of opiates in shock by blood pressure reduction, and its reversal by naloxone. I have two questions. Have you considered this, and if so, have you tried naloxone or nalorphine, as to whether they modify the effect of captopril on shock?

A.M. LEFER: That is a very fascinating area you mentioned. We have tried naloxone in hemorrhagic shock. Mark Curtis in my laboratory has done these experiments and in fact it does protect in shock. We have not considered however, whether there would be an interaction between converting enzyme inhibition and endorphin or enkephalin production. That is another possibility and I don't believe anybody has studied it. The studies that you mentioned by Holiday and Faden, in *Science* and *Nature*, did not address that question. I gather that you think perhaps that there is an interaction with converting enzyme inhibitor and the generation of endorphins in the hypophysis or the brainstem. That indeed is a very interesting possibility.

F. FYHRQUIST: Or the metabolic breakdown.

A.M. LEFER: Yes, that is another area of interest.

SECTION IV

Hypertension

Doris N. McKinstry, *Chairperson*

The Renin-Angiotensin-Aldosterone System: A Normal Mechanism for Maintaining Blood Pressure in Normotensive and Hypertensive Subjects[1]

G. A. MacGregor, N. D. Markandu, J. E. Roulston, J. C. Jones and J. J. Morton[2]
Department of Medicine, Charing Cross Hospital, Medical School London W6 8RF, United Kingdom

SUMMARY

Captopril given for 5 days caused a fall in blood pressure in normotensive subjects. The percentage fall in blood pressure was greatest (19.6%) on a low sodium diet (10 mmol/day), least (11%) on a high sodium diet (350 mmol/day) and intermediate (16.5%) on a normal sodium intake (120 mmol/day). Captopril also caused a marked fall in plasma aldosterone on all 3 sodium intakes, and an increase in sodium excretion on the normal and low sodium diets. Captopril given to patients with essential hypertension on their normal diet caused a similar percentage fall in blood pressure as compared to the normotensive, for a given plasma renin activity. These results suggest that the renin-angiotensin-aldosterone system is a normal mechanism for maintaining blood pressure and aldosterone secretion in normal man, and it plays no greater role in the maintenance of blood pressure in essential hypertension, for normal levels of plasma renin activity.

[1] These studies were supported by the Wellcome Trust, the Medical Research Council and the National Kidney Research Foundation.
[2] The Medical Research Council, Blood Pressure Unit, Glasgow.

INTRODUCTION

There is uncertainty about the importance of the renin-angiotensin-aldosterone system in maintaining blood pressure in normal man. The system has generally been regarded as a defense mechanism that maintains blood pressure only under conditions of volume depletion or sodium restriction (Haber, 1976). Work with saralasin (Posternak et al., 1977; MacGregor et al., 1979), a competitive inhibitor of angiotensin II and teprotide (Haber, 1976), a parenteral inhibitor of angiotensin converting enzyme, has shown that the renin-angiotensin-aldosterone system does maintain blood pressure, particularly in the upright position, in normal man on a low sodium intake. Neither of these inhibitors of the renin-angiotensin system caused any fall in blood pressure in normal subjects studied on their normal diet. These findings confirmed the view previously held that this system was important only under conditions of volume depletion. However, saralasin is a potent agonist (MacGregor and Dawes, 1976; Hollenberg et al., 1976; Anderson et al., 1977; Case et al., 1976) and would, therefore, underestimate the role of the renin-angiotensin-aldosterone system; in addition, both saralasin and teprotide block only the immediate effects of circulating angiotensin II. A third factor that may have underestimated the role of this system in blood pressure maintenance was that both of these experiments were done after the subjects had rested for some time before the infusion of saralasin or converting enzyme inhibitor.

In view of the uncertainty of the importance of this system in normotensive man, it is not surprising that there is considerable controversy over its role in the maintenance or causation of elevated blood pressure in essential hypertension (Case et al., 1977). With the recent development of an oral inhibitor of angiotensin converting enzyme, captopril, long-term inhibition of the formation of angiotensin II is now possible (Ondetti et al., 1977; Ferguson et al., 1977; and Gavras et al., 1978). In this study we have compared the effect of captopril given to normal subjects for 5 days on 3 different sodium intakes to the effect in patients with essential hypertension studied on their normal diet. The results suggest that the renin-angiotensin-aldosterone system is a normal mechanism for maintaining blood pressure and plays no greater role in essential hypertension in maintaining blood pressure than it does in normotensive man with comparable levels of plasma renin activity.

METHOD

Normotensive Subjects

Normotensive, healthy, male subjects, whose diastolic pressure was less than 80 mm Hg after a 2 mo observation period, were included in the study. All subjects gave their informed consent. Mean age of the subjects was 22 yr, range 21–26. Subjects were studied on 3 sodium balances, on each balance for 3 wk, with an interval between each balance study when they ate their normal diet. The low sodium diet was 10 mmol/day (n = 5), the normal sodium intake was 120 mmol/day (n = 8) and the high sodium diet was 350 mmol/day (n = 7). The high sodium diet was achieved by supplementing the diet with slow sodium (CIBA). Potassium intake was kept constant at 115 mmol/day. During the three weeks of each diet, continuous 24-hr urine collections were made. The subjects were allowed to go about their normal activities throughout the study and were not admitted to a hospital. They were discouraged from vigorous exercise. Captopril was started on the eighth day of the balance study. Blood pressure measurements were made 2 hr after the dose of 25 mg. The subjects were allowed to pursue their usual activities during these 2 hr. The dose of captopril was 25 mg three times for the first day, increasing to 3 doses of 50 mg on the second day, 100 mg thrice daily for the next 2 days, and 1 dose of 100 mg on the morning of the fifth day. Daily measurements of blood pressure and 24-hr urinary and sodium excretion were made. All blood pressures were measured by nurses using semiautomatic ultrasound sphygmomanometers (Arteriosonde 1217) (George et al., 1975). The mean value of five readings at 1 to 2 min intervals (supine, sitting and standing), were taken as the blood pressure in that position. Blood pressure measurements were made at the same time of day (between 10:00 a.m. and 12:00 midday) in the same room by the same nurse on the same arm. Pulse rate was measured on a Cambridge 3048 pulse monitor with ECG display and recording. Plasma angiotensin II (Dusterdieck et al., 1971), plasma renin activity (PRA) (Roulston and MacGregor, 1978), plasma aldosterone (James and Wilson, 1978), blood urea, electrolytes and plasma creatinine were measured on the fifth and eighth day of the balance 2 hr after the first dose of captopril, and on the fourth and fifth day during treatment with captopril. Plasma noradrenalin was measured on the eighth day of the balance before captopril and on the fifth day of captopril treatment. All blood was taken without stasis after sitting upright for 5 to 10 min.

Hypertensive Patients

Patients with uncompleted essential hypertension were included in the study. All had been referred by local general practitioners and were included in the study only if, after an initial observation period of 1 to 2 mo when they received no treatment, their sitting diastolic pressure was between 100 and 120 mm Hg. The mean age of the patients was 52 yr, range 41–65, and included 10 men and 8 women. The patients were excluded if there was evidence of renal failure, ischemic heart disease or cerebrovascular disease, or if they were taking oral contraceptives or any other drug. Informed consent was obtained from each patient.

After an initial assessment period of 1 to 2 mo, all 18 patients were observed weekly for a further 4 wk while receiving no treatment. Twelve patients were then treated with a matching placebo 3 times a day for 1 mo and seen every week. All 18 patients then received captopril 25 mg three times a day. Observations were made 2 hr after the first 25 mg dose in 15 patients. After 1 wk of captopril, the dose was increased to 50 mg thrice daily during the second week, 100 mg three times daily during the third week, and 150 mg thrice daily during the fourth week. If the supine diastolic pressure was less than 85 mm Hg the dose was not increased but was kept at the same amount as that for the previous week. Two patients had diastolic pressure below 85 mm Hg during the third week of treatment while taking captopril 100 mg thrice daily.

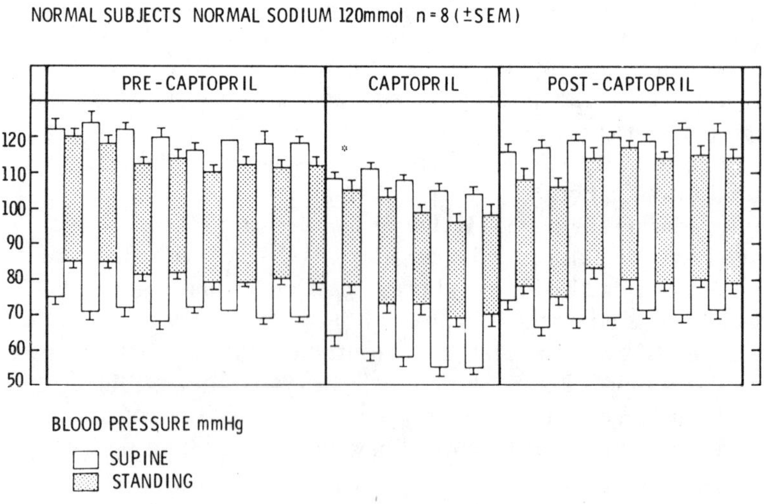

Fig. 1. Daily average blood pressure before, during and after captopril in 8 normal subjects on a normal sodium intake of 120 mmol.

During the next week, however, the diastolic pressure was greater than 85 mm Hg so that the dose was increased to 150 mg thrice daily in the fifth week of treatment. For the analysis of treatment effects the blood pressure in this fifth week for these 2 patients was handled as though it was the pressure in the fourth week. In the other 16 patients the diastolic pressure did not fall below 85 mm Hg and the dose of captopril was increased each week.

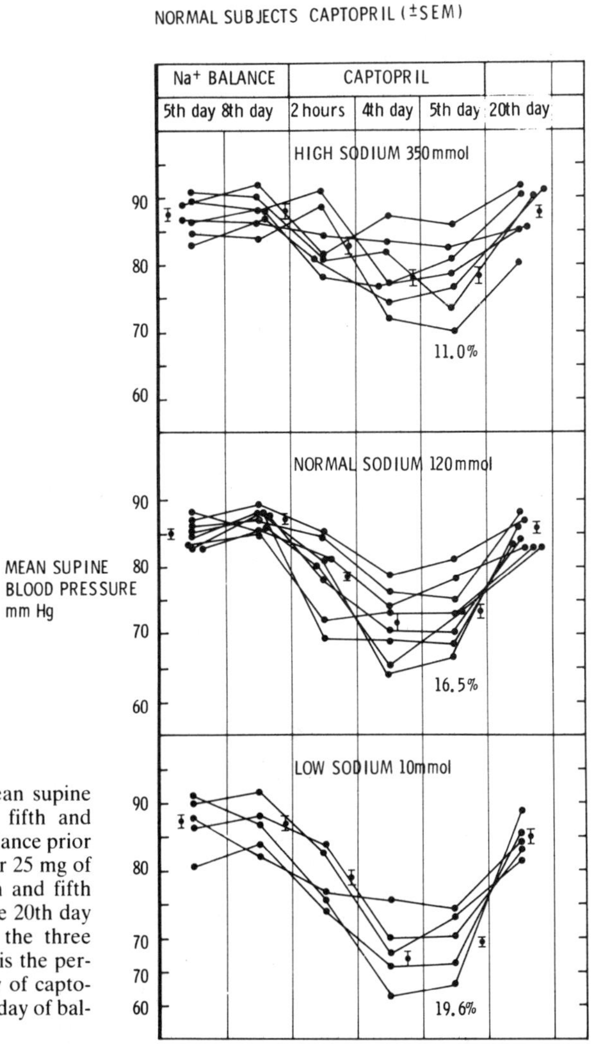

Fig. 2. Changes in mean supine blood pressure on the fifth and eighth day of sodium balance prior to captopril, at 2 hr after 25 mg of captopril, on the fourth and fifth days of captopril and the 20th day of sodium balance for the three sodium intakes. Shown is the percentage fall on fifth day of captopril compared to eighth day of balance prior to captopril.

All patients were seen by the same nurse in the blood pressure unit in the same room at the same time of day and on the same day of the week for weekly visits. Patients were instructed to take their tablets 2 to 3 hr before a visit and 1 hr before or 2 hr after a meal. All blood pressures were measured in the same manner as that used for normotensive subjects. Blood samples for PRA, plasma aldosterone, urea, electrolytes and creatinine concentrations were taken before active treatment, 2 hr after the first dose and at weekly intervals during treatment. Routine hematological and biochemical examinations were done before active treatment and during the second and fourth week of treatment. All hypertensive patients ate their normal diet and all studies were done

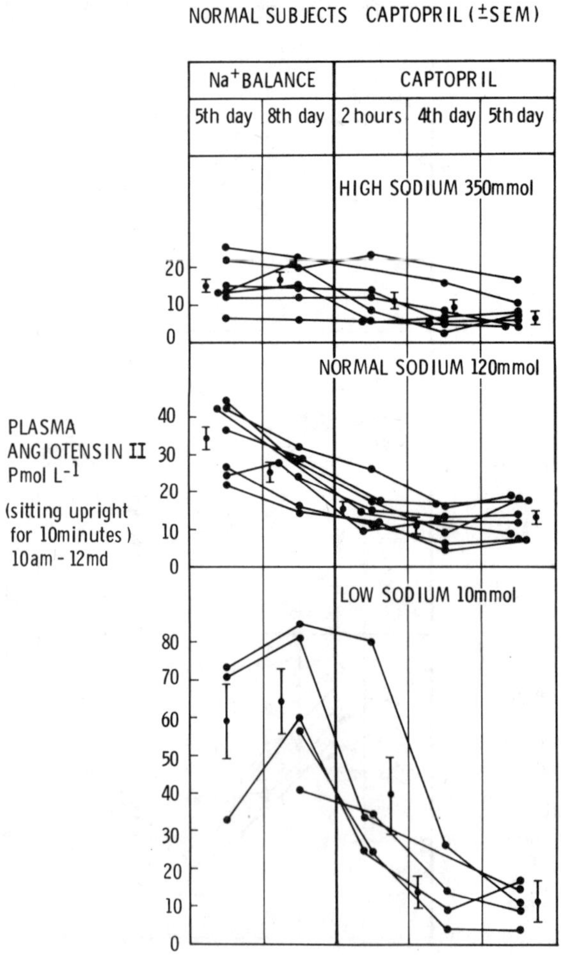

Fig. 3. Changes in plasma angiotensin II on the fifth and eighth days of sodium balance prior to captopril, 2 hr after 25 mg of captopril, and the fourth and fifth days of captopril.

as outpatients. Urinary collections for measurement of 24-hr sodium excretion were made before active treatment.

Statistical analysis for the hypertensive patients were performed using the University of London computer and the Northwestern University statistical package for social sciences. For the normotensive subjects, statistical analysis was done by Student's t-test and, for correlation, the method of least squares was used. Mean blood pressure was calculated as the diastolic pressure $+(\frac{1}{3} \times$ pulse pressure). Results are reported as the mean + the standard error of the mean.

RESULTS

Normotensive Subjects

Normal Sodium Intake (120 mmol/day)

Eight normotensive subjects were studied on a sodium intake of 120 mmol per day. Captopril caused a significant fall in blood pressure within 2 hr of the first dose ($P < 0.001$), and the decrement in pressure was greater by the fifth day (Figs. 1 and 2). The fall in standing, sitting and postexercise blood pressure was similar to that in supine blood pressure. Blood pressure returned to levels similar to the precaptopril values in the 7 days after stopping captopril (Fig. 1). The percentage fall in mean supine blood pressure with captopril by the fifth day on the normal sodium intake was 16.5%. Mean plasma angiotensin II before captopril was 25.3 ± 2.3 pmol/1. There was a significant fall at 2 hr in angiotensin II (Fig. 3, $P < 0.01$), and by the fifth day the mean level of angiotensin II was 13.3 ± 1.7 pmol/1. Plasma renin activity was 3.3 ± 0.5 ng h^{-1} ml^{-1} before captopril, rose within 2 hr of the first dose ($P < 0.001$) and by the fifth day, had reached a level of 41.4 ± 7.1 ng h^{-1} ml^{-1} (Fig. 4). Mean plasma aldosterone before captopril was 549 ± 64 pmol/l, falling within 2 hr of the first dose ($P < 0.001$), and by the fifth day reaching a mean level of 158 ± 14 pmol/l (Fig. 5). There was a mean loss of sodium during captopril of 142 mmol. Sodium balance did not come back to normal within 7 days of stopping captopril (Fig. 6). There was a fall in urinary aldosterone during captopril therapy (Fig. 6). All subjects lost weight during captopril treatment; by the fifth day, the mean weight loss was 1.15 kg. Mean plasma noradrenalin before captopril was 392 ± 39 pg/ml, and by the fifth day of captopril had risen to 712 ± 125 pg/ml ($P < 0.05$). There were no significant changes in pulse rate, urinary potassium excretion, blood urea, electrolytes and plasma creatinine during captopril treatment. Plasma potassium rose from 3.9 ± 0.1 mmol/l to 4.1 ± 0.1 mmol/l with captopril, but this increase was not significant.

High Sodium Intake (350 mmol/day)

Seven normotensive subjects were studied on the high sodium intake. Mean 24-hr urinary sodium excretion by the eighth day of the high sodium diet was 334 ± 21 mmol (Fig. 7). There was no significant

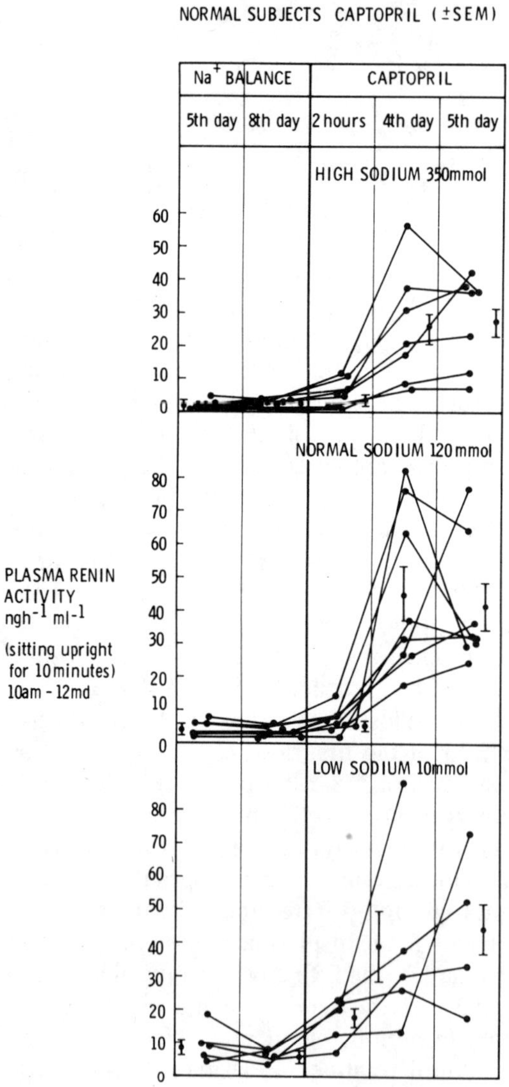

Fig. 4. Changes in plasma renin activity on the fifth and eighth days of sodium balance prior to captopril, 2 hr after 25 mg of captopril, and the fourth and fifth days of captopril.

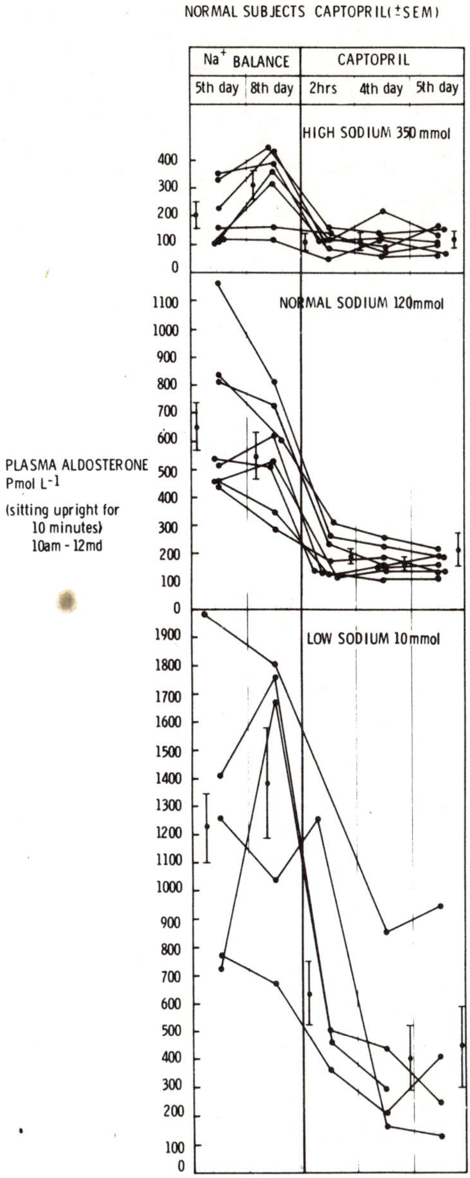

Fig. 5. Changes in plasma aldosterone on the fifth and eighth days of sodium balance prior to captopril, 2 hr after 25 mg of captopril, and the fourth and fifth days of captopril.

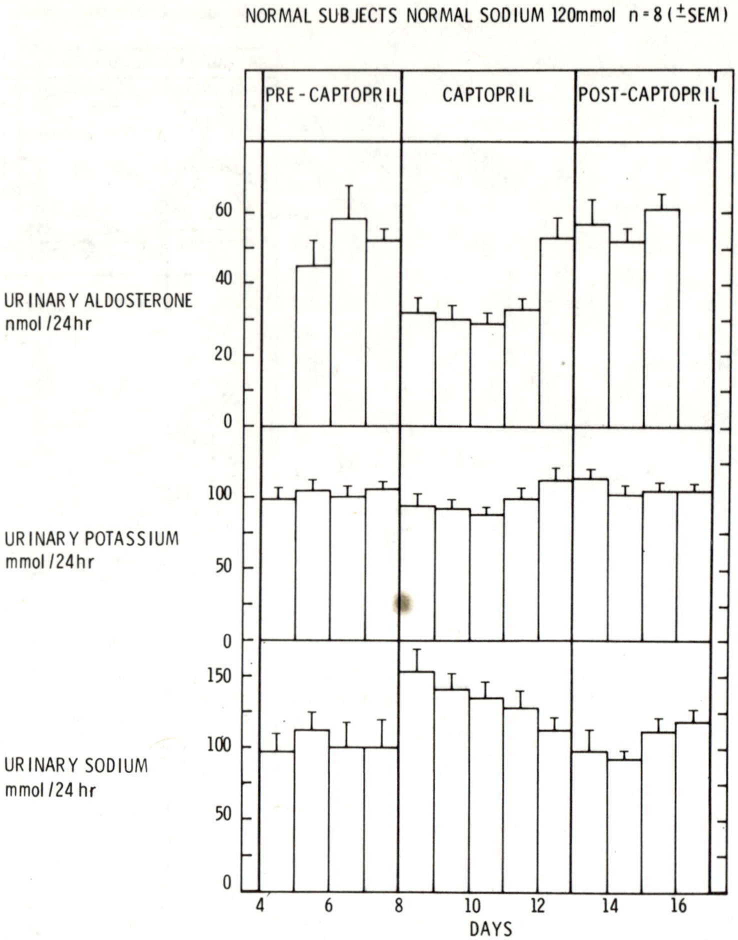

Fig. 6. Changes in 24-hr urinary aldosterone, potassium and sodium 4 days before captopril, during the 5 days of captopril and 4 days after captopril, during normal sodium intake.

change in weight with the high sodium diet. Plasma angiotensin II, renin activity and aldosterone had fallen as a result of the high sodium diet (Figs. 3, 4, 5). Captopril caused a significant fall in blood pressure within 2 hr ($P < 0.01$) (Fig. 2). This fall was less than that on the normal sodium diet but the difference was not significant. Percentage fall in mean supine blood pressure by the fifth day of captopril, as compared to the control, was 11%. Mean plasma angiotensin II before captopril was lower (as compared with normal salt intake) on the high sodium diet, 17.0 ± 2.6 pmol/l, and fell by the fifth day of captopril therapy

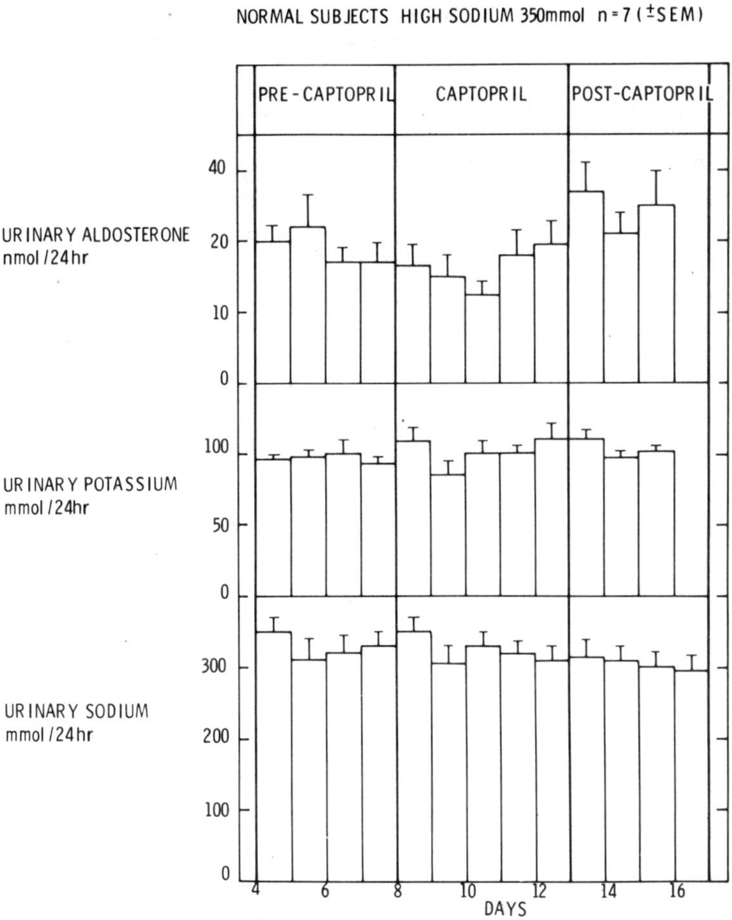

Fig. 7. Changes in 24-hr urinary aldosterone, potassium and sodium 4 days before captopril, during the 5 days of captopril and 4 days after captopril, during high sodium intake.

to 7.9 ± 3.2 pmol/l ($P < 0.001$) (Fig. 3). Plasma renin activity was also lower in this group as compared with that seen in the normal sodium group. The mean PRA before captopril was 2.7 ng h^{-1} ml^{-1}, rising by the fifth day of captopril to 27.9 ($P < 0.001$) (Fig. 4). Mean plasma aldosterone in subjects on the high sodium diet was also lower. The mean value for captopril was 318 ± 53 pmol/l, falling within 2 hr of the first dose ($P < 0.001$), and by the fifth day of captopril it was 115 ± 16 pmol/l (Fig. 5). There was no detectable change in sodium balance (Fig. 7). Urinary aldosterone showed a slight fall (Fig. 7). Plasma noradrenalin rose from a mean value of 361 ± 62 pg/ml to 562 ± 71 pg/ml by the fifth day of captopril treatment ($P < 0.05$). Plasma potassium rose from 3.7 ± 0.15 mmol/l to 4.0 ± 0.1 mmol/l during captopril therapy ($P < 0.01$).

Low Sodium Intake (10 mmol/day)

Five normotensive subjects were studied on a sodium intake of 10 mmol/day. By the eighth day of the low sodium diet, the 5 normotensive subjects had lost a mean of 150.6 mmol of sodium, and 1.5 ± 0.5 kg

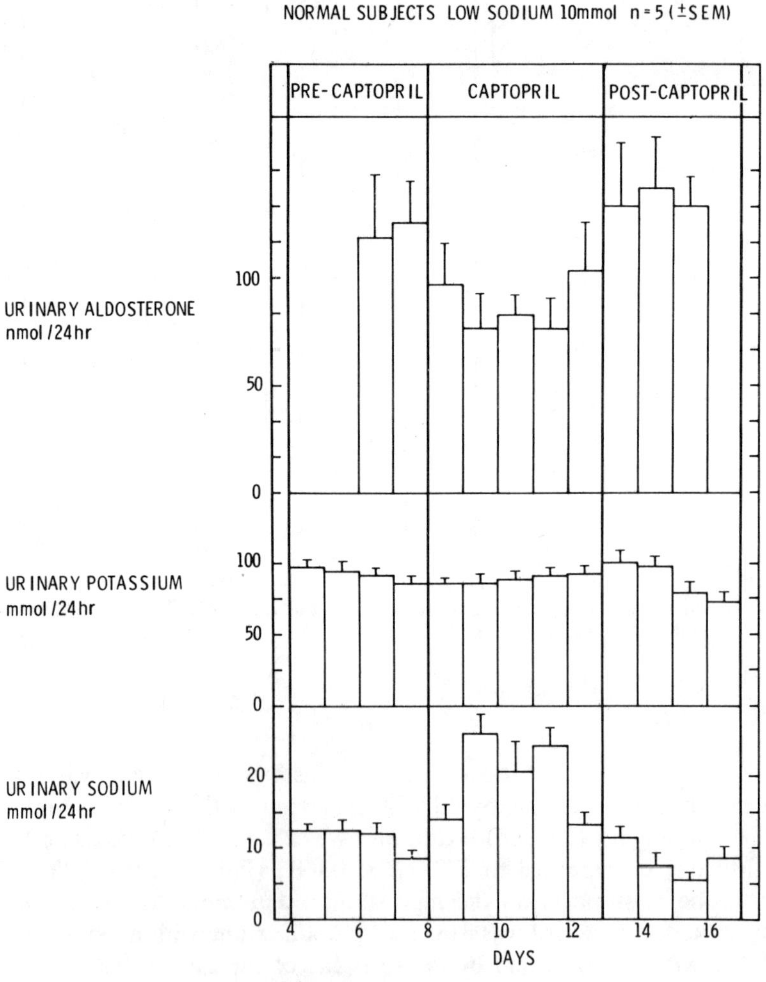

Fig. 8. Changes in 24-hr urinary aldosterone, potassium and sodium 4 days before captopril, during the 5 days of captopril and 4 days after captopril, during low sodium intake.

in weight. Mean 24-hr urinary sodium excretion was 8.6 ± 2.0 mmol. Plasma angiotensin II, renin activity and aldosterone had risen (Figs. 3, 4, 5). Blood pressure fell significantly within 2 hr of the first dose of captopril ($P < 0.001$) (Fig. 2). The fall in blood pressure by the fifth day was significantly greater than the fall seen in subjects on the high sodium intake ($P < 0.01$). The percentage fall in mean supine blood pressure was 19.6%. There was no evidence of postural hypotension. One subject felt light-headed on standing up suddenly on the fourth day of captopril therapy. Mean plasma angiotensin II on the low sodium

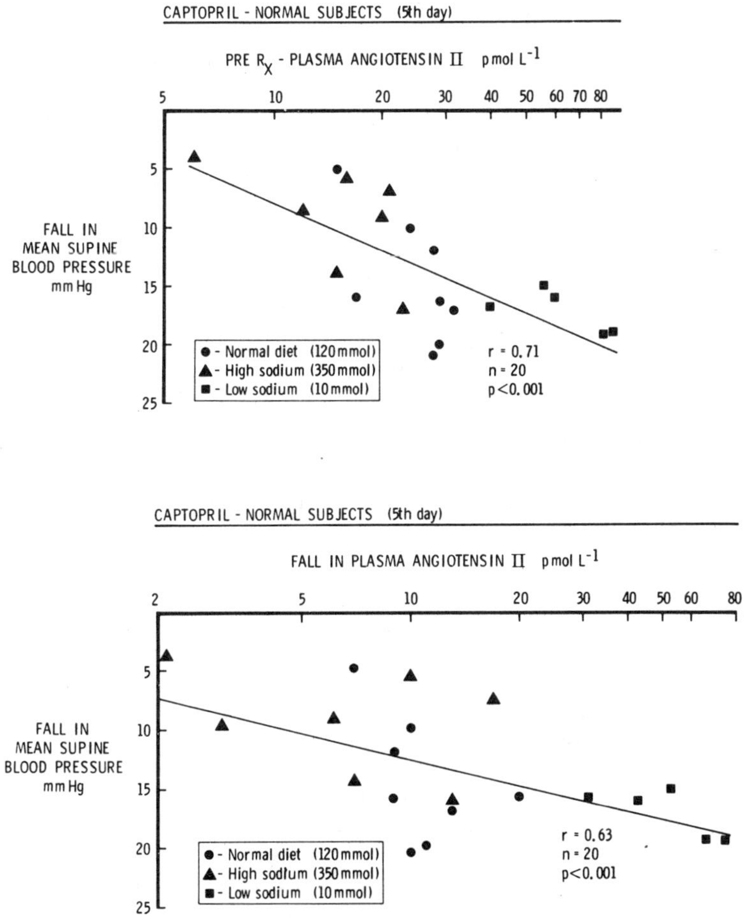

Fig. 9. Relationship between pretreatment plasma angiotensin II, fall in angiotensin II with captopril (fifth day) and the fall in mean supine blood pressure with captopril (fifth day).

diet was higher at 64.4 ± 9.3 pmol/l, falling to a mean of 11.0 ± 2.5 pmol/l by the fifth day ($P < 0.001$) (Fig. 3). Plasma renin activity was higher, with a mean value of 6.2 ± 0.9 before captopril rising to 44.6 ± 3.0 ng h^{-1} ml^{-1} by the fifth day of captopril (Fig. 4).

Plasma aldosterone was higher on the low sodium diet, with a mean value of 1395 ± 252 pmol/l before captopril, falling by the fifth day of captopril to 483 ± 173 pmol/l ($P < 0.001$) (Fig. 5). Mean loss of sodium during captopril was 43.6 mmol and urinary aldosterone fell (Fig. 8). Plasma noradrenalin rose from a mean value before captopril of 438 ± 65 to 679 ± 153 pg/ml by the fifth day of treatment. Plasma potassium rose from 3.9 ± 0.05 mmol/l to 4.0 ± 0.11 with captopril (N.S.).

Plasma angiotensin II fell to approximately the same values on captopril by the fifth day, irrespective of the diet or starting level of angiotensin II (Fig. 3). Plasma aldosterone, while showing a roughly equal percentage fall for each diet, showed much greater falls in absolute terms with captopril for the low sodium diet as compared to the normal diet and high sodium diet (Fig. 5). However, the absolute value of plasma aldosterone reached by the fifth day of captopril was higher on the low sodium diet, as compared to the normal diet, which was higher as compared to the high sodium diet (Fig. 5). There was a significant correlation between the fall in the blood pressure and the initial $\log_{10}$ plasma renin activity (r = 0.71, n – 20, $P < 0.001$), the initial $\log_{10}$ plasma angiotensin II and the fall $\log_{10}$ angiotensin II with captopril (Fig. 9). There was also a significant correlation between the decrement in plasma aldosterone that occurred with captopril and the initial plasma renin activity, initial angiotensin II and the fall in angiotensin II that occurred: (r = 0.66, r = 0.82, r = 0.78, respectively, n = 20, $P <$ 0.001). These correlations are not strictly statistically justified, as 5 of the same subjects were studied on the 3 different diets and the diets were not achieved by chance. Captopril was well tolerated in these normal subjects and no significant side effects were encountered in the 5 days of treatment. All subjects were able to go about their normal activities during the 5 days of treatment.

Patients with Essential Hypertension

The mean supine blood pressure in the 18 patients before the start of the trial, but after the initial assessment period of 1 to 2 mo, was 178 ± 4.5 mm Hg systolic and 113 + 2.9 mm Hg diastolic. Four weeks of observation in 6 patients and 4 weeks of treatment with placebo in the 12 others made no significant difference with respect to the initial or control mean blood pressure. Two hours after the first dose of captopril (25 mg) in the 15 patients who were studied, there was a significant reduction in blood pressure. This fall in blood pressure after the first dose was not significantly different from the average supine blood pressure at one week when the dose was 25 mg thrice daily, or during the second and third weeks of treatment at 50 mg, 100 mg, respectively. During the fourth week of treatment at 150 mg thrice daily, the mean supine blood pressure was significantly lower than 2 hr after the first dose of captopril (Fig. 10). Changes in sitting, standing and postexercise blood pressure showed no significant difference from the changes in supine blood pressure. The percentage fall in mean blood pressure on the maximum dose of captopril, as compared with control values, was 14.5%.

Plasma Renin Activity

Plasma renin activity before active treatment was within the normal range in 13 patients and below the normal range in 5. The mean 24-hr urinary sodium excretion in the 18 patients before receiving captopril was 150.6 mmol/l, range 90–198 mmol. Plasma renin activity rose two

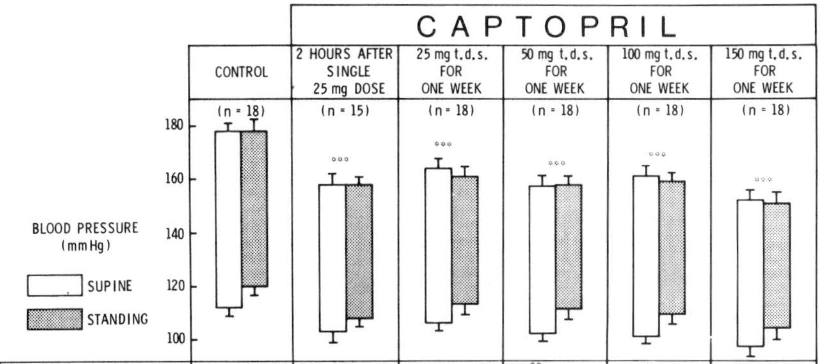

	CONTROL	2 HOURS AFTER SINGLE 25 mg DOSE	25 mg t.d.s. FOR ONE WEEK	50 mg t.d.s. FOR ONE WEEK	100 mg t.d.s. FOR ONE WEEK	150 mg t.d.s. FOR ONE WEEK
PLASMA RENIN ACTIVITY (ng/ml/hr)	0.97 (± .12)	1.83** (± .34)	2.20*** (± .36)	2.76** (± .63)	2.79** (± .57)	3.02* (± 3.6)
PLASMA ALDOSTERONE (pmol/L)	303 (± 31)	201*** (± 34)	225** (± 30)	227** (± 28)	198** (± 13)	207** (± 20)
PLASMA K+ (mmol/L)	3.91 (± .07)	3.91 (± .12)	4.07 (± .11)	4.19** (± .09)	4.17** (± .08)	4.10* (± .07)
WEIGHT (kg)	76.4 (± 3.0)	76.3 (± 3.0)	75.8* (± 3.0)	75.5** (± 3.0)	75.1*** (± 3.1)	75.1*** (± 3.0)
PULSE RATE (Supine)	80 (± 2.5)	79 (± 2.7)	76 (± 2.1)	75 (± 2.4)	78 (± 3.6)	75* (± 2.3)

Fig. 10. Mean systolic and diastolic, supine and standing blood pressures, plasma renin activity, aldosterone, potassium, weight and pulse rate, before and during treatment with increasing doses of captopril. Significance of difference:

$*P < 0.05$, $**P < 0.01$, $***P < 0.0001$.

(Blood pressure, plasma renin activity and aldosterone were not significantly different with increasing doses of captopril compared to values two hours after a single 25 mg dose (n = 15), except for blood pressure at 150 mg t.i.d. (n = 15, $P < 0.01$).

hours after the first dose of captopril and did not change significantly from this value during further treatment in these 15 patients. Values in the table, apart from the 2-hr value, are shown for all 18 patients. The fall in blood pressure that occurred with captopril in the 4th week of treatment was significantly correlated with the logarithm of the pretreatment plasma renin activity: $r = 0.59$, $P < 0.005$, $n = 18$. This correlation was significant for the falls in supine, upright, sitting, and postexercise systolic/diastolic pressures, as well as the mean blood pressure 2 hr after the first dose and mean pressure in the second, third and fourth weeks. However, there was no significant correlation for the first week of treatment.

Plasma Aldosterone Concentration

Two hours after the first dose of captopril, the plasma aldosterone concentration fell significantly (Table on Fig. 10). No further significant change occurred with continued therapy. There was no significant correlation between the pretreatment aldosterone and the fall in blood pressure that occurred with treatment.

Plasma Potassium

Plasma potassium concentration rose significantly during the second week of treatment (Table on Fig. 10).

Weight

The mean weight had fallen significantly by the end of the first week of treatment, and this weight loss increased significantly in the third week. The mean weight in the third and fourth week of treatment was not significantly different (Table on Fig. 10).

There was no significant change in pulse rate with captopril, except during the fourth week when it fell significantly (Table on Fig. 10).

Other Blood Tests

Plasma sodium, plasma chloride, blood urea, plasma creatinine and urate concentrations showed no significant change in each week of treatment. However, plasma bicarbonate concentration showed a significant fall by the fourth week of captopril, 25.1 ± 0.42 to 24.2 ± 0.29 mmol/l, respectively, $P < 0.02$. Blood sugar, alanine transferase, bilirubin, alkaline phosphatase, calcium, phosphate, cholesterol and triglyceride concentrations showed no significant changes in the second and fourth weeks of treatment as compared with control values. The mean hemoglobin concentration showed a significant fall by the fourth week of captopril—i.e., from 14.6 ± 0.28 to 13.81 ± 0.29 g/dl, $P < 0.001$. This was associated with a significant decrease in packed cell volume. White cell count did not change significantly. No patient developed or had proteinuria during the trial.

Side Effects

Captopril was well tolerated in this short-term trial and no patient had to be withdrawn because of side effects. Two patients complained of headaches on the second, third, and fourth weeks of treatment; one patient had a transient urticarial rash on both hands and arms 2 wk after starting captopril. This rash disappeared within 3 days, with continuation of captopril therapy, and did not return. One patient lost taste on the fourth week of treatment, necessitating stopping captopril during the fifth week of treatment; taste was recovered within 2 wk of stopping captopril. No other side effects were recorded.

Subsequent to the completion of this trial, a further 18 patients have been treated with captopril in a similar fashion. A comparison has been made between the 36 essential hypertensives, who had completed 4

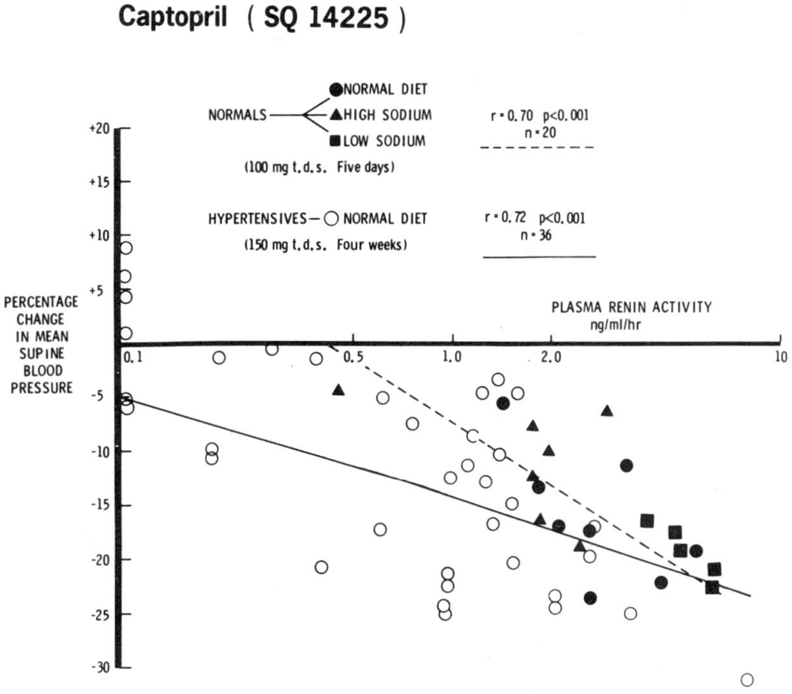

Fig. 11. Percentage change in mean supine blood pressure with captopril versus log initial plasma renin activity. ●Normotensive normal diet. ▲Normotensive high sodium. ■Normotensive low sodium diet (captopril given for 5 days). ○Hypertensives normal diet (captopril given for 4 wk).

weeks of captopril therapy at a dose of 150 mg 3 times a day, and the normotensive subjects, who were treated for 5 days only up to 100 mg 3 times a day (Fig. 11). The two correlation lines for the fall in blood pressure in percentage terms against the $\log_{10}$ of the plasma renin activity for normotensive and hypertensive subjects were not significantly different, but the residual variance was such that they are not strictly comparable. This is partly due to the higher plasma renin activity in the normal subjects.

DISCUSSION

Our results show that captopril caused a fall in blood pressure in normotensive subjects irrespective of sodium intake. The fall in blood pressure was related to the initial level of angiotensin II, plasma renin activity and the fall in angiotensin II that occurs with captopril. There is controversy about how captopril lowers blood pressure. It is clearly a potent inhibitor of converting enzyme, as judged by the fall in angiotensin II, the rise in plasma renin activity and the fall in plasma aldosterone that occurred. The angiotensin converting enzyme is also partially responsible for the breakdown of bradykinin. Inhibition of this enzyme could lead to an accumulation of bradykinin, a potential vasodilator. However, normotensive subjects and hypertensive patients did not have symptoms of raised bradykinin concentration, and recent work has shown that there is not an increase in plasma bradykinin with captopril (Matthews et al., 1979) or with teprotide (SQ 20,881), the parenteral converting enzyme inhibitor (Hulthen and Hokfelt, 1978). Significant correlations have also been shown between the fall in angiotensin II and the blood pressure fall in captopril in patients with hypertension (Fagard et al., 1979 and Atkinson et al., 1979) and salt depleted dogs (Morton et al., 1979). Angiotensin II has many other actions besides those of direct vasoconstriction of the arterioles and these may also contribute to the fall in blood pressure that occurs with captopril. Our results, showing a close relationship in these normal subjects between the fall in blood pressure with captopril and the initial plasma renin activity, initial plasma angiotensin II and the fall in angiotensin II, would suggest that perhaps part of the mechanism whereby captopril lowers blood pressure is through reduction of angiotensin II. If this is correct, our results indicate that the renin-angiotensin-aldosterone system is a normal mechanism for maintaining blood pressure. The degree of maintenance of the blood pressure would appear to these normal subjects to depend on the level of angiotensin II which, in its turn, will be dependent on the sodium intake.

Captopril also caused a fall in plasma aldosterone in all normal subjects on all three diets. This indicates that angiotensin II plays an important role in aldosterone secretion, irrespective of sodium intake, in normal man. Nevertheless, the plasma aldosterone level on captopril was still related to the dietary intake of sodium. Plasma aldosterone levels were highest on the low sodium diet, illustrating that other factors as well as changes in angiotensin II influence aldosterone secretion with changes in sodium intake. Captopril also caused in these normal subjects an increase in sodium excretion on the normal diet and the low sodium diet, but there was no detectable change on the high sodium diet. This increase in sodium excretion with captopril could be directly due to the fall in angiotensin II, or the fall in aldosterone, or some other factor, such as increased bradykinin or some other vasodilator in the kidney. The direct changes in sodium balance provide circumstantial evidence that the renin-angiotensin-aldosterone system may be important in the control of sodium balance in normal subjects on a low and normal sodium intake, but less important on a high sodium intake. The rise in plasma noradrenalin that occurred with captopril on all three diets is difficult to interpret, particularly in view of the fact that there was no significant change in pulse rate with captopril. It could indicate an increase in sympathetic tone with angiotensin converting enzyme blockade, perhaps related to the fall in blood pressure that occurred.

The results in hypertensive patients show that captopril is an effective antihypertensive drug in unselected patients with moderate essential hypertension. The fact that the main fall in blood pressure with captopril occurred two hours after the first 25 mg dose suggests that increasing the dose of captopril does not greatly increase its blood pressure lowering effect, although it may increase the duration of action. In this trial, we did not study the duration of action of captopril, but the results do indicate that a randomized trial of different doses of captopril is necessary, as well as an investigation of the duration of action of captopril when given long-term.

When comparing the changes with captopril in hypertensive patients and normotensive subjects several differences must be borne in mind: the normal subjects were much younger and were all male; and they did not receive captopril in the same way as hypertensives or for the same length of time. Nevertheless, the percentage fall in blood pressure that occurred with captopril in normal subjects was quite similar to that in patients with essential hypertension for similar values of plasma renin activity. This would suggest that, whatever the mechanism of action of captopril, it has the same blood pressure lowering effect in normotensive subjects and hypertensive patients for a given plasma renin activity. If part of the mechanism of action of captopril is through the lowering of plasma angiotensin II, then our results necessarily imply that the renin-angiotensin system plays no greater role in patients with essential hypertension in the maintenance of blood pressure than in normotensive subjects for a given value of plasma renin activity or angiotensin II (MacGregor et al., 1979). The renin-angiotensin-aldosterone system can, therefore, only be the direct cause of the high blood pressure if the level of angiotensin II is inappropriately high for the sodium balance, or there is an increase in the sensitivity to the various actions of angiotensin II.

ACKNOWLEDGMENTS

We are very grateful to Professor P. Sever, St. Mary's Hospital Medical School, London for the measurements of plasma noradrenalin.

REFERENCES

Anderson, G.H., Streeten, D.H.P., and Dalakos, T.H.G. 1977. Pressor response to 1-sar-8-ala-angiotensin II (saralasin) in hypertensive subjects. Circulation 40:243–250

Atkinson, A.B., Brown, J.J., Fraser, R., Leckie, B., Lever, A.F., Morton, J.J., and Robertson, J.I.S. 1979. Captopril in hypertension with renal artery stenosis and in intractable hypertension; acute and chronic changes in circulating concentrations of renin, angiotensins I and II and aldosterone, and in body composition. Clin Sci 57:139s–143s

Case, D.B., Wallace, J.M., Keim, H.J., Weber, M.A., Drayer, J.I.M., White, R.P., Sealey, J.E., and Laragh, J.H. 1976. Estimating renin participation in hypertension: superiority of converting enzyme inhibitor over saralasin. Am J Med 790–796

Case, D.B., Wallace, J.M., Keim, H.G., Weber, M.A., Sealey, J.E., and Laragh, J.H. 1977. Possible role of renin in hypertension as suggested by renin-sodium profiling and inhibition of converting enzyme. New Engl J Med 296:641–646

Dusterdieck, G., and McElwee, G. 1971. Estimation of angiotensin II concentration in human plasma by radioimmunoassay. Some applications to physiological and clinical states. Eur J Clin Invest 2:32–38

Fagard, P., Amery, A., Lionen, P., and Reybrouck, T. 1979. Haemodynamic effects of oral converting enzyme inhibitor captopril in hypertensive patients: comparison with saralasin. Clin Sci 57:131s–134s

Ferguson, R.K., Turini, G.A., Brunner, H.R., Gavras, H., and McKinstry, D.N. 1977. A specific orally active inhibitor of angiotensin-converting enzyme in man. Lancet I:775–778

Gavras, H., Brunner, H.R., Turini, G.A., Kershaw, G.R., Tifft, C.P., Cuttelod, S., Gavras, I. Vukovich, R.A., and McKinstry, D.N., 1978. Antihypertensive effect of the oral angiotensin converting enzyme inhibitor SQ 14,225 in man. New Engl J Med 298:991–995

George, C.F., Lewis, P.J., and Petrie, A. 1975. Clinical experience with use of ultrasound sphygomomanometer. Brit Heart J 37:804–807

Haber, E. 1976. The role of renin in normal and pathological cardiovascular homeostasis. Circulation 54:849–861

Hollenberg, N.K., William, G.H., Burger, B.M., Ishikawa, I., and Adams, D.F. 1976. Blockade and stimulation of renal, adrenal, and vascular angiotensin II receptors with 1-sar-8-angiotensin II in normal man. J Clin Invest 57:39–46

Hulthen, U.L., and Hokfelt, B. 1978. The effect of the converting enzyme inhibitor SQ 20,881 on kinins, renin-angiotensin-aldosterone and catecholamines in relation to blood pressure in hypertensive patients. Acta Medica Scandinavia 204:497

James, V.H.T., and Wilson, G.A. 1976. Determination of aldosterone in biological fluids. In *Methodological Developments in Biochemistry,* ed. E. Reid, Vol. 5, 149–158. North Holland: Elsevier

MacGregor, G.A., and Dawes, P.M. 1976. Agonist and antagonist effects of 1-sar-8-ala-angiotensin II in salt loaded and salt depleted man. Brit J Clin Pharm 3:483–487

MacGregor, G.A., Markandu, N.D., and Roulston, J.E. 1979. Does the renin-angiotensin system maintain blood pressure in both hypertensive and normotensive subjects? A comparison of propranolol, saralasin and captopril. Clin Sci 57:145s–148s

Matthews, P.G., McGrath, B.P., and Johnston, C.I. 1979. Hormonal changes with long-term converting-enzyme inhibition by captopril in essential hypertension. Clin Sci 57:135s–138s

Morton, J.J., Brown, J.J., Casal-Stenzel, J., Lever, A.F., Robertson, J.I.S., and Tree, M. 1979. Changes in the blood levels of angiotensin I, II and III in dogs infused at various doses with the converting enzyme inhibitor SQ 14,225: effect of inhibitor on blood pressure. Clin Sci 56:4p

Ondetti, M.A., Rubin, B., and Cushman, D.W. 1977. Design of specific inhibitors of angiotensin-converting enzyme: new class of orally active antihypertensive agents. Science 196:441–444

Posternak, L., Brunner, H.R., Gavras, H., and Brunner, D.B. 1977. Angiotensin II blockade in normal man: Interaction of renin and sodium in maintaining blood pressure. Kidney Int'l 2:197–203

Roulston, J.E., and MacGregor, G.A. 1978. Measurement of plasma renin activity by radioimmunoassay after prolonged cold storage. Clin Chim Acta 88:45–48

Clinical Pharmacological Studies with Captopril

Hans R. Brunner, Gustave A. Turini, Bernard Waeber and Haralambos Gavras
Division of Nephrology and Hypertension
University of Lausanne
Lausanne, Switzerland
Thorndike Memorial Laboratories
Boston City Hospital
Boston, Massachusetts 02118

INTRODUCTION

The function of a hormone can best be demonstrated by removal of the organ that produces it. In the case of angiotensin II, removal of the producing organ is not possible since angiotensin is merely the final product of renin secretion by the juxtaglomerular cells of the kidney. Removing the kidney is not an ideal method to examine the vasopressor hormone, angiotensin II, since it inevitably entails other modifications that are not directly related to elimination of circulating angiotensin II. The development of inhibitors of angiotensin converting enzyme that prevent the transformation of angiotensin I to angiotensin II has provided a new tool to study the role of angiotensin II while preserving the kidney intact.

The first substance found to specifically inhibit angiotensin converting enzyme was a component of a viper venom (Ferreira, 1965). Several peptides exercising similar action have subsequently been synthetized (Ondetti et al., 1971). Unfortunately, the therapeutic value of such peptides was limited, since they had to be administered parenterally. Finally, in 1976, the synthesis of a nonpeptidic substance with a chemical configuration permitting it to specifically attach to the active site of angiotensin converting enzyme led to the creation of several molecules. D-3-mercapto-2-methylpropanoyl-L-proline, captopril (SQ 14,225), proved to be the most active of these compounds, and was shown to be a competitive, orally active inhibitor of converting enzyme (Ondetti et al., 1977).

CLINICAL PHARMACOLOGY (NORMALS)

The pharmacologic properties of captopril were tested for the first time in 14 normotensive volunteers in December 1976 (Ferguson et al., 1977). A dose-response curve for pressor responses to intravenous angiotensin I was established for each subject prior to the first oral administration of captopril. After ingestion of a single dose of captopril, the maximum dose of angiotensin I that had been used previously was reinjected several times over the next several hours. Increasing doses of captopril (1, 2.5, 5, 10 and 20 mg) were tested in the same way. Both the degree and the duration of converting enzyme inhibition appeared to be dose-dependent. Thus, 1 mg of captopril produced only partial inhibition whereas higher doses afforded almost complete blockade of angiotensin I induced pressor responses, confirming the efficacy of this new inhibitor in man. Twenty milligrams of captopril was sufficient to completely block the response to angiotensin I (up to 160 or even 320 ng/kg of body weight) from 15 min to over 2 hr. Consequently, increasing the dose of captopril above 20 mg may only prolong the duration of the blockade, and this has to be kept in mind when treating hypertensive patients.

CLINICAL PHARMACOLOGY (PATIENTS)

Figure 1 illustrates the time course of the blood pressure lowering effect of three different single doses of captopril observed in 6 hypertensive patients on three consecutive days. With 25 mg, mean blood pressure decreased from a control of 130 ± 8 to 125 ± 11 mm Hg at 15 min and 111 ± 7 mm Hg at 60 min after administration of captopril. It returned to baseline levels of 127 ± 8 mm Hg at 360 min. Blood pressure reduction induced by 100 mg showed a similar pattern, whereas 200 mg clearly had a prolonged antihypertensive effect. At 6 hr after a 200 mg dose, blood pressure was still reduced by 9 ± 4%. Thus, increasing doses did not produce a greater blood pressure fall but prolonged its duration. Indeed, blood pressure reduction expressed in percent appears even slightly smaller with 200 mg as compared with 25 mg, but this is probably due to the lower starting blood pressure of 121 ± 6 mm Hg as opposed to 130 ± 8 mm Hg (Brunner et al., 1979).

The blood pressure reduction induced by captopril in hypertensive patients is usually rapid; the lowest blood pressures are usually reached within 2 hr following administration of the drug. Early studies in a group of 14 patients suggested that the blood pressure drop obtained during the first 2 hr predicted very well the antihypertensive effect obtained when captopril was administered for a longer period of time. Thus, in this group of patients, the blood pressures obtained 2 hr after the first administration of captopril were the same as the mean blood pressure obtained after 4 to 6 days of treatment (Brunner et al., 1979).

Converting enzyme inhibition by captopril reduces blood pressure without change in heart rate. Furthermore, there is no orthostatic syndrome induced by captopril. When 12 patients, treated for at least 1 mo with captopril, were changed from a supine to upright position, their systolic blood pressure remained unchanged at 135 ± 6 mm Hg, whereas their diastolic blood pressure increased slightly from 85 ± 2 to 93 ± 3 mm Hg. This blood pressure behavior, together with the slight increase in pulse rate when assuming the upright position, parallels closely what one might expect when untreated normotensive volunteers assume the upright position (Brunner et al., 1979).

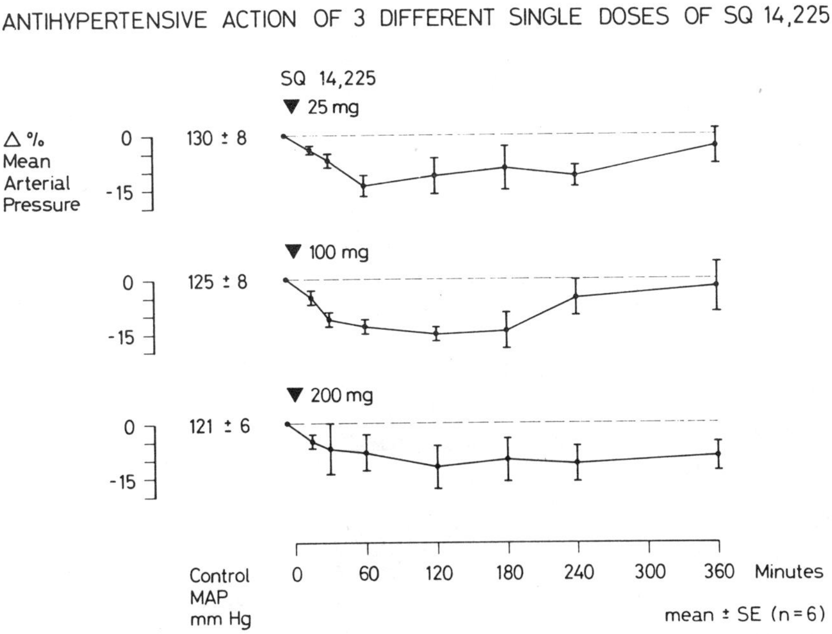

Fig. 1. Magnitude and duration of antihypertensive effect obtained with 3 different doses of captopril administered on 3 subsequent days. Note that starting mean arterial pressure (MAP) is different each time. From Brunner et al., (1979).

No gain in body weight was observed during blood pressure reduction by captopril. In 8 patients in whom a metabolic study was done, weight remained constant at 63.5 ± 4.5 kg during the first 3 days of treatment, while blood pressure fell from 172/110 ± 9/5 to 144/96 ± 6/4 mm Hg (Fig. 2). These results are corroborated by measurements of 24 hr urinary sodium excretion which, rather than decreasing as one might expect with the blood pressure reduction, tended to increase from 74 ± 15 on the last control day to 97 ± 18 mEq per 24 hr on the third day of treatment. During the same period, 24 hr urinary excretion of aldosterone fell from 13.6 ± 3 to 5.3 ± 2 μg per 24 hr. Twenty-four hour urinary excretion of potassium decreased slightly from 47 ± 6 to 43

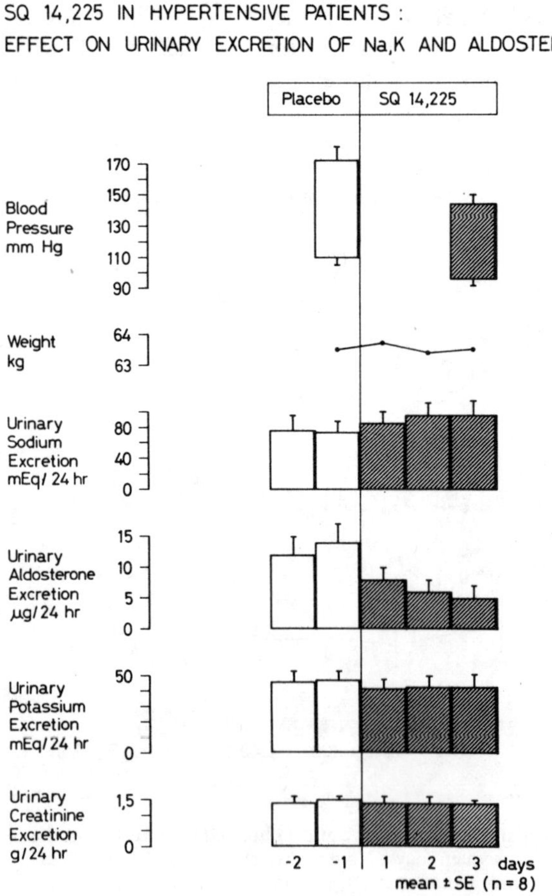

Fig. 2. Metabolic studies conducted in 8 patients during the placebo phase and the first 3 days of treatment with captopril. From Brunner et al., (1979).

± 8 mEq per 24 hr while creatinine excretion remained unchanged (Brunner et al., 1979). The lack of sodium retention may be explained by the reduced aldosterone levels. Moreover, antagonizing intrarenal effects of the renin system may also counteract sodium retention. Mimran and coworkers (1979) have recently observed that captopril increases renal blood flow and simultaneously lowers filtration fraction. Such findings suggest that captopril indeed diminishes constriction of the efferent arterioles of the glomeruli. Because sodium depletion has been shown to have a potentiating effect on blood pressure reduction induced by angiotensin blockade (Brunner et al., 1974; Gavras et al., 1974; Gavras et al., 1976), the lack of sodium retention can be expected to enhance the therapeutic efficacy of captopril.

EFFECT ON PRESSOR HORMONES

Figure 3a illustrates the behavior of plasma catecholamine levels during converting enzyme inhibition by captopril in 12 patients. While blood pressure decreased significantly, neither heart rate nor plasma norepinephrine or epinephrine changed significantly. The right panel shows 12 patients under captopril treatment. Seven of them had essential hypertension and 5 had hypertension associated with chronic renal failure. In hospital, angiotensin converting enzyme inhibition for 4 to 6 days induced a drop in blood pressure from 182/115 ± 6/4 to 150/96 ± 6/4 mm Hg. This blood pressure reduction was associated with a marked increase in plasma renin activity, probably due to the absence of a negative feedback normally exerted by angiotensin II on renin release. Despite this increase in renin, plasma aldosterone dropped significantly, which again reflects the blockade of angiotensin II production. Plasma converting enzyme activity was greatly reduced by the administration of captopril.

INTERACTION WITH DIURETICS

Figure 3b illustrates some other important aspects of converting enzyme inhibition as a therapeutic modality, i.e., its interaction with changes in sodium balance. All 12 patients needed a diuretic to normalize their arterial pressure in addition to captopril. During previous treatments consisting of at least beta-blockers and diuretics for all patients, their arterial pressure was 178/115 ± 6/2 mm Hg. When the diuretics were

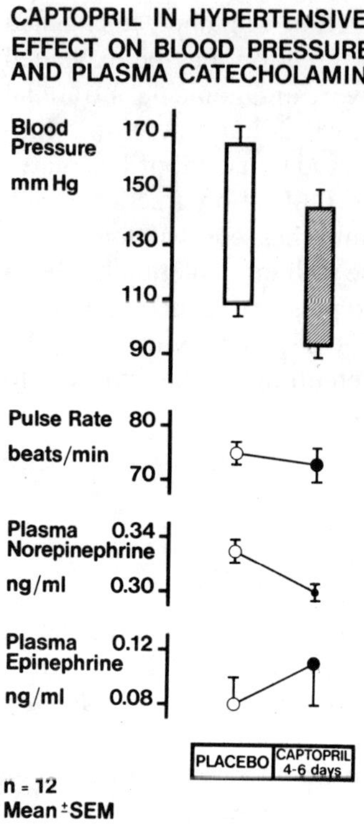

Fig. 3a. Effect of captopril on blood pressure, pulse rate, and plasma catecholamines.

added to captopril, the additional sodium reduction potentiated the hypotensive effect of the converting enzyme inhibition and arterial pressure dropped to 132/88 ± 5/3 mm Hg. The doses of diuretics used with captopril were less than or equal to those used in association with the beta-blockers previously, when arterial pressure was poorly controlled.

The sodium reduction produced in these patients by diuretics led to a further increase in renin that, at least theoretically, could have attenuated the hypotensive effect. Converting enzyme inhibition renders this compensatory response of the renin-angiotensin system ineffective and allows maximum benefit from sodium depletion. Moreover, renin-angiotensin blockade makes it possible to quantitate precisely the amount of sodium that needs to be subtracted in order to normalize pressure. After neutralizing the compensatory reaction of the renin-angiotensin system in response to sodium subtraction, the sodium factor becomes the main determinant of arterial pressure as was previously observed in binephrectomized patients (Onesti et al., 1975).

EFFECT OF CAPTOPRIL AND DIURETICS ON PLASMA RENIN, ALDOSTERONE AND ANGIOTENSIN CONVERTING ENZYME ACTIVITY

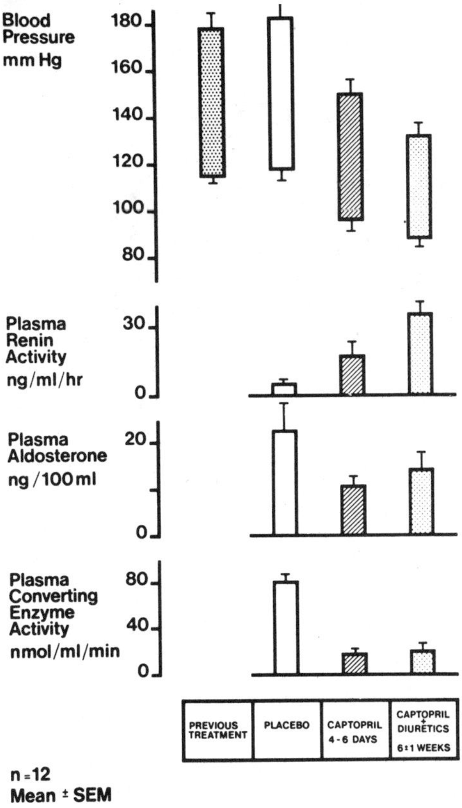

Fig. 3b. Effect of captopril alone and in combination with diuretics on blood pressure, plasma renin activity, plasma aldosterone and angiotensin-converting enzyme activity. Blood pressure on previous therapy is depicted as a basis for the evaluation of the efficacy of captopril and diuretics.

THERAPEUTIC EFFICACY

Chronic converting enzyme inhibition, whether alone or associated with diuretics, has been used successfully for the long-term treatment of hypertensive patients. Thus, in our initial group of 16 patients who have been treated over a period of one year, blood pressure, which initially had been at 176/113 ± 6/4, was still at 129/88 ± 4/3 mm Hg after one year. Accordingly, no escape from the antihypertensive effect of captopril has been observed, though increased salt ingestion or decreasing renal capacity to excrete sodium may mimic the development of treatment resistance. These factors, however, can always be corrected by increasing the amount of diuretics administered or reducing sodium intake.

SIDE EFFECTS

In a majority of the patients, treatment with captopril is tolerated extremely well. Unfortunately, several side effects have become apparent in some patients. Skin rashes occur in approximately 10% of all patients. In many, this phenomenon is transient and reducing the dose without interrupting the treatment is sufficient to eliminate the problem. Nevertheless, in some patients, treatment has to be stopped because the rash persists. Loss of taste is another side effect that has been seen in about 5% of patients. Again, in some patients this problem disappears spontaneously while the treatment is continued, but in others captopril has to be discontinued. Proteinuria has been observed by some investigators, but this probably rarely presents a major problem. In contrast, leukopenia or even agranulocytosis has to be taken much more seriously. In about 0.5% of the patients treated so far, decreased white cell counts have been observed during the first 3 mo of captopril therapy. While a cause-effect relationship between captopril and leukopenia has not been established beyond any doubt because of the complex medical problems of the involved patients and the use of strong concomitant therapies, it nevertheless cannot be discarded at the present time. Recently, a few cases of pemphigus have also been observed in patients treated with captopril. In general, the pattern of side effects observed resembles very closely that reported for patients treated with penicillamine. This may not be surprising, since both drugs share the characteristic of a mercapto group. Even if there exist side effects that have been observed with captopril, it appears that all are probably related to the compound per se and not to converting enzyme inhibition.

PLASMA RENIN RELATIONSHIP

Oral administration of captopril has made it possible to decrease arterial pressure in a majority of hypertensive patients. To date, most investigators using this renin-angiotensin system blocker have observed an antihypertensive effect not only when plasma renin activity was normal or elevated, but also when it was relatively low (Brunner et al., 1979; Bravo and Tarazi, 1979; Case et al., 1978; Gavras et al., 1978). Figure 4 shows the relationship between the decrease in diastolic pressure measured one hour after the first dose of captopril and control plasma renin activity in 32 hypertensive patients. Five patients had renovascular hypertension, 10 essential hypertension and 2 primary hyperaldosteronism. Elevated blood pressure was associated with chronic renal failure in 15 of the 32 patients; 7 of the 15 had serum creatinine levels between 1.5 and 7.4 mg% and 8 were under chronic hemodialysis. All of these patients were given a constant diet containing 100 mEq of sodium and 60 to 80 mEq of potassium per day throughout the hospitalization. Immediately preceding captopril administration, a placebo was admin-

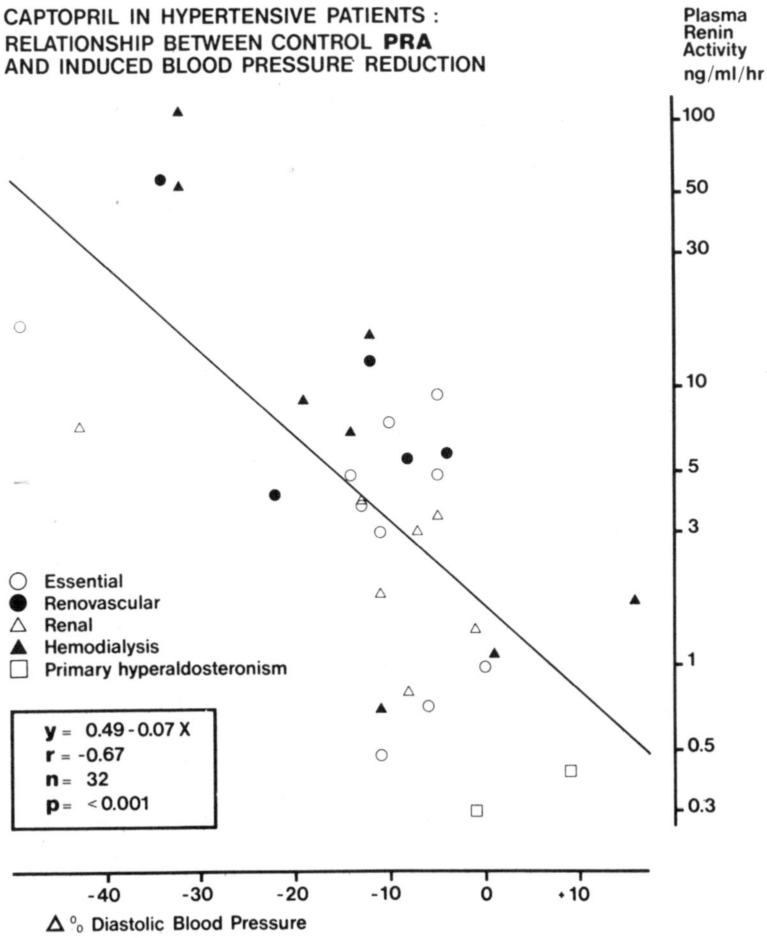

Fig. 4. Correlation between control plasma renin activity and induced diastolic blood pressure reduction 1 hr following the first dose of captopril in 32 patients with different types of hypertension. From Brunner et al, (1980).

istered for 3 days. The blood pressure response to converting enzyme inhibition correlated significantly with control plasma renin activity ($r = 0.67$, $P < 0.001$). These results suggest that the renin angiotensin system also plays a role when spontaneous renin secretion is "normal" or "low." If the effect of captopril is due exclusively to blockade of angiotensin II production, even "low" renin levels may be capable of actively participating in controlling arterial pressure, e.g., by way of increased sensitivity of the arteriolar receptors to angiotensin II. Nevertheless, it is still possible that a renin independent mechanism is also involved in the antihypertensive action of captopril. It should, however, be pointed out that arterial pressure underwent no change after converting enzyme inhibition in the 2 patients presenting with primary hyperaldosteronism and the lowest renin levels.

The correlation between control plasma renin activity and the blood pressure drop induced by converting enzyme inhibition was further analyzed in 16 patients. Indeed, the change in blood pressure observed during the first 2 hr following captopril administration correlated significantly with the control plasma renin activity ($r = -0.68$, $P < 0.01$). However, no significant relationship was found between the control plasma renin activity and the blood pressure change observed in the

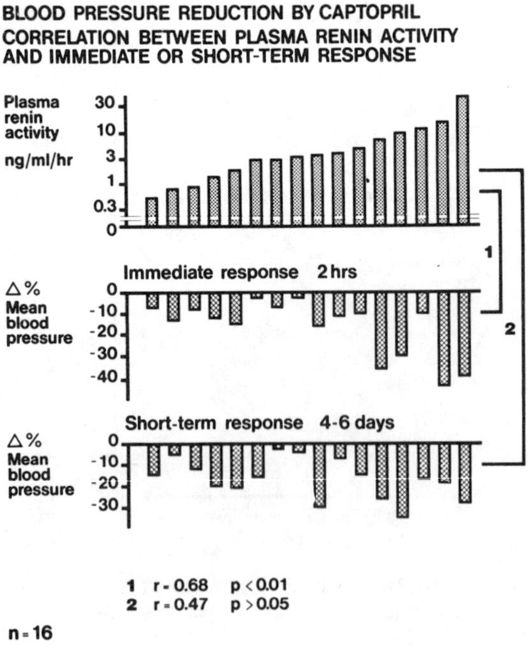

Fig. 5. Relationship between control plasma renin activity of 16 hypertensive patients and the captopril-induced reduction in mean blood pressure measured 2 hr and 4 to 6 days after starting converting enzyme inhibition.

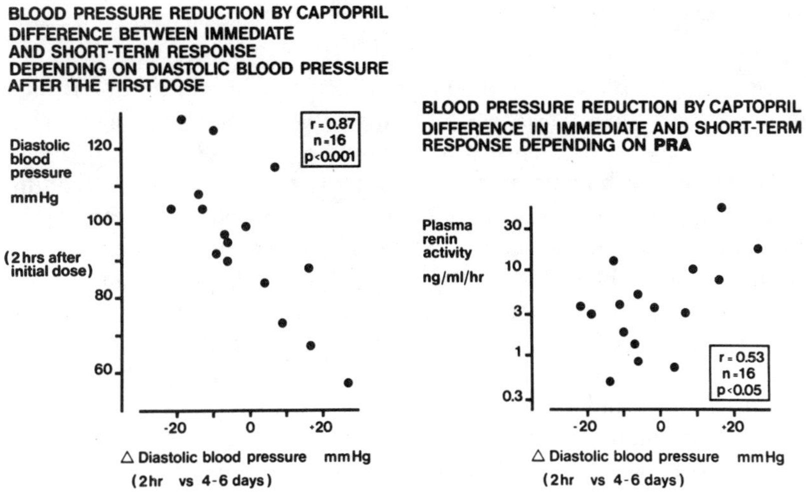

Fig. 6. Left panel: Relationship between the captopril-induced immediate (2 hr) diastolic blood pressure reduction and the changes in diastolic pressure which occurred during the following 4 to 6 days of converting enzyme blockade.
Right panel: Relationship between the control plasma renin activity (PRA) and the changes in diastolic blood pressure which occurred between 2 hr and 4 to 6 days following initiation of converting enzyme blockade.

same patients over the first 4 to 6 days of captopril treatment ($r = -0.47$, $P > 0.05$) (Fig. 5). What was the reason for this? Those patients having the greatest drop in pressure during the first 2 hr after captopril administration tended to slightly increase their blood pressure during the following days, whereas those who exhibited the slightest blood pressure reduction during the first 2 hr had further lowering of their blood pressure during the following days. In fact, there was a close correlation between diastolic pressure exhibited 2 hr after captopril administration and the blood pressure change during the following 4 to 6 days, such that the lower the diastolic blood pressure initially the greater the blood pressure increase during the following days ($r = -0.87$, $P < 0.001$). Not surprisingly, the lowest diastolic blood pressures immediately following captopril administration were observed in the patients with the highest renins. Therefore, the blood pressure change during the following days also correlated with the initial plasma renin activity ($r = 0.53$, $P < 0.05$) (Fig. 6).

BRADYKININ

There exists a potential problem of specificity when interpreting results obtained with angiotensin converting inhibitors. Indeed, angiotensin converting enzyme is identical with kininase II, which is a part of the physiological mechanism responsible for the inactivation of circulating bradykinin. The cardiovascular effects of this hormone are complex. As well as direct vasodilation, it induces catecholamine secretion by the adrenal medulla (Feldberg and Lewis, 1964) and releases prostaglandins from various organs (Ferreira et al., 1973; Giff et al., 1972). Thus, the antihypertensive effect of converting enzyme inhibitors could be heightened both by an accumulation of endogenous bradykinin due to converting enzyme inhibition and by way of enhanced actions of prostaglandins. In man, plasma bradykinin measurements obtained after inhibition of the renin angiotensin system following teprotide administration are contradictory. Both normal (Hulthen and Hokfelt, 1978) and elevated (Williams and Hollenberg, 1977) levels of bradykinin have been observed. This may be suggestive of methodological problems. With captopril, it was recently shown that plasma bradykinin levels remained unchanged after long-term converting enzyme inhibition in hypertensive patients (Johnston et al., 1979). Thus, it seems that, for the moment, the quantitation of this hormone is of limited use in evaluating its possible vasoactive role after converting enzyme inhibition; in addition, it is not known whether plasma bradykinin levels reflect the concentration of this peptide on the vascular receptors. In any event, there seems to be no correlation between the proportionate decrease in pressure induced by captopril in hypertensive patients and the degree to which plasma bradykinin levels increase. However, the fact that plasma kinins do not increase despite the administration of an angiotensin converting enzyme inhibitor is not incompatible with the fact that the converting enzyme is identical to kininase II. When purified from human lung tissue, this enzyme actually has a specific affinity for bradykinin 14 times less than that for angiotensin I (Nishimura et al., 1977). In other words, total inhibition of the degradation of bradykinin in this way should necessitate larger doses of teprotide or captopril than inhibition of angiotensin I conversion.

There are several clinical hypotheses against a major role for bradykinin in the antihypertensive effects of converting enzyme inhibitors. In normal subjects without sodium depletion and in hypertensive patients with primary hyperaldosteronism, captopril does not modify arterial pressure. Blockade of the renin-angiotensin system by captopril produces neither tachycardia nor an increase in plasma catecholamines despite the drop in pressure. No tendency toward orthostatic hypotension has been noted in patients chronically treated with this drug. These results are in contrast with the pulse acceleration and large pressure drops noted orthostatically in patients with hyperbradykininism (Streeten et al., 1972). It is also interesting that teprotide has not reduced arterial pressure in nephrectomized patients (Case et al., 1979). It should be pointed out, however, that recently one group reported a blood pressure drop in nephrectomized patients with captopril (Man in't Veld, 1980), although this observation has not been confirmed by another group of investigators (Leslie et al., 1980).

DOSAGE REGIMEN

Most investigators have administered captopril 3 to 4 times daily (usually not exceeding 450 mg/day), which would seem reasonable based on the rather short half-life of the drug. Nevertheless, for practical reasons, we have preferred to treat the majority of our patients with twice daily administration of captopril 50 to 200 mg. In order to ensure that the twice daily dose of captopril really does provide 24-hour control of arterial pressure, blood pressure was recorded in 17 outpatients throughout one day using a portable and semiautomatic gauge (portometer, Remler Corp., San Francisco). The resulting profile shows blood pressure values that are, on average, within the normal range from morning to night while the patients go about their usual activities (Fig. 7). Even

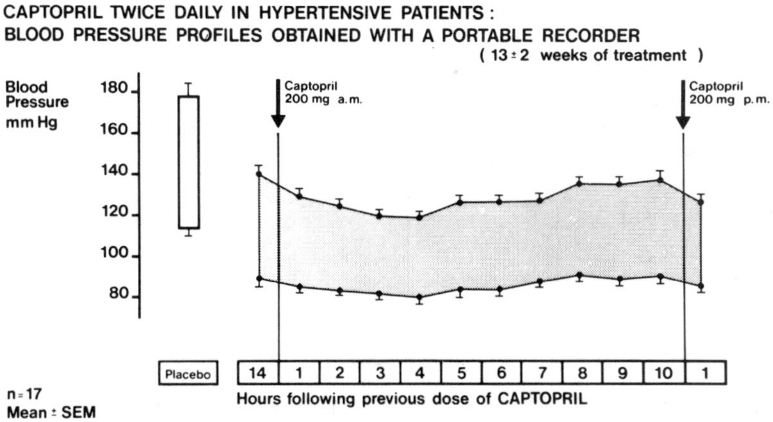

Fig. 7. Blood pressure profiles measured with a portable recorder in 17 ambulatory patients on long-term twice daily administration of captopril.

14 hr after the evening dose of captopril, arterial pressure averaged 140/89 ± 4/4 mm Hg. After the morning dose, pressure dropped a bit further, reaching 138/91 ± 4/4 mm Hg in the evening just before the second dose.

MECHANISM OF ACTION

These excellent results are surprising since continuous blockade of the renin-angiotensin system does not seem essential. In fact, Figure 8 shows that, in patients with normal renal function treated with captopril for several weeks, plasma converting enzyme activity regained its control value before the morning dose of captopril. One hour after ingestion of 200 mg of captopril, renin-angiotensin system blockade again became apparent through a decrease in converting enzyme activity comparable to that measured in the hospital 1 hr after the first administration of an

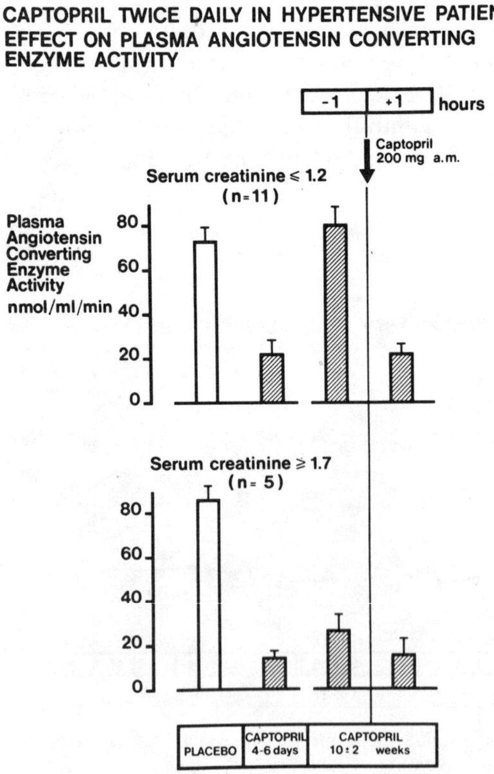

Fig. 8. Plasma converting-enzyme activity at initiation of captopril treatment and after long-term therapy with twice daily administration of the drug. From Waeber et al., (1980).

equivalent dose of captopril. In patients with chronic renal failure, however, plasma activity of the converting enzyme remained partially inhibited 14 hr after the last dose of captopril, probably due to renal retention of the drug. The fact that any modification of the activity of this enzyme in the plasma corresponds to an appropriate change in renin and plasma aldosterone levels (Waeber et al., 1980) suggests that these changes in plasma converting enzyme activity reflect the degree of angiotensin I to angiotensin II conversion taking place mainly in the pulmonary circulation. For the time being, the dichotomy between the degree of control of arterial pressure and the intermittent absence of renin angiotensin system blockade observed in patients chronically treated with captopril has not been definitively explained. It is possible, for example, that vascular sensitivity to angiotensin II may be diminished under these circumstances. Also, the possibility exists that vascular renin may play an important role in the maintenance of arterial pressure and that converting enzyme activity occurring in the cells of extrapulmonary vessels may still be inhibited, although there are no longer signs of blockade of this enzyme in the plasma (see Antonaccio et al., this volume). A patient with severe hypertension and primary hyperparathyroidism studied recently may provide a clue to this puzzle. Blockade of angiotensin converting enzyme induced an impressive fall in blood pressure. However, during the third or fourth hour following each dose of captopril, blood pressure increased again to very high levels. When captopril was administered 3 times daily, systolic pressure could be kept within reasonable limits but diastolic blood pressures still rose intermittently to 110 or even 120 mm Hg. Only when verapamil, a calcium antagonist, was administered 3 times daily together with captopril could the blood pressure be normalized throughout the day. Verapamil alone was not able to control blood pressure. The rise of blood pressure occurring within a few hours following each dose of captopril was probably related to progressive resumption of plasma converting enzyme activity. The premature blood pressure increase that distinguished this case from all our other hypertensive patients might have been mediated by enhanced responsiveness of the arterial walls to the progressively rising angiotensin II levels, and this in turn could have been due to higher calcium concentrations at the receptor site or within vascular smooth muscle cells.

OTHER INDICATIONS

Converting enzyme inhibition has been used to treat several other hypertensive conditions, such as high blood pressure in patients on main-

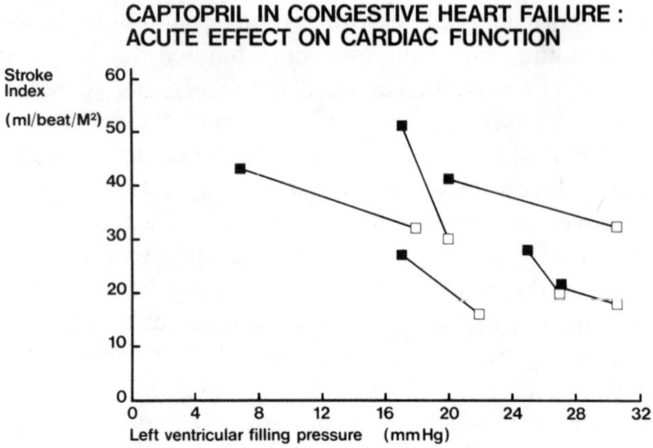

Fig. 9. Relationship between stroke index and left ventricular filling pressure of 6 normotensive patients with congestive heart failure before (open symbols) and one to three hours after (i.e., at time of maximum effect of) captopril administration (closed symbols). From Turini et al., (1979).

tenance hemodialysis and hypertensive crises (Tifft et al., 1979). Another interesting field in which converting enzyme inhibition has provided promising results is in the treatment of normotensive patients with congestive heart failure (Turini et al., 1979). Figure 9 summarizes some results obtained in 6 patients with congestive heart failure and depicts the relationship for each patient between stroke index and pulmonary capillary wedge pressure before and after captopril administration. In every patient, a decrease in pulmonary capillary wedge pressure was accompanied by an increase in stroke index. Simultaneously, mean arterial pressure fell from 73 ± 5 to 55 ± 7 mm Hg.

CONCLUSION

In conclusion, captopril is the most useful of converting enzyme inhibitors currently available for the treatment of arterial hypertension, since it has the advantage of being orally effective. Chronic blockade of the renin angiotensin system seems extremely effective as antihypertensive therapy. It allows recognition of those patients who need additional sodium subtraction to normalize their pressure, and furthermore allows quantification of the sodium excess in each patient. Diuretics are probably particularly effective during blockade of the renin system, possibly because aldosterone levels and renal vascular resistance remain low. In view of the results obtained with angiotensin converting enzyme inhibitors in man, it seems that these agents owe their sustained antihypertensive activity primarily to their ability to block endogenous production of angiotensin II.

REFERENCES

Bravo, E.L., and Tarazi, R.C. 1979. Converting enzyme inhibition with an orally active compound in hypertensive man. Hypertension 1:39

Brunner, H.R., Gavras, H., Laragh, J.H., and Kennan, R. 1974. Hypertension in man. Exposure of the renin and sodium components using angiotensin II blockade. Circ Res 34, 35 (Suppl 1):35

Brunner, H.R., Gavras, H., Waeber, B., Kershaw, G.R., Turini, G.A., Vukovich, R.A., McKinstry, D.N., and Gavras, I. 1979. Oral angiotensin-converting enzyme inhibitor in long term treatment of hypertensive patients. Ann Intern Med, 90:19

Brunner, H.R., Gavras, H., Waeber, B., Textor, S.C., Turini, G.A. and Wauters, F.P. 1980. Clinical use of an orally acting converting enzyme inhibitor: captopril. Hypertension 2:558

Case, D.B., Atlas, S.A., Laragh, J.H., Sealey, J.E., Sullivan, P.A., and McKinstry, D.N. 1978. Clinical experience with blockade of the renin-angiotensin aldosterone system by an oral converting-enzyme inhibitor (SQ 14,225, captopril) in hypertensive patients. Prog Cardiovasc Dis 21:195

Case, D.B., Wallace, J.M., and Laragh, J.H. 1979. Comparison between saralasin and converting enzyme inhibitor in hypertensive disease. Kidney Int 15:107

Feldberg, W., and Lewis, G.P. 1964. The action of peptides on adrenal medulla. Release of adrenaline by bradykinin and angiotensin. J. Physiol (Lond) 171:98

Ferguson, R.K., Turini, G.A., Brunner, H.R., Gavras, H., and McKinstry, D. 1977. A specific orally active inhibitor of angiotensin-converting enzyme in man. Lancet (I) 775

Ferreira, S.H. 1965. A bradykinin-potentiating factor (BPF) present in the venom of *Bothrops jararaca.* Brit J Pharmacol Chemother 24:163

Ferreira, S.H., Moncada, S., and Vane, J.R. 1973. Prostaglandins and the mechanism of analgesia produced by aspirin-like drugs. Brit J Pharmacol 49:86

Gavras, H., Brunner, H.R., Laragh, J.H., Sealey, J.E., Gavras, I., and Vukovich, R.A. 1974. The use of an angiotensin converting enzyme inhibitor to identify and treat vasoconstrictor and volume factors in hypertensive patients. New Engl J Med 291:817

Gavras, H., Ribeiro, A., Brunner, H.R., and Gavras, I. 1976. Reciprocal relationship between renin dependency and sodium dependency in essential hypertension. New Engl J Med 295:1278

Gavras, H., Brunner, H.R., Turini, G.A., Kershaw, G.R., Tifft, C.P., Cuttlelod, S., Gavras, I., Vukovich, R.A., and McKinstry, D.N. 1978. Antihypertensive effect of the oral angiotensin converting-enzyme inhibitor SQ 14,225 in man. New Engl J Med 298:291

Hulthen, L., and Hökfelt, B. 1978. The effect of the converting enzyme inhibitor SQ 20,881 on kinins, renin-angiotensin-aldosterone and catecholamines in relation to blood pressure in hypertensive patients. Acta Med Scand 204:497

Johnston, C.I., Millar, J.A., and McGrath, B.P. 1979. Long term effects of captopril (SQ 14,225) on blood pressure and hormone levels in essential hypertension. Lancet 2:493

Leslie, B.R., Case, D.B., Sullivan, J.F., and Vaughan, E.D., Jr. 1980. Absence of blood pressure lowering effect of captopril in anephric patients. Brit Med J 280:1067–1068

Man in't Veld, A.J., Schicht, I.M., Derkx, F.H.M., De Bruyn, J.H.B., and Schalekamp, M.A.D.H. 1980. Effects of an angiotensin converting enzyme inhibitor (captopril) on blood pressure in anephric subjects. Brit Med J 280:288–290

McGiff, J.C., Terragno, N.A., Malik, K.U., and Lonigro, A.J. 1972. Release of a prostaglandin E-like substance from canine kidney by bradykinin. Cir Res 31:36

Mimran, A., Brunner, H.R., Turini, G.A., Waeber, B., and Brunner, D. 1979. Effect of captopril on renal vascular tone in normal subjects and in patients with essential hypertension. Clin Sci 57:421S

Nishimura, K., Yoshida, N., Howada, K., Ueda, E., and Kokubu, T. 1977. Purification of angiotensin I converting enzyme from human lung. Biochim Biophys Acta (Amst) 483:398

Ondetti, M.A., Williams, N.J., Sabo, E.F., Pluščec, J., Weaver, E.R., and Kocy, O. 1971. Angiotensin converting enzyme inhibitors from the venom of *Bothrops jararaca:* Isolation, elucidation of structure and synthesis. Biochemistry (Wash) 10:4033

Ondetti, M.A., Rubin, B., and Cushman, D.W. 1977. Design of specific inhibitors of angiotensin converting enzyme: new class of orally active antihypertensive agents. Science 196:441

Onesti, G., Kim, K.E., Greco, J.A., del Guerico, E.T., Fernandez, M., and Swartz, C. 1975. Blood pressure regulation in end-stage renal disease and anephric man. Cir Res 36 (Suppl 1), 145–152

Streeten, D.H.P., Kerr, L.P., Kerr, C.B., Prior, J.C., and Dalakos, T.G. 1972. Hyperbradykininism: a new orthostatic syndrome. Lancet 2:1048

Tifft, C.P., Gavras, H., Kershaw, G.R., Gavras, I., Brunner, H.R., Liang, C.S., and Chonobian, A.V. 1979. Converting enzyme inhibition in hypertensive emergencies. Ann Int Med 90:43

Turini, G.A., Brunner, H.R., Gribic, M., Waeber, B., and Gavras, H. 1979. Improvement of chronic congestive heart failure by oral captopril. Lancet 1:1213–1216

Waeber, B., Brunner, H.R., Brunner, D.B., Curtet, A.L., Turini, G.A., and Gavras, H. 1980. Discrepancy between antihypertensive effect and angiotensin converting enzyme inhibition by captopril. Hypertension 2:236–242

Williams, G.H., and Hollenberg, N.K. 1977. Accentuated vascular and endocrine response to SQ 20,881 in hypertension. New Engl J Med 297:184

Captopril in Severe Treatment Resistant Hypertension

Roger K. Ferguson, Harris Koffer and Peter H. Vlasses
Division of Clinical Pharmacology, Department of Medicine
Jefferson Medical College, Thomas Jefferson University
Philadelphia, Pennsylvania 19107

INTRODUCTION

Regardless of etiology, severe treatment resistant hypertension can be defined as a diastolic blood pressure of 120 mm Hg or greater that either is not controlled by conventional drug therapy, or is controlled by a regimen not tolerated by the patient. An estimated 3% of treated hypertensives fall into this category (Andersson et al., 1979). Patients with resistant essential hypertension have been shown to have increased peripheral vascular tone despite treatment (Andersson et al., 1979). Treatment resistant blood pressure may complicate the malignant phase of essential hypertension or occur in secondary causes of hypertension such as renal disease. Since prolonged unresponsive blood pressure elevations are associated with significant morbidity and mortality, the control of blood pressure in these patients is of prime importance. Bilateral nephrectomy and the use of potent agents such as guanethidine or minoxidil have been reserved for the management of such patients, but each of these has inherent disadvantages.

The renin-angiotensin-aldosterone system appears to play an important role in the pathophysiology of various types of treatment resistant hypertension (Laragh et al., 1972). The development of the orally active angiotensin converting enzyme inhibitor, captopril, has provided a novel approach to the lowering of blood pressure in such patients and the initial results are promising. The purpose of this text is to review the published literature and to assess the value of captopril in the management of various types of refractory hypertension.

Table 1. Reports of captopril (Cap) use in severe and/or resistant hypertension.

	Type of Hypertension (No.)	No. of Responders	Dose/day of Cap	Comments
Treatment Resistant				
Cody et al. (1978)	Essential (4) Renovascular (4)	6	1000 mg	6 of 8 responded to cap alone; hemodynamic measurements
McCaa et al. (1979)	Essential (9)	9	up to 450 mg	All failed standard triple therapy; all required a diuretic
Atkinson et al. (1979)	Essential (6) Renovascular (7)	13	up to 450 mg	Cap alone in renovascular group; and cap + diuretic effective for essential hypertensives
Ferguson et al. (1980)	Essential (11) Renovascular (4)	13	up to 600 mg	8 responded to cap + diuretic; 5 responded to cap + diuretic + propranolol
Dialysis/Chronic Renal Failure				
Brunner et al. (1978a)	Chronic renal failure (7)	7	400 mg	5 patients required diuretic; all patients had low or normal PRA
Vaughan et al. (1980)	Dialysis resistant (2)	2	10–40 mg	Both patients with high PRA
Malignant				
Zappacosta et al. (1980)	Essential (1)	1	225 mg	Diuretic required
Oberfield et al. (1979)	Essential (1)	1	225 mg	Diuretic + propranolol required
Lopez-Ovejero et al. (1979)	Scleroderma (2)	2	600 mg	Both patients with high PRA
Other				
Brunner et al. (1980)	Primary aldosteronism (2)	1	200 mg	Pretreatment with spironolactone required
Aurell et al. (1979)	Renin-producing tumor (1)	1	450 mg	Resolution of aldosteronism

CLINICAL OVERVIEW

Captopril has been used in the management of various types of hypertension, such as essential, renovascular, hemodialysis resistant, malignant, and that secondary to chronic renal failure, primary hyperaldosteronism, scleroderma or renin-producing tumors. Reports on the use of captopril in such patients are summarized in Table 1. These patients had failed trials of various antihypertensive drug combinations; in some cases, these combinations included minoxidil. All drugs were stopped for variable periods of time before the initiation of captopril. Captopril was started in patients with little risk, even though in some cases the effects of other drugs were still present. The initial blood pressure response in this situation was greater than that which occurred with subsequent captopril administration. Occasionally, upon stopping other drugs such as clonidine, withdrawal reactions occurred that masked the initial response to captopril (Vlasses et al., 1979).

In patients who responded to captopril alone, the maximum lowering of blood pressure was usually evident within 60 min; the higher the blood pressure, the greater the response. The total daily dose of captopril required to effectively control blood pressure when given alone in these patients ranged from 40 to 1000 mg and was divided into three or four daily doses. Brunner et al. (1979) have demonstrated that increasing the dose of captopril prolongs its duration of action without increasing the magnitude of its depressor response. Patients who do not respond initially to captopril are unlikely to respond to higher doses. Currently, the maximum recommended daily dose is 400 mg in patients with normal renal function, and some investigators believe that twice daily dosing is adequate (Brunner et al., 1980). Concomitant diuretic therapy and renal disease alter the dosage requirements of captopril (see below).

ESSENTIAL AND RENOVASCULAR HYPERTENSION

Cody et al. (1978) have demonstrated that the depressor effect of captopril alone in patients with renovascular or essential hypertension is associated with a decrease in total peripheral resistance. Patients with resistant essential or renovascular hypertension with high plasma renin activity often responded to captopril alone; the doses necessary to achieve adequate duration of blood pressure reduction were often quite high. For the majority of patients, especially those with low or normal plasma renin activity (PRA) (Fig. 1), it was necessary to add a diuretic or restrict sodium intake to achieve optimal control (Ferguson et al., 1980). In this regard, Cody et al. (1978) demonstrated a further decline in total peripheral resistance and blood pressure with sodium restriction. Lower doses of captopril are adequate when given with a diuretic and the relation between dose and duration of response is maintained (Koffer et al., 1980; Larochelle et al., 1979). In our experience (Fig. 2), many patients previously uncontrolled on three or more drugs could be controlled with the addition of a diuretic to captopril therapy.

When diuretics are administered alone, a compensatory rise in renin and aldosterone occurs, which may limit their effectiveness. Captopril may enhance the effectiveness of diuretics by blocking this response. In the case of volume depletion, a patient's blood pressure may be primarily supported by the actions of angiotensin II. Administration of captopril to such patients may result in sudden symptomatic hypotension (Case et al., 1978); when this has occurred, the blood pressure could be restored with saline infusion. To avoid the rare possibility of hypotension, captopril should be initiated without concomitant diuretic therapy. If the depressor response to captopril is not adequate, diuretics may then be safely added (Case et al., 1978).

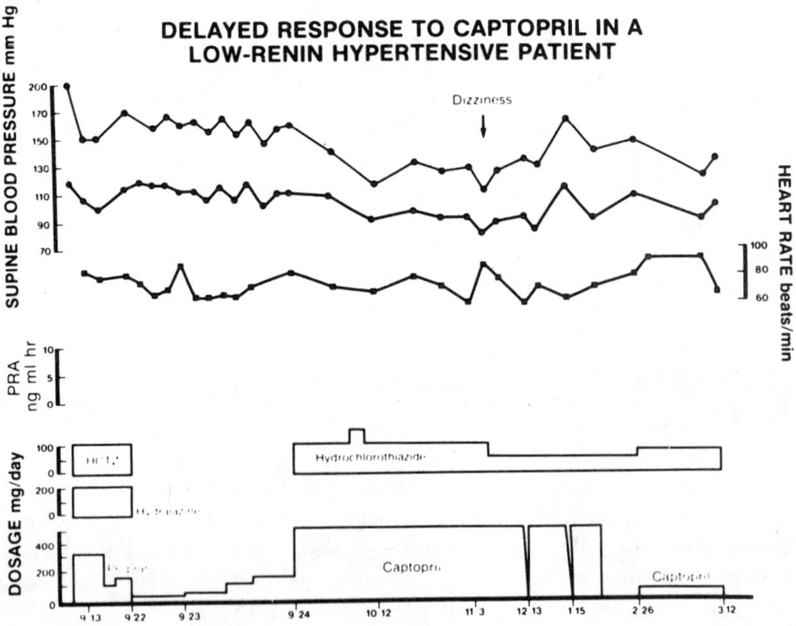

Fig. 1. A 34-year-old Black woman with low-renin essential hypertension had no blood pressure response to captopril alone, but was controlled upon addition of a diuretic. Subsequently, it was possible to lower the dose of captopril without loss of effect.

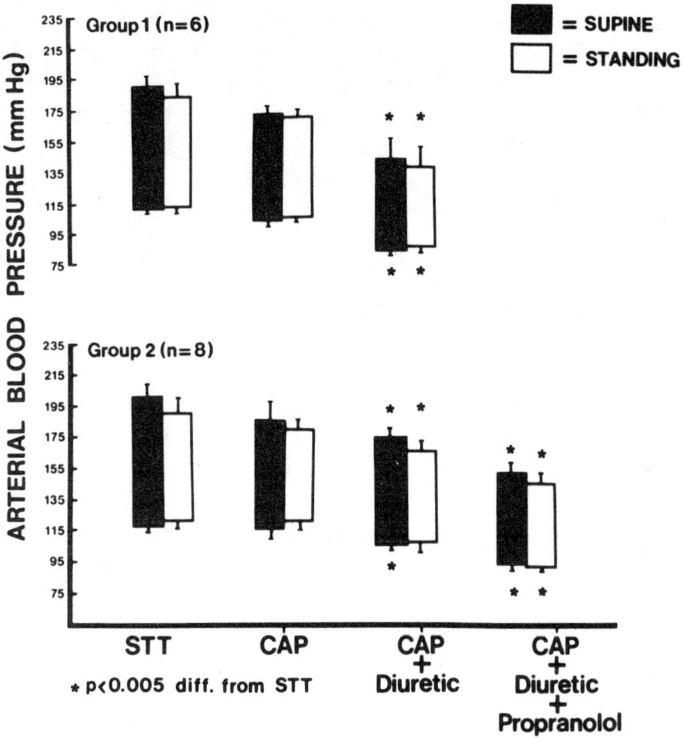

Fig. 2. Control of blood pressure with captopril (CAP) plus diuretic (Group 1) or CAP, diuretic and propranolol (Group 2) in hypertensive patients previously resistant to standard triple therapy (STT = diuretic, propranolol, hydroalazine). (Reprinted with permission of the *American Heart Journal.*)

In a certain proportion of patients, the combination of captopril and diuretic will not adequately control blood pressure. The addition of a beta-adrenergic blocking agent (Fig. 2) to the regimen of many of these patients has led to a reduction in or a normalization of blood pressure and was usually well tolerated. This is of interest since propranolol is thought by many to produce its antihypertensive effect primarily by suppressing renin (Buhler et al., 1973). A few patients were resistant even to this combination; the reason for this resistance is unclear though some of the patients had consistently low PRA. Addition of reserpine, guanethidine and prazosin was attempted in these patients without success.

Long-term follow-up of responders to captopril therapy reveals no clear evidence of tolerance to its antihypertensive effects. While marked increases in blood pressure may result from the abrupt discontinuation of some antihypertensive drugs, this has not been noted with captopril either after short- or long-term therapy (Cody et al., 1978; Vlasses et al., in press); blood pressure increased gradually without the occurrence of symptoms. This rise in blood pressure following captopril withdrawal is also associated with a gradual increase in total peripheral resistance, plasma angiotensin II and aldosterone, and a decrease in PRA.

RENAL FAILURE AND DIALYSIS RESISTANT HYPERTENSION

Some of the patients in the studies described above had renal insufficiency. There was no apparent relationship between the response to captopril and the presence of renal failure. Patients responded well to captopril regardless of pretreatment PRA, suggesting that their renin level even when low or normal contributes measurably to the maintenance of blood pressure (Brunner et al., 1978a). Captopril has not produced detrimental effects on renal hemodynamics (Mimran et al., 1979), and with rare exception has not produced deterioration of renal function. However, captopril is eliminated primarily by the kidneys and its dosage should be reduced in patients with poor renal function (Rommel et al., 1980).

Hypertension in patients with end-stage renal disease can usually by controlled by dialysis with or without antihypertensive drugs; however, some patients remain resistant. In these patients renin secretion from the diseased kidneys is disproportionately stimulated by volume depletion and the elevation in angiotensin II contributes to the resistance (Davies et al., 1973). On this basis, captopril has been found useful in controlling blood pressure in such patients (Brunner et al., 1978a; Vaughan et al., 1979).

MALIGNANT HYPERTENSION

Malignant hypertension is life threatening if not successfully treated. Patients with this disorder may be refractory to combinations of antihypertensive drugs. Though exceptions occur, PRA is usually increased in response to renal ischemia and inhibition of converting enzyme would be expected to be useful.

Captopril has been reported to be effective in two patients with malignant hypertension whose blood pressure could not be adequately controlled by conventional parenteral and oral therapy (Oberfield et al., 1979; Zappacosta et al., 1980). Both patients had high PRA before treatment. Captopril in combination with a diuretic was sufficient to normalize their blood pressure. Similarly, captopril has proved effective in patients with hypertensive crisis associated with scleroderma (Lopez-Ovejero et al., 1979). However, in two black men with malignant hypertension, we unsuccessfully attempted to substitute captopril in combination with furosemide for nitroprusside therapy. The PRA of these patients while on nitroprusside and furosemide was low and captopril did not significantly alter serum angiotensin II and plasma aldosterone concentrations. Subsequently, one of these patients was controlled by a three drug regimen. In this regard, Gavras evaluated the diagnostic and therapeutic value of the intravenous converting enzyme inhibitor teprotide in malignant hypertension.* After acute therapy with this drug, those patients with normalization of blood pressure were then treated with captopril, nonresponders with diuretics and intermediate responders with the combination of captopril and diuretics. Blood pressure responded favorably to the chosen regimen for each group. On this basis, it would appear that captopril can play an important role in the therapy of malignant hypertension in patients with normal or elevated PRA, but would be of little benefit in patients with low PRA.

OTHER CAUSES OF RESISTANT HYPERTENSION

Though not expected, captopril has benefited some patients with primary hyperaldosteronism whose blood pressure was resistant to other drug therapy (Brunner et al., 1980). Captopril was not effective as a single agent in these cases, but rather enhanced the effectiveness of spironolactone. The action of captopril in these patients is unclear and its use remains empiric. Captopril has also been used successfully to treat hypertension associated with a renin producing tumor (Aurell et al., 1979).

ADVERSE EFFECTS

To date, captopril has been generally well tolerated (Atkinson and Robertson, 1979). Because it apparently does not cross the blood-brain barrier, central side effects, seen with many other antihypertensive agents, are noticeably absent. A common problem is a maculopapular or morbilliform, pruritic skin rash which usually improves after dose reduction or discontinuation of the drug. Other side effects have in-

* H. Gavras, personal communication.

cluded hypotension, usually in patients who were volume depleted prior to receiving captopril, and standing tachycardia, which should be kept in mind in treating patients with compromised cardiovascular systems (Ferguson et al., 1980). Some patients have reported temporary ageusia during captopril therapy, the mechanism of which is unknown (Vlasses and Ferguson, 1979). Proteinuria, rises in serum creatinine, reversible renal failure, nephrotic syndrome, leukopenia and agranulocytosis in certain patients have been associated with captopril but further investigation is required (Atkinson and Robertson, 1979; van Brummelen et al., 1980; Amann et al., 1980). Investigators have postulated that accumulation of captopril and its metabolites in patients with decreased renal function may predispose to several of the above side effects. For this reason, as mentioned earlier, the dose of captopril should be minimized in patients with renal failure.

CONCLUSIONS

Captopril is a welcome addition to the therapeutic armamentarium of hypertension. Administered alone or in combination with a diuretic or a diuretic and a beta-adrenergic blocking agent, captopril has controlled the blood pressure of many patients previously resistant to conventional antihypertensive drug therapy. Whether the side effects of captopril will outweigh its benefits in these situations remains to be evaluated. Nonetheless, decreasing peripheral vascular resistance by inhibiting the effects of the renin-angiotensin-aldosterone system appears to be a valid therapeutic mode in these resistant cases.

REFERENCES

Amann, F.W., Buhler, F.R., Conen, D., Brunner, F., Ritz, R., and Brunno, S. 1980. Captopril-associated agranulocytosis. Lancet 1:150

Andersson, O., Hansson, L., and Silvertsson, R. 1979. Primary hypertension refractory to triple drug treatment. A study on central and peripheral hemodynamics. Circulation 58:615–622

Atkinson, A.B., and Robertson, J.I.S. 1979. Captopril in the treatment of clinical hypertension and cardiac failure. Lancet 2:836–839

Atkinson, A.B., Brown, J.J., Fraser, R., Leckie, B., Lever, A.F., Morton, J.J., and Robertson, J.I.S. 1979. Captopril in hypertension with renal artery stenosis and in intractable hypertension; acute and chronic changes in circulating concentrations of renin, angiotensins I and II and aldosterone, and in body composition. Clin Sci 57:139s–143s

Aurell, M., Rudin, A., Tisell, L.E., Kindblom, L.G., and Sandberg, G. 1979. Captopril effect on hypertension in patient with renin-producing tumor. Lancet 2:149–150

Brunner, H.R., Gavras, H., Turini, G.A., Waeber, B., and Wauters, J.P. 1978a. Long-term angiotensin blockade in hypertensive patients with chronic renal failure. Clin Res 26(Abs):459A

Brunner, H.R., Waeber, B., Wauters, J.P., Turini, G., McKinstry, D., and Gavras, H. 1978a. Inappropriate renin secretion unmasked by captopril (SQ 14225) in hypertension of chronic renal failure. Lancet 2:704–707

Brunner, H.R., Gavras, H., Waeber, B., Kershaw, G.R., Turini, G.A., Vukovich, R.A., McKinstry, D.N., and Gavras, I. 1979. Oral angiotensin-converting enzyme inhibitor in long-term treatment of hypertensive patients. Ann Intern Med 90:19–23

Brunner, H.R., Waeber, B., Turini, G.A., Wauters, J.P., and Gavras, H. 1980. Clinical use of captopril in patients with hypertension or congestive heart failure. In *Pharmacology and Clinical Use of Angiotensin I Converting Enzyme Inhibitors/Drug Development and Evaluation,* vol. 4, eds. F. Gross and R.K. Liedtke, pp. 57–73. New York: Gustav Fischer Verlag

Buhler, F.R., Laragh, J.H., Vaughan, E.D., Brunner, H.R., Gavras, H., and Baer, L. 1973. Antihypertensive action of propranolol. Am J Cardiol 32:511–522

Case, D.B., Atlas, S.A., Laragh, J.H., Sealey, J.E., Sullivan, P.A., and McKinstry, D.N. 1978. Clinical experience with blockade of the renin-angiotensin-aldosterone system by an oral converting-enzyme inhibitor (SQ 14225, Captopril) in hypertensive patients. Progr Cardiovasc Dis 21:195–206

Cody, R.J., Tarazi, R.C., Bravo, E.L., and Fouad, F.M. 1978. Haemodynamics of orally-active converting enzyme inhibitor (SQ 14225) in hypertensive patients. Clin Sci Mol Med 55:453–459

Davies, D.L., Schalekamp, M.A., Beevers, D.G., Brown, J.J., Briggs, J.D., Lever, A.F., Medina, A.M., Morton, J.J., Robertson, J.I.S., and Tree, M. 1973. Abnormal relation between exchangeable sodium and the renin-angiotensin system in malignant hypertension and in hypertension with chronic renal failure. Lancet 1:683

Ferguson, R.K., Vlasses, P.H., Koplin, J.R., Shirinian, A., Burke, J.F., and Alexander, J.C. 1980. Captopril in severe treatment-resistant hypertension. Am Heart J 99:579–587

Koffer, H., Vlasses, P.H., Ferguson, R.K., Weis, M., and Adler, A.G. 1980. Captopril in diuretic-treated hypertensive patients. J A M A, 244:2532–2535

Laragh, J.H., Baer, L., Brunner, H.R., Buhler, F.R., Sealey, J.E., and Vaughan, E.D. 1972. Renin, angiotensin and aldosterone system in pathogenesis and management of hypertensive vascular disease. Am J Med 52:633–652

Larochelle, P., Genest, J., Kuchel, O., Boucher, R., Gutkowska, Y., and McKinstry, D. 1979. Effect of captopril (SQ 14225) on blood pressure, plasma renin activity and angiotensin I converting enzyme activity. Can Med Assoc J 121:309–316

Lopez-Ovejero, J.A., Saal, S.D., D'Angelo, W.A., Cheigh, J.S., Stenzel, K.H., and Laragh, J.H. 1979. Reversal of vascular and renal crises of scleroderma by oral angiotensin-converting-enzyme blockade. N Engl J Med 300:1417–1419

Man in't Veld, A.J., Wenting, G.J., and Schalekamp, M.A. 1979. Does captopril lower blood pressure in anephric patients? Br Med J 2:1110

McCaa, C.S., Langford, H.G., Cushman, W.C., and McCaa, R.E. 1979. Response of arterial blood pressure, plasma renin activity and plasma aldosterone concentration to long-term administration of captopril in patients with severe, treatment-resistant malignant hypertension. Clin Sci 57:371s–373s

Mimran, A., Brunner, H.R., Turini, G.A., Waeber, B., and Brunner, D. 1979. Effect of captopril on renal vascular tone in patients with essential hypertension. Clin Sci 57:421s–423s

Oberfield, S.E., Case, D.B., Levine, L.S., Rapaport, R., Rauh, W., and New, M.I. 1979. Use of the oral angiotensin I-converting enzyme inhibitor (captopril) in childhood malignant hypertension. J Ped 95:641–644

Rommel, A.J., Pierides, A.M., and Heald, A. 1980. Captopril elimination in chronic renal failure. Clin Pharmacol Ther 27(Abs):282

van Brummelen, P., Willemze, R., Tan, W.D., and Thompson, J. 1980. Captopril-associated agranulocytosis. Lancet 1:150

Vaughan, E.D., Carey, R.M., Ayers, C.R., and Peach, M.J. 1979. Hemodialysis-resistant hypertension: Control with an orally active inhibitor of angiotensin-converting enzyme. J Clin Endocrinol Metab 48:869–871

Vlasses, P.H., and Ferguson, R.K. 1979. Temporary ageusia related to captopril. Lancet 2:526

Vlasses, P.H., Koplin, J.R., Shirinian, A., and Ferguson, R.K. 1979. Clonidine-propranolol withdrawal syndrome. Clin Res 27(Abs):661A

Vlasses, P.H., Koffer, H., Ferguson, R.K., Green, P.J., and McElwain, G.E. Abrupt discontinuation of captopril after chronic therapy in hypertensive patients. Clin Exp Hypertension, in press

Zappacosta, A.R., Vlasses, P.H., and Ferguson, R.K. 1980. Captopril in refractory malignant hypertension. Drug Intell Clin Pharm, 14:373–374

Comparison of Captopril with Placebo in the Treatment of Essential Hypertension

John C. Alexander and Joseph H. Meyer
The Squibb Institute for Medical Research
Princeton, New Jersey 08540

INTRODUCTION

Captopril, the first of a new class of orally active antihypertensive agents, inhibits the enzyme peptidyl dipeptide hydrolase which converts angiotensin I to angiotensin II and inactivates bradykinin. A number of studies have been published that suggest captopril is effective and safe in milder forms of essential hypertension (Atkinson and Robertson, 1979; Friedlander, 1979a,b; Jenkins and McKinstry, 1979; Johnston et al., 1979a,b; Klaus et al., 1979; Liedtke and Riedl, 1979; MacGregor et al., 1979a,b; McNeil et al., 1979; Röckel et al., 1979; Seedat, 1979; Sharpe, 1979; and Whitworth et al., 1979). Since placebo controlled studies have not been conducted in essential hypertension, a multicenter study was organized to demonstrate the efficacy and safety of captopril. A secondary purpose was to determine the therapeutic advantage of adding a diuretic agent to the regimen of the patients in either group.

PATIENTS AND METHODS

Males or nonpregnant females 18 years of age or older with a primary diagnosis of essential hypertension were eligible for enrollment. Patients receiving treatment for hypertension were to be considered for the study only after treatment had been discontinued and replaced with a placebo for at least 2 wk of the scheduled 8 wk lead-in period. The supine diastolic blood pressure (SDBP) of previously untreated patients at the initial visit, and of all patients at the last two visits of the placebo lead-in period, was to be 100 mm Hg or greater.

The patients selected could not receive other vasoactive drugs during the study. If any of the following conditions were present, the patient was not accepted for enrollment: 1) serious allergy, or a history of drug hypersensitivity, or recurrent dermatoses; 2) significant gastrointestinal disease that could impair the absorption of medication taken orally; or 3) creatinine clearance < 50 ml/min/1.73 m^2, or serum creatinine level twice the upper limit of normal.

Informed consent was obtained from each patient enrolled in the study.

The trial was divided into three periods (Fig. 1): a placebo lead-in period of 2 to 8 wk to establish the baseline blood pressure and to wash out previous antihypertensive medication (if any); a dose titration period of 4 wk during which the dose of either captopril or placebo was increased until the SDBP was ≤ 90 mm Hg or until the maximum dose of the drug was reached; and an 8 wk maintenance period during which hydrochlorothiazide (HCTZ) was added to the regimen if the SDBP

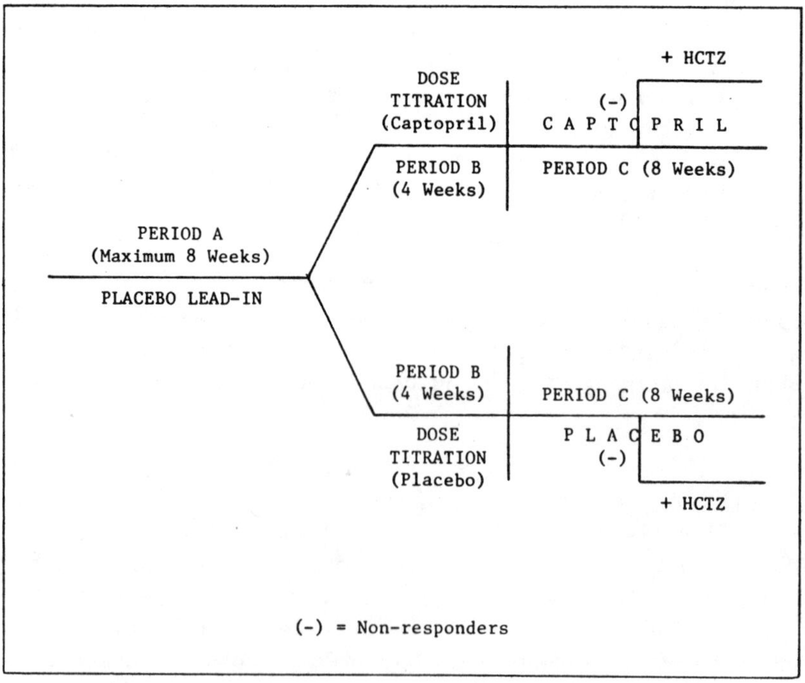

Fig. 1. Design of the double-blind comparison of captopril with placebo.

was not normalized ($\leqslant$ 90 mm Hg). If the SDBP was normalized, the patient was to continue on the same regimen as that used during the final week of the dose titration period.

Before entry into the dose titration period, patients were assigned to receive captopril or placebo according to a randomization scheme constructed so that 60% of the patients received captopril and 40% received placebo.

Captopril or matching placebo was administered t.i.d. at least 1 hr before or 2 hr after meals; hydrochlorothiazide was administered b.i.d. The total daily dose of captopril or the matching placebo was increased as necessary in the following sequence: 75, 150, 300, and 450 mg or placebo equivalent. When necessary, HCTZ was added at a dosage of 50 mg/day, which could be increased to 100 mg/day.

Each patient was seen at approximately the same time of day at each visit. The visit was made between 3 and 8 hr after the last dose of drug. Physical activity was kept to a minimum on the day of the visit to ensure basal conditions. Smoking and ingestion of beverages containing alcohol or caffeine were avoided for at least 2 hr before a visit. The blood pressure was recorded with a standard mercury sphygmomanometer, and the diastolic level was taken as the disappearance of the audible pulse beat (Korotkoff V). The value recorded was the average of two consecutive readings in which the diastolic pressure did not differ by more than 5 mm Hg. Supine blood pressure was recorded after 10 min of rest, and erect blood pressure after 2 min in the standing position.

The usual intake of salt was maintained throughout the study. Physical examinations, laboratory tests and electrocardiograms were performed periodically during the study.

Patient visits were scheduled weekly until the last 4 wk of the maintenance period when they were seen every 2 wk. At each visit, blood pressure was measured, unused tablets counted and patients questioned concerning any new symptoms.

A variety of statistical procedures was employed. A general linear model (GLM) approach was used to compare the two treatments at fixed time points with respect to changes from baseline for blood pressure and heart rate. Comparisons of the proportion of favorable responders and patients requiring concomitant diuretic treatment in each group were carried out using the Mantel-Haenszel procedure for combining results from several studies. For overall response, Cochran's test for a linear trend in proportions was used.

Table 1. Comparison of sex, age and weight in captopril group with those in placebo group before treatment.

Treatment	Sex	Number Patients	Age (yr) Range	Age (yr) Mean	Weight (kg) Range	Weight (kg) Mean
Captopril	M	72	23–71	47.9	50.0–137.6	85.2
	F	53	22–75	49.1	36.6–155.5	72.9
Placebo*	M	43	21–74	53.9	45.1–123.0	84.1
	F	29	27–72	48.7	21.0– 88.2	66.4

*The weights of one man and one woman in the placebo group were not reported.

Table 2. Comparison of blood pressure before treatment in captopril group with those in placebo group.

Systolic Pressure					Diastolic Pressure				
Captopril			Placebo		Captopril			Placebo	
Mean (mmHg)	170		174		Mean (mmHg)	110		111	
Range (mmHg)	Number	%	Number	%	Range (mmHg)	Number	%	Number	%
<150	17	13.6	7	9.7	100–104	36	28.8	18	25.0
150–169	48	38.4	24	33.3	105–114	57	45.6	34	47.2
170–189	35	28.0	21	29.2	⩾115	32	25.6	20	27.8
⩾190	25	20.0	20	27.8					

RESULTS

A total of 329 patients was enrolled by 16 investigators from the U.S. and overseas. Seventy-six (23.1%) were dropped from the evaluation because their baseline (blood pressure at end of lead-in) SDBP during the placebo lead-in period was less than 100 mm Hg. Data on 56 other patients were not included for a variety of reasons, including dropouts (20), marked blood pressure elevation during the lead-in period (13), complicating concomitant illness (9), and the remaining for a variety of protocol deviations.

The remaining 197 patients form the basis of this report. The comparison of the response to the two treatments is based on the results of the 125 patients randomized to receive captopril and the 72 patients assigned to placebo.

Patients' Characteristics Before Treatment

Patients assigned to receive treatment with captopril and those assigned to received placebo were very similar in age, sex and weight (Table 1).

The baseline pressure data were also nearly identical for both treatment groups (Table 2). The majority of the patients had moderate hypertension, SDBP on placebo between 105 and 114 mm Hg. Most patients (73% of those assigned to captopril and 65% of those assigned to placebo) were receiving antihypertensive therapy prior to the study. The mean baseline serum creatinine concentrations were 1.0 mg% for both groups. There was also a similar distribution of patients by WHO classification and Keith-Wagener grade in the two treatment groups.

Direct Comparison of Efficacy

After a lead-in period of about 6 wk for both groups, patients were assigned to receive either captopril or placebo alone for 4 wk. The dose of the medication was increased at weekly intervals unless the patient's blood pressure had normalized (SDBP $\leq$ 90 mm Hg).

The mean supine diastolic blood pressures of the patients in the two treatment groups for the first 4 wk of treatment are shown in Table 3. At each week of the four weekly visits, the efficacy of captopril was superior to placebo ($P < 0.05$). The adjusted mean decrease in SDBP at week 4 was about 15% for captopril and 10% for placebo. More than half (56%) of the decrease ultimately achieved with captopril alone occurred by the end of 1 wk of treatment and 63% after 2 wk. Surprisingly, in the placebo group only 58% of the total 4 wk decrease was found after 2 wk.

Table 3. Supine diastolic blood pressures after weeks 1, 2, 3 and 4 of treatment.

	Captopril				Placebo			
	N	Pre Mean	Post Mean	Adjusted Change*	N	Pre Mean	Post Mean	Adjusted Change*
Week 1	125	109.7	99.9	− 8.4%	72	110.3	103.9	−5.5%
Week 2	125	109.7	99.2	− 9.4%	72	110.3	103.7	−5.6%
Week 3	123	109.6	96.4	−12.9%	71	110.3	103.4	−7.0%
Week 4	119	109.4	94.7	−14.9%	65	110.0	100.9	−9.6%

*Adjusted for investigators and pretreatment levels

Table 5. Statistical results at week 4 geometric mean pre- and posttreatment levels and adjusted percent change.

	Captopril				Placebo				
	N	Pre Mean	Post Mean	Adjusted Change[a]	N	Pre Mean	Post Mean	Adjusted Change[a]	F-Value
Supine									
Systolic BP	119	167.6	146.8	−13.9%	65	171.9	162.4	−6.6%	27.25 P <0.001
Diastolic BP	119	109.4	94.7	−14.9%	65	110.0	100.9	−9.6%	15.75 P <0.001
Mean BP	119	129.0	112.2	−14.4%	65	130.8	121.6	−8.3%	24.73 P <0.001
Heart Rate	119	76.6	74.7	−3.1%	65	77.0	73.8	−4.7%	0.81 NS
Standing									
Systolic BP	119	164.4	143.3	−13.6%	65	164.3	158.3	−3.6%	43.95 P <0.001
Diastolic BP	119	112.1	97.2	−13.2%	65	110.7	104.4	−5.5%	29.54 P <0.001
Mean BP	119	129.7	112.7	−13.3%	65	128.7	122.5	−4.7%	43.21 P <0.001
Heart Rate	118	82.5	82.0	−1.2%	64	82.5	80.1	−3.5%	1.62 NS

[a] Adjusted for investigators and pretreatment levels.
NS = Not Significant

Table 4. Overall blood pressure response to either captopril or placebo by dose.

		Week 4			
Drug and Dose (mg)	N	Normalized	Responders	Nonres-ponders	Treatment Failures (Cumulative)
Captopril					
75	13	12	0	1	0
150	22	13	3	6	0
300	28	13	6	8	1
430	57	12	20	25	0
Total	120	50	29	40	1
Placebo	68	11	14	40	3

All but one of the patients on captopril received a total daily dose of 75 mg during the first week of treatment. During the second week, 26 patients continued to receive 75 mg and 96 patients had their dose of captopril increased to 150 mg. As can be seen in Table 4, about half of the patients were receiving the maximal permitted dose of captopril, 450 mg/day, at week 4.

Also in Table 4, the categorization of responses on the basis of supine diastolic blood pressure are listed:

a) Normalized: Pressure at 90 mm Hg or less
b) Responder: Not normalized, but at least a 10% decrease from baseline
c) Nonresponder: Not normalized, and less than 10% decrease from baseline.

In addition, some patients were discontinued from the study due to a lack of response. These patients were classified as Treatment Failures. Based on Cochran's Chi-square Test over these four response categories, a statistically significant difference between captopril and placebo was found at week 4 ($P < 0.01$). At this point, with all patients still on monotherapy, only 16.2% of those treated with placebo had achieved normalization of blood pressure compared with 41.7% of those given captopril.

Comparisons (Table 5) of the geometric mean changes in the various blood pressure parameters after 4 wk of treatment all favored captopril ($P < 0.001$). A similar percentage decrease (13–14%) after captopril was found regardless of the blood pressure determination selected, supine or standing, diastolic, systolic, or mean pressure.

Table 6. Statistical results at week 8 and 12 geometric mean pre and post treatment levels and adjusted percent change.

	Captopril[a]				Placebo[a]				
	N	Pre Mean	Post Mean	Adjusted Change[b]	N	Pre Mean	Post Mean	Adjusted Change[b]	F-Value
				Week 8					
Supine									
Systolic BP	106	168.6	138.7	−18.3%	52	173.9	153.5	−11.4%	13.94 $P < 0.001$
Diastolic BP	106	110.0	88.9	−19.5%	52	110.6	96.9	−12.4%	24.85 $P < 0.001$
Mean BP	106	129.7	105.6	−19.0%	52	131.9	116.0	−11.9%	22.33 $P < 0.001$
Heart Rate	106	76.9	75.8	−1.1%	52	76.5	74.9	−1.8%	0.14 NS
				Week 12					
Supine									
Systolic BP	96	168.0	134.3	−20.6%	49	173.9	154.7	−11.4%	27.36 $P < 0.001$
Diastolic BP	96	109.4	8.85	−21.5%	49	110.7	96.9	−12.5%	30.59 $P < 0.001$
Mean BP	96	129.1	102.1	−21.1%	49	132.0	116.4	−12.1%	33.24 $P < 0.001$
Heart Rate	96	77.2	75.4	−2.8%	49	77.2	76.4	−1.9%	0.19 NS

[a] Some patients also received concomitant diuretic therapy.
[b] Adjusted for investigators and pretreatment levels.
NS = Not Significant

Table 7. Frequency distributions of overall response categories.

	Week 8*				Week 12**			
	Captopril Group		Placebo Group		Captopril Group		Placebo Group	
Category	n	%	n	%	n	%	n	%
Normalized	65	61	16	29	76	78	13	24
Responders	26	24	12	21	10	10	15	28
Nonresponders	15	14	24	43	10	10	21	39
Treatment Failures (Cumulative)	1	1	4	7	2	2	5	9
Total	107		56		98		54	

* 45% of patients in the captopril group and 83% of patients in the placebo group were also receiving diuretic.
**58% of patients in the captopril group and 90% of patients in the placebo group were also receiving diuretic.

Maintenance Period

After the 4 wk titration period, investigators were permitted to add diuretic as concomitant therapy if the blood pressure had not normalized. The proportion of patients requiring diuretic therapy was 48 of 106 (45%) for the captopril group and 43 of 52 (83%) for the placebo group at week 8, and 56 of 96 (58%) for captopril and 44 of 49 (90%) for placebo at week 12. For a variety of reasons, some patients were not seen at either week 8 or 12. Significantly more placebo-treated patients required treatment with diuretic ($P < 0.001$).

Statistically significant differences in favor of the group originally assigned to captopril ($P < 0.01$) were also found in the geometric mean decreases in supine blood pressure at week 8 and week 12 (Table 6). The mean percentage decrease in SDBP at the end of week 12 was 21.5% for those who received captopril (alone or in combination with a diuretic) compared with 12.5% for the group assigned to placebo (alone or in combination with a diuretic).

By the end of week 12, 78% of the patients assigned to receive captopril and 24% of patients in the placebo group had achieved normalization of their supine diastolic blood pressures (Table 7).

For the 40 (41.6%) patients who received captopril alone for 12 wk, the full antihypertensive effects were apparent at the end of week 4. For those who failed to normalize on captopril alone (56 patients), the addition of a diuretic resulted in a further 12% decrease in SDBP (Table 8).

This compares to an additional 4% decrease when a diuretic was given to the patients treated with placebo.

No significant change in heart rate was found in either treatment group at any time point.

Table 8. Mean supine diastolic blood pressures for the four treatment groups.

Treatment Group	Diuretic By Week 12	N	Week	Mean SDBP	Change From Baseline
Captopril	No	40	0	107	
			4	88	−18%
			8	87	−19%
			12	86	−20%
	Yes	56	0	111	
			4	99	−11%
			8	91	−19%
			12	86	−23%
Placebo	No	5	0	104	
			4	89	−14%
			8	90	−14%
			12	94	−10%
	Yes	44	0	111	
			4	102	− 9%
			8	97	−13%
			12	97	−13%

SIDE EFFECTS

A total of 228 patients was included in the evaluation of safety; 141 were assigned to captopril and 87 to placebo. As mentioned previously, a number of these patients were not discussed in the efficacy section because of protocol violations. Seventy-five of the patients assigned to captopril and 62 patients who received placebo had HCTZ added at some time during the study.

Dizziness, headache and malaise were the reactions reported most often (Table 9). The percentage of patients who reported each of these reactions was similar during treatment with captopril alone and placebo alone. The incidence of rashes and alterations in taste perception during treatment with captopril was higher than that in the placebo group.

Combination treatment with captopril and diuretic was well tolerated.

Table 9. Comparison of percentage of patients with most frequent adverse reactions by treatment.

	Treatment Group			
Adverse Reaction	Captopril	Captopril + HCTZ	Placebo	Placebo + HCTZ
Headache	12.1%	1.3%	14.9%	—
Malaise	9.2%	6.7%	9.2%	3.2%
Rash	7.8%	2.7%	2.3%	3.2%
Dizziness	5.7%	17.3%	4.6%	12.9%
Nausea/Vomiting	4.3%	4.0%	1.2%	6.5%
Taste Alteration	2.1%	4.0%	—	3.2%
Total Patients	141	75	87	62

Adverse reaction is counted for the treatment being received when it first occurred. The "Total Patients" for each treatment group indicates the number of patients who received the specified treatment.

Neither tachycardia nor orthostatic hypotension was reported. The incidence of dizziness was somewhat higher on combination therapy than on captopril alone.

The percent of patients for whom "severe" untoward events were reported was nearly identical among patients taking captopril alone and among those taking placebo alone (3.5 and 3.4%, respectively). Patients taking either captopril or placebo plus hydrochlorothiazide had more "serious" adverse reactions (10.7% of captopril patients and 11.3% of placebo patients).

A number of patients did not complete the study due to adverse reactions; 7.0% of 129 patients assigned to captopril and 5.6% of 72 patients who received placebo. Rash and taste alteration were the most frequent reasons for discontinuation of captopril.

No medically significant changes in laboratory test results were observed in patients treated with captopril. A slight increase occurred in the mean serum urea nitrogen and potassium values and a slight decrease in hemoglobin and white blood cell count.

DISCUSSION

In this placebo controlled, double-blind multicenter study, captopril was clearly effective in lowering blood pressure in patients with essential hypertension.

It is well known that blood pressure frequently falls when patients are under close medical supervision and after administration of placebo. Any interpretation of efficacy of a new drug thus requires that controls be employed to demonstrate the true pharmacologic effects. In this study, approximately 36% of patients previously diagnosed as having hypertension responded to treatment with placebo (26% during the 2 to 8 wk lead-in period). The mean diastolic blood pressure continued to fall in the group assigned to placebo. After 4 wk of placebo treatment, the mean supine diastolic blood pressure decreased about 10 mm Hg (9.6%). About 10% of patients normalized on placebo alone during the study.

Despite the placebo response, the results with captopril were superior. The data at the end of week 4 provide a direct comparison of monotherapy with captopril and placebo. The SDBP was normalized in 42% of the patients randomized to receive captopril and only 15% of those treated with placebo. The mean percentage decrease in SDBP at the end of week 4 was 14.9% with captopril compared to 9.6% with placebo ($P < 0.001$).

The dose-response data suggest that in patients with essential hypertension most of the antihypertensive effect can be achieved with a daily dosage of captopril of 75 or 150 mg. If a satisfactory blood pressure has not been obtained after treatment with 150 mg of captopril per day, diuretic should be added before increasing the dose of captopril further.

The superiority of captopril compared to placebo was even more evident after the 8 wk maintenance period (during which hydrochlorothiazide was added if normalization of the blood pressure had not been accomplished). By the end of week 12, 58% of the patients in the captopril group and 90% of the patients in the placebo group required concomitant therapy with HCTZ ($P < 0.01$). Addition of diuretic resulted in an additional 12% decrease in SDBP below that achieved with captopril alone; this compares to the very small additional decrease of 4% when a diuretic was given to patients treated with placebo. The additive effects of combination therapy with captopril and diuretic can be explained by the effects of diuretic on the renin-angiotensin system. Treatment with a diuretic generally results in an increase in plasma renin activity, which blunts the hypotensive effect of the diuretic. The effect of an increase in plasma renin activity would be blocked by captopril.

At the end of 12 wk of treatment, 78% of the patients in the captopril group and 24% of the patients in the placebo group had a SDBP of 90 mm Hg or less. The mean decrease in SDBP was 20% for the patients in the captopril group and 10% for those in the placebo group ($P < 0.001$).

Treatment with captopril alone or in combination with a diuretic was well tolerated. Drug related side effects included taste alteration and skin rash, which were usually mild and transient.

SUMMARY AND CONCLUSION

A total of 197 patients with essential hypertension was included in this double-blind comparison study of the efficacy and safety of captopril and placebo (125 were randomized to captopril and 72 to placebo).

After treatment for 4 wk, the blood pressure of 42% of the patients treated with captopril and 16% of the patients treated with placebo had normalized. The mean decrease in supine diastolic blood pressure (SDBP) was 15% for captopril and 10% for placebo ($P < 0.001$).

During the next 8 wk of treatment, diuretic was given concomitantly to all patients whose pressure had failed to normalize on monotherapy. By the end of week 12, diuretic had been added to 58% of patients treated with captopril and 90% of those given placebo. Normalization of the blood pressure was achieved for 78% of the patients in the captopril group and only 24% of the patients randomized to placebo. The mean decrease in SDBP at the end of the study was 20% for the patients in the captopril group and 10% for those in the placebo group.

Side effects of captopril were essentially limited to rash and taste alteration, which were usually mild and transient.

In this three month study, captopril was safe and effective for the treatment of essential hypertension.

REFERENCES

Atkinson, A.B., and Robertson, J.I.S. 1979. Captopril in the treatment of clinical hypertension and cardiac failure. Lancet 2:836–839

Friedlander, D.H. 1979a. Captopril and propranolol in mild and moderate essential hypertension. Preliminary Report. N Z Med J 90:146–149

Friedlander, D.H. 1979b. Comparison of captopril with propranolol in the treatment of essential hypertension. Med J Aust (Suppl) 2:30–32

Jenkins, A.C., and McKinstry, D.N. 1979. Review of clinical studies of hypertensive patients treated with captopril. Med J Aust (Suppl) 2:32–37

Johnston, C.I., McGrath, B.P., and Matthews, P.G. 1979a. Comparison of the hormonal effects of captopril (SQ 14225) and hydrochlorothiazide in the treatment of essential hypertension. Med J Aust (Suppl) 2:17–22

Johnston, C.I., McGrath, B.P., Millar, J.A., and Matthews, P.G. 1979b.

Long-term effects of captopril (SQ 14225) on blood pressure and hormone levels in essential hypertension. Lancet 2:493–496

Klaus, D., Lederle, R., and Braun, B. 1979. Captopril in hypertension. Effect on blood pressure, plasma renin, plasma aldosterone and converting enzyme. In *Pharmacology and Clinical Use of Angiotensin I Converting Enzyme Inhibitors,* eds. Franz Gross and Rainer Liedtke, pp. 92–106. New York: Gustav Fischer Verlag

Liedtke, R., and Riedl, L. 1979. Comparative monitoring of biochemical parameters during treatment with captopril and other antihypertensive agents. In *Pharmacology and Clinical Use of Angiotensin I Converting Enzyme Inhibitors,* eds. Franz Gross and Rainer Liedtke, pp. 107–122. New York: Gustav Fischer Verlag

MacGregor, G.A., Markandu, N.D., and Roulston, J.E. 1979a. Does the renin-angiotensin system maintain blood pressure in both hypertensive and normotensive subjects? A comparison of propranolol, saralasin and captopril. Clin Sci (Suppl) 57:1455–1485

MacGregor, G.A., Markandu, D.U., Roulston, J.E., and Jones, J.C. 1979b. Essential hypertension: effect of an oral inhibitor of angiotensin-converting enzyme. Brit Med J 2:1106–1109

McNeil, J.J., Anderson, A., Christophidis, N., Mendelsohn, F.A.P., Coghlan, J., and Louis, W.J. 1979. Comparison of captopril and hydrochlorothiazide in the treatment of moderate hypertension. Med J Aust (Suppl) 2:22–24.

Röckel, A., Brand, A., Heidbreder, E., Stürmer, G., and Heidland, A. 1979. A clinical comparison of a converting enzyme inhibitor and of hydrochlorothiazide—effects on urinary and salivary kallikrein excretion. In *Pharmacology and Clinical Use of Angiotensin I Converting Enzyme Inhibitors,* eds. Franz Gross and Rainer Liedtke, pp. 82–91. New York: Gustav Fischer Verlag

Seedat, Y.K. 1979. Comparison of captopril with propranolol in the treatment of mild and moderate essential hypertension. S A Med J 56:983–986

Sharpe, D.N. 1979. Comparison of captopril (SQ 14225) with hydrochlorothiazide in the treatment of mild and moderate essential hypertension. Med J Aust (Suppl) 2:24–26

Whitworth, J.A., Walter, N.M.A., and Kincaid-Smith, P. 1979. Clinical trial of an orally administered converting enzyme inhibitor (captopril). Med J Aust (Suppl) 2:27–30

Discussion

Z.P. HOROVITZ: I have a question for Dr. Ferguson but before that I want to make a statement because I am afraid that Dr. MacGregor feels that Squibb doesn't love him and I would like to make a point which I think is valid. We don't like John Laragh, Bob McCaa and Hans Brunner any more than you, but I think one of the reasons they got captopril earlier is that the Swiss regulatory authorities and the FDA have a very much more realistic view on starting clinical trials than the U.K. Medicines Committee and I think that created part of the problem. I understand the Medicines Committee is trying to change that and liberalize the requirements for Phase I patient studies and that would be a good thing if it happens. The question is to Dr. Ferguson; Roger, you talked about standing tachycardia in three patients and I wasn't sure if you mentioned if those patients had a beta blocker withdrawn at any time before they went on to captopril?

R.K. FERGUSON: I should emphasize that of those three patients who developed an infarct each had severe atherosclerotic disease and confirmed or probable renal artery stenosis on the basis of atherosclerotic disease. One of the patients could not take and two did not require a beta blocker and so none of the three patients had beta blockers before the event. Subsequently we did add a small dose of beta blocker (20–40 mg/day of propranolol) to control marked standing tachycardia in those patients in whom it occurs.

F.W. WOLFF: I'd like to address my question entirely to the adverse effects of this very exciting agent, in the hope that these adverse effects will receive fullest possible publication in the forthcoming issue and that a full discussion of this point would prevent later untoward situations and any enormous delay of the introduction of this really very exciting drug. My first question refers to loss of taste. Was this an abnormality in taste or were in fact some taste measurements taken? Was it a specific abnormality of taste? I know that d-penicillamine gives rise to taste loss. I believe in hepatitis there is taste loss and I remember that the antimalarials are sometimes associated with taste loss and perhaps sometimes a very bitter taste. Now my question is again was this a complete taste loss or an abnormality of taste?

D.N. MCKINSTRY: Dr. MacGregor, can you respond to that question with respect to your patient population and then perhaps Dr. Alexander can address the question from the standpoint of our overall clinical experience.

G.A. MACGREGOR: I am a bit reluctant to answer the question the way it's been asked. But nevertheless I will attempt to. When we were doing the trial using large doses of captopril up to 150 mg t.i.d., we found we had 6 patients out of a total of 45 who lost taste and that was within the second month of therapy. Three of those patients had quite severe loss of taste and 2 of them completely lost taste and lost 4–6 kilos each because of loss of appetite. This was not just a mild amount of taste loss. It was complete loss of taste and we have tape recordings of these patients as to exactly how they felt. As you may know it is extremely difficult to do proper tests of taste and we attempted to do these but because it is not our primary interest to look at the side effects of drugs we did not pursue it. Having now switched to a much lower dose of captopril with equal efficacy in more severe patients, because most of our patients now are on long-term treatment and in those treatment resistant patients we have had no cases of loss of taste in the last 9 months.

F.W. WOLFF: Thank you. The next part of the question—and I am sorry to trouble you—is related to the agranulocytosis problem. We have been told that 15 patients out of 3000, an incidence of 0.5%, developed agranulocytosis. We are also told at the same time that many of these people have other drugs. Well, any of us who have been in drug research for a number of years know that this is very often the case. People are on numerous drugs at one time. However, an incidence of 0.5% of agranulocytosis is an absolutely unacceptable figure. I have heard this may be associated with the sulfhydryl radical which is of course essential to the action of this drug. I hope we will have a very full explanation of what this agranulocytosis amounted to, an explanation for it, and perhaps like metiamide, second generations of the drug may be developed where this does not occur. Could we have a response to the agranulocytosis problem?

J.C. ALEXANDER: It is difficult to give a precise incidence because it depends on the patient population at risk for the problem. We think the patient population at risk are those patients that have been treated under an "emergency" protocol within the last six months. These are the patients with severe hypertension receiving multiple drugs and having a variety of concomitant diseases, especially autoimmune diseases such as lupus. It is encouraging that in our moderate hypertension controlled studies in about 1500 patients we did not see a case. If the incidence is 0.5%, then the probability of seeing agranulocytosis in 1500 patients should be quite high. This condition occurs within the first three months of treatment. It is very rapidly reversible. We have a number of studies underway to get a better understanding of the problem. You have to remember that this is something that we have just observed in our studies, and we have not had time to fully explore all the ramifications of it.

F.W. WOLFF: The final part of this question refers to the proteinuria. Is this real proteinuria? Is there evidence of renal toxicity? Are studies going into renal toxicity adequate? Is there an explanation for this proteinuria? How severe is it?

J.C. ALEXANDER: Approximately 1 to 3% of patients treated with captopril for three months or more have developed significant proteinuria; nephrotic syndrome developed in about one fourth of these cases. The majority of cases of proteinuria occurred in patients with clinical evidence or history of renal disease. Where a biopsy was performed, the pathologic finding was membranous glomerulopathy. The proteinuria usually decreased or disappeared, even though captopril was sometimes continued at the same dose. However, some patients had persistent proteinuria for six months or more. Other parameters of renal function such as BUN and serum creatinine were seldom altered. Since patients with hypertension may have proteinuria, the presence or absence of urinary protein should be determined before starting therapy and periodically during therapy. Consideration should be given to the benefit vs. risk of continuing captopril therapy in any patient who develops unexplained proteinuria exceeding 1 g/24 hr, or proteinuria that is increasing progressively.

F. GROSS: Is anything known about whether this leukopenia is of toxic or allergic origin?

J.C. ALEXANDER: We do not know with certainty the mechanism of the neutropenia. Our hematology consultants who reviewed the cases thought it was a toxic effect. Two patients have been rechallenged with the drug without recurrence of the agranulocytosis.

G.A. MACGREGOR: I would like to comment about agranulocytosis. I am very concerned about it. On the other hand, as you know, all the reports apart from one (who was treated with 600 mg/day) have been in patients with renal failure. They have all been on what I regard as large amounts of captopril. Captopril and its metabolites accumulate in renal failure. I believe that by using lower doses there may be less toxicity, particularly in patients with renal failure. In our experience, the drug at a lower dose, e.g., 12.5 mg–25 mg t.i.d., is as effective in lowering blood pressure in the majority of patients as the large amounts currently being used.

F. SPINELLI: Blood pressure in patients treated with captopril appears to become dependent on extracellular volume. Does this imply a risk of overdose, especially in patients with the risk of severe hypotension?

J.C. ALEXANDER: Patients on diuretics and especialy those in whom diuretic therapy was recently instituted, as well as those subject to severe dietary salt restriction or undergoing dialysis, may occasionally experience a precipitous reduction of blood pressure within the first three hours after receiving the initial dose of captopril. An exaggerated hypotensive effect may also occur with one or two subsequent doses, but less frequently and with less severity. This marked response is attributed to the fact that angiotensin II plays an important role in the maintenance of blood pressure under conditions of sodium/volume depletion. The possibility of hypotensive effects can be minimized by either discontinuing the diuretic or increasing the salt intake approximately one week prior to initiation of treatment with captopril. Alternatively, provide medical supervision for at least three hours after the initial dose. If hypotension occurs, the patient should be placed in a supine position, and if necessary, receive an intravenous infusion of normal saline. This transient hypotensive response is not a contraindication to further doses which can be given without difficulty once the blood pressure has increased after volume expansion.

D.N. MCKINSTRY: Dr. MacGregor, I have a comment rather than a question regarding your studies in normal subjects. We have studied captopril in approximately 175 normal subjects. Most of them, admittedly, were single-dose studies, with subjects on their usual salt intake. Our experience has been that, in sodium replete subjects, either in the supine or standing position, we see no significant fall in blood pressure. We did one study in young healthy males in which we gave 100 or 200 mg 3 times daily for 10 consecutive days after they had been brought into sodium balance during a 5 day lead-in period in a metabolic ward, either on 100 mEq or 10 mEq sodium diet. We did, indeed, see a significant fall in standing blood pressure in normal subjects who had been on the 10 mEq sodium diet. The average daily decrease in mean blood pressure was about 10 to 15% compared with either their pretreatment blood pressure or with a parallel placebo control group. However, in the sodium replete subjects who received the same relatively high doses of captopril for 10 days, we did not see a significant fall in either standing or supine blood pressure. The only real difference I can see between your study and those we have conducted is that our subjects were confined to our clinical pharmacology unit in our studies rather than being ambulatory in the sense of going about their usual daily activities. However, our subjects were certainly ambulatory within the unit. I was wondering if you have any possible explanations for the differing results in sodium replete subjects in your study as compared with our studies.

G.A. MacGregor: I think there are two important things: firstly, the activity of the normal subjects. Our normal subjects were medical students who were going about their normal activities, except for the fact that all their food and drink was being obtained from the metabolic ward kitchen. Secondly, methodology in measuring blood pressure. The changes in blood pressure we are looking for are small, i.e. in the order of 10 to 15%. All blood pressures are measured in the same room by the same nurse, and at the same time of day for daily measurements. All the measurements are done on Arteriosondes with five recordings in each position, lying, sitting, and standing, and post-exercise measurements. In other studies that we have done in normal subjects, double-blind with a placebo and an active drug, we have shown under similar conditions no change in blood pressure with a placebo.

B.B. Levine: I have worked for many years in the field of drug hypersentivitity. I'd like to express my opinion that many of these reactions, except for the taste reaction, are most likely allergic in mechanism. Of course we wouldn't know whether this is the case until we do the appropriate immunological studies, which I assume as yet have not been done; however, I base my opinion on several facts: First, many mercaptans are sensitizers. We have shown d-penicillamine to be a very potent sensitizer in allergic contact dermatitis in guinea pigs. Many dermatologists over the years have found a variety of complex mercaptans to be sensitizers. Secondly, the clinical manifestations observed (rash, proteinuria, agranulocytosis) as well as the occurrence of these reactions several weeks or several months after the initiation of drug therapy are consistent with allergic mechanisms.

Z.P. Horovitz: Dr. Levine and I have had this discussion before and he's visited us and discussed the early results on captopril. At the moment we can't quite agree with him, although he may be more of an expert than we are. We have done a number of studies actually designed to look for an allergic reaction in animals and in patients' blood. We have not been able to find any sign of an allergic reaction. We have never been able to raise any antibodies to captopril, although I realize that's not a completely positive answer. We have been looking for antibodies for another reason, to try to develop a radioimmunoassay. I think what might explain the time factor other than an allergic response is that if you remember the protocols the time factor was also related to protocol dosage increase. So with the protocols we used, because of the quick dosage increase, we don't know whether the side effects were allergic type response or just drug overdose. It could be either one simply because of the very quick rise in dosage.

C.I. JOHNSTON: I would just like to make a comment on whether or not captopril lowers blood pressure in normals. I personally don't regard it as a very important question. It just depends on the condition of the study. It is quite clear that the acute hypotensive effect of captopril is very dependent on how much angiotensin II is reduced and how much your blood pressure at that time is dependent on angiotensin. I think if you have been running around like a medical student going into the wards and running back to get to bed to lie down and get your blood pressure taken, it is not surprising it falls a bit when Dr. MacGregor gives them captopril. Whereas, if subjects have been lying quietly in a room for 2 hr, the renin system is turned off and captopril is unlikely to affect blood pressure.

G.A. MACGREGOR: I agree with you that lying flat in a room for 2 hr will reduce the dependency of blood pressure on angiotensin II. Nevertheless, we do not treat high blood pressure by laying patients flat for 2 hr in a room. Our experiments were designed to see if the effect of captopril given over a period of a few days was the same in normotensive subjects as in patients with essential hypertension, both groups going about their normal activities.

R.K. FERGUSON: That's true. It was interesting to me that your plasma angiotensin levels in normals, which were 35 pg/ml, are in fact about 50% higher than ours.

G.A.MACGEGOR: Again, this reflects the difference in the way the samples were collected. Our normals, who are young, have all blood measurements done sitting upright, and these values are after ten minutes of sitting upright. It is not surprising, therefore, that the values of angiotensin II are higher than in normals lying flat.

R.K. FERGUSON: You say that if the patient's blood presure is due in part to angiotensin's effect and you block or remove that, then as long as the patient or the volunteer can compensate by other mechanisms, it's not a great problem. One place where it worries you is in the patients with congestive heart failure where it seems, at least in some cases, they cannot compensate, which can be a problem.

G.A. MACGREGOR: I was surprised that Dr. Ferguson did not mention weight in his studies of treatment-resistant hypertensives who were given captopril and a diuretic. Diuretics lower blood pressure through the loss of sodium and water opposed by a reactive rise in angiotensin II. If the rise in angiotensin II is blocked by captopril the blood pressure fall with the diuretic will be largely dependent on the amount of sodium and water lost or, in the short term, the weight loss. In similar patients to the ones he described we found that in some patients already on captopril we had to remove several kilos of weight, i.e., several liters of sodium and water, with large amounts of furosemide before we could control the blood pressure. I wondered what his experience was. The second question I would like to ask is why he found it necessary to add in propranolol to the combination of captopril and a diuretic.

R.K. FERGUSON: They didn't put on weight because for the most part our patients were already pretty well dried out from diuretic therapy and sodium restriction and there was very little change in weight by adding captopril; if anything it decreased slightly, though not a great deal. In the studies that I showed in our patients, about half were not controlled by captopril and diuretics although they certainly had a significant improvement in their blood pressure. The protocol allowed us to see what happened if we added propranolol and even other drugs. We added the propranolol and in *that* group of patients we could get a further small decrease in diastolic blood pressure, perhaps 4 to 5 mm, which was a surprise to us. Nonetheless, it seemed real, because when we subsequently withdrew the propanolol in some patients the blood pressure again increased.

D.N. MCKINSTRY: Dr. Ferguson and Dr. Alexander, correct me if I am wrong, but I think several of our investigators have reported to us that they see either a so-called biphasic or triphasic action of captopril in lowering blood pressure in some patients. They report an immediate fall in blood pressure manifesting itself within the first day or so of captopril therapy. The blood pressure then tends to increase towards pretreatment levels with continued captopril treatment. If a patient responding in this manner is continued on captopril monotherapy without the addition of other drugs, a secondary fall in blood pressure is observed in 6 to 10 days. Dr. Ferguson, I wonder how often the third drug, propranolol, might have been added during that period when some sort of reactive process was occurring after the initial fall in pressure and, if propranolol had not been added but the patient had simply continued captopril and diuretic, whether you might not have eventually achieved the same therapeutic result.

R.K. FERGUSON: I think that in some of our patients that is possible because we did not wait forever to try adding propranolol. But in other patients, that was not truc because we did wait a long enough period of time for this effect to occur. As I mentioned we probably got a greater blood pressure lowering initially than subsequently because the effects of the other drugs were still present. We went immediately from standard triple therapy to captopril. I suspect what you ask could be true in some patients, however.

M.J. ANTONACCIO: I think in animals there is a very good reason why propranolol might further lower blood pressure. After having already given captopril in combination with a diuretic, Dr. Ferrone has found, in SHR for instance, that captopril alone will increase cardiac output and decrease peripheral resistance. If you give the SHR a diuretic you get a further increase in cardiac output and further decrease in peripheral resistance. There is also a further fall in pressure because the resistance goes down relatively more than cardiac output goes up. Then, if you give propranolol, you reduce the cardiac output more than resistance and blood pressure might fall further.

SECTION V

The A. N. Richards Memorial Lecture

John H. Laragh

The Renin System in Hypertension: A Research Journey

John H. Laragh
Division of Cardiology, Cornell Medical Center
The New York Hospital, New York 10021

INTRODUCTION

I am pleased to be chosen by the Physiological Society of Philadelphia to deliver the annual A.N. Richards Memorial Lecture. I am expecially honored to be in the company of the many distinguished scientists who have given this lecture before me. At the beginning of my own research career A.N. Richards was one of my heroes. I was interested in the role of the kidneys in heart failure and how aldosterone governed sodium balance. I enjoyed reading the pioneering work of Richards and his group and of their use of micropuncture to define mechanisms of renal tubular electrolyte transport.

In this lecture I will review a research odyssey of more than 20 years that has led me and my colleagues to the view that all clinical states of high blood pressure are not alike and that they may be better understood and treated in reference to a biochemical profile (Laragh et al., 1979). The main features of that profile consist of a measure of the patient's sodium intake together with the measured values of a cascade of hormones (Laragh and Sealey, 1973)—renin, angiotensin and aldosterone. All together, we have learned these hormones compose a biologic control system that exerts constant cybernetic control over the body's balance of sodium and potassium and its blood pressure. This control system is unique in that it seems designed to protect tissue perfusion by first rapidly defending blood pressure, via angiotensin-vasoconstriction. This is followed by a slower-acting limb that restores flow, via aldosterone induced renal sodium retention, an effect which in turn osmotically expands the volume of the blood and extracellular fluids.

The significance of this new viewpoint is that it brings coherence and rationality to the understanding, and hence management, of one of man's most common group of disorders, one that has been called his "silent killer." Heretofore, "essential" hyptertension, the most common form, has been generally considered to be a single entity with a single cause, course and outlook, and so the possible etiologies within this category of hypertension have eluded definition. Thus, effective treatment programs perforce have been blind processes of trial and error.

My research journey began with an interest in the hormonal and renal mechanisms of salt (i.e., sodium and potassium) metabolism and their involvement in fluid retaining states such as congestive heart failure. We know that the intake-output balance of sodium, with its capacity for retaining water, determines the volume of the vascular fluid and so contributes hydraulically to blood pressure and flow. I was concentrating in my early work on the role of aldosterone, the recently discovered sodium-retaining adrenocortical hormone, which acts in the renal tubule to enhance distal tubular sodium transport thus establishing an electrochemical gradient to drive the concurrent urinary excretion of potassium. Our clinical studies of the behavior of aldosterone were made possible by its recent isolation and synthesis and by our development of a laborious, but accurate, double isotope dilution assay that permitted measurement of the minuscule quantities of hormone secreted daily (Laragh et al., 1960b).

OVERSECRETION OF ALDOSTERONE IN MALIGNANT HYPERTENSION

With this orientation and the methodology, we began our research into high blood pressure with a survey type of study that examined aldosterone secretion rates in the whole gamut of clinical hypertension. In this effort we expected to verify, and we did, a pattern of excessive aldostereone secretion, with depleted potassium, in patients with primary aldosteronism. This rare form of hypertension is sodium-volume dependent. It is cured by surgical removal of an offending adrenocortical tumor. We also established the presence of normal or nearly normal rates of aldosterone secretion in patients with common essential hypertension. However, in this study, it came as a great surprise to us to discover massive oversecretion of aldosterone in patients with malignant hypertension (Laragh et al., 1960b). The values were much higher than in any other form of clinical hypertension. The finding excited us because it was the first biochemical abnormality of potential pathophysiologic relevance ever identified among these patients. Its impressive degree also suggested that it might be involved in causation. Why was there this massive oversecretion? Did malignant hypertension resemble primary aldosteronism? Would it respond to the same treatment? No. Actually, we learned that even total adrenalectomy in these patients failed to reverse the disease; but surgical explorations showed us that the glands were not adenomatous but were bilaterally enlarged. We therefore reasoned that this adrenal hyperplasia of malignant hypertension must be a response to a blood borne chemical messenger originating elsewhere in the body. We suspected a renal signal because the kidney is the most damaged organ in malignant hypertension and because in research of another era the kidneys had been implicated both structurally and functionally in hypertension. Thus, in 1898, Tigerstedt and Bergmann (1898) had produced hypertension in rabbits by injecting saline extracts of minced rabbit kidneys, and had called this pressor agent renin. This work was not generally confirmed. Then, in 1934 Goldblatt, in his famous experiment (Goldblatt et al., 1934), raised the blood pressure of the dog by clamping one of its renal arteries. However, after another wave of enthusiastic research, renin had again fallen into disrepute as a relevant substance for hypertensive phenomena because it was not possible to relate its blood levels to either Goldblatt or human forms of renal hypertension. Nonetheless, we suspected that abnormal renin secretion might be linked to the abnormal aldosterone secretion of human malignant hypertension.

Renin, the enzyme secreted by the kidneys, was not available for experiment or direct measurement (it has only recently been purified), but some of its actions were known. It has no direct physiologic effects, but acts in the bloodstream enzymatically to cleave the decapeptide angiotensin I from a plasma globulin substrate of hepatic origin. This product is inactive, but upon meeting converting enzymes in a passage through the lung it is further cleaved to the pressor octapeptide angiotensin II. These two angiotensins had been isolated and synthesized. Theoretically, then, angiotensin II's formidable pressor qualities could account for a considerable portion of the elevated blood pressure in malignant hyptertension.

EXPOSING THE RENIN-ANGIOTENSIN-ALDOSTERONE "AXIS"

Accordingly, we decided to find out whether release of excess renin from the kidneys could explain the aldosteronism (Laragh et al., 1960B) we had discovered. Angiotensin II had just become available for clinical trial. During 1958 and 1959 we infused it into human volunteers and were excited to observe a prompt and large rise in their aldosterone secretion rates (Laragh et al., 1960a). This response was direct and specific for angiotensin and was not just related to the state of vasoconstriction or blood pressure, because no other pressor substance had this effect (Laragh et al., 1960a). We studied vasopressin, serotonin, epinephrine, norepinephrine and a number of their synthetic analogues.

These experiments in normal subjects exposed a specific biochemical signal system between three organs—the kidneys, the blood and the adrenal cortex—via renin to angiotensin to aldosterone (Goldblatt et al., 1934). We realized that we had revealed a coherent messenger system that we believed exerted bifold control over blood pressure and fluid volume by regulating vasoconstriction and sodium balance. Actually, we learned that Gross (1958) had suspected a renal-adrenal interaction from his animal studies. This research had already implicateed a renal-adrenal interplay in the control of sodium balance but not in blood pressure regulation.

RENIN EXCESS EXPLAINS THE PATHOGENESIS OF MALIGNANT HYPERTENSION

At the same time our experiments allowed us to develop an explanation for the pathophysiology of human malignant hypertension (Fig. 1).In this hypothesis (Laragh, 1960) we proposed that, when the kidneys develop a critical degree of damage, the malignant syndrome is initiated by the inappropriate release of large excesses of renin into the bloodstream. This creates excesses of angiotensin, which raise pressure by vasoconstricting arterioles and which also stimulate aldosterone oversecretion. Aldosterone, acting on the kidney, attempts to cause sodium retention, an effect which would restore blood volume and renal flow and thereby normally would "turn off" the initial signal for renin release. However, the kidneys, because they are damaged, cannot respond fully to the command of aldosterone. Instead, a vicious circle (i.e., short-circuit) is initiated in which there is more renin secretion, more angiotensin, more aldosterone and higher and higher pressures, with hypokalemia (Fig. 1).

There is now considerable evidence to indicate that such an excess of angiotensin and aldosterone in the blood is critically related to the causation not only of severe hypertension but also to fibrinoid and necrotizing arteriolitis and the fatal outcome characteristic of untreated malignant hypertension. Thus, in practically every animal species studied, administration of either renin or angiotensin with aldosterone, in the presence of available dietary sodium, produces severe vasculitis, often with death of the animals, in a few days (Masson et al., 1962; Gavras et al., 1974a).

Equally convincing evidence for this vasculotoxic role of renin-angiotensin-aldosterone excess in malignant hypertension has come from the human experience. A number of groups (Vertes et al., 1969; Gleadle et al., 1969; Onesti et al., 1968) have observed that a total nephrectomy will reverse malignant hypertension with biopsies showing healing of the vasculitis as blood pressure and flow are returned to normal. Actually, a similar sequence sometimes occurs in the severe hypertension of unilateral renal disease, too, where the hypertension and attendant vasculitis, hypokalemia and alkalosis can be corrected by a nephrectomy or repair of the renal vasculature.

Thus, it became apparent that among hypertensive patients who have overt oversecretion of aldosterone there are two discrete syndromes.

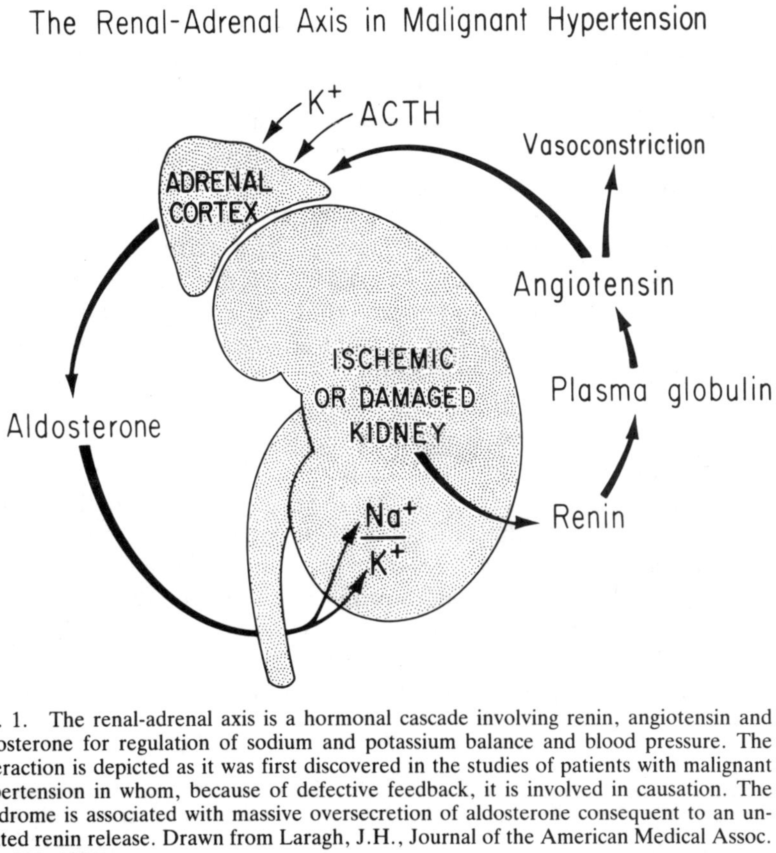

Fig. 1. The renal-adrenal axis is a hormonal cascade involving renin, angiotensin and aldosterone for regulation of sodium and potassium balance and blood pressure. The interaction is depicted as it was first discovered in the studies of patients with malignant hypertension in whom, because of defective feedback, it is involved in causation. The syndrome is associated with massive oversecretion of aldosterone consequent to an unabated renin release. Drawn from Laragh, J.H., Journal of the American Medical Assoc. 174, 293, 1960.

The first, primary hyperaldosteronism or Conn's syndrome, is the expression of autonomous aldosterone oversecretion in an otherwise healthy subject. The aldosterone excess induces sodium retention, volume expansion and hypertension with the consequent suppression of renin secretion. It also causes renal potassium wastage, with muscle weakness and polyuria. These abnormalities are entirely correctable by removal of the oversecreting adrenal adenoma.

The second hypertensive syndrome with hyperaldosteronism is malignant hypertension. In contrast to the first, it is characterized clinically by severe hypertension and an accelerated pace, and pathologically by adrenal hyperplasia and fibrinoid vasculitis. There is either bilateral or unilateral renal disease with excessive secretion of renin as the basic for the accelerated hypertension.

The benign course of primary hyperaldosteronism as compared with malignant hypertension can be accounted for in terms of differences in behavior of the proposed renal-adrenal control system. A primary adrenal oversecretion of aldosterone, as in primary aldosteronism, works to cause sodium volume excess and thereby to suppress renin release from a normally functioning kidney. In contrast, malignant hypertension is initiated by inappropriate renin secretion and so is characterized instead by intense vasoconstriction from too much angiotensin plus the consequences of too much aldosterone.

Accordingly, in these two hypertensive syndromes, for a similar rise in blood pressure different mechanisms are at work to explain the gross differences in vascular damage. In primary aldosteronism, hypertension occurs with overfilling of the circulation from a *sodium volume* excess and renin is consequently low. In contrast, malignant hypertension is accompanied by intense *vasoconstriction* from a renin excess. There is poor systemic flow with acrocyanosis, hemoconcentration and generalized tissue ischemia. Aldosterone oversecretion, elicited by angiotensin, contributes to the syndrome by driving sodium retention linked to kaliuresis. It seemed likely to us that the presence of the renin induced vasoconstriction was the difference that explained the much greater target organ damage of malignant hypertension as compared with the low renin state of primary aldosteronism. In our future studies of essential hypertension, we pursued and developed this concept to two different final mechanisms in hypertension, one in which a *volume* factor predominates as opposed to a largely *vasoconstriction* factor in the other.

HOW THE RENIN-ANGIOTENSIN-ALDOSTERONE AXIS FUNCTIONS

It is our view that the renin "axis" is a biological control system designed to maintain arterial blood pressure and sodium balance simultaneously by responding to changes in renal perfusion (Laragh and Sealey, 1973; Laragh et al., 1972) (Fig. 1). Renin is secreted in response to a whole variety of factors that reduce arterial blood pressure and renal perfusion, such as shock, hemorrhage or heart failure, or which reduce flow in the distal renal tubule, as in sodium depletion. Renin, with no physiologic action its own, acts enzymatically on a plasma protein to release the inactive decapeptide angiotensin I, which is then rapidly hydrolyzed to the active octapeptide, angiotensin II, by pulmonary, plasma or tissue converting enzymes. Angiotensin II, the most potent natural pressor known, raises blood pressure quickly by constricting the arterioles. At the same time it stimulates the secretion of aldosterone, which acts over the next several hours to cause renal sodium retention and potassium excretion. The retained sodium leads secondarily to water retention, expanding the extracellular fluid volume. By restoring flow, further support is given to the blood pressure. Acting in concert, angiotensin II and aldosterone restore systemic blood flow and renal perfusion, thereby compensating the system. Since the stimulus for renin secretion is reduced renal perfusion, the renin induced increases in pressure and in flow provide the negative feedback signals that shut off renin release. These signals involve baroreceptor mechanisms in the afferent renal arterioles, as well as a sodium load receptor in the mucula densa portion of the renal tubules; there is now good evidence that beta-adrenergic activity negotiates an efferent part of the baroreceptor communication (Buhler et al., 1972).

COULD ANGIOTENSIN II BE A VECTOR IN ESSENTIAL HYPERTENSION?

Following this discovery of the renin-aldosterone system for control of blood pressure and sodium balance (Laragh et al., 1960a), and the demonstration of its crucial participation in malignant (Laragh et al., 1960a; Laragh, 1960) and subsequently in other unusual forms of clinical hypertension (Laragh et al., 1972), investigators continued to doubt a role for renin in "esential" hypertension. Perhaps this was because of a general failure in the past to relate plasma renin levels to blood pressure levels in this large group, and also because many such hypertensive patients were known to exhibit ostensibly "normal" renin levels.

However, because of the powerful pressor actions of angiotensin and aldosterone, our research group felt that this control system must participate in some way, at least, in some forms of essential hypertension. One key question was whether so-called "normal" circulating levels of these hormones could be sufficient to contribute to the hypertension.

A research series changes traditional thought on this question. The evidence began with studies in which we infused angiotensin or norepinephrine continuously into normal human volunteers for up to 11 days (Ames, 1965) (Fig. 2). With angiotensin, we found that, as the infusion proceeded, less and less angiotensin was needed to raise the

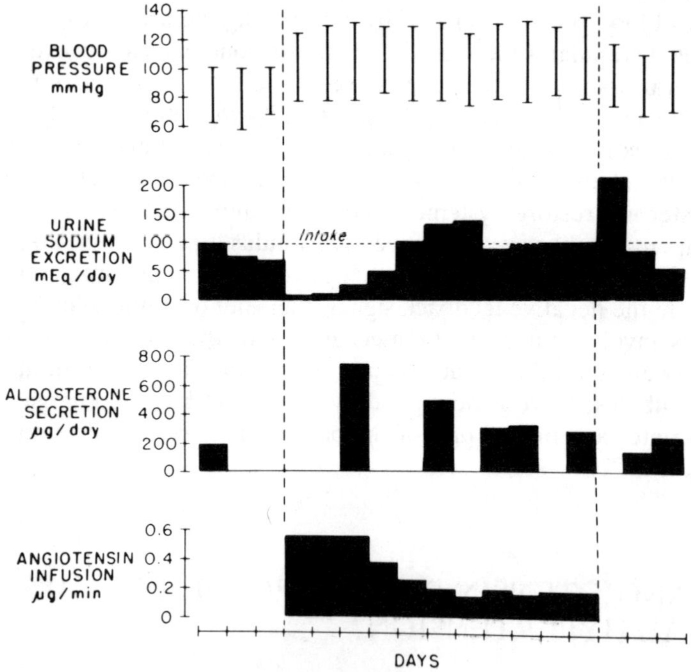

Fig. 2. The prolonged continuous angiotensin infusion in a normal subject for 11 days. The dose of angiotensin was adjusted to keep a mildly pressor response. Angiotensin II induced a marked and selective increase in the adrenal cortical secretion of aldosterone together with consequent sodium retention. As sodium was retained angiotensin became *more* pressor. Because of this increasing pressor sensitivity to angiotensin the dose was serially reduced to a point where aldosterone secretion returned to control levels. Thus the pressor sensitivity to angiotensin increased as sodium retention progressed. The results indicate that angiotensin, unlike norepinephrine, can produce and then sustain hypertension in diminishingly small amounts. Redrawn from Ames R.P., Borkowski A.J., Sicinski A.M., and Laragh J.H. J. Clin. Invest. 44: 1171–1186, 1965.

pressure, because of activation of aldosterone and resultant sodium retention and weight gain, so that very small amounts became pressor by the fourth or fifth day (Fig. 2). Norephinephrine exhibited the opposite features so that more and more hormone was needed to keep the pressure elevated as sodium balance became negative. All this meant to us that angiotensin fit the requirements for an active pressor agent over the long term even in relatively small amounts. In contrast, circulating norephinephrine levels failed to fit the requirements of this model as a vector of long-term hypertension.

RENIN IN "PROFILE" WITH SODIUM

Recognizing that renin and aldosterone secretion fluctuate widely but predictably in normal people according to the state of their sodium balance, in 1967 we described procedures called *renin-sodium profiling* (or *renin profiling* for short) and *aldosterone profiling* to evaluate the normalcy of the system in hypertensive patients (Laragh et al., 1972). In this approach hormone measurements are referenced against the concurrent 24-hr rate of urinary sodium excretion, used as an index of intake and of balance. The values from patients are compared to curves obtained from studies on normal subjects (Fig. 3). Venous blood for renin assay is collected in seated ambulatory patients so that this test also assesses the postural stimulus to renin secretion, which, for the most part, we know to be neurogenically mediated. The second essential for evaluating the system is the use of a reliable assay for plasma renin activity, sensitive enough to fully explore and discriminate the subnormal range. The requirements for a simple, reliable renin assay (Sealey and Laragh, 1977) and the physiologic protocol for performing renin-sodium profiling (Laragh and Sealey, 1977) in human subjects have recently been updated.

RENIN PROFILES IN THE SPECTRUM OF HYPERTENSIVE DISEASES

Following exposure of the active renin participation in malignant hypertension, other research defined particular abnormalities of the renin control system which appear to play a causal or participatory role in three and possibly four forms of hypertensive disease (Laragh et al., 1972). These four conditions are: malignant hypertension (excess renin and aldosterone), primary aldosteronism (excess aldosterone and low

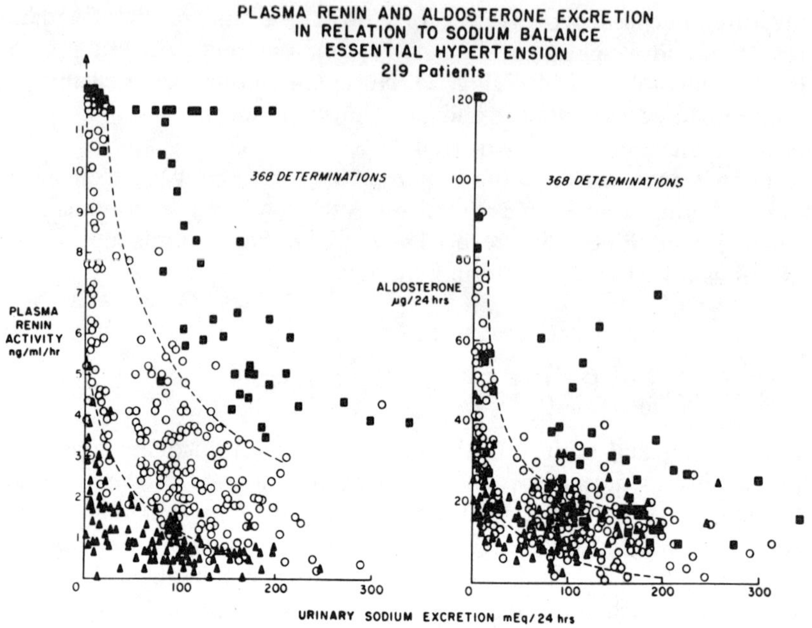

Fig. 3. The relation of the noon ambulatory plasma renin activity and of the corresponding daily urinary aldosterone excretion (right panel) to the concurrent daily rate urinary sodium excretion. The dotted lines define the normal channel derived from the study of normotensive people. Two hundred nineteen patients with untreated essential hypertension were studied, some on several occasions at different levels of sodium intake. Triangles indicate low renin, open circles indicate normal renin, squares indicate high renin essential hypertension. Three major subgroups are defined by the appropriateness or normalcy of the plasma renin activity to the rate of sodium excretion which is used as an index of dietary intake and of sodium balance. Additional abnormal subgroups are defined when aldosterone (right panel) is included in the analysis. From Brunner et al., New Engl. J. Med. 286: 441, 1972.

renin), renovascular hypertension (excess renin) (Vaughan et al., 1973), and oral contraceptive hypertension (excess renin substrate) (Laragh et al., 1967).

ESSENTIAL HYPERTENSION DIFFERENTIATED

Moreover, in using the renin-sodium profiling technique to study *essential hypertension*, by far the most common form, we developed evidence suggesting that this category is a biochemically heterogeneous group (Brunner et al., 1972). Patients with this condition fell into three major physiologic subgroups exhibiting either a low (about 30%), normal

(55%) or high (15%) renin profile (Fig. 3). Many studies throughout the world have described similar distributions.

RENIN SUBGROUPS RELATED TO NATURAL HISTORY OF ESSENTIAL HYPERTENSION

Our companion clinical data (Brunner et al., 1972) indicated that patients within the three renin subgroups of essential hypertension might have different disorders, because their different renin levels were related to differences in their clinical course and their prognosis. Low renin patients were found to suffer fewer heart attacks and strokes than did normal or high renin patients. The validity of this observation has at times been questioned, but most such reports have not tested the proposal critically (Laragh, 1978). The fact stands that low renin patients, as a group, are significantly older by a least 9 years than patients in the other two subgroups, even though they are often more hypertensive. In view of the fact that low renin patients have higher blood pressure than do the medium renin patients, a logical explanation for the low renin group's greater longevity is that they enjoy a measure of protection from the cardiovascular damage that is more often associated with a high renin level and its attendant vasoconstriction.

Subsequently both clinical and experimental evidence that excessive renin activity, directly or indirectly, is *vasculotoxic* and so predisposes to vascular injury has grown (Gavras et al., 1974a; Laragh, 1978; Brunner et al., 1973a). Thus, patients with malignant or renovascular hypertension and high plasma renin levels (Laragh et al., 1972) or with renin-secreting tumors (Orjavik et al., 1975) have more striking vascular damage; in animals, injections of renin are extremely damaging to the cardiovascular system, especially the heart, brain and kidneys (Gavras et al., 1974a). Moreover, various acute clinical situations associated with high renin levels (e.g. renal artery graft closure, renal trauma and saralasin rebound [Keim et al., 1976]), are accompanied by stroke, encephalopathy or heart attack. At the same time, other studies in humans have established that a plasma renin measurement truly reflects the degree of its active vasoconstriction (Brunner et al., 1974; Case et al., 1976b, 1977). It is such vasoconstriction with impaired microcirculatory flow and tissue ischemia that might cause or predispose to target organ damage. All of this has led us to believe that renin levels could be useful clinically for exposing differences in risk that reflect real differences in pathophysiology.

PHARMACOLOGICAL PROOF OF RENIN INVOLVEMENT IN ESSENTIAL HYPERTENSION

Having applied renin-sodium profiling to reveal and characterize the biochemical and epidemiological heterogeneity among patients with essential hypertension, the stage was set to perform more definitive studies to expose and quantify renin system involvement in so-called essential hypertension.

THE BETA-ADRENORECEPTOR–RENIN SECRETION LINKAGE

A giant stride forward was taken in 1972 when Bühler and associates first reported (Bühler et al., 1972) that the antihypertensive action of the beta-adrenergic receptor blocker, propranolol, is directly related to the pre-existing plasma renin levels and also to the degree of which the drug reduces the renin level. Moreover, the induced fall in the renin levels caused by the drug preceded the fall in blood pressure. This research showed that propranolol was most effective in lowering blood pressure in patients with high renin levels, significantly effective in patients with "normal" or medium renin levels, and —of equal theoretical importance—it was ineffective or even *pressor* in hypertensive patients with low plasma renin levels (Fig. 4). To my knowledge this report provided the very first evidence, albeit circumstantial, for the pressor participation of the renin system in major fractions of patients with essential hypertension, while also showing that a major mechanism for the antihypertensive action of propranolol is mediated via renin suppression. The study showed that monotherapy with a beta receptor blocker could be effective in appropriately selected subgroups of patients, since, in them, propranolol by itself produced striking effects without the need to add other agents to the regimen.

This study verified our suspicion that a "normal" renin level in hypertensive patients is in fact abnormal since the expected physiologic response of the renin control system to a high blood pressure would be a suppression or a "turnoff" of renin secretion in an effort to bring the blood pressure back down (Laragh and Sealey, 1973) (Fig. 1). This failure of renin to turn itself off could be pathologically involved in these patients.

While this research strongly implicated renin suppression in the antihypertensive action of propranolol it certainly did not establish it as the *only* mechanism by which beta-blockers might lower pressure. How-

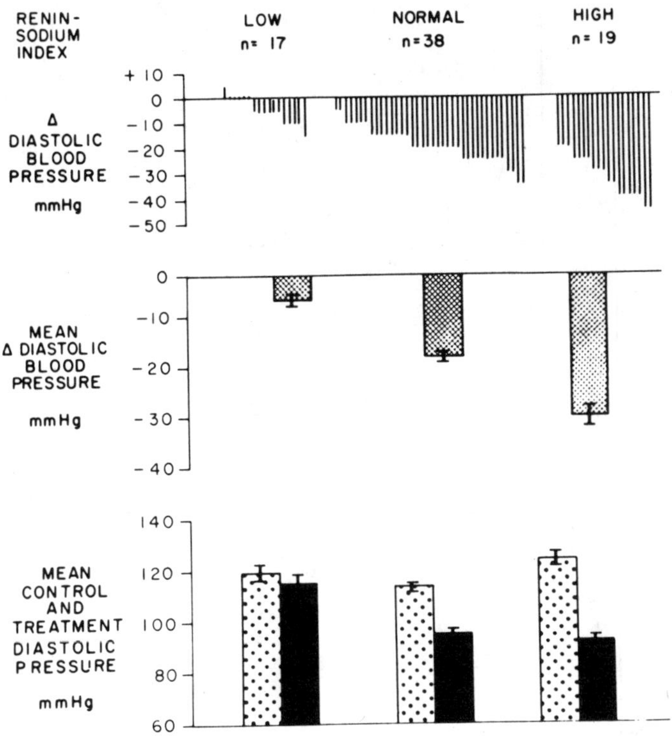

Fig. 4. Propranolol-induced changes in diastolic blood pressure in 74 patients. Absolute changes in diastolic pressure in each individual patient are depicted in the upper section. Other results are expressed as mean ± SEM for the low, normal and high renin subgroups. In the lower section, the dotted bars represent control blood pressures and the solid bars pressure after propranolol administration. Propranolol produced dramatic improvement in blood pressure in high and in many normal renin patients and conversely no effect in the low renin patients. The changes in blood pressure were also closely related to induced decrements in plasma renin. The studies indicate a major antirenin action of propranolol and in so doing, for the first time they suggested major role for renin in sustaining the high blood pressure of most patients with "essential" hypertension. From Bühler et al., Amer. J. Cardiol. 32: 511, 1973.

ever, as of today, alternate theories, invoked to explain the main antihypertensive action of beta blockers, are not supported by either clinical observations or investigative data. For example, the theory that beta blockers lower blood pressure by lowering cardiac output must account for an apparent inconsistency. Cardiac output is lowered in all patients treated with beta blockers, and yet blood pressure may either be unaltered, rise or decrease greatly. Thus this cardiac theory must then invoke some secondary mechanism, related to variable reactive changes in peripheral resistance consequent to the reduction in cardiac output

Table 1. The selective effectiveness of antihypertensive drugs.

Beta Blockers are Preferentially Effective in High and Normal Renin Patients	*Diuretics* are Preferentially Effective in Low Renin Patients
Bühler et al. N Engl J Med 287:1209, 1972	Crane et al. Am J Med Sci 260:311, 1970
Bühler et al. Am J Cardiol 32:511, 1973	Spark et al. Ann Intern Med 75:831, 1971
Bühler et al. Am J Cardiol 36:652, 1975	Crane et al. Am J Med 52:457, 1972
Castenfors et al. Acta Med Scand 193:189, 1973	Carey et al. Arch Intern Med 130:849, 1972
Pettinger et al. N Engl J Med 292:1214, 1975	Adlin et al. Arch Intern Med 130:855, 1972
Karlberg et al. Am J Cardiol 27:642, 1976	Vaughan et al. Am J Cardiol 32:523, 1973
Hollifield et al. N Engl J Med 295:68, 1976	Castenfors et al. Acta Med Scand 193:189, 1973
Weidman et al. Klin Wochenschr 54:765, 1976	Distler et al. Dtsch Med Wochenschr 99:864, 1974
Menard et al. Am J Med 60:886, 1976	Douglas et al. JAMA 227:518, 1974
Stumpe et al. Am J Med 60:853, 1976	Karlberg et al. Am J Cardiol 37:642, 1976
MacGregor et al. Clin Sci Mol Med 50:18p, 1976	MacGregor et al. Clin Sci Mol Med 50:18p, 1976
Boerth, Pediatr Res 10:328, 1976	
Moore et al. Lancet II:67, 1976	
Bahr et al. Clin Pharm Ther 20:130, 1976	
Zech et al. Postgrad Med J 53:(Suppl 3):134, 1977	
Philipp et al. Dtsch Med Wochenschr 102:569, 1977	
Beta Blockers Can Actually *Raise* Pressure in Low Renin Patients	*Diuretics* Can Actually *Raise* Pressure in High Renin Patients
Drayer et al. Am J Med 60:897, 1976	Baer et al. Ann Intern Med 86:257, 1977

(Hanssan, 1973; Tarazi and Dustan, 1972). However, such postulated reactive changes in arteriolar tone would still have to be shown to be independent of falls in resistance produced by the induced reduction of angiotensin II levels. Another alternate theory, that beta blockers act on the central nervous system, has been proposed but has received little support (Conway et al., 1978; Pickering, 1979), especially since some beta blockers do not enter the central nervous system and since higher dosages of beta blockers are more often *pressor* than depressor (Atsmon et al., 1972).

Some opposition to the renin theory of propranolol's antihypertensive action also developed when some investigators could not demonstrate a renin lowering effect using certain other beta blockers. The details of this controversy have been summarized elsewhere (Pickering, 1979). But at least one such claim has been retracted (Amery et al., 1976) as it has become generally appreciated that beta-blocking drugs as a species all exhibit a marked suppressing action on renin secretion. The failure to demonstrate renin suppression with a beta blocker in some attempts was very likely due to the use of renin methods that inadvertently activated the inactive form of plasma renin (prorenin) which is not suppressed by beta-blockade therapy (Atlas et al., 1977).

At the same time, more clinical and experimental evidence has accumulated that strengthens the view that the major antihypertensive action of beta blockers is mediated via renin suppression. This includes the lack of antihypertensive action or the *pressor* effect of beta-adrenergic blockers in low renin states such as low renin essential hypertension (Drayer et al., 1976; Bühler et al., 1975) (Table 1), in primary aldosteronism and in anephric humans. Furthermore, a group of animal experiments where more rigid control is possible also show that propranolol and other beta blockers are in fact effective in lowering pressure in relation to the height of the control renin level, and are quite ineffective in lowering pressure or even pressor in hypertensive models where renin in low or is absent (Chiu and Sommer, 1979; Gaina et al., 1979; Lydtin and Sommerfeldt, 1972). Moreover, propranolol is ineffective in lowering blood pressure in experimental angiotensin II infusion hypertension (Young and Diepstra, 1978).

SARALASIN

Convincing complementary evidence for the participatory role of renin in essential hypertension has emerged from subsequent studies using other pharmocological probes that specifically block either angiotensin II action at the receptor site (saralasin) or the formation of angiotensin

II (converting enzyme inhibitors). The broad experience from many research groups using acute infusions of saralasin (Baer et al., 1977; Brunner et al., 1973; Case et al., 1976a; MacGregor and Dawes, 1976; Marks et al., 1976; Streeten et al., 1975) found that only about 15 to 20% of patients with essential hypertension have a significant renin factor as revealed by a depressor response to this agent.

The fact that saralasin testing identified a smaller fraction (about 15–20%) with a putative renin participation than did the propranolol experience (about 50%), we believe to be adequately explained by the fact that saralasin is in itself a weak angiotensin-like agent and acts as an agonist particularly in situations where receptors are partially or completely vacant (that is, in medium or low renin patients) (Case et al., 1976s). This causes a gross underestimation of the true renin participation in the blood pressure level (Fig. 5). For this reason in 1973 we sought for and found another type of antirenin pharmocologic probe which had no agonistic activity. Indeed, it was at this time that I first got in touch with Dr. Zola Horovitz at The Squibb Institute for Medical Research to begin our continuing productive interaction. They had the compound, SQ 20,881, we were looking for, Dr. Horovitz was most cooperative and, fortunately too, nobody else at that time seemed interested.

CONVERTING ENZYME INHIBITORS

Teprotide

In our studies from 1973–77, the differences between estimates of renin system participation in essential hypertension obtained with beta-receptor blockers and saralasin were resolved. This was possible because we had found that the nonapeptide converting enzyme inhibitor (SQ 20,881 or teprotide) had no intrinsic pressor properties of its own, but produced virtually complete blockage of angiotensin II generation (Case et al., 1977; Gavras et al., 1974b). This first converting enzyme inhibitor, teprotide, had to be given intravenously. Using this agent, we demonstrated (Case et al., 1976b, 1977; Gavras et al., 1974b) broad participation of the renin system in essential hypertension as revealed by an acute depressor response to the intravenous infusion in 89 patients. Thus, all high renin patients and 90% of the medium renin patients, i.e., altogether about 70% of patients with essential hypertension, revealed a significant to large renin factor operating to sustain their hypertension (Fig. 6). The striking parallelism of the depressor responses within the spectrum of patients with essential hypertension to propran-

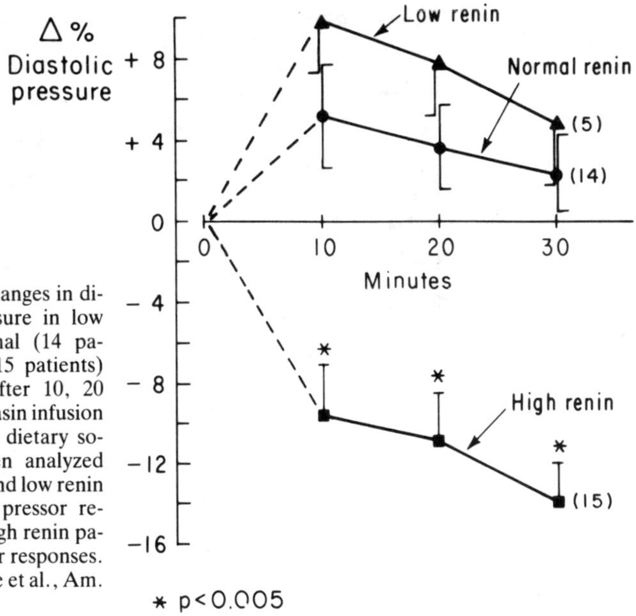

Fig. 5. Percent changes in diastolic blood pressure in low (5 patients), normal (14 patients) and high (15 patients) renin subclasses after 10, 20 and 30 min of saralasin infusion while on a normal dietary sodium intake. When analyzed by group, normal and low renin patients exhibited pressor responses whereas high renin patients had depressor responses. Redrawn from Case et al., Am. J. Med. 60, 1976.

olol, to saralasin or to converting enzyme inhibitor identifies their common mode of action largely related to an antirenin effect. In fact, the degree of response to all the three types of blocking drugs was predictable from the baseline renin sodium profile measurement. Moreover, the converse is true. All three types of antirenin drugs were relatively or absolutely ineffective in the 30 percent or so of patients with low renin essential hypertension and they were also ineffective in the other low renin states of primary aldosteronism and anephric subjects. These patients respond instead to diuretics or other means of volume depletion, evidence which suggests that sodium volume excess is instead an important mechanism in maintenance of their hypertension.

CAPTOPRIL

The practical translation of all of this research has become a reality with the development by Squibb scientists of captopril, the first orally active converting enzyme inhibitor. In an ingenious piece of biochemical pharmacology this compound was modeled after the nonapeptide, fitted to the active center of the enzyme and built for oral absorption (Ondetti et al., 1977). As one might expect, the experience to date using oral captopril (Brunner et al., 1979) retraces what we had described using

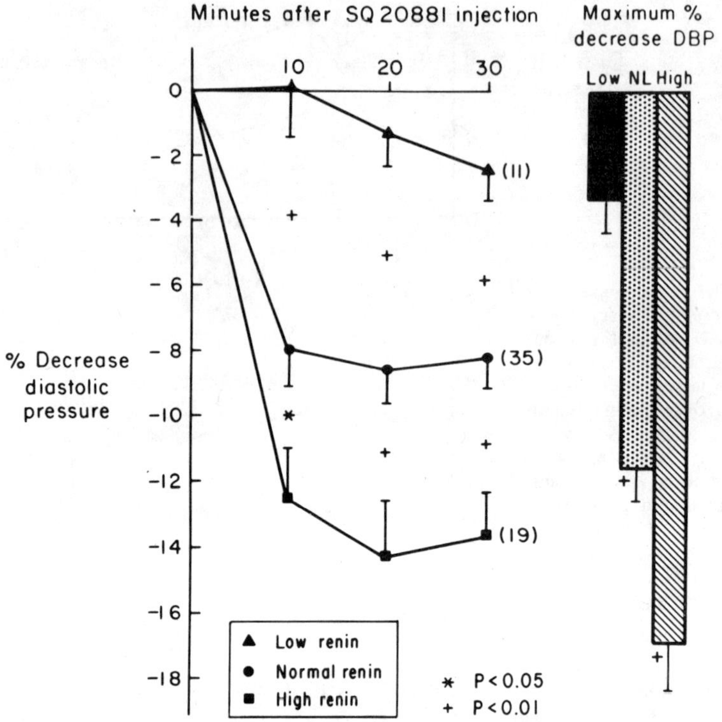

Fig. 6. Depressor responses to converting enzyme blockade in 65 seated hypertensive patients. The left panel represents the time course of the diastolic pressure changes, comparing the value of the low renin (triangle), normal renin (circle), and high renin (square) patients. The panel at the right compares the maximal percent decreases in diastolic blood pressure (DBP) of the three subgroups. The data show a striking antihypertensive effect in high renin patients, a lesser but still impressive action in most normal renin patients and no significant antihypertensive action in low renin patients. From Case, Wallace, Keim, Weber, Sealey, Laragh. N. Engl. J. Med. 296: 641–646, 1977.

teprotide (Case et al., 1977; Gavras et al., 1974b). In clinical evaluations to date (Case et al., 1978; Brunner et al., 1979), captopril has proven to be an extremely effective and potent antihypertensive agent that lowers blood pressure according to the height of the control renin level. Moreover it lowers blood pressure with fewer side effects, suggesting that maybe we are striking even closer to the causal process. While we need to learn more about captopril action it already seems clear that the blood pressure lowering effect may occur in stages (Fig. 7). There is an immediate response occurring by 90 min after an oral dose, often followed by a rebound of several days or more before the final antihypertensive benefit is re-achieved. This may be due to a transient

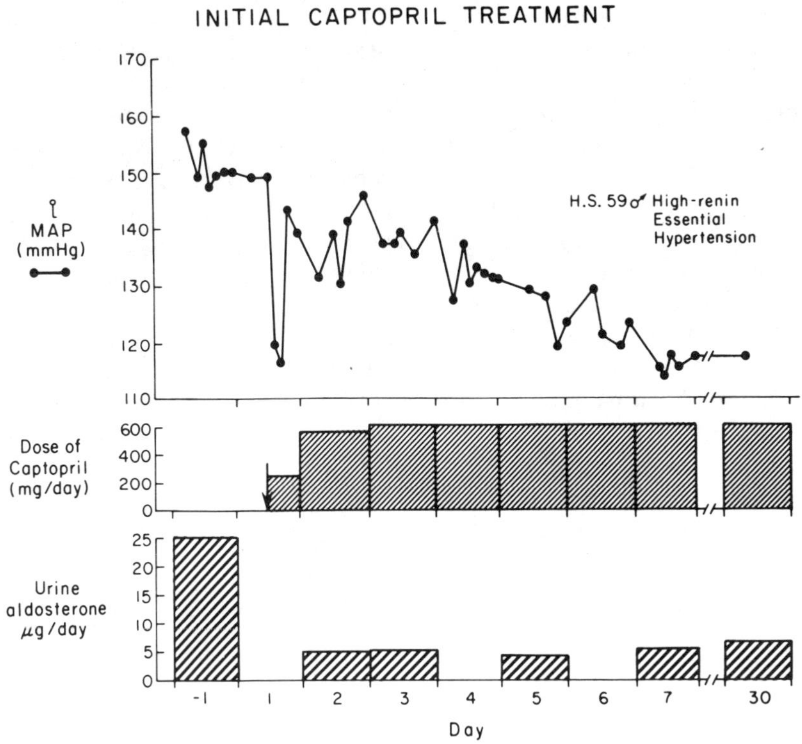

Fig. 7. Course of blood pressure response in a 59-year-old man with high renin essential hypertension. Mean arterial pressure (MAP) fell strikingly after the first dose but then returned toward control levels during the first 5–6 days before reaching a stable level of correction representative of the long-term blood pressure. This triphasic response occurred despite an increase in captopril dosage to 600 mg/day and a decline in urine aldosterone excretion to 20% of the pretreatment level. The aldosterone suppression was maintained for 30 days. From Laragh, Case, Atlas, Sealey, *Hypertension*, 1980 2:586–593.

baroreceptor system opposition to the drug (Laragh et al., 1980). Undoubtedly, one of the long-term benefits of the new agent derives from the more slowly expressed reduced aldosterone secretion following the converting enzyme blockade. The delayed response pattern illustrated (Fig. 7) is of considerable clinical relevance since it tells the physician to restrain himself from adding a second drug until a week or so of treatment with captopril by itself.

One of the great conceptual charms of captopril-type therapy is the fact that as the blood pressure is lowered, flow to vital organs is maintained or even increased. This is in contrast to diuretic treatment or even to beta blockade, where arterial flow is reduced to a greater or lesser extent as part of the tradeoff for reducing pressure. A second attractive feature is the fact that quite unlike other vasodilators (e.g., hydralazine, minoxidil), there is no reflex tachycardia or cardiac stimulation. The reason for the latter poses a key research question possibly linking angiotensin action to the baroreceptor system. Another good feature seems to be the fact that the captopril–converting enzyme blockade is not complete but just operates to dampen the renin system. Thus, in response to physiologic stimuli such as posture or exercise renin rises enough to override the block and stimulate enough angiotensin and aldosterone to achieve homeostasis. Indeed, if the blockade were absolutely complete captopril-type therapy might prove dangerous.

All of this work now allows us to say, with certitude, that a properly performed plasma renin measurement is, after all, a true measure of the degree of active reniin vasoconstriction. Thus, this test becomes the first

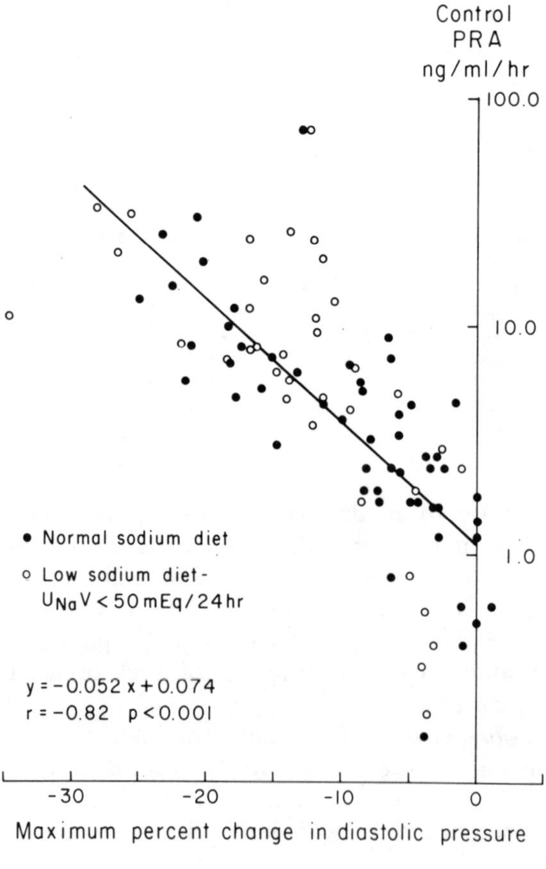

Fig. 8. Relation between pretreatment level of plasma renin activity (PRA) and the maximum induced change in diastolic pressure by the nonapeptide SQ20881 (teprotide) in 89 seated hypertensive patients. A dose of 1 mg was given to patients on normal sodium intake (closed circles) and to those after mild dietary sodium depletion (open circles). Significant falls in diastolic blood pressure (75%) occurred with PRA values of 2 ng/ml/hr or greater. From Case et al., N. Engl. J. Med. 296: 641–646, 1977.

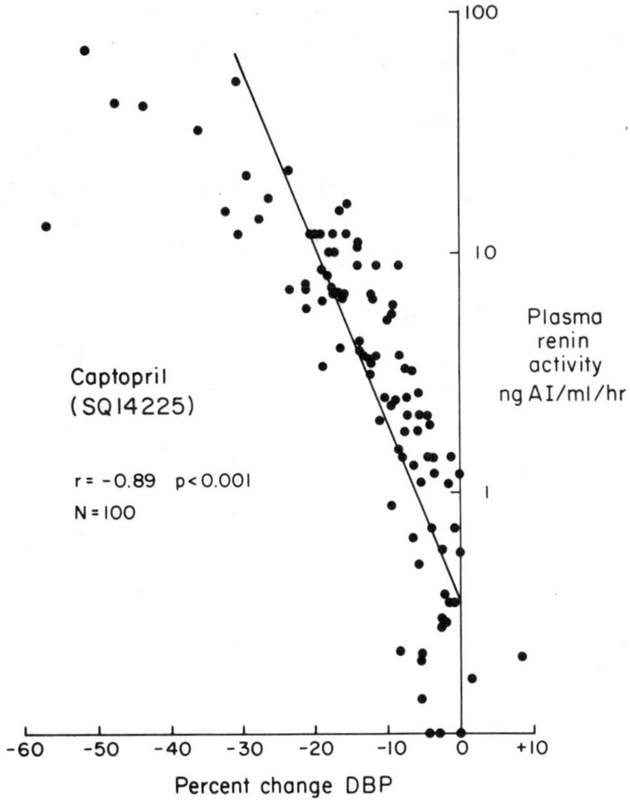

Fig. 9. Relationship between the pretreatment level of plasma renin activity and the precent changes is diastolic blood pressure (DBP) 90 min after the first oral dose of captopril (SQ 14225). Courtesy of Dr. D.B. Case.

reliable measurement of vasoconstrictor activity in patients. The simultaneously collected 24 hr urinary sodium measurements determine the appropriateness of this vasoconstriction for the sodium volume status of the patient. The correlations between the control plasma renin activity and the degree of blood pressure reductioon produced by either the nonapeptide, teprotide or captopril are extremely impressive. They are derived from our research beginning in 1973 and now include 89 patients given the nonapeptide (Fig. 8) and 100 patients given captopril (Fig. 9). If you will allow me to mix my metaphors, these data are the icing on the cake for our research journey that exposed the major renin system involvement in human hypertension.

A PRACTICAL INTERLUDE: HOW TO MEASURE PLASMA RENIN ACTIVITY

One of the reasons for the long delay in appreciating the role of the renin system in human hypertension has surely been methodological. Actually, it is unfortunate but true that renin assays widely applied over the past 10 years or so (Table 2) have contained large sources of error that render much of the clinical data obtained with their usage inaccurate or incorrect. The errors built into these three methods resulted from an unawareness at the time of their development of the presence in plasma of large amounts of inactive renin (i.e. prorenin) (Sealey and Laragh,

Table 2. Three inaccurate or insensitive renin assays.

Method	Faults	Resulting Problem
1. Skinner (1967)	Acid treatment activates inactive plasma renin	Results can be from 1 to 20 times too high with inconsistency in this error. Patients with PRA = 0 (i.e. anephric or primary aldosteronism) will appear to have active plasma renin. Renin does not appear to fall with beta-blockade.
2. Brown, Davies, Lever, Robertson and Tree (1964)	Same	Same
3. Haber, Koerner, Page, Kliman and Purnode (1972)	1. No pH control during incubation—as pH rises renin is destroyed 2. Incomplete angiotensin—ACE inhibition by BAL + 8-OH-quinoline 3. BAL also inhibits renin activity	Generally less accurate, especially so for defining low-renin values

Solution:

The renin assay used should: control pH at its optimum for human renin (5.5 to 6.0) during the incubation and use of EDTA plus either DFP or PMSF to inhibit angiotensinases and allow for prolonged incubations (18 hr) for the low values. (Sealey and Laragh, 1976)

1975), which in two of the three methods (Table 2) routinely is converted to active renin by acid treatment, thereby misleading the active renin measurement in a variable way and sometimes by a factor of 20 or more.

The third unreliable method which has been rather widely used lacks the sensitivity and consistency to identify low renin patients (Table 2). This results from no pH control, poor angiotensinase inhibition and the use of B.A.L. as an angiotensinase inhibitor—a compound now known to inhibit renin activity (Ryan and Weinberger, 1980). Fortunately, all these problems are easily solvable (Sealey and Laragh, 1977) (Table 2) so that, in future research, broader and fuller definition of the role of active renin in normal physiology and also in diagnosis will be possible.

DISSECTION OF VASCONSTRICTION AND VOLUME FACTORS

The reciprocity between vasoconstriction and volume factors in blood pressure support dramatically demonstrated in these studies has broad applications. Thus, for example, when renin vasoconstriction is opposed with a vasodilator drug such as hydralazine, minoxidil or nitroprusside, the renin system reacts and induces, via aldosterone activation, salt and water retention. In the same way, diuretics, uniquely effective in low renin patients, are likely to be ineffective or even pressor in high renin patients, presumably by triggering a reactive renin overshoot in response to volume depletion (Gavras et al., 1974b). This seesaw relationship explains why diuretic resistant patients respond to the addition of an antirenin drug and why some medium or high renin patients need a diuretic to cope with reactive sodium retention. However, many patients will respond to a single agent providing it is the correct one for their physiologic lesion. A worldwide literature has accumulated to verify the different selectivity of beta-receptor blockers and of diuretics when given to different subgroups of patients as defined by renin testing (Table 1).

This vasoconstriction-volume seesaw or reciprocity helps explain why many insufficiently analyzed patients do not respond to trial and error therapy: the agent prescribed may be inappropriate to deal with the factors behind their hypertension—diuretics are usually ineffective in vasoconstriction dependent hypertension while renin blockage is ineffective for volume dependent hyptertension (Table 1).

THE VASOCONSTRICTION-VOLUME INTERPLAY IN TWO FORMS OF EXPERIMENTAL RENAL HYPERTENSION

Concurrently, on another research front, renin system blockade in animal and human forms of renovascular hypertension has demonstrated the same interplay between vasoconstriction and volume factors that is also apparent in essential hypertension (Laragh, 1973). The 1 kidney clipped with the opposite kidney untouched hypertension model is renin dependent and is corrected by angiotensin antibody or blockade (Brunner et al., 1971). This form resembles surgically curable human renovascular hypertension, since nephrectomy or vascular repair is curative. On the other hand, the 1 kidney clipped hypertensive animal with the opposite kidney removed exhibits a low renin level and is instead using volume to support the hypertension. But this form can be converted to renin dependency by dietary sodium depletion (Gavras et al., 1973). Only then does the blood pressure fall with antirenin system drugs. The human counterpart of this second form is a patient with bilateral kidney disease and reduced renal mass. They are similar because they exhibit medium or low renin levels and respond to sodium depletion but not to antirenin drugs. These studies provide a basis for the value of the peripheral renin profile as a primary screening test for surgically curable renovascular hypertension (unilateral hypersecretion of renin), because renin hypersecretion is a characteristic of this form. Practically, this blood test often allows the physician to defer or eliminate the necessity for the more invasive tests of pyelography and arteriography until the need is definitely established by a positive renin profile followed then by renal vein renin studies.

THE VASOCONSTRICTION-VOLUME HYPOTHESIS

When all these research findings are put together, a construction emerges that we have come to call the Vasoconstriction-Volume Hypothesis (Laragh, 1973). This serves as a useful framework for analyzing, explaining and treating all hypertensive patients. The hypothesis sees hypertension as a bipolar spectrum ranging from a predominant excess of arterial vasoconstriction to a predominant excess of effective volume. The volume excess form is identifiable by a low renin level and verifiable by correction of the hypertension by diuresis alone. All hypertension is characterized by vasoconstriction and an increased peripheral resistance. However, volume dependent forms are considered to have less vasoconstriction and more flow for a given degree of blood pressure elevation. The reason for the vasoconstriction of low renin forms remains a research problem.

In viewing hypertension as a spectrum, there are intermediate forms in which relative excesses of one or the other may overlap and jointly contribute to high blood pressure (Fig. 10). Malignant hypertension and primary aldosteronism are the polar extremes of the spectrum, and the other clinical entities fall at intermediate points. Renin profiling can generally suggest the likelihood of the particular type of hypertension within this spectrum. Just as important, the exceptions to the rule, who do not respond as expected to antirenin or antivolume therapy, are framed for further study.

More research is needed to define fully the accompanying physiologic differences among patients within the various renin subgroups. For example, we do know that the plasma volume is usually proportionally lower in hypertensive than in normotensive people (Tarazi, 1976), but we need to know conclusively if there are volume differences among equally hypertensive patients with differing renin profiles. However, all available data are in keeping with what the vasoconstriction-volume model predicts, that low renin patients are relatively hemodiluted and volume expanded. Thus, in all reported studies, tabulated elsewhere (Laragh et al., 1979; Laragh, 1973), patients with low renin hypertension have relatively lower blood urea, hematocrit and hemoglobulin values, and their total or central blood volumes are almost without exception as high or higher than those of normal-renin patients. The converse is

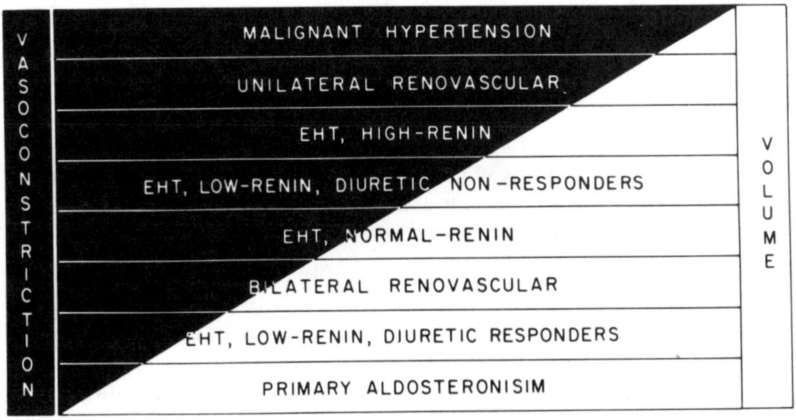

Fig. 10. The Vasoconstriction-Volume Spectrum in Hypertension. Depicting individual patients at any given sodium intake. From Laragh J.H., Letcher R.L., and Pickering T.G. JAMA 241: 151–156, 1979.

also true: when hypervolemia is present in hypertensive patients renin values are considerably reduced. Moreover, high renin patients have the lowest plasma volumes, averaging 92% of control values (Laragh et al., 1979). In further studies of these relationships, sodium balance should be rigorously controlled, and greater attention should be paid to errors inherent in space measurement while also recognizing that, so far, the arterial volume is an unmeasureable fraction of total blood volume.

NEW DIAGNOSTIC AND TREATMENT PLAN

Information created in this research voyage presages the onset of a new era of more predictable and selective drug therapy in essential hypertension. The new work certainly does not refute the old hard-earned empiric data on treatment. Rather, it provides a second generation of knowledge that will better inform clinical practice while bringing us closer to full understanding. Based on this research, we have recently set forth in detail the methods of a complete, integrated system for diagnosis and for the simpler, more specific and more predictable treatment of all hypertension (Laragh et al., 1977, 1979). In it (Fig. 11), the renin-sodium profile and a baseline plasma-potassium measurement provide the basis for 1) definitive diagnosis or adrenocortical hypertension, 2) definitive diagnosis of surgically curable renovascular hypertension, 3) the acquisition of baseline data about the relative participation of vasoconstriction and volume factors that enable more specific therapy, 4) the evaluation of the pace, severity and outlook of the underlying disorder and 5) identification of patients whose hypertension cannot be explained by our current state of knowledge. The plan also defines an optional procedure for empirical use when a renin profile is not easily available.

The new system differs in concept from previous approaches such as "stepped care" because these older systems have assumed all hypertension is alike and therefore a single additive recipe system is used for all. Thus, "stepped care" makes no provision for systematic drug substitution or dose reduction. In contrast, our new system seeks to identify those large subgroups of different people in whom long term monotherapy with either an antivasoconstriction or antivolume drug is possible. Among the advantages of the new system is that, for major fractions of all hypertensive patients, it enables, for the long-term commitment, therapy with one drug instead of two, or two instead of three of more, with the attraction that the right type of drug is applied to contain the physiological lesion, i.e., antivolume therapy with a diuretic in patients

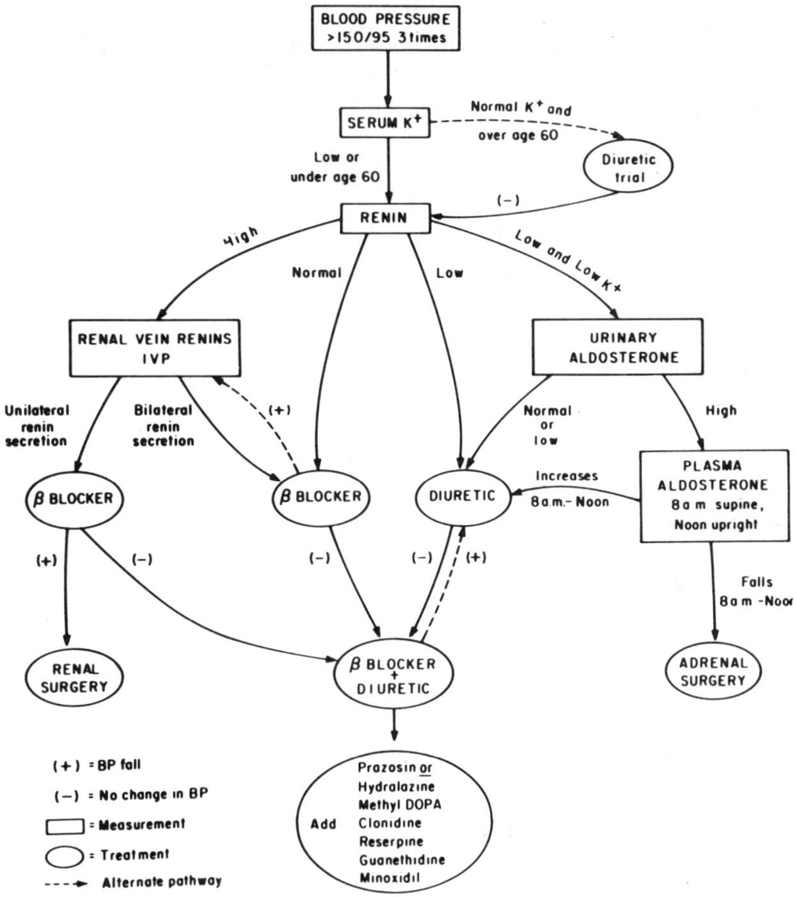

Fig. 11. Flow sheet for complete diagnosis and treatment of hypertension. Low-renin patients with hypokalemia (serum K+ <3.5) are evaluated for adrenocortical disease. Normokalemic low renin patients receive a diuretic first. The medium and high-renin patients are started on a beta-blocker. Whenever either course is only partly corrective, the two agents are combined, followed by a reverse subtraction trial. Altogether this nromalizes some 85% of patients. For the residual resistant group, other less palatable drugs are superimposed in traditional trial by error fashion. Also unlike "stepped care", drug subtraction and dose reduction is performed whenever pressure is normalized, to achieve the fewest drugs in the smallest dose.
Those medium or high renin patients with a good response to beta-blocker alone based on other clinical features too, are evaluated for surgically curable renovascular hypertension. Only if the renal vein renin study reveals ipsilateral hypersecretion with ischemia is angiography necessary.
When renin profiling is not available this basic approach can be partially applied, albeit more slowly, by serially evaluating the response to antirenin or antivolume therapy.
Redrawn from Laragh, J.H., Letcher, R.L., and Pickering, T.G. JAMA 241, 151–166, 1979, (38).

with low renin levels or antirenin therapy with a β-blocker in patients with high and normal renin levels (Fig. 4 and Table 1). It also diagnoses or excludes surgically curable forms and reduces the use of and aids in selection of patients for pyelography and arteriography.

CONCLUDING REFLECTIONS

The research story that I have told creates new perceptions that have practical application. In the broadest sense the work indicates that essential hypertension is probably not a single disorder, as has long been thought. In fact, high blood pressure may be like having an elevated body temperature. Both can have many different causes, and therefore will respond or fail to respond to particular treatments. This concept explains the long known heterogeneity in pharmacological responsiveness observed among different patients. When differences in drug responsiveness are related to differences in biochemical and pathophysiological features, the patterns that emerge help explain accompanying differences in clinical course and in susceptibility to cardiovascular damage observed among patients who have similar blood pressure elevations.

This heterogeneity probably explains that phenomenon—well-known to practicing physicians—in which some patients with alarming elevations of blood pressure can disdain all therapy and yet go on to live a long and merry life while others, with lesser elevations, succumb prematurely to heart attack or stroke. In other words, the risk is certainly not equally distributed among hypertensive patients. With all this information in hand, it does not seem likely that a single genetic or acquired defect can explain the spectrum now labeled as essential hypertension.

We now know that the renin-angiotensin-aldosterone control system actively participates in most human hypertension. The observed patterns of its participation support the hypothesis that essential hypertension is a spectrum that can be sorted out according to whether or not there is—at one extreme—excessive vasoconstriction, as in the high and medium renin patients, or—at the other extreme—a relative absence of vasoconstriction (low renin patients). In this relationship, participation of excessive sodium volume in the hypertensive condition is inversely related to the plasma renin activity. The concept that these low renin patients have overfilling with better microcirculatory flow and less vasoconstriction fits with the fact that they are older than the rest despite an equal or greater degree of hypertension, suggesting per se that they are relatively protected from stroke or heart attack.

This new knowledge, together with companion pharmacological advances, opens the door to a new era of more specific diagnosis and treatment. Renin profiles can be used to separate out clinically similar patients who have high as opposed to low renin values. This separation is the basic first step for definitive diagnosis or exclusion of the surgically curable renovascular or adrenocortical forms. At the same time, it provides at the beginning a guide for applying simpler and more specific therapy for major subgroups. This will allow separation of low renin patients best treated for the long term with a diuretic alone as antivolume monotherapy. High or medium renin patients are treated instead with an antirenin agent as the single first step.

Present evidence suggests that beta-receptor blocking drugs lower blood pressure largely by suppressing renal renin secretion. Accordingly, the spectrum of their effectiveness resembles that observed with the use of the orally active converting enzyme inhibitor, captopril, with the latter being more potent because of its more complete block of renin activity. Both types of drugs lower pressure according to the control renin level and both are especially effective in patients with high or medium renin values. The newest antirenin therapy in appropriate patients has the conceptual advantage, not offered by diruetics, of improving organ flow while lowering the blood pressure. Moreover, the greater specificity of this form of treatment is accompanied by fewer unwanted side effects. This in itself may mean that we are striking closer to the causal mechanism.

In future research we need to learn why the renin system participates, apparently inappropriately, in so much of the hypertensive spectrum. That is to say, what and where is this signal coming from? This framework per se perhaps suggests that hypertension is a disorder of a regulatory control system. The fact that correction of renin excess by captopril produces no unwanted tradeoffs or impairment of other functions implies that the system is, in fact, behaving inappropriately rather than just contributing its physiologic share to the elevated pressure. We also need to understand the basis for the sodium volume excess in volume dependent low renin hypertension,—a situation in which the renin system turn off does in fact appear to be quite appropriate.

Beyond these questions we need to learn more about a possibly important link between renin system behavior and the autonomic and central nervous systems. This potential linkage is well illustrated by the action of beta-receptor blockers on renin release and by the absence of an autonomic reflex response (i.e., reflex tachycardia) to angiotensin II depletion. We also need to know more about prostaglandin mediation, angiotensin or catecholamine induced changes of vascular tone. Finally, in future research, the role of prorenin (Sealey and Laragh, 1975) and the nature of its conversion step to active renin, possibly involving renal kallikrein (Sealey et al., 1978a,b) deserves special attention as an important new control point in the system.

SUMMARY

It is good to report that a research journey over the past 20 years has exposed the renin-angiotensin-aldosterone control system and defined its involvement in human hypertension. This knowledge opens the door to a new era in the treatment of hypertension, which is now viewed as a common manifestation of what we have come to realize is a spectrum of pathophysiologic disorders. Like all advances in medicine, this new era is characterized by more diagnosis and less guess. We are beginning to treat the hypertensive lesion rather than the sphygmomanometer. The biochemical heterogeneity among hypertension patients is matched by differences in their drug responsiveness and in their clinical course and risk of cardiovascular events. Accordingly, it is no longer necessary to treat all patients with a single recipe. Moreover, we are coming closer to defining who is at risk so that we can improve on the present strategy of treating every patient for life in order to improve the outcome for only a few.

Exposure of the renin-aldosterone control system that regulates blood pressure and flow and its involvement in malignant hypertension was followed by definition of its diagnostically useful involvement in surgically curable renovascular and adrenocortical forms, and its involvement in oral contraceptive hypertension.

In this setting, the definition of a series of drugs, first beta-receptor blockers, then saralasin and converting enzyme inhibitors, whose main action can be traced to suppression of the renin factor has enabled new understanding and a new era of treatment just as the advent of sodium volume suppressive drugs gave impetus to an earlier advance. We now know that renin induced vasoconstriction is involved in the blood pressure elevation of most patients, while in a minor subfraction sodium-volume retention is the final culprit. Renin-sodium profiling defines this involvement by revealing the degree of renin participation and its appropriateness to the salt balance.

Using renin profiling, we can solve for the pathogenic factor in the vasoconstriction-volume equation that regulates blood pressure in all of us, and thereby plan simpler and more specific therapy, often with one drug being sufficient. At the same time, surgically curable forms are diagnosed or excluded and an impression is gained about pace and severity of the disorder. Moreover, those patients who do not fit into the framework are separated for further study.

We still need to learn why renin secretion so often participates in hypertensive disease and why it fails to turn itself off. And it is obvious too, from the 10 to 15% of patients who still elude satisfactory control by any analysis, that we still have a way to go. Meanwhile, the creation of this new knowledge brings us closer to a final solution and provides the understanding essential for better diagnosis and treatment of this entire group of diseases.

REFERENCES

Amery, A., Lijnen, P., Fagard, R. and Reybrouck, T. 1976. Plasma renin activity vs. concentration. N Engl J Med 295:1198c

Ames, R. P., Borkowski, A. J. Sicinski, A. M., and Laragh, J. H. 1965. Prolonged infusions of angiotensin II and norepinephrine and blood pressure, electrolyte balance, aldosterone and cortisol secretion in normal man and in cirrhosis with ascites. J Clin Invest 44:1171–1186

Atlas, S. A., Sealey, J. E., Laragh, J. H., et al. 1977. Plasma renin and "prorenin" in essential hypertension during sodium depletion, beta-blockade and reduced arterial pressure. Lancet 2:785

Atsmon, A., Blum, E., Steiner, M., et al. 1972. Further studies with propranolol in psychotic patients. Psychopharmacologia 27:249

Baer, L., Parra-Carrillo, J. Z., Radichevich, I., and Williams, G.S. 1977. Detection of renovascular hypertension with angiotensin II blockade. Ann Int Med 86:275

Brunner, H. R., Kirshman, J. D., Sealey, J. E., and Laragh, J.H. 1971. Hypertension of renal origin: Evidence for two different mechanisms. Science 174:1344–1346

Brunner, H. R., Laragh, J. H., Baer, L., Newton, M. A., Goodwin, F. T., Krakoff, L. R., Bard, R. H., and Bühler, F.R. 1972. Essential hypertension: Renin and aldosterone, heart attack and stroke. N Engl J Med 286:441–449

Brunner, H. R., Gavras, H., and Laragh, J. H. 1973a. Angiotensin II blockade in man by sar[1]-ala[8]-angiotensin II for understanding and treatment of high blood pressure. Lancet ii:1045

Brunner, H. R., Sealey, J. E., and Laragh, J. H. 1973a. Renin as a risk factor in essential hypertension: more evidence. Am J Med 55:295–302

Brunner, H. R., Gavras, H., Laragh, J. H., and Keenan, R. 1974. Hypertension in man. Exposure of the renin and sodium components using angiotensin II blockade. Circ Res 34 & 35: Suppl I:I–35–I–45

Brunner, H. R., Gavras, H., Walker, B., Kershaw, G. R., Turini, G. A., Vukovich, R. A., McKinstry, D. N., and Gavras, I. 1979. Oral angiotensin-converting enzyme inhibitor in long-term treatment of hypertensive patients. Ann Int Med 90:19

Bühler, F. R., Laragh, J. H., Baer, L., Vaughan, E. D., Jr., and Brunner, H. R. 1972. Propranolol inhibition of renin secretion. A specific approach to diagnosis and treatment of renin-dependent hypertensive disease. N Engl J Med 287:1209–1214

Bühler, F. R., Burkhard, F., and Lütold, B. E. 1975. Antihypertensive beta blocking action as related to renin and age: A pharmacologic tool to identify pathogenic mechanisms in essential hypertension. Am J Cardiol 36:653

Case, D. B., Wallace, J. M., Keim, H. J., Sealey, J. E., and Laragh, J. H. 1976a. Usefulness and limitations of saralasin, a weak competitive agonist of angiotensin II, for evaluating the renin and sodium factors in hypertensive patients. Am J Med 60:825

Case, D. B., Wallace, J. M., Keim, H. J., Weber, M. A., Drayer, J. I. M., White, R. P., Sealey, J. E., and Laragh, J.H. 1976b. Estimating renin participation in hypertension. Superiority of converting enzyme inhibitor over saralasin. Am J Med 61:790–796

Case, D. B., Wallace, J. M., Keim, J. H., Weber, M. A., Sealey, J. E., and Laragh, J. H. 1977. Possible role of renin in hypertension as suggested by renin-sodium profiling and inhibition of converting enzyme. New Engl J Med 296:641–646

Case, D. B., Atlas, S. A., Laragh, J. H., Sealey, J. E., Sullivan, P. A., and McKinstry, D. N. 1978. Clinical experience with blockade of the renin-angiotensin-aldosterone system by an oral converting-enzyme inhibitor (SQ 14,225, captopril) in hypertensive patients. Prog in Cardiovasc Dis 21:195–206

Chiu, P. J. S., and Sommer, E. J. A. 1979. Effect of Na^+ restriction on blood pressure responses to β-adrenoreceptor blockade in spontaneously hypertensive rats (SHR). Fed Proc 38:2372 (Abs)

Conway, J., Greenwood, D. T., and Middlemiss, D. N. 1978. Central nervous actions of beta-adrenoreceptor antagonists. Clin Sci Mol Med 54:119

Drayer, J. I. M., Keim, H. J., Weber, M. A., Case, D. B., and Laragh, J. H. 1976. Unexpected pressor responses to propranolol in essential hypertension. Am J Med 60:897

Gavras, H., Brunner, H. R., Vaughan, E. D., Jr., and Laragh, J.H. 1973. Angiostensin-sodium interaction in blood pressure maintenance of renal hypertensive and normotensive rats. Science 180:1369–1372

Gavras, H., Brunner, H. R., and Laragh, J. H. 1974a. Renin and aldosterone and the pathogenesis of hypertensive vascular damage. Prog Cardiovasc Dis 17:39–49

Gavras, H., Brunner, H. R., Laragh, J. H., Sealey, J. E., Gavras, I., and Vukovitch, R. A. 1974b. An angiotensin converting enzyme inhibitor to identify and treat vasoconstrictor and volume factors in hypertensive patients. New Engl J Med 291:817–821

Gleadle, R. I., Brown, J. J., Curtis, J. R. et al 1969. Plasma renin concentration and the control of blood pressure in patients with chronic renal failure: The effect of haemodialysis. In D.N.S. Kerr (ed), *Proceedings of the European Transplant Association*, Stockholm, London: Pitman Medical, Vol. 6:131

Goldblatt, H. J., Lynch, R. F., Hanzal, R. F., et al 1934. Studies on experimental hypertension. Production of persistent elevation of systolic blood pressure by means of renal ischemia. J Exp Med 59:347–379

Gross, F. 1958. Renin und Hypertension, physiologische oder pathologische Wirkstuffe? Klin Wochenschr 36:693–705

Hansson, L. 1973. Beta-adrenergic blockade in essential hypertension. Acta Medica Scand (Suppl) 550:1

Iaina, A., Goldfarb, D., and Eliahou, H. E. 1979. Unexpected pressor effect of propranolol and angiostensin II in bilaterally nephrectomized rats. Life Sciences 25:1153

Keim, H. J., Drayer, J. I. M., Case, D. B., Lopez-Ovejero, J.A., Wallace, J. M., and Laragh, J. H. 1976. A role for renin in rebound hypertension and encephalopathy after infusion of Sar[1]-ala[8]-angiotensin II. N Engl J Med 295:1175–1177

Laragh, J. H. 1960. The role of aldosterone in man: Evidence for regulation of electrolyte balance and arterial pressure by renal-adrenal system which may be involved in malignant hypertension. JAMA 174:293–295

Laragh, J. H. 1973. Vasoconstriction-volume analysis for understanding and treating hypertension: The use of renin and aldosterone profiles. Am J Med 55:261–274

Laragh, J. H. 1978. Renin as a predictor of hypertensive complications: Discussion. Ann NY Acad Sci 304:165–177

Laragh, J. H., and Sealey, J. E. 1973. the renin-angiotensin-aldosterone hormonal system and regulation of sodium, potassium, and blood pressure homeostasis. American Physiological Soc *Handbook of Physiology—Renal Physiology*, eds. J Orloff and R W Berliner, Chapter 26:831–903. Baltimore: Waverly Press, Inc.

Laragh, J. H., and Sealey, J. E. 1977. Renin-sodium profiling: Why, how and when in clinical practice. Cardiovasc. Med 2:1053–1075

Laragh, J. H., Angers, M., Kelly, W. G., and Lieberman, S. 1960a. Hypotensive agents and pressor substances. The effect of epinephrine, norepinephrine, angiotensin II and others on the secretory rate of aldosterone in man. JAMA 174:234–240

Laragh, J. H., Ulick, S., Januszewicz, W., Deming, Q. B., Kelly, W. G., and Lieberman, S. 1960b. Aldosterone secretion and primary and malignant hypertension. J Clin Invest 39:1091–1106

Laragh, J. H., Sealey, J. E., Ledingham, J. G. G., and Newton, M. A. 1967. Oral contraceptives. Renin, aldosterone, and high blood pressure. JAMA 201:918–922

Laragh, J. H., Baer, L., Brunner, H. R., Bühler, F. R., Sealey, J. E., and Vaughan, E. D., Jr. 1972. Renin, angiotensin and aldosterone system in pathogenesis and management of hypertensive vascular disease. Am J Med 52:633–652

Laragh, J. H., Letcher, R. L., and Pickering, T. G. 1979. Renin profiling for modern diagnosis and treatment of hypertension. JAMA 24:151–156

Laragh, J. H., Case, D. B., Atlas, S. A., and Sealey, J. E. 1980. Captopril compared with other anti-renin system agents in hypertensive patients, its use to identify and treat the renin factor. Hypertension in press

Lydtin, H., and Sommerfeldt, H. 1972. The effect of d- and dl-propranolol on blood pressure and cerebral norepinephrine in rats with DOCA-hypertension. Int J Clin Pharmacol 64:328

MacGregor, G. A., and Dawes, P. M. 1976. Angiotensin II blockade in patients with essential hypertension. Aust NZJ Med 6:53

Marks, L. S., Maxwell, M. H., and Kaufman, J. J. 1977. Renin, sodium, and vasodepressor response to saralasin in renovascular and essential hypertension. Ann Int Med 87:176

Masson, G.M.C., Mikasa, A., and Yasuda, H. 1962. Experimental vascular disease elicited by aldosterone and renin. Endocrinology 71:505–512

Ondetti, M. A., Rubin, B., and Cushman, D. W. 1977. Design of specific inhibitors of angiotensin-converting enzyme: new class of orally active antihypertensive agents. Science 196:441

Onesti, G., Swartz, C., Ramirez, O., *et al* 1968. Bilateral nephrectomy for control of hypertension in uremia. Trans Am Soc Artif Intern Organs 14:361–364

Orjavik, O. S., Aas, M., Fauchold, P., *et al* 1975. Renin secreting tumors with severe hypertension. Acta Med Scand 197:329–336

Pickering, T. G. 1979. Beta-blocking drugs in hypertension. NY State J Med Dec:2054

Ryan, N., and Weinberger, M. H. 1980. Effect of sulfhydryl reagents on the enzymatic activity of human renin: Implications for renin assay. Clin Chim Acta 106:135–143

Sealey, J. E., and Laragh, J. H. 1975. "Prorenin" in human plasma? Methodological and physiological implications. Circ Res 35 & 37 Suppl I:10–16

Sealey, J. E., and Laragh, J. H. 1977. How to do a plasma renin assay. Cardiovasc Med 2:1079–1092

Sealey, J. E., Atlas, S. A., and Laragh, J. H. 1978a. Linking the kallikrein and renin systems via activation of inactive renin. New data and hypothesis. Am J Med 65:994–1000

Sealey, J. E., Atlas, S. A., and Laragh, J. H. 1978b. Human urinary kallikrein converts inactive to active renin and is a possible physiological activator of renin. Nature 275:144–145

Streeten, D. H. P., Anderson, G. H., Jr., Freiberg, J. M., and Dalakos, T. G. 1975. Use of an angiotensin II antagonist (saralasin) in the recognition of "angiotensinogenic" hypertension. N E J Med 292:657

Tarazi, R. C. 1976. Hemodynamic role of extracellular fluid in hypertension. Circ Res 38 (Suppl 2) 73–83

Tarazi, R. C., and Dustan, H. P. 1972. Beta adrenergic blockade in hypertension. Am J Cardiol 29:633

Tigerstedt, R., and Bergman, P. G. 1898. Niere und kreislauf. Scand Arch Physiol 8:223–271

Vaughan, E. D., Jr., Bühler, F. R., Laragh, J. H., Sealey, J.E., Baer, L., and Bard, R. H. 1973. Renovascular hypertension: Renin measurements to indicate hypersecretion, and contralateral suppression, estimate renal plasma flow and score for surgical curability. Am J Med 55:402–414

Vertes, V., Cangiano, J. L., Berman, L. B. et al 1969. Hypertension in end-stage renal disease N E J Med 280:978–981

Young, D. B., and Diepstra, G. R. 1978. Lack of antihypertensive effects of propranolol in experimental angiotensin II hypertension. Physiologist 21:132

SECTION VI

Symposium Summary

Franz Gross

Symposium Summary

Franz Gross
University of Heidelberg, 6900 Heidelberg, West Germany

Dr. McKinstry, ladies and gentlemen. Being in Philadelphia, I cannot resist remembering my first visit to this city, 27 years ago, on the occasion of a satellite symposium following the International Congress of Physiology in Montreal. I think, in fact, it was the Physiological Society of Philadelphia which sponsored the symposium. Newton Richards was Honorary President, an impressive old gentleman, and we respectfully remembered that he had been the first to puncture glomeruli in frogs. The meeting was organized by Professor Carl Schmidt, at that time one of the leading pharmacologists, not only in the United States, but worldwide. But my reason for mentioning all this is that on that occasion Marthe Vogt, working then with John Gaddum in Edinburgh, demonstrated noradrenaline in the hypothalamus and other brain areas for the first time. A remarkable event in the history of pharmacology and neuroendocrinology. It may be that we have witnessed another outstanding event in pharmacology during these past two days of our symposium.

Yesterday, Dr. Horovitz, in his introduction, distinguished between questions which had been settled and those for which we still don't know the answer. Well, I am afraid his wish that our meeting help settle open problems has been fulfilled in part only. There are still quite a few unanswered questions, and it will take much more time to find a solution for them. Despite the fact that Dr. Cushman gave a remarkable example of a systematic approach to drug development in a field that, in the beginning, did not look promising at all from a therapeutic point of view, we still do not know what would be the most effective way to block the converting enzyme, to what degree it should be blocked to get a satisfactory therapeutic effect, what will be all the repercussions of such a blockade on the renin-angiotensin system and other enzyme substrate systems, whether the various converting enzymes in different tissues are similarly inhibited, and so on.

Remarkable progress has been made since Dr. Cushman reported on converting enzyme inhibition and inhibitors at the Jacques Genest Festival of 1972 at Mont Gabriel in the Laurentian Mountains, and the step from the snake venom peptides, such as teprotide, to the nonpeptidic, orally active inhibitor, captopril, is one of the few examples of successful drug design, at least until now. It is also an excellent example of systematic research invested in a field which did not at all look like it would yield results rapidly that would be applicable to therapeutics—a research philosophy not rated very high in drug industry today, where targeted research and aiming at quick application is in the foreground.

It is reassuring to see that 27 companies supported our meeting; I hope this does not mean a corresponding number of captopril followers, as has been the case in the field of beta-blockers. Of course, there will be many modifications, and the sulfhydryl group, which is under some suspicion today, might disappear, but we should not forget Dr. Cushman's statement that "if it were not for the sulfhydryl group, we wouldn't have captopril".

Dr. Rubin gave a review of his extensive animal studies and demonstrated in this way the predictive value of such experimental studies for therapeutic efficacy. I think we should not be surprised that it takes time for the blood pressure to return to elevated levels after withdrawal of long-term treatment. I should be more interested in learning whether a sudden interruption of treatment causes a rebound effect, but this has obviously not been observed so far. It is remarkable that we don't know what other mechanisms, besides the inhibiton of the converting enzyme, are involved in the lowering of blood pressure: Is it a direct vasodilation, an interference with the sympathetic system, an effect on the central renin-angiotensin system, an accentuation of bradykinin, the release of prostaglandins, a reduction of aldosterone secretion, or an effect on the arteriolar renin-angiotensin system? More questions than answers.

On point should be kept in mind: converting enzyme is an activator of angiotensin and inactivates bradykinin, and since the lung is a major site for this activation or inactivation, circulating angiotensin II may obviously have a function in the systemic circulation, while bradykinin has none.

What about the central effects of the renin-angiotensin system and the inhibition of angiotensin I conversion? Does peripherally administered captopril pass the blood-brain barrier and interfere with the control of the central renin-angiotensin system? Obviously not, as Dr. Ganten and his associates have shown, but it may be that more lipophilic derivatives do. Experimental procedures are of significance, the dosages used, the duration of treatment. Intraventricularly administered captopril may lower the blood pressure in spontaneously hypertensive rats, but it is not clear whether this antihypertensive effect is related to central mechanisms. The converting enzyme activity in the brain may increase under the influence of captopril, and the same may happen in plasma, but the practical significance of these findings is not yet clear, and the same holds true for the increase in renin activity at various sites of the central nervous system, such as the neurohypophysis, the hypothalamus, or the medulla. Somewhat confusing, but not at all hopeless, are all these changes in the central nervous system. The same, of course, applies to the various peptides that Dr. Ganten offered in reply to Dr. Horovitz's questions, the peptides on which captopril may work: substance P, the endorphins, or even the enkephalins, all possible candidates for the converting enzyme, which may not be too selective in choosing its reaction partners.

Discussing all these data, one should not forget to distinguish between enzyme concentration and enzyme activity. I was surprised not to see one Michaelis-Menten constant or a Lineweaver-Burk plot during all of our symposium. In this context, Dr. Fyhrquist showed the increase in converting enzyme activity in various tissues as well as in plasma during captopril treatment, and consequently the decrease in angiotensinogen, when plasma renin activity and angiotensin I rise.

For once, we turned from rats to cats, in which the renin system has been poorly studied so far. Dr. Conway demonstrated the time differences in quick sympathetic reflexes and the slower response in renin release, and that inhibition of converting enzyme does not affect baroreceptor reflexes. Also, centrally administered converting enzyme inhibitor was ineffective under these conditions. Hence, the renin-angiotensin system is a system which interacts with the sympathetic nervous system, but is not a back-up system for it—an important conclusion from a pathophysiological point of view.

Then we entered the kinin field, and I hope all of you realized how much more complex this system is than the renin-angiotensin system. Oscar Carretero gave an excellent introductory review, hinting at the various pitfalls and possibilities of erroneous interpretation or even of creating artifacts. He argued by *exclusion* criteria, and in the end little remained of what would speak in favor of a participation of bradykinin in the blood pressure lowering effect of converting enzyme inhibitors. Consequently, I think it is not fair to blame bradykinin for being responsible for the rash. Of course, we admit that intrarenal blood distribution may change, and Colin Johnston showed that captopril had some effect in addition to that of saralasin on blood pressure, renal blood flow and renovascular resistance, but the inhibition of the activation of an endogenously produced substance, such as angiotensin II, and the administration of a competitive antagonist to angiotension II are different things. Anyhow, Dr. Thurston also gave some evidence for the fact that captopril may exert some of its activities by delayed bradykinin inactivation. However, I dare predict that it is at best marginal.

And of course the prostaglandins had to be discussed in the context, as they participate in the regulation of blood flow in the kidney, perhaps also in other organs, their release being affected by bradykinin. Oscar Carretero was reluctant to rule out completely the prostaglandins as participants in the overall action of captopril, while Dr. Antonaccio did not have much belief in them and presumes that they don't play a significant role. I fully agree with him in this respect. But *what* plays a significant role apart from the converting enzyme?

Dr. Horovitz complained that we looked too much at the possible additional sites of action instead of at the basic mechanisms of action, but I think the two run closely parallel, since the mechanism depends quite a bit upon the site. And, in addition, is the transfer of the renin system to the arteriolar wall not a change of site which has little to do with the mechanism of action, because the converting enzyme is inhibited not only in the kidney, but also in the vessel wall? Here I should like to know a little more about the pharmacokinetics of captopril, e.g., whether it is concentrated in those parts of the arteries where Dr. Antonaccio and his associates have so nicely demonstrated the converting enzyme by histofluorescent methods.

The results reported during these two days were in various respects hard to compare with one another. Short-term and long-term studies, various modes of administration, different animal species and various drugs—clean ones and dirty ones (such as aprotinin [Trasylol])—have been used as tools. No wonder that the results differ, and it is good that, in most cases, it is possible to find a reasonable interpretation, e.g., that stroke-prone spontaneously hypertensive rats may respond differently from simple spontaneously hypertensive rats, and so on. Excuses—sorry: explanations—are always at hand.

Even if we accept the hypothesis that the delayed kinin inactivation may be of some significance for the acute antihypertensive effect of captopril, this is obviously not so for the prolonged administration of the drug. We have to realize that we know little about the physiological significance of the kallikrein-kinin system, particularly of the renal kallikrein-kinin system, and that we do not have reliable methods by which to turn it on or off, as we can do with the renin-angiotensin system. In my opinion, it would be advisable to forget, for the time being, the kallikrein-kinin system in relation to the converting enzyme inhibitors, or at least to put it into a drawer until we shall know more about it; otherwise we shall have too many speculations, and hypotheses may be created, which will keep many of us busy with demonstrating that they are wrong.

That captopril is a stronger blood pressure lowering drug than saralasin is not surprising, the latter having some pressor activity itself, and in addition captopril acts on an endogenously produced substance. We also heard from Dr. Zimmerman that converting enzyme inhibitors may prevent the intrarenal conversion of angiotensin I, without causing an increase in renal blood flow. What the additional mechanism may be, which makes captopril the more effective compound, is not yet settled. Again I should like to stick out my neck and say: It is not bradykinin.

Dr. Antonaccio complemented well Dr. Rubin's studies on the antihypertensive effects of captopril and confirmed in rats the synergistic effect of diuretics. I always wonder why the system which has to be inactivated or antagonized has to be stimulated first to make the inhibitor become more effective. If renin and angiotensin are responsible for high blood pressure, why increase their activity or concentration first, before preventing the conversion into the more active angiotensin II? It is as if I gave a poor chap a lot of money so that I can steal more from him. Anyhow, we heard that arterial renin can be increased similarly as renal renin by sodium depletion, and that obviously renin can be synthetized in the vascular wall, not only in the arteries, but also in the veins, a still neglected part of the circulatory system. Do we really need such an additional pool of renin to explain the action of captopril, and does the renin-angiotensin system contribute to vascular tone, or does it have some other function? More work for the future.

On the second day, we turned from biochemistry and pharmacology to physiology and pathophysiology and, of course, clinical implications, perhaps too little to possible clinical complications, since each progress or new development, and especially a new drug, charges its toll. Hemodynamics do not look nearly as complicated for captopril as for other blood pressure lowering substances. Total peripheral resistance is consistently reduced without affecting cardiac output or heart rate, nearly an ideal hemodynamic situation. Dr. Bravo arrived just in time to bring us his message from Cleveland and to present a nice diagram of the relationship between blood volume, total peripheral resistance, cardiac output, and heart rate, the two remaining variables being baroreceptor activity and venous compliance. Just before, Bob McCaa had given an impressive review of what can be done with dogs to elucidate the regulation of aldosterone release, and correctly he mentioned that opinions were more controversial in 1963 than five years earlier, when the relationship between the renin-angiotensin system and aldosterone was claimed for the first time; no doubt, we have a remarkable ability to complicate matters all the time. But Bob started with a suitable animal model, the intact conscious dog, and convincingly demonstrated the significance of angiotensin for the release of aldosterone and the effect of captopril with all its consequences: the fall in pressure, the enhanced sodium excretion, and the increase in plasma renin activity. The fact that the elevated excretion of sodium is independent of pressure and of aldosterone secretion demonstrates that angiotensin II may have a direct effect on the renal handling of sodium. In the end, he also joined the party of those who doubt that bradykinin contributes substantially to the effect of captopril. I think he is on the right side.

The kidney, of course, deserves special attention, and Dr. Zimmerman clearly demonstrated that inhibition of converting enzyme increases renal blood flow. He also showed that angiotensin II helps to reabsorb sodium in the distal tubules. More and more it becomes clear that the reduction in arteriolar tone and the subsequent increase in renal blood flow contribute to the overall hemodynamic effect of captopril, but it is also remarkable that the effect of captopril is limited, that only about 50% of angiotensin I are prevented from conversion to angiotensin II, and that there is a maximum inhibition within the kidney, which is not identical to complete inhibition. Let us keep this important finding in mind, and let us not forget that there are other vascular beds, where the situation may be different, as was shown for the femoral artery.

The turn to the heart revealed some other fascinating observations, on which Dr. Lefer reported. First, the hemorrhagic-shock model, in which the renin-angiotensin system is of course maximally stimulated; no wonder that captopril improves the situation when blood is reinfused. But that it acts as a lysosomal stabilizer, that it prevents proteolysis in pancreatic homogenates, that it seems to have direct effects on beta-glucuronidase and on cathepsin D—these are real surprises. Is captopril even a dirty substance, because it can do too much, as we said of Trasylol yesterday? Fortunately, in contrast to Trasylol, captopril has no effect on bradykinin levels in shock; hence, it is innocent in this respect. And it also does not seem to influence the myocardial depressant factor, which has some local significance in Philadelphia and which is prevented by saralasin—a separation between the antagonist and the converting enzyme inhibitor.

With respect to the treatment of congestive heart failure, we witness a fundamental change during these days, to which Dr. Chatterjee has made a major contribution: the restitution of the hemodynamic situation is a much more physiological approach in the treatment of heart failure than the stimulation of a failing heart. The reduced total peripheral resistance and pulmonary capillary wedge pressure as well as the diminished right atrial pressure are effects caused by captopril that support the re-establishment of the deranged circulation. The increase in cardiac output, together with the drop in pulmonary capillary wedge pressure, and the decrease in heart rate are essential features of captopril, and it is remarkable that these are not initial effects only, but that they are maintained during prolonged administration. Furthermore, the increased venous compliance may contribute to reducing preload, perhaps on the basis of an inhibition of the renin-angiotensin system in the venous walls—but I should refrain from further speculation along these lines.

The various other indications that Dr. Stritar mentioned from the thick files in the Squibb offices are of course interesting, but to a major degree anecdotal. I should like to issue a word of warning not to make captopril a panacea: it is an old experience in industry that drugs which do too many different things are not good drugs. Most of the more or less rare diseases which respond to captopril respond to it because of the reduced total peripheral resistance and the improved hemodynamic situation. One should also be careful in trying captopril in desperate cases. We just heard from Dr. Alexander what may happen. The results may be rewarding in some patients, but failures may occur in others, and all the good friends will then quickly blame the drug and not consider the severe underlying disease.

Now we come to the clinical pharmacology section. Dr. MacGregor started his talk by referring to John Laragh's tales of last evening. He, too, believes that the renin-angiotensin system contributes to the maintenance of blood pressure under normal sodium intake and in normotensive subjects. It may be that normotensive medical students are a special subgroup of human beings, especially in Britain. Nevertheless, Dr. MacGregor presented an excellent study, just the type of study we need. Angiotensin II levels fell significantly in all his three groups, to the same low values, independent of sodium intake. The percentage fall in aldosterone was similar in all the three groups of subjects. Plasma angiotensin II and fall in blood pressure were well correlated, although Dr. MacGregor admitted that such correlations do not mean too much. Sodium excretion under normal and low sodium diet increased significantly. All these are data to think about when assessing the value of captopril. In my opinion, his study was one of the best I have seen for a long time, although I doubt that the renin-angiotensin system may be compared with the sympathetic system in the way he tried to do it. Dosage finding, he stressed, is important; I myself think one should not administer more than 25 mg, perhaps twice daily, but certainly not three times 50 mg, as was just mentioned for the last study that was presented. The duration of action has to be established in essential hypertension, and the average patient has to be considered, not only the exceptional patient who has many complications. One of the most important conclusions drawn from this study is that, whenever angiotensin II is above normal, without corresponding sodium loss, the renin-angiotensin system may be the cause for, or may at least contribute substantially to, high blood pressure.

I apologize for commenting only shortly on the contributions of the last three speakers, but I had to summarize (what they had said) during the coffee break. Dr. Brunner presented valuable data on the dose-response relationship and on basic clinical pharmacology. It is remarkable that no resistance to captopril develops, and that he could give it over a period of three years. He seems to be a close friend of Squibb's, receiving the drug for such a long time, whereas Dr. MacGregor in Britain had the greatest difficulty in obtaining it. An interesting observation has been that the inhibition of the converting enzyme in the plasma did not last the whole day, and that the activity varies, but we don't know what happens in tissues. Just another example showing that what we measure in plasma does not give us any information at all about what may happen at the target site. Plasma levels in various respects are unsatisfactory indicators.

The separation of various degrees of hypertension with respect to their response to captopril may not have been necessary. It is perhaps a little artificial, because there is not much difference between the moderate and the mild cases. On the other hand, Dr. Ferguson gave us some impressive examples of what can be done in resistant cases. "Resistant", in his patients, did not only mean resistant to other antihypertensives, but, in some cases, also to captopril. Uneasy feelings would be created if resistance to captopril did not exist, because it exists for each of the drugs we have today. It is important to know that captopril acquired a position in the treatment of malignant hypertension, and we should keep in mind that quite remarkable results can be obtained with it. Dr. Ferguson was the first to mention also adverse reactions; allergic reactions, tachycardia during standing are certainly effects which are not too severe and do not cause much concern. However, the adverse reactions mentioned subsequently by Dr. Alexander may not be disregarded, and we have to watch carefully what happens in this respect and have to monitor captopril closely. Once more, I think that the results of the study that Dr. Alexander presented, the double-blind study, are most encouraging, especially with respect to the fact that the pressure could be well controlled over a prolonged period. Nevertheless, the place of captopril—i.e., where it will fit in best with the large group of antihypertensive drugs—has still to be found.

I mentioned already the side effects. I should like to say that, during a treatment of about six months, an incidence of proteinuria or of renal complications in the order of between 0.7 and 1% is rather high, and of course we still have to see what may happen to the various other adverse reactions, such as taste alterations, skin rashes and especially those on the bone marrow. All the same, this should not discourage us from making further use, wide use, of captopril.

Before concluding, I should like to come back to the two types of questions asked by Dr. Horovitz yesterday, at the beginning of our meeting: those which have been solved and those which are still open. I think everybody will agree to two facts:

1. Captopril and teprotide inhibit kininase II, the converting enzyme, effectively, and
2. Converting enzyme inhibitors lower high blood pressure and are very effective antihypertensive agents.

However, it is still not settled how far the causal relationship goes between these two facts. Is blood pressure lowering exclusively the consequence of the enzyme inhibition? Many observations are in favor of such an assumption, but there remain various unanswered questions—more work for the future.

If you permit that I guess at what is the crucial effect of captopril for its therapeutic usefulness, I should like to say: The reduction in total peripheral resistance without an influence on cardiac output; perhaps also the reduction in pulmonary capillary wedge pressure and right atrial pressure in cases where these are high. It may be that the reduced arteriolar tone, which can be obtained by suppressing the formation of angiotensin II either systemically or locally in the vasculature, is responsible for the beneficial effect of captopril. In several years we may have enough data at hand to prove this hypothesis.

And this brings me to the end and to Dr. Page's letter which was quoted yesterday by Dr. Horovitz. He wrote to the Journal of the American Medical Association: "I think the time has come when the cliché 'the mechanisms and cause of essential hypertension are unknown' can justifiably be altered to 'it is highly probable that the core of the mechanism of essential, renovascular, and malignant hypertension is angiotensin.' The evidence from blocking the converting enzyme and from peptide blocking agents all but conclusively shows that when angiotensin is unable to function, blood pressure returns to normal or near normal, especially if salt has been depleted."

I have great admiration for Irv Page, and I fully agree with him as regards renovascular and malignant hypertension, but essential hypertension is a different disease, which, despite the converting enzyme inhibitors, will remain enigmatic for some time to come. The renin-angiotensin system is primarily a physiological system for the regulation of sodium balance, and angiotension II is a physiological factor, which, when its formation gets out of control, may become of pathological significance—in which way, we still do not fully understand. The story of the converting enzyme inhibitors is not over, it is just beginning.

Index

WITHDRAWN